Vibrational States

Vibrational States

S. CALIFANO

Institute of Chemical Physics
University of Florence

A Wiley–Interscience Publication

JOHN WILEY & SONS

London · New York · Sydney · Toronto

Library of Congress Cataloging in Publication Data:

Califano, S.
Vibrational states.

'A Wiley–Interscience publication.'
1. Vibrational spectra. I. Title.
QC454.V5C34 539'.6 75-19268

ISBN 0 471 12996 8

Printed by J. W. Arrowsmith Ltd., Winterstoke Road, Bristol BS3 2NT

*To my father and to my wife
who continuously transmitted
to me their love for science
and culture.*

Preface

This book originates from a course that the author has given for some years at the University of Florence for chemistry and physics students. The content has been modified and adapted to fulfil the need for a general theoretical introduction to vibrational spectroscopy at a graduate level. The book also includes topics of a more advanced character for the benefit of young research workers who face an immense literature from which the important new developments are not easily extracted.

The content is limited to the theoretical treatment of the vibrations of isolated molecules. To keep the book a reasonable length, important aspects of modern vibrational spectroscopy, for example crystal and high resolution spectroscopy, band shapes and vibronic couplings, are not discussed at all. Care has been taken, however, to maintain the theoretical treatment at a very general level so that a reader who masters this exposition should have no difficulties in extending his interest to other areas of vibrational spectroscopy. For this reason, coordinate transformations have been treated at length and the methods of quantum field and group theory widely used. The role of off-diagonal operators in the construction of symmetry coordinates has been emphasized and a detailed treatment of redundancies is given.

My generation has received an immense benefit from the books by G. Herzberg, *Infrared and Raman Spectra*, and by E. B. Wilson jr., J. C. Decius and P. C. Cross, *Molecular Vibrations*. I believe that it is not sufficient to specify that I have constantly drawn from these sources, to give them all the credit they deserve. It is actually hard to imagine a new book on vibrational spectroscopy which does not reflect their influence. I am also indebted to other books in relation to specific chapters, such as the book by W. Heitler *The Quantum Theory of Radiation* with reference to Chapter 3, the book by R. McWeeny *Symmetry* for Chapter 5 and the famous article of H. H. Nielsen in the *Handbuch der Physik* for Chapter 9.

This book would have never appeared without the constant pressure that my wife exerted on me. She convinced me that an incomplete book is better than nothing and compelled me to ignore my natural aversion for writing.

It is my pleasure to acknowledge the help I received from Professors N. Neto, V. Schettino and G. Taddei who gave me many helpful suggestions concerning both the content and the form of the manuscript. I am also deeply indebted to Professor Stuart Walmsley for his patience in correcting the manuscript. Finally I wish to express my appreciation to Mrs Lella Marzocchi for typing the text with care and patience.

Preface

Contents

Chapter 1
Infrared and Raman Spectroscopy

1.1 Introduction

Molecular vibrational spectroscopy deals with all processes of interaction of the electromagnetic radiation with matter which produce a variation in the vibrational motions of the molecules. The study of these vibrations is of paramount importance for the understanding of several physical properties of molecular compounds which depend upon the vibrational states of the molecules, and represents a powerful tool for the investigation of molecular structure.

The interaction of the electromagnetic radiation with an assembly of molecules can be analysed by measuring either the absorption of infrared radiation or the scattering of light, two phenomena that, although governed by completely different interaction mechanisms, give rise to the same types of transitions between vibrational states. In order to establish some basic terminology for our discussion, we recall that the electromagnetic radiation is characterized by the wavelength $\tilde{\lambda}$ or by the frequency $\tilde{\nu}$, connected by the well-known relation

$$\tilde{\lambda}\tilde{\nu} = c \tag{1.1.1}$$

where c is the velocity of light. Different units are currently used in different regions of the spectrum for these quantities. In the region of interest to us, the wavelength is best expressed in units of microns ($1\ \mu = 10^{-4}$ cm) and the frequency $\tilde{\nu}$ in units of cycles/s (hertz). The frequency $\tilde{\nu}$ is, however, seldom used in vibrational spectroscopy. In its place it is customary to utilize the quantity $\nu = 1/\tilde{\lambda}$, representing the number of waves per centimetre and thus expressed in units of reciprocal centimetres (cm^{-1}) or 'wavenumbers'.

Recently the term 'Kayser' has been introduced to replace wavenumber but, the recommendation for the use of this unit is almost systematically ignored in the spectroscopic literature. In the same way the introduction of other units such as the 'fresnel' for $\tilde{\nu}$ (1 fresnel $= 10^{12}$ cycles/s $=$ 1 terahertz) does not seem to have great popularity. For this reason we shall continue in this book to use wavenumbers and microns as units for ν and $\tilde{\lambda}$ respectively. We notice also that, although ν and $\tilde{\nu}$ differ by a factor c ($\nu = \tilde{\nu}/c$), the word 'frequency' is generally utilized for both quantities.

When a molecule undergoes a vibrational motion, it may absorb radiation of frequencies falling in the infrared region of the electromagnetic spectrum, which extends roughly from the lower frequency limit of the visible spectrum ($\nu \simeq 10{,}000\ \mathrm{cm}^{-1}$) to the upper frequency limit of the microwave spectrum ($\nu \simeq 10\ \mathrm{cm}^{-1}$). The absorption of infrared radiation is a very selective phenomenon for a molecule in the sense that it is localized in several small frequency intervals (absorption bands) which are specific to the compound studied and correspond to different types of vibrational motion.

In addition, when a sample of a molecular compound is irradiated with monochromatic light of a frequency that cannot be absorbed directly by the molecules (for instance visible light for non-coloured compounds), the light scattered by the sample will contain, in addition to the exciting frequency, frequencies differing from this by amounts equal to vibrational frequencies of the molecules. This phenomenon takes the name of *Raman* effect and constitutes a powerful method, complementary to infrared spectroscopy, for the study of molecular vibrations. The Raman spectrum presents the same degree of specificity and selectivity as the infrared spectrum and also consists of bands characteristic of the compound studied.

There is a large amount of information in the infrared and Raman spectra of a polyatomic molecule, and the task of the spectroscopist is essentially to unravel the information in terms of molecular structures. An enormous amount of work has been accumulated in the last fifty years on the interpretation of the vibrational spectra of molecular compounds and a great amount of significant information is extracted from these spectra and currently utilized for the understanding of molecular properties.

The information is localized at the molecular level and is carried to a great extent by the frequencies of the molecular vibrations. The vibrational frequencies depend upon the masses of the atoms, the molecular geometry and the forces binding the atoms together in the molecule. Once the masses and the geometry are known, it is therefore possible to evaluate the binding forces from the vibrational frequencies. Conversely, if the forces are known the spectrum can be calculated for possible molecular conformations and thus be used to elucidate the molecular structure. The theoretical treatment of molecular vibrations that permits the use of the spectral information for a quantitative evaluation of the binding forces involves a considerable amount of classical and quantum mechanics and the handling of some advanced mathematical techniques. This book is essentially concerned with this aspect of vibrational spectroscopy and thus little space is devoted to other applications or to the experimental problems. The importance of these other aspects, however, should not be underestimated. By far the most popular use of the information derived from the vibrational spectra is the analysis of the chemical constitution of the molecules. Although well grounded theoretically, this 'chemical spectroscopy' has been developed on an empirical basis, through the correlation between chemical constitution and characteristic vibrational frequencies. It has long been recognized that the occurrence of chemical groups in a molecule is associated with the appearance of characteristic infrared and Raman bands in specific and well defined regions of the spectrum with frequencies which depend upon the nature of the group and its position relative to other groups. The correlation between chemical constitution and vibrational bands has now reached such a high level of sophistication that it constitutes an almost independent branch of vibrational spectroscopy in which theoretical concepts, structural arguments and experimental findings are amalgamated in a unitary and self-consistent treatment.

Information concerning the intermolecular forces is also carried by the molecular vibrational frequencies, since they are very sensitive to the molecular environment and change from the gaseous to the liquid and to the solid state. These changes are normally small for Van der Waals forces but furnish direct evidence of intermolecular interactions. For strong directional intermolecular forces the frequency variations can be very large. In the case of hydrogen-bonded systems, shifts of several hundred wavenumbers are observed for some bands involving the motion of the hydrogen-bonded groups. In recent years the use of vibrational frequencies for the determination of intermolecular potentials in solids has assumed a particular relevance and has been put on a rigorous and quantitative basis.

Important information is also carried by the intensity of the infrared and Raman bands, concerning essentially the distribution of charges in the molecules and its variation with the nuclear displacements. The intensity of infrared bands is actually related to the variation of the molecular dipole moment and the intensity of the Raman bands to the variation of the molecular polarizability in the vibrations. The manipulation of this information in terms of molecular properties is not simple and in our opinion has not completely reached its definitive assessment. For this reason we shall present in this book the basic quantum-mechanical treatment of infrared and Raman intensities but we shall not enter into the discussion of the recent developments of the theory. This is an area of rapid development and any detailed treatment is likely to become rapidly obsolete. In addition, the connection between infrared intensities and molecular structure is discussed in detail in a book by Gribov[1] entitled *Intensity Theory for Infrared Spectra of Polyatomic Molecules* and the theory of Raman intensities in two recent books on the Raman effect edited by Szymanski[2] and Anderson[3] respectively. It would be difficult to add anything important to what is contained in these books and thus we recommend their reading to the student interested in this argument.

There are still other pieces of information in the vibrational spectra that are not yet properly utilized or fully understood, for instance those relative to the lifetime of the vibrational states and to the energy transfer processes between these states. Considerable progress has been made in recent years on these topics, but they remain for the moment of a very specialistic character and are thus better treated in review articles than in standard texts.

1.2 Infrared spectra

In principle the infrared spectrum of a chemical compound can be observed either in absorption or in emission, since the molecules can absorb the same radiations that they can emit. In practice, however, the absorption spectrum is much more convenient for laboratory measurements and thus this is the type of spectrum that is normally utilized, except in special cases (astrophysical problems, emission of flames, etc.).

In an infrared absorption measurement, the radiation emitted from a source at high temperature is passed through the sample, analysed by a spectrograph and sent on to a suitable detector. The signal produced by the detector is then amplified and fed to a recording system. In modern instruments the ratio I/I_0, where I and I_0 are the intensities of the radiation after and before passing through the sample, is plotted versus the frequency (in cm^{-1}) in a continuous graph. Normal infrared spectrographs of the best commercial type have resolving powers of the order of 0·1 cm^{-1} and cover the spectral region from 5000 to 200 cm^{-1}. Far-infrared spectrographs and interferometers are also available to extend the range up to about 10 cm^{-1} with comparable resolution.

The infrared spectrum of a substance in the gas phase at low pressure can be considered, to a very good approximation, as characteristic of the isolated molecules and free from distortions introduced by intermolecular interactions. When the gas spectrum is measured under low resolution the spectral pattern observed consists of a certain number of broad bands corresponding to the different molecular vibrations that can produce absorption of infrared radiation. If, however, the spectrum is measured under sufficient resolution, the bands may show a fine structure consisting of several sharp lines. These lines correspond to transitions between rotational levels and originate from the fact that changes in the vibrational energy of a molecule may be accompanied by simultaneous changes of the rotational energy. The rotational lines appear as a fine structure of the vibrational bands simply because the separation between rotational levels is much smaller than the separation between vibrational levels. The separation between two successive rotational levels is large enough to be observed only if the moments of inertia of the molecule are relatively small. For this reason the rotational structure of vibrational bands is observed for small molecules only. For molecules of larger dimensions the spacing of the rotational lines is too small to be resolved, even under high resolution, and thus only the envelope of the rotational structure is observed in the spectrum. The analysis of the rotational structure of vibrational bands permits quantitative information to be obtained on the molecular structure as well as on the binding forces in the molecule. The basic theory of molecular rotations and the calculation of rotational energies is not included in this book although data obtained from vibro-rotational and rotational spectra will often be utilized. The reader interested in learning the theory is referred to the book[4] by G. M. Barrow, *Molecular Spectroscopy*, for the basic theory and to the book[5] by H. C. Allen and P. C. Cross, *Molecular Vib-rotors*, for a more advanced treatment.

In the infrared spectra of liquids the rotational fine structure is completely smeared out by the high rate of molecular collisions and by the random effect of the intermolecular interactions. The intermolecular interactions also produce frequency shifts and intensity variations of the vibrational bands, whereas the deformations in the molecular symmetry due to the combined effect of the collisions and of the intermolecular interactions may cause the appearance in the spectrum of bands which are absent for symmetry requirements in the gas spectrum.

The infrared spectra of solids, especially at very low temperatures, show very sharp vibrational bands because of the absence of rotational structure and of broadening due to the collisions. Since the intermolecular forces couple together the vibrations of the molecules in the crystal, the vibrational bands are split into several components depending upon the number of molecules in the unit cell and their site symmetry. Since the site symmetry is normally lower than the symmetry of the isolated molecule, bands inactive by symmetry in the gas phase may appear with noticeable intensity.

Many features of the infrared spectra of polyatomic molecules may be explained classically. According to classical electromagnetic theory, a periodic variation of the dipole moment of the molecule gives rise to absorption or emission of radiation of frequency equal to the frequency of the oscillation of the dipole moment. A polyatomic molecule behaves, as far as its vibrational properties are concerned, as a system of coupled oscillators and thus possesses a finite number of proper frequencies of vibration, which are the only simple periodic motions that the system can perform (normal vibrations). In the harmonic approximation, any complex motion of the molecule is then a combination of some of these normal vibrations with appropriate amplitudes and phases. During a normal vibration of the molecule the atoms are periodically displaced from their equilibrium positions and correspondingly the charges within the molecule are displaced with the same period of oscillation. This periodic displacement of the charges may or may not produce a periodic variation of the molecular dipole moment, depending upon the actual form of the vibrational motion. The net result is then that the molecule will absorb or emit all the radiations of frequency equal to those normal vibrations which produce a dipole moment variation. In order to obtain a dipole moment variation during a normal vibration it is not necessary for the molecule to possess a permanent dipole moment. Consider, for instance, the two normal vibrations of the linear molecule CO_2 shown schematically in Figure 1.2.1. In the first type of vibration (Figure 1.2.1a) the two oxygen atoms move in phase away from the carbon atom in a half-cycle of the vibration and in phase toward the carbon atom in the next half-cycle. The dipole moment that was zero by symmetry for the equilibrium configuration, remains constantly zero during the vibration. Because of the absence of a dipole moment oscillation this vibration cannot interact with the electromagnetic radiation

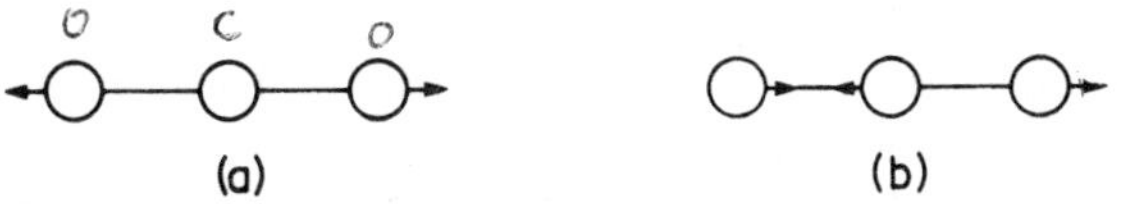

Figure 1.2.1. Schematic representation of two of the normal vibrations of CO_2. The arrows represent the direction in which the atoms move in a half-cycle of the vibration. After reaching the maximum displacements the atoms invert their directions of motion

and thus cannot give rise to an absorption band in the infrared. In the second type of vibration, the oxygen atoms move together (Figure 1.2.1b) toward the right while the carbon atom moves toward the left in a half-cycle. In the next half-cycle the direction of motion of the atoms is inverted. During the vibration the molecular symmetry is lost, each bond being alternately shortened and lengthened with respect to the other. As a result the centre of the positive charges in the molecule is periodically displaced from the centre of the negative charges and a periodic variation of the dipole moment is thus produced.

Vibrations associated with a dipole moment variation are called *infrared active* since they give rise to infrared absorption, whereas vibrations for which the dipole variation is zero by symmetry are called *infrared inactive* since they are unable to produce absorption of infrared radiation.

In the harmonic approximation the vibrational motions of the molecule contain only the frequencies of the fundamental modes of vibration. The fact that the infrared spectra of molecular compounds contain, in addition to the infrared active fundamentals, a number of weak bands with frequencies roughly equal to $2\nu_i, 3\nu_i, \ldots$, as well as to $\nu_i + \nu_k$, $\nu_i - \nu_k$, $\nu_i + 2\nu_k$, etc., does not contradict the classical interpretation but only shows that the harmonic approximation is insufficient. As a matter of fact, the occurrence of vibrational motions of frequencies $n\nu_i$ (*overtone vibrations*) and of frequencies $n\nu_i \pm m\nu_k$ (*combination* or *difference vibrations*) can also[6] be explained classically by dropping the harmonic approximation and considering anharmonic motions. As for the fundamentals, overtones, combinations and difference vibrations will give rise to infrared bands whenever a dipole variation is produced during the vibrations.

The band structure of the infrared spectra is therefore well reproduced by the classical theory. However, as soon as one tries to explain the details of the spectra, the classical interpretation fails. For instance the vibro-rotational fine structure of the bands cannot be accounted for, since classically the rotational energy is a continuous function of the angular velocity which can assume all possible values. In the next section we shall see that the intensity of Raman bands is also incorrectly predicted by the classical theory.

All these difficulties are eliminated in the quantum-mechanical treatment in which only discrete values are possible for the rotational and vibrational energies. Transitions between these energy levels occur with absorption or emission of radiation according to Bohr's rule

$$\nu = (E_{vj} - E_{v'j'})/hc \tag{1.2.1}$$

where h is Planck's constant and E_{vj}, $E_{v'j'}$ are two possible vibro-rotational energy levels of the molecule.

We consider now pure vibrational transitions, i.e. transitions in which the rotational energy does not change. In this case the energy levels, assuming harmonic motions of the nuclei, are given by the expression

$$E_v = hc \sum_i \nu_i(v_i + \tfrac{1}{2}) \tag{1.2.2}$$

where the ν_i are the classical normal frequencies of vibration of the molecule and the v_i are quantum numbers that can assume only integer values.

From Bohr's rule and from expression (1.2.2) it follows that the transition from the lowest energy level of the molecule to a higher level in which only the quantum number v_k is equal to one, all others being still equal to zero, gives rise to absorption of the classical vibrational frequency ν_k. In the same way, transitions to levels in which only one quantum number is multi-excited give rise to overtones and transitions to levels in which two (or more) quantum numbers are excited together, give rise to combination bands. The inverse transitions from higher levels to the lowest give rise, instead, to emission of radiation of the same frequencies.

In quantum mechanics a radiative transition can, however, take place only if the integral

$$\int \psi_{v_k} \mu \psi_{v'_k} \, d\tau \qquad (1.2.3)$$

is different from zero, where ψ_{v_k}, $\psi_{v'_k}$ are the wavefunctions associated with the two states and μ is the dipole moment. It turns out that, when ψ_{v_k} and $\psi_{v'_k}$ are harmonic oscillator wavefunctions, the integral (1.2.3) is always zero unless $v'_k = v_k \pm 1$. In the harmonic approximation, therefore, the only transitions that can take place with absorption ($\Delta v_k = +1$) or emission ($\Delta v_k = -1$) of radiation are those between levels differing by one unit in one quantum number only. The absorption or emission frequencies are then exactly the classical normal frequencies of the molecule.

The fact that the quantum-mechanical frequencies of vibration of the molecule are identical to these found in the classical treatment is of extreme importance in vibrational spectroscopy since it permits the calculation of the harmonic frequencies of vibration of the molecules using standard methods of classical mechanics.

When the harmonic approximation is abandoned and anharmonic motions of the nuclei are taken into account, overtones and combination transitions also become possible in the quantum treatment. In this case, however, relation (1.2.2) becomes more complex.

1.3 Raman spectra

When the spectrum of the radiation scattered by a sample illuminated with monochromatic light is observed, it is found that it consists of a very strong line at the frequency ν_0 of the incident light, as well as of a series of weaker lines at both higher and lower frequencies. The important fact for the utilization of Raman spectra in vibrational spectroscopy is that the frequencies of these new lines differ from ν_0 by amounts $\Delta\nu$ equal to the vibrational frequencies of the molecules of the sample. Another important feature is that to each band displaced from ν_0 toward lower frequencies there corresponds another band, with much lower intensity, displaced by the same amount toward higher frequencies. The Raman bands occurring at frequencies lower than ν_0 are called *Stokes* whereas those occurring at higher frequencies are called *anti-Stokes*.

In a Raman experiment the light scattered by the sample is normally observed at 90° from the direction of the incident light, although observations at 180° are often utilized. The scattered light is collected by a condensing lens and focused on the entrance slit of a spectrograph. To reduce the background, double or even triple monochromators are utilized. For many years the line at 4358 Å of the mercury arc has been the principal light source for Raman spectroscopy. Since the development of high-power lasers, the mercury arc has been almost completely replaced by these more convenient sources. The choice of the exciting frequency is dictated by the need to avoid direct absorption from the sample and to utilize lines of high intensity and high frequency. Today the most used laser lines are those at 6338 Å of the Helium–Neon laser and at 4880 or 5145 Å of the Argon ion laser. Many other laser sources are, however, available and especially with the development of tunable dye lasers there is practically no limitation in the choice of the exciting line.

The classical theory of Raman scattering is based on the idea that the electromagnetic field of the incident light induces in the molecules a variable dipole moment which oscillates with the frequency of the radiation and thus acts as an emitter of radiation in all directions. The dipole moment induced in a molecule by the electric vector $\mathbf{E} = \mathbf{E}_0 \cos(2\pi\nu_0 t)$ of the incident radiation is given by

$$\boldsymbol{\mu} = \boldsymbol{\alpha}\mathbf{E}_0 \cos(2\pi\nu_0 t) \tag{1.3.1}$$

where $\boldsymbol{\alpha}$ is the molecular polarizability, $\mathbf{E}_0$ the amplitude of the electric field oscillation and ν_0 the frequency of oscillation.

According to classical electromagnetic theory, a dipole oscillating with frequency ν emits radiation of this frequency, with an intensity given by[7]

$$I = \frac{16\pi^4\nu^4}{3c^2}|\mu|^2 \tag{1.3.2}$$

Thus, if the molecular polarizability were a constant quantity, by substitution of (1.3.1) into (1.3.2) and by taking $\nu = \nu_0$ we would obtain

$$I = \frac{16\pi^4\nu_0^4}{3c^2}\alpha^2 E_0^2 \cos^2(2\pi\nu_0 t) \tag{1.3.3}$$

The molecular polarizability is, however, not a constant for a vibrating molecule since it depends upon the molecular conformation which changes with the displacements of the nuclei during the vibrations. If a normal vibration of frequency ν_k is excited, the polarizability will oscillate with frequency ν_k according to a relation of the type

$$\boldsymbol{\alpha} = \boldsymbol{\alpha}_0 + \boldsymbol{\alpha}_k \cos(2\pi\nu_k t + \varphi_k) \tag{1.3.4}$$

where φ_k is a phase factor, $\boldsymbol{\alpha}_0$ a constant and $\boldsymbol{\alpha}_k$ the maximum change in $\boldsymbol{\alpha}$ when the molecule vibrates. Using (1.3.4) the induced dipole moment (1.3.1) then becomes

$$\boldsymbol{\mu} = \boldsymbol{\alpha}_0\mathbf{E}_0 \cos(2\pi\nu_0 t) + \boldsymbol{\alpha}_k\mathbf{E}_0 \cos(2\pi\nu_0 t)\cos(2\pi\nu_k t + \varphi_k)$$

i.e.

$$\boldsymbol{\mu} = \boldsymbol{\alpha}_0\mathbf{E}_0 \cos(2\pi\nu_0 t) + \tfrac{1}{2}\boldsymbol{\alpha}_k\mathbf{E}_0\{\cos[2\pi(\nu_0 + \nu_k)t + \varphi_k] + \cos[2\pi(\nu_0 - \nu_k)t - \varphi_k]\} \tag{1.3.5}$$

Therefore, when the molecule vibrates with frequency ν_k, the induced dipole oscillates not only with frequency ν_0 but also with frequencies $(\nu_0 + \nu_k)$ and $(\nu_0 - \nu_k)$.

By use of (1.3.5) the intensity of the scattered light is then given by

$$I = \frac{16\pi^4}{3c^2}E_0^2\left\{\nu_0^4\alpha_0^2\cos^2(2\pi\nu_0 t) + (\nu_0 + \nu_k)^4\frac{\alpha_k^2}{4}\cos^2[2\pi(\nu_0 + \nu_k)t + \varphi_k]\right.$$
$$\left. + (\nu_0 - \nu_k)^4\frac{\alpha_k^2}{4}\cos^2[2\pi(\nu_0 - \nu_k)t - \varphi_k]\right\} + \text{cross-terms} \tag{1.3.6}$$

where the cross-terms can be neglected since the power they irradiate averages to zero on time intervals of sufficient length. The first term in (1.3.6) is the intensity scattered at the frequency ν_0 of the incident radiation (*Rayleigh scattering*) and is a coherent type of scattering since it is in phase with the incident light. The second and the third terms represent the intensity of the radiation scattered at the anti-Stokes frequency $(\nu_0 + \nu_k)$

and at the Stokes frequency $(\nu_0 - \nu_k)$ respectively. Since there is a phase factor which will be different from molecule to molecule, the scattered radiation will be incoherent.

The classical theory thus correctly predicts the occurrence of the Stokes and anti-Stokes lines but leads to an incorrect ratio of their intensities. From (1.3.6) it follows that the ratio of the intensities of the Stokes and anti-Stokes lines should be

$$\frac{I_{\text{Stokes}}}{I_{\text{anti-Stokes}}} = \frac{(\nu_0 - \nu_k)^4}{(\nu_0 + \nu_k)^4} \tag{1.3.7}$$

and thus should be less that the unity, whereas experimentally it is found that the Stokes lines are much stronger than the anti-Stokes lines. This inconsistency is eliminated in the quantum theory of the Raman effect. Since the complete quantum theory is rather complex we shall give here only a phenomenological interpretation in terms of discrete energy levels and of photons. The complete theory will be discussed in detail in Chapter 3.

In terms of the corpuscular theory of light the Rayleigh scattering corresponds to an elastic collision process between the photon and the molecule, whereas Raman scattering corresponds to an inelastic collision in which the photon either loses a quantum of vibrational energy (Stokes lines) or acquires a quantum of vibrational energy (anti-Stokes lines). This picture accounts immediately for the higher intensity of the Stokes lines. In an assembly of molecules at not very high temperature, the number of molecules excited to the vibrational level E_{v_k} is always a small fraction of the total number of molecules in the lowest vibrational level. The number of elementary scattering processes in which the molecules give the energy $h\nu_k$ to the photons is thus small with respect to the number of processes in which the molecules acquire the energy $h\nu_k$ from the photons.

A deeper insight into the scattering of light can be obtained through a reasonable model of the elementary perturbation process that involves a photon and a molecule. A photon of energy $h\nu_0$ in the visible or ultraviolet region of the spectrum, when approaching a molecule, sets up a perturbation of its electronic wavefunctions, because only electrons are light enough to follow the fast-changing field of the photon. The wavefunctions of the perturbed system then acquire a mixed character and become linear combinations of all possible wavefunctions of the unperturbed molecule with time-dependent coefficients. Through this perturbation the photon recognizes the stationary states of the molecule and the molecule measures the energy of the photon.

For a very small time interval the identity of the photon is lost and its energy becomes spread out over the entire molecule, being indistinguishable from all the changes in the kinetic and potential energy of the electrons caused by the perturbation. We can formally regard the molecule as having attained a non-stationary energy level of higher energy by considering the perturbation energy as belonging to the molecule. This formal description has nothing to do with the concept of energy levels used to describe the absorption process. For this reason we shall speak of a *virtual* level in order to specify that it is introduced into the discussion only for a modellistic description of the perturbation process.

As soon as the photon discovers that the molecule has no stationary levels of energy $h\nu_0$, it leaves the unstable situation in which it has been trapped. In terms of the same modellistic representation we can formally consider the photon as being emitted by the perturbed molecule which jumps back to one of its stationary states. In this process the photon can recover the entire energy and thus the molecule returns to its original state, that in the great majority of the cases is the ground state. This gives rise to Rayleigh scattering since the frequency of the photon remains the same and all that can happen to it is a change in the direction of propagation.

With a much lower probability, however, the photon can loose part of its energy in the interaction process and thus leave the molecule with a lower energy $h\nu$. Since the molecule must return to a stationary state, the difference $h\nu_0 - h\nu$ must correspond to a quantum of vibrational or rotational energy. In the same way, the photon can leave the molecule also with a higher energy if by chance it finds the molecule in an excited vibrational or rotational energy level and the molecule jumps after the interaction process, to the ground level.

It is important to notice that the absorption of the photon $h\nu_0$ and the emission of the photon $h\nu$ are truly simultaneous events and cannot be separated in time one from the other. In our description, however, we have established a sequential occurrence of events in which the emission of the photon from the virtual state follows the absorption process. We might equally well split up the scattering into perfectly equivalent sequential events in which the emission of the photon precedes the absorption process. Neither of these paths can in principle be preferred and thus both must be considered together. This is shown schematically in Figure 1.3.1 where the virtual levels are indicated by broken lines and the stationary levels by full lines. Arrows are used to indicate transitions among them.

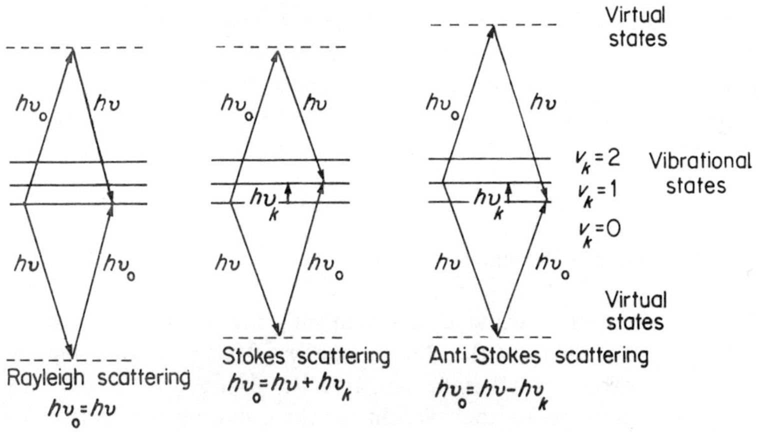

Figure 1.3.1 Schematic representation of the Raman effect

As the frequency of the incident radiation approaches an electronic absorption of the molecule, the intensity of the Raman bands is strongly enhanced. Raman spectra obtained with exciting frequencies close to absorption bands of the compound studied are called '*resonance* Raman spectra' in order to emphasize the fact that a resonant denominator occurs in the quantum-mechanical description and that this causes large deviations from the intensity relation (1.3.2). When ν_0 coincides with an absorption frequency of the molecule we have the fluorescence phenomenon and at first sight it might seem that the Raman effect goes into fluorescence with continuity as the exciting frequency goes to an absorption frequency of the molecule. Fluorescence and the Raman effect should, however, not be confused. In the case of fluorescence the incident photon is truly absorbed by the molecule, which reaches an excited stationary state with a well defined lifetime. After some time the molecule returns to a lower energy state and, if this is different from the initial one, irradiates light of frequency lower than the exciting frequency. The two steps are then really sequential in time and in fact fluorescence can be quenched by adding to the system a species that can take away by collision the excitation energy from the excited molecule in the time interval between absorption and emission. This is not possible in the case of the Raman effect, since there is no delay between the annhilation of the incident photon and the creation of the scattered photon. The fact that the scattering intensity is much greater when ν_0 is close to an absorption frequency simply reflects a greater efficiency of the perturbation in these conditions.

The general formulation of the quantum theory of scattering is not simple. Under some specific conditions, that are actually fulfilled by the large majority of molecules, the quantum-mechanical analogue of the polarizability, i.e.

$$\alpha_{v_k v'_k} = \int \psi_{v_k} \alpha \psi_{v'_k} \, d\tau \tag{1.3.8}$$

can be utilized, as shown by Placzek.[8] These conditions are:

(a) The ground electronic state must be non-degenerate.
(b) The Born–Oppenheimer approximation must be valid.
(c) The exciting frequency must be less than any electronic transition of the molecule although much larger than any vibrational frequency.

Although the theory can be modified to include the cases in which these conditions are not satisfied, we shall here limit ourselves to the standard treatment.

When the harmonic oscillator wavefunctions are used in (1.3.8), the integral is always zero unless $v'_k = v_k \pm 1$. In the harmonic approximation, therefore, only fundamentals are allowed in the Raman spectrum, a result that parallels the one obtained for the infrared spectrum. Anharmonic motions of the nuclei must be taken into account for the Raman activity of overtones and combination bands.

1.4 The polarization of Raman bands

In the previous section we have seen that only the quantum treatment gives a correct explanation of all features of the Raman spectra. However, valuable indications on the handling of the experimental data can also be obtained from the classical treatment, since these indications retain their validity in the quantum-mechanical formulation.

Owing to the fact that the classical treatment is much simpler and that it furnishes a convenient basis for the introduction of the quantum-mechanical interpretation, we shall utilize the classical concepts of induced dipole and of molecular polarizability for the discussion of the polarization of Raman bands, a problem of considerable practical interest in the interpretation of Raman spectra.

We begin by noticing that the relationship

$$\boldsymbol{\mu} = \boldsymbol{\alpha} \mathbf{E} \tag{1.4.1}$$

between the induced dipole and the electric field vector is a vectorial relation since both $\boldsymbol{\mu}$ and $\mathbf{E}$ are vectors. The *molecular polarizability* $\boldsymbol{\alpha}$ is thus a second-order tensor and then equation (1.4.1) written in full has the form

$$\begin{pmatrix} \mu_X \\ \mu_Y \\ \mu_Z \end{pmatrix} = \begin{pmatrix} \alpha_{XX} & \alpha_{XY} & \alpha_{XZ} \\ \alpha_{YX} & \alpha_{YY} & \alpha_{YZ} \\ \alpha_{ZX} & \alpha_{ZY} & \alpha_{ZZ} \end{pmatrix} \begin{pmatrix} E_X \\ E_Y \\ E_Z \end{pmatrix} \tag{1.4.2}$$

where the components of $\boldsymbol{\mu}$, $\mathbf{E}$ and $\boldsymbol{\alpha}$ are referred to a space-fixed reference system (X, Y, Z).

As in all tensors relating to physical quantities in present theories, $\boldsymbol{\alpha}$ is symmetric, i.e. $\alpha_{RS} = \alpha_{SR}$.

When the molecular orientation is fixed in space we expect from (1.4.2) different band intensities for different directions of observation, when linearly polarized light, such as that of a laser beam, is used as exciting line. Consider for instance the case, sketched in

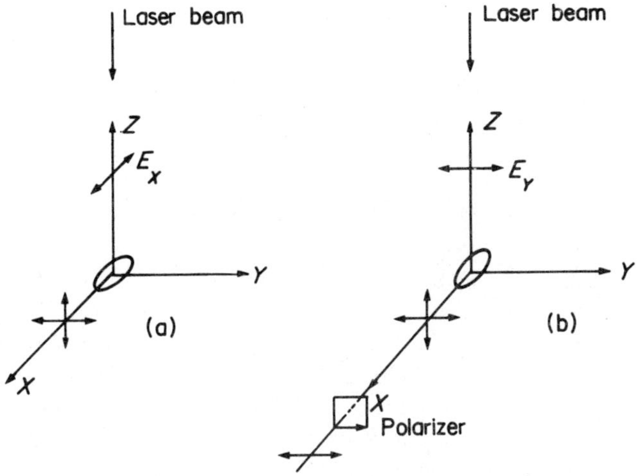

Figure 1.4.1 Schematic representation of the experiments for measuring I_T (obs $\|$), I_T (obs $\perp$) and $I_{\|}$ (obs $\perp$)

Figure 1.4.1(a), in which the laser beam propagates along the Z-axis with the electric vector parallel to X, the scattered light being observed in the X-direction. In this case we have for the exciting beam

$$E = E_X \neq 0 \qquad E_Y = E_Z = 0$$

and thus the intensity of the light collected in the X-direction per unit solid angle is given by

$$I_T(\text{obs } \|) = \frac{2\pi^3 \nu^4}{c^3}(\mu_Z^2 + \mu_Y^2) = \frac{2\pi^3 \nu^4}{c^3}(\alpha_{ZX}^2 + \alpha_{YX}^2)E_X^2 \tag{1.4.3}$$

where the symbol $I_T(\text{obs } \|)$ means that we are observing the total scattered light in a direction parallel to that of the electric vector of the incident beam.

Consider now the case in which the laser beam is still propagating along the Z-axis but the electric vector oscillates in the Y-direction (Figure 1.4.1b). In this case we have

$$E = E_Y \neq 0 \qquad E_X = E_Z = 0$$

and thus

$$I_T(\text{obs } \perp) = \frac{2\pi^3 \nu^4}{c^3}(\mu_Z^2 + \mu_Y^2) = \frac{2\pi^3 \nu^4}{c^3}(\alpha_{ZY}^2 + \alpha_{YY}^2)E_Y^2 \tag{1.4.4}$$

where the symbol $I_T(\text{obs } \perp)$ means that we observe the total scattered light in a direction perpendicular to that of the electric vector of the incident beam. By insertion of a polarizer after the sample we can collect only the part of the scattered light which is polarized in the direction of E_Y. The intensity in this case will be indicated by the symbol $I_{\|}(\text{obs } \perp)$

and is given by

$$I_{\parallel}(\text{obs} \perp) = \frac{2\pi^3\nu^4}{c^3}\mu_Y^2 = \frac{2\pi^3\nu^4}{c^3}\alpha_{YY}^2 E_Y^2 \tag{1.4.5}$$

Using the relation[7]

$$I_0 = \frac{c}{8\pi}E^2 \tag{1.4.6}$$

equations (1.4.3), (1.4.4) and (1.4.5) become

$$\begin{aligned} I_{\mathrm{T}}(\text{obs} \parallel) &= \frac{16\pi^4\nu^4 I_0}{c^4}(\alpha_{ZX}^2 + \alpha_{YX}^2) \\ I_{\mathrm{T}}(\text{obs} \perp) &= \frac{16\pi^4\nu^4 I_0}{c^4}(\alpha_{ZY}^2 + \alpha_{YY}^2) \\ I_{\parallel}(\text{obs} \perp) &= \frac{16\pi^4\nu^4 I_0}{c^4}(\alpha_{YY}^2) \end{aligned} \tag{1.4.7}$$

A unique feature of Raman spectra is that band polarization is also observed in liquids and gases where the molecules are randomly oriented. This is due to the fact that the squares of the polarizability components average to different values over all angles. In order to extend equations (1.4.7) to liquids and gases, we need to introduce another reference system (x, y, z), attached to the molecule and rotating with it.

If $\mathbf{T}$ is the transformation matrix from the space-fixed to the molecule-fixed reference system, we have

$$\boldsymbol{\mu}' = \mathbf{T}\boldsymbol{\mu} \qquad \mathbf{E}' = \mathbf{T}\mathbf{E} \tag{1.4.8}$$

where $\boldsymbol{\mu}'$ and $\mathbf{E}'$ are the dipole moment and the electric vector in the rotating system. We notice that $\mathbf{T}$ is an orthogonal matrix and that its elements are the direction cosines of the rotating system with respect to the fixed system of axes.

From (1.4.1) and (1.4.8) we obtain

$$\boldsymbol{\mu}' = \mathbf{T}\boldsymbol{\mu} = \boldsymbol{\alpha}'\mathbf{E}' = \boldsymbol{\alpha}'\mathbf{T}\mathbf{E} \tag{1.4.9}$$

$$\boldsymbol{\mu} = \mathbf{T}^{-1}\boldsymbol{\alpha}'\mathbf{T}\mathbf{E}$$

i.e.

$$\boldsymbol{\alpha} = \mathbf{T}^{-1}\boldsymbol{\alpha}'\mathbf{T} \tag{1.4.10}$$

or

$$\boldsymbol{\alpha}' = \mathbf{T}\boldsymbol{\alpha}\mathbf{T}^{-1} \tag{1.4.11}$$

Among the possible molecule-fixed systems, it is convenient to choose the particular one in which $\boldsymbol{\alpha}$ is diagonal and is thus of the form

$$\begin{pmatrix} \alpha_{xx} & 0 & 0 \\ 0 & \alpha_{yy} & 0 \\ 0 & 0 & \alpha_{zz} \end{pmatrix} \tag{1.4.12}$$

The axes (x, y, z) in which the polarizability is diagonal are called the *principal axes* of the polarizability ellipsoid since the set of equations (1.4.9) to (1.4.11) defines in general the transformation of the equation of an ellipsoid from a generic set of axes to the principal axes.

An important property of transformations like (1.4.11) is that they leave unaltered the trace of the tensor. We can then define a quantity $\bar{\alpha}$ that is invariant under a rotation of the axes, by means of the relation

$$\bar{\alpha} = \tfrac{1}{3}(\alpha_{XX} + \alpha_{YY} + \alpha_{ZZ}) = \tfrac{1}{3}(\alpha_{xx} + \alpha_{yy} + \alpha_{zz}) \tag{1.4.13}$$

We shall call $\bar{\alpha}$ the *mean value* or the *isotropic part* of the polarizability. We can also define a quadratic quantity γ^2 that is invariant under a rotation of axes through the relation

$$\begin{aligned}\gamma^2 &= \tfrac{1}{2}[(\alpha_{XX} - \alpha_{YY})^2 + (\alpha_{YY} - \alpha_{ZZ})^2 + (\alpha_{ZZ} - \alpha_{XX})^2 + 6(\alpha_{XY}^2 + \alpha_{YZ}^2 + \alpha_{ZX}^2)] \\ &= \tfrac{1}{2}[(\alpha_{xx} - \alpha_{yy})^2 + (\alpha_{yy} - \alpha_{zz})^2 + (\alpha_{zz} - \alpha_{xx})^2]\end{aligned} \tag{1.4.14}$$

We shall call γ^2 the *anisotropy* of the ellipsoid since γ^2 is a direct measure of the departure of the ellipsoid from a sphere. The importance of $\bar{\alpha}$ and γ^2 is that the average values of the components of α over all angles are easily expressed in terms of these quantities.

As discussed before, equations (1.4.7) refer to a molecule fixed in space. In order to obtain the corresponding equations for a liquid or for a gas with random molecular orientations, equations (1.4.7) must be multiplied by the number of molecules and the average value of the squares of the components of $\boldsymbol{\alpha}$ must be computed for all possible orientations of the principal axes with respect to the fixed axes system.

If we consider a generic element of equation (1.4.10), this will have the form

$$\alpha_{RS} = \sum_i \alpha_{ii} T_{Ri} T_{Si} \qquad \begin{array}{c} i = x, y, z \\ R, S = X, Y, Z \end{array} \tag{1.4.15}$$

where T_{Ri} and T_{Si} are direction cosines of the molecule-fixed reference system with respect to the space-fixed reference system.

The average value of the square of a component of $\boldsymbol{\alpha}$ is then

$$\overline{\alpha_{RS}^2} = \sum_i \alpha_{ii}^2 \overline{T_{Ri}^2 T_{Si}^2} + 2\sum_{ij} \alpha_{ii}\alpha_{jj} \overline{T_{Ri} T_{Si} T_{Rj} T_{Sj}} \tag{1.4.16}$$

since the α_{ii} are invariant under a rotation of axes.

In order to calculate these averages use can be made of the orthogonality properties of the **T**-matrix, summarized in the relations

$$\begin{aligned} T_{Xi}^2 + T_{Yi}^2 + T_{Zi}^2 &= 1 \\ T_{Xi}T_{Xj} + T_{YiYj} + T_{Zi}T_{Zj} &= 0 \end{aligned} \tag{1.4.17}$$

By squaring these relations and by averaging we obtain

$$\begin{aligned} 3\overline{T_{Ri}^4} + 6\overline{T_{Ri}^2 T_{Rj}^2} &= 1 && \text{(a)} \\ 3\overline{T_{Ri}^2 T_{Rj}^2} + 6\overline{T_{Ri} T_{Si} T_{Rj} T_{Sj}} &= 0 && \text{(b)} \end{aligned} \tag{1.4.18}$$

It is thus evident that once $\overline{T_{Ri}^4}$ has been obtained, all other averages can be calculated from (1.4.18). Using polar coordinates we have

$$\overline{T_{Ri}^4} = \overline{\cos^4 \theta_{Ri}} = \frac{1}{4\pi}\int_0^{\pi}\int_0^{2\pi} \cos^4 \theta_{Ri} \sin \theta_{Ri}\, d\theta\, d\varphi = \tfrac{1}{5} \tag{1.4.19}$$

and therefore

$$\begin{aligned} \overline{T_{Ri}^2 T_{Rj}^2} &= \tfrac{1}{15} && \text{(a)} \\ \overline{T_{Ri}^2 T_{Si}^2} &= \tfrac{1}{15} && \text{(b)} \\ \overline{T_{Ri} T_{Si} T_{Rj} T_{Sj}} &= -\tfrac{1}{30} && \text{(c)} \end{aligned} \tag{1.4.20}$$

Using (1.4.19) and (1.4.20), equation (1.4.16) yields

$$\overline{\alpha_{RR}^2} = \tfrac{1}{5}\sum_i \alpha_{ii}^2 + \tfrac{2}{15}\sum_{ij} \alpha_{ii}\alpha_{jj} = \tfrac{1}{45}(45\bar{\alpha}^2 + 4\gamma^2) \qquad \text{(a)}$$

$$\overline{\alpha_{RS}^2} = \tfrac{1}{15}\sum_i \alpha_{ii}^2 - \tfrac{1}{15}\sum_{ij} \alpha_{ii}\alpha_{jj} = \tfrac{1}{15}\gamma^2 \qquad \text{(b)} \qquad (1.4.21)$$

As discussed before, equations (1.4.7) can be extended to liquids and gases by multiplying by the number N of molecules and averaging over all orientations. We then have, by use of (1.4.21),

$$I_{\mathrm{T}}(\text{obs} \parallel) = \frac{16\pi^4\nu^4 N I_0}{c^4} \tfrac{2}{15}\gamma^2 \qquad \text{(a)}$$

$$I_{\mathrm{T}}(\text{obs} \perp) = \frac{16\pi^4\nu^4 N I_0}{c^4} \frac{(45\bar{\alpha}^2 + 7\gamma^2)}{45} \qquad \text{(b)} \quad (1.4.22)$$

$$I_{\parallel}(\text{obs} \perp) = \frac{16\pi^4\nu^4 N I_0}{c^4} \frac{(45\bar{\alpha}^2 + 4\gamma^2)}{45} \qquad \text{(c)}$$

For 90° scattering we can define an experimentally measurable quantity ρ_l that represents the ratio of the intensities scattered perpendicular and parallel to the direction of **E**. Obviously ρ_l is given by

$$\rho_l = \frac{I_{\mathrm{T}}(\text{obs} \perp) - I_{\parallel}(\text{obs} \perp)}{I_{\parallel}(\text{obs} \perp)} = \frac{3\gamma^2}{45\bar{\alpha}^2 + 4\gamma^2} \qquad (1.4.23)$$

and is called the *depolarization ratio*. The depolarization ratio is a very useful experimental parameter for the identification of the vibrations associated with Raman bands.

Consider first the case of Rayleigh scattering. According to equation (1.3.6) what matters in this case is the constant part $\boldsymbol{\alpha}_0$ of the tensor $\boldsymbol{\alpha}$. We shall indicate with the symbol α_{ij}^0 the components of $\boldsymbol{\alpha}_0$ and with the symbols $\bar{\alpha}_0$ and γ_0^2 the derived invariants occurring in (1.4.23). We notice that $\bar{\alpha}_0$ can never be equal to zero since this would require $\alpha_{xx}^0 = \alpha_{yy}^0 = \alpha_{zz}^0 = 0$ and thus no scattered radiation would appear. However, the anisotropy γ_0^2 can be equal to zero if the polarizability ellipsoid is a sphere, i.e. if $\alpha_{xx}^0 = \alpha_{yy}^0 = \alpha_{zz}^0$. This situation occurs when the molecule has a very high symmetry and its physical properties are thus isotropic. In this case $\rho_l = 0$ and then the Rayleigh line is completely polarized. For non-isotropic molecules γ_0 is different from zero and then obviously $\rho_l < \frac{3}{4}$.

Consider now the case of Raman scattering. According to equation (1.3.6) the part $\boldsymbol{\alpha}_k$ of the scattering tensor responsible for the scattering process is the part that changes periodically during a normal vibration. We shall use the symbol α_{ij}^k for the components of $\boldsymbol{\alpha}_k$ and the symbols $\bar{\alpha}_k$ and γ_k^2 for the corresponding derived quantities in (1.4.23).

The main difference between Rayleigh and Raman scattering is that $\bar{\alpha}_k$ can be equal to zero since the components α_{ii}^k may be zero or even negative. The polarization ratio ρ_l can thus vary between the limits

$$0 \leqslant \rho_l^{\text{Raman}} \leqslant \tfrac{3}{4}$$

The cases in which $\bar{\alpha}_k$ or γ_k^2 are equal to zero can be obtained by symmetry considerations. The proper use of symmetry in the interpretation of molecular quantities is through the use of group theory. This is done in detail in Chapters 5, 6 and 7. Here we shall be satisfied with giving some general results of practical use in the assignment of Raman bands to normal vibrations of the molecules.

The first important result of the application of group theory is that $\bar{\alpha}_k$ is different from zero only for normal vibrations in which the symmetry of the molecule of equilibrium is preserved. Such vibrations are called *totally symmetric* and an example is given in Figure 1.2.1(a) for CO_2. Therefore, for totally symmetric vibrations $\rho_l < \frac{3}{4}$. Raman bands for which $\rho_l < \frac{3}{4}$ are called *polarized.* For non-totally symmetric vibrations $\rho_l = \frac{3}{4}$ and the Raman bands in this case are called *depolarized.*

A depolarization ratio can also be defined when non-polarized exciting radiation is used. We shall not discuss this case, which is of no practical interest for Raman spectroscopy with a laser source. A discussion of the depolarization ratio for non-polarized exciting radiation can be found in References 2, 3, 6, 7.

The concept of depolarization ratio has been derived here in terms of the classical treatment. Placzek[8] has shown, however, that the quantum-mechanical treatment leads to the same conclusions when a component of the polarizability tensor is replaced by the quantum analogue

$$(\alpha_{ij})_{mn} = \int \psi_m \alpha_{ij} \psi_n \, d\tau \tag{1.4.24}$$

References

1. L. A. Gribov, *Intensity Theory for Infrared Spectra of Polyatomic Molecules*, Consultant Bureau, 1964.
2. *Raman Spectroscopy* (ed. H. A. Szymanski), Plenum Press, 1967.
3. *The Raman Effect* (ed. A. Anderson), Marcel Dekker, 1967, Vols. I and II.
4. G. M. Barrow, *Molecular Spectroscopy*, McGraw-Hill, 1962.
5. H. C. Allen and P. C. Cross, *Molecular Vib-rotors*, Wiley, 1963.
6. G. Herzberg, *Molecular Spectra and Molecular Structure*, D. Van Nostrand Co., 1945.
7. E. B. Wilson Jr., J. C. Decius and P. C. Cross, *Molecular Vibrations*, McGraw-Hill, 1955.
8. G. Placzek, *Handbuch der Radiologie* (ed. E. Marx), Akademische Verlagsgesellschaft, 1934, Vol. VI.

Chapter 2

The Classical Treatment of Molecular Vibrations

2.1 The mechanical model

The study of the motion of a polyatomic molecule in space is a very complex problem. Not only is the molecule free to translate and rotate but in addition all its basic components, nuclei and electrons, are in continuous movement each relative to the others.

In order to solve the dynamical problem of a molecular system it is first of all necessary to build up a suitable model for the molecule. Without entering into a discussion on the significance of physical models and on their relationships with physical theories,[14] we wish to point out that the validity of a model is judged by comparing '*a posteriori*' its physical properties with experimental quantities which can be considered 'inherent' to the model itself.

The body of information accumulated until now on the molecular structure already gives us an indication of the model. We know, for instance, that in a molecule the binding electrons hold the nuclei at approximately fixed distances and with approximately fixed valence angles. Since in this book we are interested in the motion of the nuclei in the molecule, we are obviously tempted to replace the binding electrons with elastic springs and therefore to build a 'ball-and-spring' model. We shall see that this model is able to reproduce correctly most of the dynamical properties of the molecules. Only at a higher level of approximation does this simple model reveal some limitations and must be replaced by a more sophisticated one.

The fact that the nuclear motions can be treated independently from the electronic motions and that the electrons are not considered as specific particles in our model is due to the difference in the relative velocities of nuclei and electrons. Since the nuclei are much heavier than the electrons, these latter move with much greater velocities. In the time interval in which a nucleus makes a small displacement relative to the others, the electrons have already traveled some thousands of times around it. As a consequence the electrons see the nuclei as practically fixed in space whereas from their side the nuclei feel the electrons as an average charge distribution within the molecule. We shall discuss this problem in more detail in Section 3.1.

Even the simple ball-and-spring model is, however, too complicated to begin with unless some other simplifications are introduced. It is convenient to assume that the atoms are point masses and that the springs holding them together are weightless and obey Hooke's Law. Although these simplifications introduce further limitations on the efficiency of the model, they permit us to apply to the molecules the complete mathematical machinery developed in the past for the harmonic motions of mechanical systems and therefore help in building the necessary basis for a more complete treatment.

2.2 The separation of molecular motions and the Sayvetz conditions

As discussed in the previous section, our molecular model involves simultaneous translation, rotation and vibration, and we need to describe this complex motion in all its details. For the interpretation of the infrared and Raman spectra of polyatomic molecules we are primarily interested in the vibrational motions of the model. Therefore, we wish to find a way of separating, as far as possible, the overall motion into translations, rotations and vibrations.

This problem has been dealt with by Casimir,[1] Eckart[2] and most generally by Sayvetz,[3] and amounts to the choice of a proper set of coordinates which realize the desired separation to the maximum possible extent.

The problem is most conveniently separated into two steps.

Let us consider the molecule as a rigid body, which can only translate and rotate, and study its motion using a space-fixed reference system (X, Y, Z). The position in space of the n atoms of the molecule is then defined by the $3n$ Cartesian coordinates X_k ($k = 1, 2, \ldots, 3n$) relative to this system. In order to describe the motion of a rigid molecule, fewer than $3n$ coordinates are needed. For a non-linear molecule only six coordinates are necessary since a rigid body possesses just three translational and three rotational degrees of freedom. Let us call α_p ($p = 1, 2, \ldots, 6$) these 'large scale' coordinates whose variation describes the overall translations and rotations of the molecule. For a linear molecule, which has two rotational degrees of freedom, only five α_ps would be needed. We might also take into account molecules with internal rotations, such as ethane, considering them to consist of two (or more) rigid parts which can rotate relative to each other and adding one (or more) α_p for each internal degree of rotation.

The equations of transformation from the Cartesian coordinates X_k to the α_ps, which can be written in the form

$$X_k = F_k(\alpha_p) \qquad \begin{array}{l} k = 1, 2, \ldots, 3n \\ p = 1, 2, \ldots, m \end{array} \tag{2.2.1}$$

are simply equations of *rigid constraint* under which the motion takes place. To obtain the Hamiltonian, upon which depend the properties of the system in both classical and quantum mechanics, it is necessary to compute only the kinetic energy of the system since the potential energy is zero.

From equation (2.2.1) the velocities are

$$\dot{X}_k = \sum_p \left(\frac{\partial F_k}{\partial \alpha_p}\right) \dot{\alpha}_p \tag{2.2.2}$$

and the kinetic energy and therefore the Hamiltonian would be

$$2H = \sum_k m_k \dot{X}_k^2 = \sum_{p,r} \sum_k F_k^p F_k^r m_k \dot{\alpha}_p \dot{\alpha}_r = \sum_{p,r} A_{pr}^0 \dot{\alpha}_p \dot{\alpha}_r \tag{2.2.3}$$

where

$$F_k^p = \left(\frac{\partial F_k}{\partial \alpha_p}\right) \quad \text{and} \quad A_{pr}^0 = \sum_k m_k F_k^p F_k^r \tag{2.2.4}$$

Once we have obtained the Hamiltonian for the rigid body, whose eigenvalues are known, we can release the constraint of rigid motion, allowing the nuclei of the molecule to vibrate with small amplitudes around their equilibrium positions. This introduces $3n - 6$ (and in general $3n - m$) new degrees of freedom that can be described by a new set of 'vibrational coordinates' q_λ ($\lambda = 1, 2, \ldots, 3n - m$) which define the deviations of the nuclei from the equilibrium positions.

The equations of transformation (2.2.1) can now be rewritten in the form

$$X_k = F_k(\alpha_p) + \sum_\lambda f_{k\lambda}(\alpha_p) q_\lambda \qquad \begin{array}{l} k = 1, 2, \ldots, 3n \\ p = 1, 2, \ldots, m \\ \lambda = 1, 2, \ldots, 3n - m \end{array} \tag{2.2.5}$$

which are now the equations of *elastic constraint* under which the motion takes place.

We note that there are $3n$ Cartesian coordinates X_k and only $3n - m$ displacement coordinates q_λ. This obviously means that there must exist m conditions relating the coefficients $f_{k\lambda}$, which are not all independent. From equation (2.2.5) the velocities are

$$\dot{X}_k = \sum_p F_k^p \dot{\alpha}_p + \sum_\lambda f_{k\lambda} \dot{q}_\lambda + \sum_{p,\lambda} f_{k\lambda}^p q_\lambda \dot{\alpha}_p \tag{2.2.6}$$

where

$$f_{k\lambda}^p = \left(\frac{\partial f_{k\lambda}}{\partial \alpha_p} \right)$$

and the kinetic energy takes the form

$$2T = \sum_{p,r} A_{pr} \dot{\alpha}_p \dot{\alpha}_r + \sum_{\lambda\mu} C_{\lambda\mu} \dot{q}_\lambda \dot{q}_\mu + 2 \sum_{p\lambda} B_{p\lambda} \dot{\alpha}_p \dot{q}_\lambda \tag{2.2.7}$$

the coefficients A_{pr}, $C_{\lambda\mu}$ and $B_{p\lambda}$ are easily obtained by insertion of equation (2.2.6) in the expression for the kinetic energy and are given by

$$A_{pr} = A_{pr}^0 + \sum_\lambda (A_{pr\lambda} + A_{rp\lambda}) q_\lambda + \sum_{\lambda\mu} A_{pr\lambda\mu} q_\lambda q_\mu \tag{2.2.8a}$$

$$C_{\lambda\mu} = C_{\lambda\mu}^0 \tag{2.2.8b}$$

$$B_{p\lambda} = B_{p\lambda}^0 + \sum_\mu B_{p\lambda\mu} q_\mu \tag{2.2.8c}$$

where

$$A_{pr}^0 = \sum_k m_k F_k^p F_k^r \qquad C_{\lambda\mu}^0 = \sum_k m_k f_{k\lambda} f_{k\mu} \qquad B_{p\lambda}^0 = \sum_k m_k F_k^p f_{k\lambda} \tag{2.2.9a}$$

$$A_{pr\lambda} = \sum_k m_k F_k^p f_{k\lambda}^r \qquad A_{rp\lambda} = \sum_k m_k F_k^r f_{k\lambda}^p \qquad B_{p\lambda\mu} = \sum_k m_k f_{k\mu}^p f_{k\lambda} \tag{2.2.9b}$$

$$A_{pr\lambda\mu} = \sum_k f_{k\lambda}^p f_{k\mu}^r \tag{2.2.9c}$$

In equation (2.2.7) the first term represents the contribution to the total kinetic energy of the translational and rotational motions, the second term represents the vibrational contribution, and the third term the energy of interaction between vibrations and rotations plus translations. The separation of vibrations from rotational and translational motions is therefore possible if we choose the coordinates α_p and the q_λs in such a way that the third term becomes as small as possible.

This can be done if we remember that we have m conditions among the coefficients $f_{k\lambda}$ and we choose them in order that

$$B_{p\lambda}^0 = \sum_k m_k F_k^p f_{k\lambda} = 0 \qquad p = 1, 2, \ldots, m \tag{2.2.10}$$

These m equations (six for a normal molecule) are generally called the *Sayvetz conditions.*

With the use of the m Sayvetz conditions, the third term of equation (2.2.7) becomes

$$2\sum_{p\lambda}\sum_{\mu} B_{p\lambda\mu} q_\mu \dot{\alpha}_p \dot{q}_\lambda \tag{2.2.11}$$

which is very small compared to the vibrational kinetic energy, since it depends on the displacement coordinates, which are small compared to the $\dot{q}$s.

To make the physical interpretation clearer, it is convenient to rewrite the Sayvetz conditions using the vector notation[4] and to specify the choice of the translational and rotational coordinates α_p.

For a normal molecule (Figure 2.2.1) we shall choose as α_ps the three Cartesian components X_0, Y_0 and Z_0 of a vector **R** which locates the centre of mass O of the molecule in the space-fixed axis system and the three Eulerian angles ϕ, θ and χ which describe the orientation in space of a rotating axis system x, y, z, whose origin is at the centre of mass.

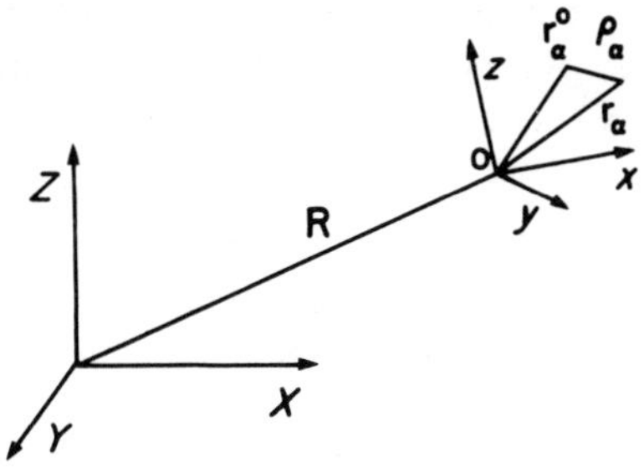

Figure 2.2.1 Choice of the fixed (X, Y, Z) and rotating (x, y, z) reference systems

Let the instantaneous position of the αth nucleus in the rotating system (see Figure 2.2.1) be given by a vector $\mathbf{r}_\alpha$ with components x_α, y_α and z_α and the equilibrium position by a vector $\mathbf{r}_\alpha^0$ with components x_α^0, y_α^0 and z_α^0.

The instantaneous displacement of the αth nucleus from the equilibrium position in the rotating system is then defined by the vector

$$\boldsymbol{\rho}_\alpha = \mathbf{r}_\alpha - \mathbf{r}_\alpha^0 \tag{2.2.12}$$

with components $\Delta x_\alpha = x_\alpha - x_\alpha^0$, $\Delta y_\alpha = y_\alpha - y_\alpha^0$, $\Delta z_\alpha = z_\alpha - z_\alpha^0$.

The total velocity $\mathbf{V}_\alpha$ of the αth nucleus is then

$$\mathbf{V}_\alpha = \dot{\mathbf{R}} + \boldsymbol{\omega} \times \mathbf{r}_\alpha + \dot{\boldsymbol{\rho}}_\alpha \tag{2.2.13}$$

where $\boldsymbol{\omega}$ is the angular velocity of the rotating system of axes.

The kinetic energy is thus

$$\begin{aligned} 2T = \sum_\alpha m_\alpha \mathbf{V}_\alpha \cdot \mathbf{V}_\alpha &= \dot{R}^2 \sum_\alpha m_\alpha + \sum_\alpha m_\alpha (\boldsymbol{\omega} \times \mathbf{r}_\alpha) \cdot (\boldsymbol{\omega} \times \mathbf{r}_\alpha) + \sum_\alpha m_\alpha \dot{\rho}_\alpha^2 \\ &+ 2\dot{\mathbf{R}} \cdot \boldsymbol{\omega} \times \sum_\alpha m_\alpha \mathbf{r}_\alpha + 2\dot{\mathbf{R}} \cdot \sum_\alpha m_\alpha \dot{\boldsymbol{\rho}}_\alpha + 2\sum_\alpha m_\alpha \boldsymbol{\omega} \times \mathbf{r}_\alpha \cdot \dot{\boldsymbol{\rho}}_\alpha \end{aligned} \tag{2.2.14}$$

The first three terms of (2.2.14) represent the pure translational, rotational and vibrational kinetic energies whereas the last three terms represent the corresponding interaction energies.

Since we have used $3n + 6$ coordinates (three components of **R**, three Eulerian angles and $3n$ components of the $\boldsymbol{\rho}_\alpha$ vectors) to describe the $3n$ degrees of freedom of the molecule, we have six conditions that we can impose among them and we can choose these six conditions in order to eliminate, or to reduce to the maximum extent, the three coupling terms of (2.2.14). The first three conditions can be chosen in vectorial notation in the form

$$\sum_\alpha m_\alpha \mathbf{r}_\alpha = \mathbf{0} \tag{2.2.15}$$

from which it follows by differentiation with respect to time

$$\sum_\alpha m_\alpha \dot{\mathbf{r}}_\alpha = \sum_\alpha m_\alpha[(\boldsymbol{\omega} \times \mathbf{r}_\alpha) + \dot{\boldsymbol{\rho}}_\alpha] = \boldsymbol{\omega} \times \sum_\alpha m_\alpha \mathbf{r}_\alpha + \sum_\alpha m_\alpha \dot{\boldsymbol{\rho}}_\alpha = \sum_\alpha m_\alpha \dot{\boldsymbol{\rho}}_\alpha = \mathbf{0} \tag{2.2.16}$$

Equation (2.2.15) is normally called *first Sayvetz condition* and specifies that during a molecular vibration the centre of mass of the molecule must remain unshifted. From (2.2.16) it follows then that in a molecular vibration there must be no linear momentum of vibration.

Introduction of equations (2.2.15) and (2.2.16) in the expression (2.2.14) of the kinetic energy causes the first two interaction terms to vanish, and therefore eliminates the coupling between translations and rotations and between translations and vibrations. Equation (2.2.14) thus reduces, by use of the scalar triple product,† to

$$2T = \dot{R}^2 \sum_\alpha m_\alpha + \sum_\alpha m_\alpha(\boldsymbol{\omega} \times \mathbf{r}_\alpha) \cdot (\boldsymbol{\omega} \times \mathbf{r}_\alpha) + \sum_\alpha m_\alpha \dot{\rho}_\alpha^2 + 2\boldsymbol{\omega} \cdot \sum_\alpha m_\alpha(\mathbf{r}_\alpha^0 + \boldsymbol{\rho}_\alpha) \times \dot{\boldsymbol{\rho}}_\alpha \tag{2.2.17}$$

We can now choose the remaining three conditions to completely define the rotating system. For this we shall require that whenever the atomic displacements in a molecular vibration tend to produce a rotation of the molecule, the rotating system reorients in order to eliminate this component of the motion. This condition is best expressed in the form

$$\sum_\alpha m_\alpha \mathbf{r}_\alpha^0 \times \mathbf{r}_\alpha = \mathbf{0} \tag{2.2.18}$$

from which it follows by differentiation with respect to time

$$\sum_\alpha m_\alpha \dot{\mathbf{r}}_\alpha^0 \times \mathbf{r}_\alpha + \sum_\alpha m_\alpha \mathbf{r}_\alpha^0 \times \dot{\mathbf{r}}_\alpha = \sum_\alpha m_\alpha(\boldsymbol{\omega} \times \mathbf{r}_\alpha^0) \times \mathbf{r}_\alpha + \sum_\alpha m_\alpha \mathbf{r}_\alpha^0 \times (\boldsymbol{\omega} \times \mathbf{r}_\alpha) + \sum_\alpha m_\alpha \mathbf{r}_\alpha^0 \times \dot{\boldsymbol{\rho}}_\alpha = \mathbf{0} \tag{2.2.19}$$

and therefore

$$\sum_\alpha m_\alpha \mathbf{r}_\alpha^0 \times \dot{\boldsymbol{\rho}}_\alpha = \mathbf{0} \tag{2.2.20}$$

Condition (2.2.20) is called the *second Sayvetz condition* and physically means that during a molecular vibration there must be no zero-order vibrational angular momentum. By use of equation (2.2.20) the expression of the total kinetic energy (2.2.14) becomes

$$2T = \dot{R}^2 \sum_\alpha m_\alpha + \sum_\alpha m_\alpha(\boldsymbol{\omega} \times \mathbf{r}_\alpha) \cdot (\boldsymbol{\omega} \times \mathbf{r}_\alpha) + \sum_\alpha m_\alpha \dot{\rho}_\alpha^2 + 2\boldsymbol{\omega} \cdot \sum_\alpha m_\alpha \boldsymbol{\rho}_\alpha \times \dot{\boldsymbol{\rho}}_\alpha \tag{2.2.21}$$

The separation of the kinetic energy into a pure translational, a pure rotational and a pure vibrational term is therefore possible only if the last term of (2.2.21), called the *Coriolis energy*, can be neglected.

It is easily seen, however, that this term is small if compared to the pure vibrational term since it depends upon both the displacement coordinates $\boldsymbol{\rho}_\alpha$ and the angular velocity $\boldsymbol{\omega}$, which are small quantities compared to the vibrational velocities $\dot{\boldsymbol{\rho}}_\alpha$. We can therefore include the Coriolis energy in the rotational part and treat the other two terms separately. Furthermore, we can ignore the translational kinetic energy from now on, since it is completely uncoupled to the other terms.

† We recall that for the scalar triple product the following relations hold:

$$\mathbf{A} \cdot (\mathbf{B} \times \mathbf{C}) = -\mathbf{B} \cdot (\mathbf{A} \times \mathbf{C}) = \mathbf{C} \cdot (\mathbf{A} \times \mathbf{B}).$$

2.3 Elementary treatment of small vibrations

As a consequence of the Sayvetz conditions discussed in the previous section, the vibrations of a molecule can be treated separately from the rotational and translational motions, using a coordinate system bound to the molecule and satisfying conditions (2.2.15) and (2.2.20).

To write the equation of motion for our molecule, we need to have the expression for the kinetic energy T and for the potential energy V.

Employing as coordinates the $3n$ displacements from the equilibrium positions of the nuclei Δx_α, Δy_α and Δz_α, namely the Cartesian components of the $\boldsymbol{\rho}_\alpha$ vectors defined in Section 2.2, the kinetic energy takes the very simple form

$$2T = \sum_\alpha m_\alpha(\Delta\dot{x}_\alpha^2 + \Delta\dot{y}_\alpha^2 + \Delta\dot{z}_\alpha^2) \tag{2.3.1}$$

where m_α is the mass of the αth atom. For the sake of simplicity, we can indicate by x_i the generic Cartesian displacement coordinate and rewrite equation (2.3.1) in the more compact form

$$2T = \sum_{i=1}^{3n} m_i\dot{x}_i^2 \tag{2.3.2}$$

in which the summation now runs from 1 to $3n$ and $m_i = m_1$ for $i = 1$, 2 and 3, $m_i = m_2$ for $i = 4$, 5 and 6 and so on.

Instead of using the simple Cartesian displacement coordinates x_i, it is often convenient to use a set of 'mass weighted' Cartesian coordinates q_i defined by the relation

$$q_i = \sqrt{m_i}x_i \tag{2.3.3}$$

in terms of which the kinetic energy takes the simpler form

$$2T = \sum_{i=1}^{3n} \dot{q}_i^2 \tag{2.3.4}$$

The analytical form of the potential energy V of the system is unfortunately not known. The only information available on it is that it must be some function of the displacement coordinates. Under these conditions, for small displacements, the best we can do is to expand the potential energy in a power series of the displacement coordinates

$$2V = 2V_0 + 2\sum_i \left(\frac{\partial V}{\partial q_i}\right)_0 q_i + \sum_{i,j}\left(\frac{\partial^2 V}{\partial q_i\,\partial q_j}\right)_0 q_iq_j + \frac{1}{3}\sum_{i,j,k}\left(\frac{\partial^3 V}{\partial q_i\,\partial q_j\,\partial q_k}\right)_0 q_iq_jq_k + \cdots \tag{2.3.5}$$

Since we are not interested in the absolute value of the potential energy (which, in any case, is unknown), but rather in its variation with the coordinates q_i, we can shift the zero of the energy scale so that the energy of the equilibrium configuration V_0 is zero. Further, since the equilibrium configuration is by definition at a minimum of the potential energy, and the q_i are all independent, it must be true that

$$f_i = \left(\frac{\partial V}{\partial q_i}\right)_0 = 0 \tag{2.3.6}$$

For small excursions of the atoms around their equilibrium positions, the higher terms of equation (2.3.5) can be neglected and it becomes

$$2V = \sum_{i,j} f_{ij}q_iq_j \tag{2.3.7}$$

In this equation, the terms f_{ij} are constants which are normally called 'force constants' since they represent the proportionality factors between the displacements of the nuclei and the restoring forces acting upon them. It must be observed that in equation (2.3.7) there are in principle $(3n) \times (3n + 1)/2$ force constants since, V being a continuous function, the order of differentiation in (2.3.5) is immaterial and thus

$$f_{ij} = f_{ji} \tag{2.3.8}$$

Once we have obtained T and V, we can write the equation of motion for the vibrating molecule. Since T is a function of the velocities only and V is a function of the displacements only, the equation of motion in Lagrangian form can be written

$$\frac{\mathrm{d}}{\mathrm{d}t}\left(\frac{\partial T}{\partial \dot{q}_i}\right) + \left(\frac{\partial V}{\partial q_i}\right) = 0 \qquad i = 1, 2, \ldots, 3n \tag{2.3.9}$$

yielding, using (2.3.4) and (2.3.7),

$$\ddot{q}_i + \sum_j f_{ij} q_j = 0 \tag{2.3.10}$$

This is a set of $3n$ simultaneous second-order differential equations, the solutions of which are the well-known harmonic oscillator functions of classical mechanics

$$q_i = A_i \cos(2\pi c v t + \phi) \tag{2.3.11}$$

where v is the frequency of vibration in cm^{-1}, ϕ is a phase factor and A_i is the maximal displacement or amplitude of the atom.

By substitution of (2.3.11) in (2.3.10) we obtain

$$\sum_j f_{ij} A_j - 4\pi^2 (cv)^2 A_i = 0 \tag{2.3.12}$$

which can be rewritten

$$\sum_j (f_{ij} - \lambda \delta_{ij}) A_j = 0 \tag{2.3.13}$$

where $\lambda = 4\pi^2 v^2 c^2$ and δ_{ij} is the Kronecker delta.

Equation (2.3.13) is a set of $3n$ simultaneous homogeneous linear equations in the $3n$ unknowns A_j, which has solutions different from the trivial one $A_j = 0$ $(j = 1, 2, \ldots, 3n)$ only if the determinant of the coefficients is equal to zero.

$$\begin{vmatrix} f_{11} - \lambda & f_{12} & f_{13} & \cdots & f_{1,3n} \\ f_{12} & f_{22} - \lambda & f_{23} & \cdots & f_{2,3n} \\ f_{13} & f_{23} & f_{33} - \lambda & \cdots & f_{3,3n} \\ \vdots & \vdots & \vdots & \ddots & \vdots \\ f_{1,3n} & f_{2,3n} & f_{3,3n} & & f_{3n,3n} - \lambda \end{vmatrix} = 0 \tag{2.3.14}$$

This is called the 'secular determinant' which, when expanded, gives rise to an algebraic equation of order $3n$ in λ. There are therefore, $3n$ values of λ, i.e. $3n$ harmonic frequencies v_k, for which the determinant vanishes. It will be shown later (Section 2.6), that six of the λs are zero and, therefore, that there are only $3n - 6$ (and in general $3n - m$) non-zero roots of the secular equation.

If the value of one of these $3n - 6$ roots, let us say λ_k, is substituted in equation (2.3.13), a solution can be found for the unknown A_j. In order to show the relation with the particular λ_k used, we shall indicate this solution by A_{jk}. It is, however, necessary to remember that from the theory of the linear systems of equations it follows that it is not possible to determine all the $3n$ A_{jk} unknown from equation (2.3.13), but only their ratios.

2.4 Normal coordinates

Expressions (2.3.4) and (2.3.7) for T and V can be rewritten in a very simple and compact form, using matrix notation. If we call $\mathbf{q}$ a column vector, with components given by the $3n$ mass-weighted Cartesian displacement coordinates q_i of the atoms, equation (2.3.4) becomes

$$2T = \tilde{\dot{\mathbf{q}}}\dot{\mathbf{q}} \tag{2.4.1}$$

and equation (2.3.7) becomes

$$2V = \tilde{\mathbf{q}}\mathbf{f}\mathbf{q} \tag{2.4.2}$$

where the symbol $\sim$ indicates the transpose vector and $\mathbf{f}$ is a square $3n \times 3n$ symmetric matrix whose elements are the force constants f_{ij} of equation (2.3.7).

Equation (2.4.1) is a sum of squares of each of the coordinate velocities. (2.4.2) is more complicated and contains all possible coordinate cross-product terms, since in general the matrix $\mathbf{f}$ has non-zero off-diagonal terms.

If the $\mathbf{f}$-matrix were diagonal, the vibrational problem would be very simple. The equations of motion (2.3.13) would reduce to

$$(f_{ii} - \lambda)A_i = 0 \qquad i = 1, 2, \ldots, 3n$$

namely to $3n$ separate equations, each associated with a given coordinate. Furthermore, the diagonal elements of the $\mathbf{f}$-matrix would be just the $3n$ vibrational frequencies (six of them would be zero), since from the preceding equation we have

$$f_{ii} = \lambda_i$$

However, the fact that the $\mathbf{f}$-matrix is not diagonal, prevents the vibrational problem from being broken down into $3n$ distinct and separate problems, each involving only one coordinate. This fact not only complicates the classical treatment but makes the solution of the quantum-mechanical problem in terms of the q_i coordinates impossible, since we know how to solve Schroedinger's equation for the harmonic oscillator only if we can write the kinetic and potential energies as a sum of separate terms, each a function of one single coordinate.

These considerations show that the Cartesian coordinates, although convenient to visualize the problem, are, from the point of view of finding an analytical solution, a bad choice. It is, therefore, legitimate to ask ourselves whether or not a better set of coordinates can be found which has the desired property of making both the kinetic and the potential energy matrices diagonal and at the same time satisfying the Sayvetz conditions.

We shall call these new coordinates '*normal coordinates*' and define a new column vector $\mathbf{Q}$ with components given by the $3n$ normal coordinates Q_k. In order to obtain both the kinetic and potential energies in diagonal form, the transformation from mass-weighted Cartesian coordinates to normal coordinates must necessarily be a linear transformation of the type

$$Q_k = \sum_i l'_{ki} q_i \tag{2.4.3}$$

which means that each normal coordinate Q_k is a linear combination of the $3n$ Cartesian coordinates q_i with coefficients l'_{ki} to be determined. In matrix notation such a transformation is simply written as

$$\mathbf{Q} = \mathscr{L}^{-1}\mathbf{q} \tag{2.4.4}$$

where for convenience we have used the symbol $\mathscr{L}^{-1}$ to indicate the transformation matrix from mass-weighted Cartesian to normal coordinates, whose elements are the coefficients l'_{ki} of equation (2.4.3). The inverse transformation from normal to Cartesian coordinates is therefore

$$\mathbf{q} = \mathscr{L}\mathbf{Q} \tag{2.4.5}$$

where $\mathscr{L}$ is the inverse of the $\mathscr{L}^{-1}$ matrix, i.e. $\mathscr{L}\mathscr{L}^{-1} = \mathbf{E}$, $\mathbf{E}$ being the unit matrix.

As discussed before, we require the Q_k to satisfy the Sayvetz conditions and to bring the kinetic and the potential energies simultaneously to diagonal form, i.e. to expressions of the type

$$2T = \tilde{\dot{\mathbf{Q}}}\dot{\mathbf{Q}} \tag{2.4.6}$$

and

$$2V = \tilde{\mathbf{Q}}\mathbf{\Lambda}\mathbf{Q} \tag{2.4.7}$$

where $\mathbf{\Lambda}$ is a diagonal matrix whose elements are the normal frequency parameters $\lambda_k = 4\pi^2c^2\nu_k^2$. Equations (2.4.6) and (2.4.7) impose severe conditions on the transformation matrix $\mathscr{L}$ and we shall find it by introduction of equation (2.4.4) into the expressions (2.4.6) and (2.4.7) of T and V or, equivalently, by introduction of equation (2.4.5) in the corresponding expressions for T and V, (2.4.1) and (2.4.2).

Considering first the kinetic energy, we have

$$2T = \tilde{\dot{\mathbf{Q}}}\tilde{\mathscr{L}}\mathscr{L}\dot{\mathbf{Q}} \tag{2.4.8}$$

and by comparison with (2.4.6)

$$\tilde{\mathscr{L}}\mathscr{L} = \mathbf{E} \tag{2.4.9}$$

which shows the first important property of the $\mathscr{L}$ matrix, namely that it is orthogonal since

$$\tilde{\mathscr{L}} = \mathscr{L}^{-1} \tag{2.4.10}$$

In the same way we obtain for the potential energy, by use of (2.4.2) and (2.4.5),

$$2V = \tilde{\mathbf{Q}}\tilde{\mathscr{L}}\mathbf{f}\mathscr{L}\mathbf{Q} \tag{2.4.11}$$

and by comparison with equation (2.4.7)

$$\tilde{\mathscr{L}}\mathbf{f}\mathscr{L} = \mathbf{\Lambda} \tag{2.4.12}$$

which shows the second important property of $\mathscr{L}$, namely that it diagonalizes the **f**-matrix.

Since $\mathscr{L}$ is orthogonal, equation (2.4.12) can be rewritten in the form

$$\mathbf{f}\mathscr{L} = \mathscr{L}\mathbf{\Lambda} \tag{2.4.13}$$

which is an eigenvalue equation. We can formally consider the matrix **f** as an operator acting on a given column $\mathscr{L}_k$ of the $\mathscr{L}$-matrix to give as a result the same column $\mathscr{L}_k$ multiplied by a constant factor λ_k, which is the corresponding kth diagonal element of the $\mathbf{\Lambda}$-matrix. For this reason the columns of $\mathscr{L}$ are called *eigenvectors* and the elements of $\mathbf{\Lambda}$ are called *eigenvalues* of the matrix **f**.

For each column $\mathscr{L}_k$ of $\mathscr{L}$ we have then an equation of the type

$$\mathbf{f}\mathscr{L}_k = \mathscr{L}_k\lambda_k$$

which can be rewritten in the form

$$(\mathbf{f} - \mathbf{E}\lambda_k)\mathscr{L}_k = 0 \qquad k = 1, 2, \ldots, 3n \tag{2.4.14}$$

The conditions of compatibility of (2.4.14) are then

$$|\mathbf{f} - \mathbf{E}\lambda_k| = 0$$

Equation (2.4.14) simply represents a different way of writing the system of simultaneous homogeneous linear equations (2.3.13). It must be observed, however, that each unknown amplitude A_{jk} (for a particular λ_k) of equation (2.3.13) is not numerically equivalent to the corresponding element $\mathscr{L}_{jk}$ of equation (2.4.14). In fact, from equation (2.4.9) it follows for the kth column of $\mathscr{L}$ that

$$\sum_j \mathscr{L}_{jk}^2 = 1 \tag{2.4.15}$$

whereas the same is not true for the corresponding A_{jk} elements.

As we have pointed out previously, the system of equations (2.3.13) does not determine the A_{jk} elements uniquely but only their ratios. For an arbitrary A_{jk} we can, however, construct a set of 'normalized' amplitudes of the type

$$\frac{A_{jk}}{[\sum_j A_{jk}^2]^{\frac{1}{2}}} \tag{2.4.16}$$

which are now numerically equivalent to the elements $\mathscr{L}_{jk}$. These considerations show more precisely the meaning of the columns of the $\mathscr{L}$-matrix, i.e. of the rows of the $\tilde{\mathscr{L}}$-matrix. If we consider equation (2.4.4)

$$\mathbf{Q} = \tilde{\mathscr{L}}\mathbf{q} \tag{2.4.17}$$

we see that for a given normal coordinate Q_k, the corresponding row of $\tilde{\mathscr{L}}$ gives the set of coefficients with which the mass-weighted Cartesian displacements take part in that normal coordinate.

In the representation of a normal coordinate it is, however, more convenient to use Cartesian displacement coordinates. Thus, if $\mathbf{M}$ is a diagonal matrix whose elements are the atomic masses, we shall use, instead of (2.4.17), the relation

$$\mathbf{Q} = \tilde{\mathscr{L}}\mathbf{M}^{\frac{1}{2}}\mathbf{x} = \tilde{\mathscr{L}}_\mathbf{x}\mathbf{x} \tag{2.4.18}$$

which is simply obtained by insertion in (2.4.17) of the relation

$$\mathbf{q} = \mathbf{M}^{\frac{1}{2}}\mathbf{x} \tag{2.4.19}$$

between mass-weighted and Cartesian displacement coordinates.

When Cartesian displacement coordinates are used, the kinetic and potential energies (2.4.1) and (2.4.2) become respectively

$$\begin{aligned} 2T &= \tilde{\dot{\mathbf{x}}}\mathbf{M}\dot{\mathbf{x}} \\ 2V &= \tilde{\mathbf{x}}\mathbf{f}_\mathbf{x}\mathbf{x} \end{aligned} \tag{2.4.20}$$

where

$$\mathbf{f}_\mathbf{x} = \mathbf{M}^{\frac{1}{2}}\mathbf{f}\mathbf{M}^{\frac{1}{2}} \tag{2.4.21}$$

and the secular equation (2.4.13) takes the form

$$\mathbf{M}^{-1}\mathbf{f}_x\mathscr{L}_x = \mathscr{L}_x\mathbf{\Lambda} \tag{2.4.22}$$

There are several numerical methods[5] for the calculation of the eigenvalues and eigenvectors of real matrices, like $\mathbf{f}$ or $(\mathbf{M}^{-1}\mathbf{f}_x)$ of equations (2.4.13) and (2.4.22), available as standard computer routines. For this reason we shall not discuss the numerical problem in this book.

2.5 Normal modes of vibration

Let us examine in more detail the nature of the solution obtained in Section 2.3 for the vibration of a polyatomic molecule. Equation (2.3.11) tells us that in a molecular vibration each atom oscillates about its equilibrium position with a frequency ν_k, a phase ϕ_k and an amplitude A_{ik}.

For each given solution $\lambda_k = 4\pi^2c^2\nu_k^2$ of the secular equation, i.e. for each normal coordinate Q_k, the molecule undergoes a simple motion in which all the nuclei move in phase with the same frequency ν_k but with different amplitudes A_{ik}. In other words, all the nuclei will pass through the equilibrium positions at the same time, reach their maximum displacement in a given direction, pass again through the equilibrium positions, reach their maximum displacement in the opposite direction and so forth.

A mode of motion of this kind is called a *normal mode of vibration* or simply a *normal vibration* and the frequency associated with it is called a *normal* or *fundamental* frequency of vibration. A normal mode of vibration can be represented by an arrow at each nucleus showing the relative displacements. In Figure 2.5.1 the normal modes of vibration of a

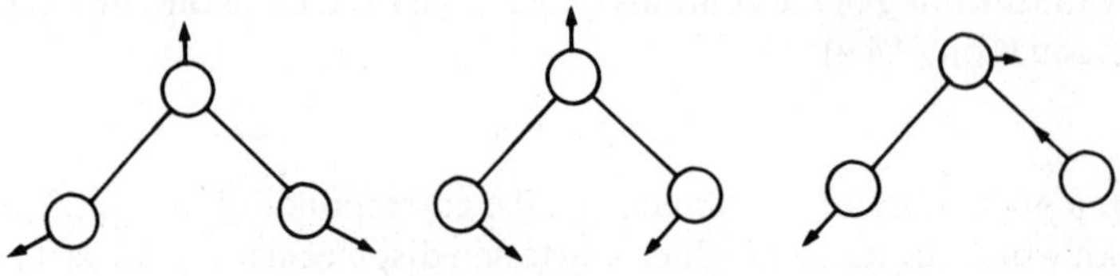

Figure 2.5.1 Schematic representation of the normal modes of a bent triatomic molecule

bent triatomic molecule XY_2, are schematically drawn. The arrows represent in direction and magnitude the atomic displacement in one half-cycle. In the second half-cycle the arrows will point in the opposite directions. The relative length of the arrows can be chosen so that they give the relative amplitudes of displacements of the individual nuclei. If this is done in the proper scale, each drawing will represent not only a molecular motion but also the normal coordinate associated with it.

Using equation (2.4.18), we may take the Cartesian components of each arrow proportional to the elements of the rows of the $\tilde{\mathscr{L}}\mathbf{M}^{\frac{1}{2}}$ matrix. Each row of the matrix gives all the components for the description of the corresponding normal coordinate. It is, therefore, obvious that in order to make a correct drawing the solution of the secular equation must be known.

In some cases it can happen that two (or more) roots of the secular equation coincide. The two (or more) corresponding normal vibrations will therefore have the same frequency. In such a case, the two (or more) vibrations are said to be *degenerate*. We shall see that the occurrence of degenerate vibrations is directly related to the existence of some particular symmetry elements in a molecule. When two vibrational frequencies ν_k and ν_l are degenerate (i.e. $\nu_k = \nu_l$), then there are only two independent sets of solutions of equation (2.3.13), A_{jk} and A_{jl}, but an infinite number of ways in which they can be chosen. Because of the homogeneity of equation (2.3.13), any linear combination $aA_{jk} + bA_{jl}$ is also a solution of (2.3.13) for the same frequency. The corresponding motion will still be a normal mode since all atoms continue to move with the same frequency. It is necessary to emphasize that this is possible only if the two motions which are superimposed have the same frequency, i.e. if it is a degenerate vibration. The sum of two motions with different frequencies does not produce a simple motion of the atoms.

A good example of degeneracy is supplied by a triangular triatomic molecule X_3. Such a molecule possesses $3 \times 3 - 6 = 3$ normal vibrations, two of which are degenerate. In Figure 2.5.2(a) a possible choice of the normal vibrations of this system is schematically represented. The two degenerate vibrations are indicated as ν_{2a} and ν_{2b}.

If the two degenerate modes are added with a phase difference of 0° and 180°, the two equivalent modes of Figure 2.5.2(b) are obtained. The reader will observe that in both cases the displacements of the nuclei are at right angles in the two degenerate modes. If the phase difference is 0° or 180° the nuclei will still move along straight lines. If, however, the phase difference is different from 0° or 180° the atoms will describe an elliptical motion around the equilibrium position. With a phase difference of 90° the ellipse degenerates into a circle as shown in Figure 2.5.2(c).

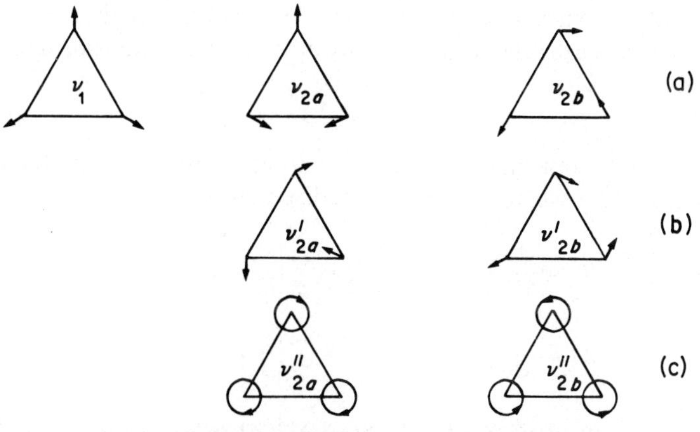

Figure 2.5.2 Normal modes of a triangular triatomic molecule X_3

2.6 The zero roots of the secular equation

In Section 2.3 we have learned how to treat the vibrations of a polyatomic molecule using Cartesian displacement coordinates. However, for a molecule with n atoms, this coordinate system includes the three translations and the three rotations in addition to the molecular vibrations. We expect, therefore, when we build up the secular equation (2.3.14) in terms of these coordinates, to obtain six zero roots corresponding to the six modes of motion with zero frequency. In order to give a more convincing proof of the existence of the zero roots, we recall the two Sayvetz conditions of equations (2.2.15) and (2.2.18) that we shall rewrite in the form

$$\sum_\alpha m_\alpha \mathbf{r}_\alpha = \sum_\alpha m_\alpha(\mathbf{r}_\alpha^0 + \boldsymbol{\rho}_\alpha) = \sum_\alpha m_\alpha \boldsymbol{\rho}_\alpha = \mathbf{0} \tag{2.6.1}$$

$$\sum_\alpha m_\alpha \mathbf{r}_\alpha^0 \times \mathbf{r}_\alpha = \sum_\alpha m_\alpha \mathbf{r}_\alpha^0 \times (\mathbf{r}_\alpha^0 + \boldsymbol{\rho}_\alpha) = \sum_\alpha m_\alpha \mathbf{r}_\alpha^0 \times \boldsymbol{\rho}_\alpha = \mathbf{0} \tag{2.6.2}$$

These two vector relations, when expanded using standard methods of vector analysis, give rise to six scalar relations of the type

$$\sum_\alpha m_\alpha \Delta x_\alpha = \sum_\alpha m_\alpha^{\frac{1}{2}} q_{\alpha x} = 0$$

$$\sum_\alpha m_\alpha \Delta y_\alpha = \sum_\alpha m_\alpha^{\frac{1}{2}} q_{\alpha y} = 0$$

$$\sum_\alpha m_\alpha \Delta z_\alpha = \sum_\alpha m_\alpha^{\frac{1}{2}} q_{\alpha z} = 0$$

$$\sum_\alpha m_\alpha (y_\alpha^0 \Delta z_\alpha - z_\alpha^0 \Delta y_\alpha) = \sum_\alpha m_\alpha^{\frac{1}{2}} (y_\alpha^0 q_{\alpha z} - z_\alpha^0 q_{\alpha y}) = 0 \tag{2.6.3}$$

$$\sum_\alpha m_\alpha (z_\alpha^0 \Delta x_\alpha - x_\alpha^0 \Delta z_\alpha) = \sum_\alpha m_\alpha^{\frac{1}{2}} (z_\alpha^0 q_{\alpha x} - x_\alpha^0 q_{\alpha z}) = 0$$

$$\sum_\alpha m_\alpha (x_\alpha^0 \Delta y_\alpha - y_\alpha^0 \Delta x_\alpha) = \sum_\alpha m_\alpha^{\frac{1}{2}} (x_\alpha^0 q_{\alpha y} - y_\alpha^0 q_{\alpha x}) = 0$$

where x_α^0, y_α^0 and z_α^0 are the coordinates of the equilibrium position of the αth atom relative to the centre of mass and

$$q_{\alpha x} = \sqrt{m_\alpha}\, \Delta x_\alpha \qquad q_{\alpha y} = \sqrt{m_\alpha}\, \Delta y_\alpha \qquad q_{\alpha z} = \sqrt{m_\alpha}\, \Delta z_\alpha$$

Since the roots of the secular equation are properties of the molecule and not of the co-ordinate system used, we shall switch to a new set of coordinates R_i which are more convenient for proving the existence of the zero roots. We shall choose the new set of coordinates in such a way that they are related to the qs by an orthogonal transformation

$$\mathbf{R} = \mathbf{Tq} \tag{2.6.4}$$

and that the first six R-coordinates describe the three translations and the three rotations.

These six coordinates, when properly normalized, have the form

$$R_1 = \frac{1}{M^{\frac{1}{2}}} \sum_\alpha m_\alpha^{\frac{1}{2}} q_{\alpha x} \qquad R_4 = \sum_\alpha \left(\frac{m_\alpha}{I_x}\right)^{\frac{1}{2}} (y_\alpha^0 q_{\alpha z} - z_\alpha^0 q_{\alpha y})$$

$$R_2 = \frac{1}{M^{\frac{1}{2}}} \sum_\alpha m_\alpha^{\frac{1}{2}} q_{\alpha y} \qquad R_5 = \sum_\alpha \left(\frac{m_\alpha}{I_y}\right)^{\frac{1}{2}} (z_\alpha^0 q_{\alpha x} - x_\alpha^0 q_{\alpha z}) \tag{2.6.5}$$

$$R_3 = \frac{1}{M^{\frac{1}{2}}} \sum_\alpha m_\alpha^{\frac{1}{2}} q_{\alpha z} \qquad R_6 = \sum_\alpha \left(\frac{m_\alpha}{I_z}\right)^{\frac{1}{2}} (x_\alpha^0 q_{\alpha y} - y_\alpha^0 q_{\alpha x})$$

where M is the mass of the molecule and I_x, I_y and I_z are the moments of inertia. The remaining $3n - 6$ coordinates have not been written in explicit form since this is not important for our purposes.

The reader will notice that the six Sayvetz conditions of equation (2.6.3) can be simply obtained by setting the six coordinates (2.6.5). equal to zero. The first three Sayvetz conditions mean, therefore, that in a molecular vibration the displacement of the nuclei from their equilibrium positions should never produce a translation of the molecule, whereas the last three Sayvetz conditions mean that the displacement of the nuclei from the equilibrium positions should not produce a rotation of the molecule (in the approximation that the instantaneous coordinates x_α, y_α and z_α can be considered equal to the equilibrium coordinates x_α^0, y_α^0 and z_α^0).

Since in a pure rotation or translation there is no variation of the interatomic distances in the molecule, there must be no contribution to the potential energy of each of the six coordinates of equation (2.6.5). Therefore if we write the potential energy in terms of the

R-coordinates in the form

$$2V = \sum_{\mu\nu} F_{\mu\nu} R_\mu R_\nu + 2\sum_{\mu i} F_{\mu i} R_\mu R_i + \sum_{ij} F_{ij} R_i R_j \qquad \begin{matrix} \mu, \nu = 1, 2, \ldots, 6 \\ i, j = 7, 8, \ldots, 3n \end{matrix} \tag{2.6.6}$$

we see that all the force constants of the type $F_{\mu\nu}$ and $F_{\mu i}$ must be zero. It follows that the secular equation in terms of the R-coordinates has the form

$$\begin{vmatrix} -\lambda & 0 & 0 & 0 & 0 & 0 & 0 & 0 & \cdot & 0 \\ 0 & -\lambda & 0 & 0 & 0 & 0 & 0 & 0 & \cdot & 0 \\ 0 & 0 & -\lambda & 0 & 0 & 0 & 0 & 0 & \cdot & 0 \\ 0 & 0 & 0 & -\lambda & 0 & 0 & 0 & 0 & \cdot & 0 \\ 0 & 0 & 0 & 0 & -\lambda & 0 & 0 & 0 & \cdot & 0 \\ 0 & 0 & 0 & 0 & 0 & -\lambda & 0 & 0 & \cdot & 0 \\ 0 & 0 & 0 & 0 & 0 & 0 & F_{7,7}-\lambda & F_{7,8} & \cdot & F_{7,3n} \\ 0 & 0 & 0 & 0 & 0 & 0 & F_{8,7} & F_{8,8}-\lambda & \cdot & F_{8,3n} \\ \cdot & \cdot & \cdot & \cdot & \cdot & \cdot & \cdot & \cdot & \cdot & \cdot \\ 0 & 0 & 0 & 0 & 0 & 0 & F_{3n,7} & F_{3n,8} & \cdot & F_{3n,3n}-\lambda \end{vmatrix} = 0 \tag{2.6.7}$$

and therefore that there are six zero roots as stated before. A more detailed discussion of the zero roots of the secular equation can be found in Reference 6.

The reader will observe that the six translational and rotational coordinates of equation (2.6.5) are true normal coordinates since for $\mathbf{R} = (R_1, R_2, \ldots, R_6)$

$$\begin{aligned} 2T &= \tilde{\dot{\mathbf{q}}}\dot{\mathbf{q}} = \tilde{\dot{\mathbf{R}}}\mathbf{T}\tilde{\mathbf{T}}\dot{\mathbf{R}} = \tilde{\dot{\mathbf{R}}}\dot{\mathbf{R}} \\ 2V &= 0 \end{aligned} \tag{2.6.8}$$

2.7 Coriolis energy and zeta matrices

In Section 2.1 we have seen that the vibrational energy of a molecule is coupled to the rotational energy by an interaction term, the Coriolis energy, and according to equation (2.2.21) this is given by

$$2T_{\text{Coriolis}} = 2\boldsymbol{\omega} \cdot \sum_\alpha m_\alpha \boldsymbol{\rho}_\alpha \times \dot{\boldsymbol{\rho}}_\alpha \tag{2.7.1}$$

In the treatment of molecular vibrations the Coriolis energy can be ignored since it is small compared to the vibrational energy. Since, however, the Coriolis coupling terms depend upon the normal modes of vibration of the molecule, they can furnish useful information on the force constants and for this reason it is worthwhile to discuss them in further detail.

The relationship between the Coriolis coupling terms and the normal modes of vibration of a molecule is most conveniently discussed in terms of a theory developed by Meal and Polo[7] and commonly used today. Equation (2.7.1) can be rewritten in the form

$$T_{\text{Coriolis}} = \Omega_x \omega_x + \Omega_y \omega_y + \Omega_z \omega_z \tag{2.7.2}$$

where ω_x, ω_y, ω_z are the components of the angular velocity of the rotating system and

$$\Omega_\sigma = \sum_\alpha m_\alpha (\boldsymbol{\rho}_\alpha \times \dot{\boldsymbol{\rho}}_\alpha) \cdot \mathbf{e}_\sigma \qquad \sigma = x, y, z \tag{2.7.3}$$

$\mathbf{e}_\sigma$ being a unit vector in the direction of the axis σ. It is convenient to rewrite equation

(2.7.3), using matrix notation, in the form

$$\boldsymbol{\Omega}_\sigma = \tilde{\mathbf{q}}\,\mathcal{M}^\sigma \dot{\mathbf{q}} \tag{2.7.4}$$

where $\mathbf{q}$ is the column vector of mass-weighted Cartesian coordinates and $\dot{\mathbf{q}}$ is the corresponding time derivative. For this, let us define three matrices $\mathcal{M}^x$, $\mathcal{M}^y$ and $\mathcal{M}^z$, each of them made of N identical 3×3 blocks along the diagonal (one block for each atom α), having respectively the form

$$(\mathcal{M}^x)_\alpha = \begin{pmatrix} 0 & 0 & 0 \\ 0 & 0 & 1 \\ 0 & -1 & 0 \end{pmatrix} \quad (\mathcal{M}^y)_\alpha = \begin{pmatrix} 0 & 0 & -1 \\ 0 & 0 & 0 \\ 1 & 0 & 0 \end{pmatrix} \quad (\mathcal{M}^z)_\alpha = \begin{pmatrix} 0 & 1 & 0 \\ -1 & 0 & 0 \\ 0 & 0 & 0 \end{pmatrix} \tag{2.7.5}$$

Performing the vector product in (2.7.3) we have

$$\begin{aligned} \Omega_x &= \sum_\alpha m_\alpha(y_\alpha \dot{z}_\alpha - z_\alpha \dot{y}_\alpha) = \sum_\alpha (q_{y\alpha}\dot{q}_{z\alpha} - q_{z\alpha}\dot{q}_{y\alpha}) \\ \Omega_y &= \sum_\alpha m_\alpha(z_\alpha \dot{x}_\alpha - x_\alpha \dot{z}_\alpha) = \sum_\alpha (q_{z\alpha}\dot{q}_{x\alpha} - q_{x\alpha}\dot{q}_{z\alpha}) \\ \Omega_z &= \sum_\alpha m_\alpha(x_\alpha \dot{y}_\alpha - y_\alpha \dot{x}_\alpha) = \sum_\alpha (q_{x\alpha}\dot{q}_{y\alpha} - q_{y\alpha}\dot{q}_{x\alpha}) \end{aligned} \tag{2.7.6}$$

and using (2.7.5) it is easily verified that (2.7.4) is equivalent to (2.7.3).

Since the mass-weighted Cartesian coordinates are related to normal coordinates by the relationship (2.4.5), we can also write

$$\boldsymbol{\Omega}_\sigma = \tilde{\dot{\mathbf{Q}}}\tilde{\mathcal{L}}\mathcal{M}^\sigma \mathcal{L}\dot{\mathbf{Q}} = \tilde{\mathbf{Q}}\boldsymbol{\zeta}^\sigma \dot{\mathbf{Q}} \tag{2.7.7}$$

where we have now defined three $\boldsymbol{\zeta}^\sigma$ matrices as

$$\boldsymbol{\zeta}^\sigma = \tilde{\mathcal{L}}\mathcal{M}^\sigma \mathcal{L} \tag{2.7.8}$$

in a normal coordinate basis.

According to Section 2.6 the normal coordinate vector $\mathbf{Q}$ has $3N - 6$ components Q_v ($3N - 5$ for a linear molecule) describing the $3N - 6$ normal modes of vibration associated with the $3N - 6$ non-zero roots of the secular equation and six components Q_r describing the three translations and the three rotations of the molecule. If we write equation (2.4.5) symbolically in the form

$$\mathbf{q} = \mathcal{L}\mathbf{Q} = \left[\begin{array}{cccc|ccc} l_{1v_1} & l_{1v_2} & \cdots & l_{1,v_{3N-6}} & l_{1r_1} & \cdots & l_{1r_6} \\ l_{2v_1} & l_{2v_2} & \cdots & l_{2,v_{3N-6}} & l_{2r_1} & \cdots & l_{2r_6} \\ & & & & & & \\ & & & & & & \\ l_{3N,v_1} & l_{3N,v_2} & \cdots & l_{3N,v_{3N-6}} & l_{3N,r_1} & \cdots & l_{3N,r_6} \end{array}\right] \left[\begin{array}{c} Q_{v_1} \\ Q_{v_2} \\ \vdots \\ Q_{v_{3N-6}} \\ \hline Q_{r_1} \\ \vdots \\ Q_{r_6} \end{array}\right] \tag{2.7.9}$$

it is easily seen that the matrix $\boldsymbol{\zeta}^\sigma$ can be partitioned into submatrices for the vibrations and the rigid molecular motions

$$\boldsymbol{\zeta}^\sigma = \begin{bmatrix} \boldsymbol{\zeta}_v^\sigma & \boldsymbol{\zeta}_{vr}^\sigma \\ \boldsymbol{\zeta}_{vr}^\sigma & \boldsymbol{\zeta}_r^\sigma \end{bmatrix} = \left[\begin{array}{c|c} \tilde{\mathcal{L}}_v\mathcal{M}^\sigma\mathcal{L}_v & \tilde{\mathcal{L}}_v\mathcal{M}^\sigma\mathcal{L}_r \\ \hline \tilde{\mathcal{L}}_r\mathcal{M}^\sigma\mathcal{L}_v & \tilde{\mathcal{L}}_r\mathcal{M}^\sigma\mathcal{L}_r \end{array}\right] \tag{2.7.10}$$

where $\mathscr{L}_v$ and $\mathscr{L}_r$ are the two parts of the $\mathscr{L}$-matrix of equation (2.7.9). The ζ_v^σ matrices are those of interest in most of the vibration–rotation treatments and we shall be essentially concerned with them.

The ζ^σ matrices have some important properties that can be summarized in a series of rules. These have been discussed by several authors[7–11] and the reader is referred to their original papers or to the books of References 12 and 13 for a detailed discussion. In this book we shall not be interested in the determination of the Coriolis ζ-coefficients from the analysis of vibro-rotational or rotational bands but rather in their use in the calculation of force constants. This problem will be discussed in detail in Chapters 4 and 8.

References

1. H. B. G. Casimir, *Dissertation*, Leyden (1931).
2. C. Eckart, *Phys. Rev.*, **47**, 552 (1935).
3. A. Sayvetz, *J. Chem. Phys.*, **6**, 383 (1939).
4. E. B. Wilson and J. B. Howard, *J. Chem. Phys.*, **4**, 260 (1936).
5. W. E. Milne, *Numerical Calculus*, Princeton University Press, 1949; E. Bodewig, *Matrix Calculus*, North Holland, 1956; B. Noble, *Numerical Methods*, Interscience, 1964; J. H. Wilkinson, *The Algebraic Eigenvalue Problem*, Oxford University Press, 1965.
6. E. B. Wilson Jr., J. C. Decius and P. C. Cross, *Molecular Vibrations*, McGraw-Hill, 1955.
7. J. H. Meal and S. R. Polo, *J. Chem. Phys.*, **24**, 1119 (1956); **24**, 1126 (1956).
8. T. Oka and Y. Morino, *J. Mol. Spectrosc.*, **6**, 472 (1961).
9. I. M. Mills and J. L. Duncan, *J. Mol. Spectrosc.*, **9**, 244 (1962).
10. T. Oka, *J. Mol. Spectrosc.*, **29**, 84 (1969).
11. L. Henry and G. Amat, *Cah. Phys.*, **14**, 230 (1960).
12. H. C. Allen and P. C. Cross, *Molecular Vib-rotors*, Wiley, 1963.
13. S. J. Cyvin, *Molecular Vibrations and Mean-Square Amplitudes*, Elsevier, 1968.
14. M. B. Hesse, *Forces and Fields. The Concept of Action at a Distance in the History of Physics*, T. Nelson and Sons, 1961.

Chapter 3
The Quantum Treatment of Molecular Vibrations

3.1 The Born–Oppenheimer approximation

In the previous chapter we have discussed the classical treatment of molecular vibrations and shown that classical mechanics provides a correct description of the vibrations of a polyatomic molecule. In conjunction with the mechanical treatment, classical electromagnetic theory can account for the general features of the vibrational spectra. It cannot, however, explain the details and fails completely in the treatment of the intensities and of the fine structure of vibro-rotational bands. The quantum-mechanical treatment not only provides a correct explanation for these problems, but also furnishes a unified picture of electronic, vibrational and rotational spectroscopy, that the classical theory is unable to achieve.

In the present chapter we shall give a concise survey of the quantum-mechanical treatment of vibrational spectroscopy covering both the vibrational problem and the interaction of the radiation field with the vibrating molecules. In doing this, we shall assume that the reader is acquainted with the basic principles and techniques of quantum mechanics and we shall limit the exposition to those features which are essential for the understanding of the vibrational spectra.

A stationary state of a polyatomic molecule is described in quantum mechanics by a wavefunction ψ and by an energy value E, an eigenfunction and eigenvalue respectively of the complete molecular Hamiltonian $\mathscr{H}$

$$\mathscr{H}\psi = E\psi \tag{3.1.1}$$

where $\mathscr{H}$ contains terms depending upon the electronic, nuclear and spin coordinates. In order to treat the vibrational problem, it is first necessary to separate the various contributions to $\mathscr{H}$ to the maximum extent, so that (3.1.1) can be divided into separate equations, one for each problem.

We shall assume right away that this separation is possible for the terms depending upon the spin coordinates and we shall therefore ignore the spin part of the Hamiltonian in the following discussion. The separation of the electronic and nuclear motions depends upon the large difference between the mass of the electrons and the masses of the nuclei. Since these latter are much heavier, the electrons have much greater velocities. To a good approximation the electronic motions can thus be treated assuming fixed positions of the nuclei in space. This approximation, known as the Born–Oppenheimer approximation,[1] corresponds in quantum mechanics to the assumption that the wavefunction $\psi(\mathbf{r}, \mathbf{R})$, where $\mathbf{r}$ and $\mathbf{R}$ are electronic and nuclear coordinates, factorizes as the product of two functions $\psi_n(\mathbf{R})$ and $\psi_e(\mathbf{r}, \mathbf{R})$, where the *nuclear* wavefunction $\psi_n(\mathbf{R})$ is a function of the nuclear coordinates only and the *electronic* wavefunction $\psi_e(\mathbf{r}, \mathbf{R})$ is a function of the electronic coordinates for a given rigid nuclear framework.

By choosing a molecular reference system located at the centre of mass of the molecule, which satisfies the first Sayvetz condition and factorizes the translational energy, we can write $\mathscr{H}$ in the form

$$\mathscr{H} = \mathscr{H}_{\mathrm{e}} + \mathscr{H}_{\mathrm{n}}$$

$$\mathscr{H}_{\mathrm{n}} = -\sum_{\alpha} \frac{\hbar^2}{2m_\alpha} \frac{\partial^2}{\partial R_\alpha^2} + \sum_{\alpha,\beta} \frac{Z_\alpha Z_\beta e^2}{R_{\alpha\beta}} \tag{3.1.2}$$

$$\mathscr{H}_{\mathrm{e}} = -\sum_{i} \frac{\hbar^2}{2m_i} \frac{\partial^2}{\partial r_i^2} + \sum_{i,j} \frac{e^2}{r_{ij}} - \sum_{i,\alpha} \frac{Z_\alpha e^2}{r_{i\alpha}}$$

where α, β label the nuclei and i, j the electrons, Z being the number of positive charges on the nuclei and $R_{\alpha\beta}$, $r_{i\alpha}$, r_{ij} being nuclear–nuclear, nuclear–electronic and electronic–electronic distances respectively. The electronic wavefunction ψ_{e} is then an eigenfunction of the electronic Hamiltonian $\mathscr{H}_{\mathrm{e}}$

$$\mathscr{H}_{\mathrm{e}}\psi_{\mathrm{e}} = E_{\mathrm{e}}\psi_{\mathrm{e}} \tag{3.1.3}$$

for each given relative nuclear conformation. By writing

$$\psi(\mathbf{r}, \mathbf{R}) = \psi_{\mathrm{e}}(\mathbf{r}, \mathbf{R})\psi_{\mathrm{n}}(\mathbf{R}) \tag{3.1.4}$$

as discussed before and by using (3.1.2) and (3.1.4) we obtain

$$(\mathscr{H}_{\mathrm{n}} + E_{\mathrm{e}})\psi_{\mathrm{n}} = E\psi_{\mathrm{n}} \tag{3.1.5}$$

which shows that E_{e} enters the effective nuclear Hamiltonian as a parameter whose variation with the internuclear distances can be treated as part of the potential energy of the nuclei. The electronic wave-equation (3.1.3) yields as eigenvalues the electronic energy levels and constitutes the basic equation of electronic spectroscopy. Hereafter, therefore, we shall limit ourselves to the eigenvalues and eigenfunctions of the nuclear Hamiltonian.

In Chapter 2 we have shown that in the classical treatment the nuclear motions can be separated, to a fairly good approximation, into vibrational and rotational motions. To the same degree of approximation, this is also true in quantum mechanics.[2] Accordingly, if an axis system is introduced which rotates with the molecule and satisfies the second Sayvetz condition, the wavefunction ψ_{n} can be written as the product of a *rotational* wavefunction ψ_r and a *vibrational* wavefunction ψ_v

$$\psi_{\mathrm{n}} = \psi_v\psi_r \tag{3.1.6}$$

and the nuclear Hamiltonian $\mathscr{H}_{\mathrm{n}}$ as

$$\mathscr{H}_{\mathrm{n}} = \mathscr{H}_v + \mathscr{H}_r \tag{3.1.7}$$

where $\mathscr{H}_r$ is the rotational Hamiltonian (including the interaction of the angular moments of rotation and vibration) and $\mathscr{H}_v$ is the vibrational Hamiltonian, which satisfies the equation

$$\mathscr{H}_v\psi_v = E_v\psi_v \tag{3.1.8}$$

The rotational Hamiltonian $\mathscr{H}_r$ and the rotational wavefunction ψ_r are discussed in great detail in the book by Allen and Cross[3] *Molecular Vib-rotors*. In this book we are concerned only with vibrational states and thus our interest will be devoted to the solutions of (3.1.8) only. In the limits of the Born–Oppenheimer and of the harmonic approximation equation (3.1.8) has simple solutions which are discussed in the next section. In some cases, however, the total wavefunction cannot be written as a simple product of the type (3.1.4) and more generally as a product

$$\psi = \psi_e \psi_v \psi_r \tag{3.1.9}$$

since the Hamiltonian contains interaction terms among the different types of coordinates that cannot be ignored. In such cases one must allow for a breakdown of the Born–Oppenheimer approximation, collecting under this name both the equations (3.1.4) and (3.1.6) as in the original work of these authors.[1]

When (3.1.9) is not valid, the wavefunction ψ is normally written as a sum of products of the type (3.1.9) multiplied by appropriate coefficients[4]

$$\psi = \sum_i c_i \psi_{ei} \psi_{vi} \psi_{ri} \tag{3.1.10}$$

Normally the breakdown of the Born–Oppenheimer approximation can be reduced to the breakdown of either (3.1.4) or of (3.1.6). The first case requires the use of a wavefunction of the type

$$\psi = \psi_v \sum_i c_i \psi_{ei} \psi_{ri} \tag{3.1.11}$$

or of the type

$$\psi = \psi_r \sum_i c_i \psi_{ei} \psi_{vi} \tag{3.1.12}$$

while the second requires a wavefunction

$$\psi = \psi_e \sum_i \psi_{vi} \psi_{ri} \tag{3.1.13}$$

Equation (3.1.11) describes a situation in which rotation–electronic interactions cause the breakdown of the Born–Oppenheimer approximation. Such interactions are of great importance in the analysis of the rotational levels but play no role in vibrational spectroscopy.

Equations (3.1.12) describes a situation in which the vibrational–electronic (vibronic) interaction is responsible for the breakdown of (3.1.4), such as in the Renner[5] and in the Jahn–Teller[5] effects. Finally, equation (3.1.13) describes a situation in which the vibro-rotational interactions are responsible for the breakdown of (3.1.6). The Coriolis interaction is the best known vibro-rotational interaction. Since this interaction takes place between states of different vibrational symmetry, it can sometimes produce alterations of the absorption spectra by making forbidden transitions appear in the infrared. This is, however, less important than the perturbations induced in the rotational structure by the Coriolis interaction. For this reason the breakdown of (3.1.6) is normally dealt with in rotational spectroscopy books. It must be pointed out, however, that Coriolis interaction constants are normally used in the calculation of accurate force constants for small molecules.

3.2 The vibrational wave equation

The vibrational Hamiltonian and the vibrational wavefunctions are written, according to the previous section, in terms of a coordinate system rotating with the molecule and satisfying the Sayvetz conditions. Among the possible sets of coordinates that one can choose, normal coordinates are by far the most convenient ones since to each normal coordinate is associated one single mode of vibration of the molecule.

The solution of the wave equation expressed in terms of normal coordinates is rather simple for molecules with non-degenerate vibrations. We shall discuss this case in part (a) of this section. For molecules with degenerate vibrations the problem is more complex. We shall see later on that only two- and three-fold degenerate vibrations can occur for polyatomic molecules. We shall discuss these cases separately in parts (b) and (c) of this section.

(A) *Non-degenerate case. The one-dimensional harmonic oscillator*

The quantum-mechanical vibrational Hamiltonian operator for a molecule is easily obtained from the classical Hamiltonian function in terms of the normal coordinates Q_i and the associated momenta

$$P_i = (\partial T/\partial \dot{Q}_i) = \dot{Q}_i \qquad H = T + V = \frac{1}{2}\sum_{i=1}^{3N-6} (P_i^2 + \lambda_i Q_i^2) \tag{3.2.1}$$

by using the postulates of quantum mechanics. It has the form

$$\mathscr{H}_v = -\frac{\hbar^2}{2}\sum_{i=1}^{3N-6}\left(\frac{\partial^2}{\partial Q_i^2} - \frac{\lambda_i}{\hbar^2}Q_i^2\right) \tag{3.2.2}$$

where $\hbar = h/2\pi$. Since the normal coordinates constitute a set of $3N - 6$ independent variables,† the wavefunction ψ_v of equation (3.1.8) can be written as the product of $3N - 6$ independent functions $\psi_{v_i}(Q_i)$, each a function of one single coordinate Q_i

$$\psi_v(Q_1, Q_2, Q_3, \ldots, Q_{3N-6}) = \psi_{v_1}(Q_1)\psi_{v_2}(Q_2)\ldots\psi_{v_{3N-6}}(Q_{3N-6}) \tag{3.2.3}$$

and the vibrational energy E_v as the sum of $3N - 6$ terms

$$E_v = E_1 + E_2 + E_3 + \ldots + E_{3N-6} \tag{3.2.4}$$

By application of the Hamiltonian operator (3.2.2) to the wavefunction (3.2.3), we obtain $3N - 6$ independent equations in one variable, of the type

$$-\frac{\hbar^2}{2}\frac{\partial^2\psi_{v_i}}{\partial Q_i^2} + \frac{\lambda_i}{2}Q_i^2\psi_{v_i} = E_{v_i}\psi_{v_i} \tag{3.2.5}$$

which is the well-known Schrödinger equation for a one-dimensional harmonic oscillator. The solution of equation (3.1.8) can thus be expressed as the product of $3N - 6$ harmonic oscillator wavefunctions and the total vibrational energy as the sum of the energies of these $3N - 6$ oscillators.

The solution of equation (3.2.5) was obtained by Schrödinger himself in terms of a variable q_i related to Q_i by the expression

$$q_i = \gamma_i^{\frac{1}{2}} Q_i \tag{3.2.6}$$

where

$$\gamma_i = 2\pi c\nu_i/\hbar = \lambda_i^{\frac{1}{2}}/\hbar \tag{3.2.7}$$

The variable q_i is dimensionless, as is easily seen by introducing (3.2.6) into the expression (2.4.7) of the potential energy. For this reason q_i is called the 'dimensionless normal coordinate'.

Using the dimensionless normal coordinate q_i, equation (3.2.5) becomes

$$\frac{d^2\psi_{v_i}}{dq_i^2} + \left(\frac{2E}{\hbar\lambda_i^{\frac{1}{2}}} - q_i^2\right)\psi_{v_i} = 0 \tag{3.2.8}$$

† $3N - 5$ for linear molecules.

and can be solved exactly. The eigenvalues are given by

$$E_{v_i} = hc\nu_i(v_i + \tfrac{1}{2}) \tag{3.2.9}$$

where ν_i is the classical frequency of vibration in cm^{-1} and v_i is a quantum number which can take any integral positive value including zero. The eigenfunctions are given by

$$\psi_{v_i} = \mathcal{N}_{v_i} H_{v_i}(q_i)\, \mathrm{e}^{-q_i^2/2} \tag{3.2.10}$$

where

$$\mathcal{N}_{v_i} = \left(\frac{\gamma_i}{\pi}\right)^{\frac{1}{4}} \frac{1}{[2^{v_i}(v_i!)]^{\frac{1}{2}}} \tag{3.2.11}$$

is a normalization factor and $H_{v_i}(q_i)$ are the Hermite polynomials of degree v_i in the variable q_i, defined as

$$H_{v_i}(q_i) = (-1)^{v_i}\, \mathrm{e}^{q_i^2} \frac{\mathrm{d}^{v_i}}{\mathrm{d}q_i^{v_i}} \mathrm{e}^{-q_i^2} \tag{3.2.12}$$

For example, the first three Hermite polynomials are, according to (3.2.12),

$$H_0(q_i) = 1 \qquad H_1(q_i) = 2q_i \qquad H_2(q_i) = 4q_i^2 - 2$$

Higher-degree polynomials are easily obtained using the recursion formula

$$H_{v_i+1}(q_i) = 2q_i H_{v_i}(q_i) - 2v_i H_{v_i-1}(q_i) \tag{3.2.13}$$

The Hermite polynomials form a set of orthogonal functions[2] in the interval $-\infty \leqslant q_i \leqslant +\infty$ of the variable q_i. The same is true for the functions (3.2.10) which obey then the orthonormality relation

$$\langle\psi_{v_i}|\psi_{v_i'}\rangle = \mathcal{N}_{v_i}\mathcal{N}_{v_i'} \int_{-\infty}^{+\infty} \mathrm{e}^{-q_i^2} H_{v_i}(q_i) H_{v_i'}(q_i)\, \mathrm{d}q_i = \begin{cases} 0 & \text{if } v_i \neq v_i' \\ 1 & \text{if } v_i = v_i' \end{cases} \tag{3.2.14}$$

In future we shall need the matrix elements of the type

$$\begin{aligned} (Q_i)_{v_i v_i'} &= \langle\psi_{v_i}|Q_i|\psi_{v_i'}\rangle \qquad &\text{(a)} \\ (P_i)_{v_i v_i'} &= \langle\psi_{v_i}|P_i|\psi_{v_i'}\rangle \qquad &\text{(b)} \end{aligned} \tag{3.2.15}$$

where Q_i and P_i are a normal coordinate and its conjugate momentum respectively. By use of (3.2.13) and (3.2.14) it is easily seen that the matrix elements (3.2.15a) are always equal to zero unless $v_i' = v_i \pm 1$. For $v_i' = v_i + 1$, we obtain from (3.2.15a)

$$\begin{aligned} (Q_i)_{v_i, v_i+1} &= \gamma_i^{-\frac{1}{2}} \mathcal{N}_{v_i}\mathcal{N}_{v_i+1} \int_{-\infty}^{+\infty} \mathrm{e}^{-q_i^2} H_{v_i} q_i H_{v_i+1}\, \mathrm{d}q_i \\ &= \gamma_i^{-\frac{1}{2}} \mathcal{N}_{v_i}\mathcal{N}_{v_i+1} \left[(v_i+1) \int_{-\infty}^{+\infty} \mathrm{e}^{-q_i^2} H_{v_i} H_{v_i}\, \mathrm{d}q_i + \frac{1}{2} \int_{-\infty}^{+\infty} \mathrm{e}^{-q_i^2} H_{v_i} H_{v_i+2}\, \mathrm{d}q_i\right] \\ &= \gamma_i^{-\frac{1}{2}} \mathcal{N}_{v_i}\mathcal{N}_{v_i+1} (v_i+1) \int_{-\infty}^{+\infty} \mathrm{e}^{-q_i^2} H_{v_i} H_{v_i}\, \mathrm{d}q_i \\ &= \gamma_i^{-\frac{1}{2}} \mathcal{N}_{v_i}\mathcal{N}_{v_i+1} (v_i+1) \frac{1}{\mathcal{N}_{v_i}^2} \end{aligned}$$

and, using (3.2.11),

$$(Q_i)_{v_i, v_i+1} = \gamma_i^{-\frac{1}{2}} \left(\frac{v_i+1}{2}\right)^{\frac{1}{2}} \tag{3.2.16}$$

Similarly we obtain for $v_i' = v_i - 1$

$$(Q_i)_{v_i,v_i-1} = \gamma_i^{-\frac{1}{2}}\left(\frac{v_i}{2}\right)^{\frac{1}{2}} \tag{3.2.17}$$

and for any other value of v'

$$(Q_i)_{v_i,v_i'} = 0 \qquad v_i' \neq v_i \pm 1 \tag{3.2.18}$$

In order to obtain the matrix elements of P_i we notice that from (3.2.12) it follows that

$$\frac{\mathrm{d}H_{v_i}}{\mathrm{d}q_i} = 2q_iH_{v_i} - H_{v_i+1} = 2v_iH_{v_i-1} \tag{3.2.19}$$

and that

$$\frac{\mathrm{d}}{\mathrm{d}q_i}[H_{v_i}\,\mathrm{e}^{-q_i^2/2}] = \mathrm{e}^{-q_i^2/2}(v_iH_{v_i-1} - \tfrac{1}{2}H_{v_i+1}) \tag{3.2.20}$$

By substitution of P_i with the operator $(-\mathrm{i}\hbar)\partial/\partial Q_i$ we obtain

$$\begin{aligned} (P_i)_{v_i,v_i+1} &= -\mathrm{i}(\hbar\lambda_i^{\frac{1}{2}})^{\frac{1}{2}}\left(\frac{v_i+1}{2}\right)^{\frac{1}{2}} \\ (P_i)_{v_i,v_i-1} &= \mathrm{i}(\hbar\lambda_i^{\frac{1}{2}})^{\frac{1}{2}}(v_i/2)^{\frac{1}{2}} \\ (P_i)_{v_i,v_i'} &= 0 \qquad v_i' \neq v_i \pm 1 \end{aligned} \tag{3.2.21}$$

(B) *Degenerate case. The two-dimensional harmonic oscillator*

If a molecule possesses two degenerate normal coordinates Q_{t1} and Q_{t2}, the secular equation has two identical roots $\lambda_{t1} = \lambda_{t2} = \lambda_t$. According to (3.2.1) the classical Hamiltonian function for a two-fold degenerate oscillator has the form

$$H_t = \tfrac{1}{2}[(P_{t1}^2 + P_{t2}^2) + \lambda_t(Q_{t1}^2 + Q_{t2}^2)] \tag{3.2.22}$$

The corresponding Hamiltonian operator is then

$$\mathscr{H}_t = -\frac{\hbar^2}{2}\left[\left(\frac{\partial^2}{\partial Q_{t1}^2} + \frac{\partial^2}{\partial Q_{t2}^2}\right) - \frac{\lambda_t}{\hbar^2}(Q_{t1}^2 + Q_{t2}^2)\right] \tag{3.2.23}$$

and has eigenfunctions

$$\psi_{v_t} = \psi_{v_{t1}}(\gamma_t^{\frac{1}{2}}Q_{t1})\psi_{v_{t2}}(\gamma_t^{\frac{1}{2}}Q_{t2}) \tag{3.2.24}$$

and eigenvalues

$$E_{v_t} = hc\nu_t[(v_{t1} + \tfrac{1}{2}) + (v_{t2} + \tfrac{1}{2})] = hc\nu_t(v_t + 1) \tag{3.2.25}$$

where

$$v_t = v_{t1} + v_{t2} \tag{3.2.26}$$

Each value of v_t is associated with a set of $v_t + 1$ degenerate energy levels and of $v_t + 1$ degenerate wavefunctions, corresponding to the different possible values of v_{t1} and v_{t2} in (3.2.26), as shown below for $v_t = 0$, 1 and 2.

v_t	v_{t1}	v_{t2}	ψ_{v_t}
0	0	0	$\psi_0(\gamma_t^{\frac{1}{2}}Q_{t1})\psi_0(\gamma_t^{\frac{1}{2}}Q_{t2})$
1	1	0	$\psi_1(\gamma_t^{\frac{1}{2}}Q_{t1})\psi_0(\gamma_t^{\frac{1}{2}}Q_{t2})$
	0	1	$\psi_0(\gamma_t^{\frac{1}{2}}Q_{t1})\psi_1(\gamma_t^{\frac{1}{2}}Q_{t2})$
2	2	0	$\psi_2(\gamma_t^{\frac{1}{2}}Q_{t1})\psi_0(\gamma_t^{\frac{1}{2}}Q_{t2})$
	1	1	$\psi_1(\gamma_t^{\frac{1}{2}}Q_{t1})\psi_1(\gamma_t^{\frac{1}{2}}Q_{t2})$
	0	2	$\psi_0(\gamma_t^{\frac{1}{2}}Q_{t1})\psi_2(\gamma_t^{\frac{1}{2}}Q_{t2})$

Although $\psi_{v_{t1}}$ and $\psi_{v_{t2}}$ furnish an adequate description of the system, a better one can be obtained by a different choice of variables. In order to clarify the reasons for this change of variables we recall from Section 2.5 that in a degenerate vibration the atoms do not move in general along straight lines but describe ellipses around the equilibrium positions. There is therefore a vibrational angular momentum associated with the degenerate vibrations, that is not explicitly quantized in the description given above.

As for the non-degenerate case of part (A), we shall express the Hamiltonian in terms of dimensionless normal coordinates and conjugate momenta, defined as

$$\begin{aligned} q_{t1} &= \gamma_t^{\frac{1}{2}}Q_{t1} \qquad p_{t1} = \gamma_t^{-\frac{1}{2}}P_{t1} \\ q_{t2} &= \gamma_t^{\frac{1}{2}}Q_{t2} \qquad p_{t2} = \gamma_t^{-\frac{1}{2}}P_{t2} \end{aligned} \tag{3.2.27}$$

Substitution of these expressions in (3.2.22) then yields

$$H_t = \tfrac{1}{2}\hbar\lambda_t^{\frac{1}{2}}[(p_{t1}^2 + p_{t2}^2)/\hbar^2 + (q_{t1}^2 + q_{t2}^2)] \tag{3.2.28}$$

from which we obtain in the usual way the Hamiltonian operator

$$\mathscr{H}_t = \tfrac{1}{2}\hbar\lambda_t^{\frac{1}{2}}\left(-\frac{\partial^2}{\partial q_{t1}^2} - \frac{\partial^2}{\partial q_{t2}^2} + q_{t1}^2 + q_{t2}^2\right) \tag{3.2.29}$$

In the two-dimensional configuration space spanned by q_{t1} and q_{t2}, we can define polar coordinates ρ and φ, through the relations

$$\begin{aligned} q_{t1} &= \rho\cos\varphi \\ q_{t2} &= \rho\sin\varphi \end{aligned} \tag{3.2.30}$$

from which it follows that

$$\begin{aligned} \rho^2 &= q_{t1}^2 + q_{t2}^2 \\ \tan\varphi &= q_{t1}/q_{t2} \end{aligned} \tag{3.2.31}$$

and thus

$$\frac{\partial}{\partial q_{t1}} = \cos\varphi\frac{\partial}{\partial\rho} - \frac{\sin\varphi}{\rho}\frac{\partial}{\partial\varphi} \qquad \frac{\partial}{\partial q_{t2}} = \sin\varphi\frac{\partial}{\partial\rho} + \frac{\cos\varphi}{\rho}\frac{\partial}{\partial\varphi} \tag{3.2.32}$$

Using (3.2.32), the Hamiltonian operator (3.2.29) becomes in polar coordinates

$$\mathscr{H}_t = -\tfrac{1}{2}\hbar\lambda_t^{\frac{1}{2}}\left[\frac{1}{\rho}\frac{\partial}{\partial\rho}\left(\rho\frac{\partial}{\partial\rho}\right) + \frac{1}{\rho^2}\frac{\partial^2}{\partial\varphi^2} - \rho^2\right] \tag{3.2.33}$$

The Schrödinger equation for the two-fold degenerate oscillator is then

$$\frac{1}{\rho}\frac{\partial}{\partial\rho}\left(\rho\frac{\partial\psi_{v_t}}{\partial\rho}\right)+\frac{1}{\rho^2}\frac{\partial^2\psi_{v_t}}{\partial\varphi^2}+\left(\frac{2E}{\hbar\lambda_t^{\frac{1}{2}}}-\rho^2\right)\psi_{v_t}=0 \tag{3.2.34}$$

Complete separation of the variables is obtained by choosing ψ_{v_t} in the form

$$\psi_{v_t}=F(\rho)\phi(\varphi) \tag{3.2.35}$$

By substitution of (3.2.35) in (3.2.34) we obtain, after division by ψ_{v_t} and multiplication by ρ^2,

$$\frac{1}{F}\rho\frac{\partial}{\partial\rho}\left(\rho\frac{\partial F}{\partial\rho}\right)+\rho^2\left(\frac{2E}{\hbar\lambda^{\frac{1}{2}}}-\rho^2\right)=-\frac{1}{\phi}\frac{\partial^2\phi}{\partial\varphi^2} \tag{3.2.36}$$

Since the left side of this equation depends only upon the independent variable ρ whereas the right side depends only upon the independent variable φ, both members must be equal to a constant that we call l^2. We then have the two equations

$$\frac{1}{\phi}\frac{\partial^2\phi}{\partial\varphi^2}=-l^2 \tag{3.2.37}$$

$$\frac{1}{\rho}\frac{\partial}{\partial\rho}\left(\rho\frac{\partial F}{\partial\rho}\right)+\left(\frac{2E}{\hbar\lambda^{\frac{1}{2}}}-\rho^2-\frac{l^2}{\rho^2}\right)F=0 \tag{3.2.38}$$

The solution of (3.2.37) is very simple, being of the form

$$\phi=e^{il\varphi} \tag{3.2.39}$$

and, in order to be single-valued, requires that l must be an integer or zero.

The solution of (3.2.38) is of the form

$$F(\rho)=e^{-\rho^2/2}\rho^{|l|}Z(\rho) \tag{3.2.40}$$

where $Z(\rho)$ is a polynomial whose form can be found by insertion of (3.2.40) in (3.2.38). We have

$$\frac{d^2Z}{d\rho^2}+\left(\frac{2|l|+1}{\rho}-2\rho\right)\frac{dZ}{d\rho}+2\left[\frac{E}{\hbar\lambda_t^{\frac{1}{2}}}-(|l|+1)\right]Z=0 \tag{3.2.41}$$

By putting $Z(\rho^2)=G(r)$ and by the change of variable $\rho^2=r$, equation (3.2.41) is converted into the differential equation for the associated Laguerre polynomials[2]

$$r\frac{d^2G}{dr^2}+(|l|+1-r)\frac{dG}{dr}+\left[\frac{1}{2}\left(\frac{E}{\hbar\lambda_t^{\frac{1}{2}}}-1+|l|\right)-|l|\right]G=0 \tag{3.2.42}$$

which, using (3.2.25), can be written in the form

$$r\frac{d^2G}{dr^2}+(|l|+1-r)\frac{dG}{dr}+(n-|l|)G=0 \tag{3.2.43}$$

where

$$n=\tfrac{1}{2}(v_t+|l|) \tag{3.2.44}$$

The solutions of (3.2.43) are the associated Laguerre polynomials[2] defined as

$$G(r)=L_n^{|l|}(r)=\frac{d^{|l|}}{dr^{|l|}}L_n(r) \tag{3.2.45}$$

where

$$L_n(r) = e^r \frac{d^n}{dr^n}(r^n e^{-r}) \tag{3.2.46}$$

are the Laguerre polynomials of degree n.

The form of the Laguerre polynomials imposes some restrictions on the possible values that the numbers l and n can assume. According to (3.2.46) $n = \frac{1}{2}(v_t + |l|)$ must be an integer, and according to (3.2.45) the associated Laguerre polynomials with $n - |l| < 0$ cannot exist. It follows then that l can taken only the values

$$l = \pm v_t, \pm(v_t - 2), \pm(v_t - 4), \ldots, \pm 1 \quad \text{or } 0 \tag{3.2.47}$$

The complete wavefunctions ψ_{v_t} of equation (3.2.35) are then

$$\psi_{v_t,l} = \mathcal{N}_{v_t,l}\, e^{-\rho^2/2} \rho^{|l|} L_n^{|l|}(\rho^2)\, e^{il\varphi} \tag{3.2.48}$$

where $\mathcal{N}_{v_t l}$ is a normalization factor given by

$$\mathcal{N}_{v_t l} = \frac{1}{\sqrt{\pi}} \frac{[(v_t - |l|)/2!]^{\frac{1}{2}}}{[(v_t + |l|)/2!]^{\frac{3}{2}}} \tag{3.2.49}$$

The associated Laguerre polynomials can be obtained using (3.2.45). The polynomials corresponding to $v_t = 0$, 1 and 2 are, for instance,

$$L_0^0 = 1 \qquad L_1^0 = 1 - \rho^2 \qquad L_1^1 = -1$$
$$L_2^0 = 2 - 4\rho^2 + \rho^4 \qquad L_2^1 = -4 + 2\rho^2 \qquad L_2^2 = 2$$

and correspondingly the wavefunctions (3.2.48) are

$$\psi_{0,0} = \frac{1}{\sqrt{\pi}} e^{-\rho^2/2}$$

$$\psi_{1,1} = \frac{1}{\sqrt{\pi}} e^{-\rho^2/2} \rho L_1^1 e^{i\varphi} \qquad \psi_{1,-1} = \frac{1}{\sqrt{\pi}} e^{-\rho^2/2} \rho L_1^1 e^{-i\varphi}$$

$$\psi_{2,2} = \frac{1}{\sqrt{\pi}} e^{-\rho^2/2} \rho^2 L_2^2 e^{2i\varphi} \qquad \psi_{2,0} = \frac{1}{\sqrt{\pi}} e^{-\rho^2/2} L_1^0 \qquad \psi_{2,-2} = \frac{1}{\sqrt{\pi}} e^{-\rho^2/2} \rho^2 L_2^2 e^{-2i\varphi}$$

We can now show that the wavefunctions (3.2.48) are also eigenfunctions of the vibrational angular momentum operator. We notice that the classical vibrational angular momentum is perpendicular to the configuration plane spanned by the normal coordinates q_{t1} and q_{t2} and its magnitude is given by

$$M = |(\mathbf{e}_1 q_{t1} + \mathbf{e}_2 q_{t2}) \mathrm{x} (\mathbf{e}_1 p_{t1} + \mathbf{e}_2 p_{t2})| = q_{t1} p_{t2} - q_{t2} p_{t1} \tag{3.2.50}$$

$\mathbf{e}_1$ and $\mathbf{e}_2$ being unit vectors in the plane defined by q_{t1} and q_{t2}. Using (3.2.32) the operators associated with p_{t1} and p_{t2} assume the form

$$\begin{aligned} &-i\hbar\left(\cos\varphi \frac{\partial}{\partial \rho} - \sin\varphi \frac{\partial}{\partial \varphi}\right) \\ &-i\hbar\left(\sin\varphi \frac{\partial}{\partial \rho} + \cos\varphi \frac{\partial}{\partial \varphi}\right) \end{aligned} \tag{3.2.51}$$

and thus, using (3.2.30) we obtain from (3.2.50) that the vibrational angular momentum operator is given in polar coordinates by

$$\mathcal{M} = -i\hbar \frac{\partial}{\partial \varphi} \tag{3.2.52}$$

It is easily seen now, by using this expression for $\mathscr{M}$, that the wavefunctions (3.2.48) are eigenfunctions of $\mathscr{M}$, since

$$\mathscr{M}\psi_{v_t,l} = \hbar l \psi_{v_t,l} \tag{3.2.53}$$

The quantum number l gives, therefore, in units of $\hbar$, the value of the vibrational angular momentum in the state $\psi_{v_t l}$.

The calculation of the matrix elements of Q_{t1} and Q_{t2}, as well as of P_{t1} and P_{t2}, is not simple as in the case of the one-dimensional harmonic oscillator. For this reason we shall not discuss them here. The reader interested in their evaluation can find a detailed discussion in Reference 6.

(C) *Degenerate case. The three-dimensional harmonic oscillator*

If a molecule possesses three degenerate normal coordinates, Q_{t1}, Q_{t2} and Q_{t3}, the secular equation has three identical roots $\lambda_{t1} = \lambda_{t2} = \lambda_{t3} = \lambda_t$. The Hamiltonian operator, in dimensionless normal coordinates, is given by the expression

$$\mathscr{H}_t = \tfrac{1}{2}\hbar\lambda_t^{\frac{1}{2}}\left[-\frac{\partial^2}{\partial q_{t1}^2} - \frac{\partial^2}{\partial q_{t2}^2} - \frac{\partial^2}{\partial q_{t3}^2} + q_{t1}^2 + q_{t2}^2 + q_{t3}^2\right] \tag{3.2.54}$$

and has eigenfunctions

$$\psi_{v_t} = \psi_{v_{t1}}(q_{t1})\psi_{v_{t2}}(q_{t2})\psi_{v_{t3}}(q_{t3}) \tag{3.2.55}$$

and eigenvalues

$$E_t = hc\nu_t(v_{t1} + \tfrac{1}{2} + v_{t2} + \tfrac{1}{2} + v_{t3} + \tfrac{1}{2}) = hc\nu_t(v_t + \tfrac{3}{2}) \tag{3.2.56}$$

where

$$v_t = v_{t1} + v_{t2} + v_{t3} \tag{3.2.57}$$

There are in this case $(v_t + 1)(v_t + 2)/2$ degenerate wavefunctions and energy levels for each value of v_t, corresponding to the different possible combinations of v_{t1}, v_{t2} and v_{t3} in (3.2.57).

As for the two-dimensional case discussed before, it is convenient to transform the Hamiltonian in polar coordinates which, for the three-dimensional space spanned by q_{t1}, q_{t2} and q_{t3}, are defined as

$$\begin{aligned} q_{t1} &= \rho \sin\theta \cos\varphi \\ q_{t2} &= \rho \sin\theta \sin\varphi \\ q_{t3} &= \rho \cos\theta \end{aligned} \tag{3.2.58}$$

The Hamiltonian transformation is long and tedious and will not be given in detail. Owing to the similarity of the problem to that of the hydrogen atom,[2] the reader should recognize in the transformed Hamiltonian the well-known expression

$$\mathscr{H}_t = -\frac{1}{2}\hbar\lambda_t^{\frac{1}{2}}\left[\frac{1}{\rho^2}\frac{\partial}{\partial\rho}\left(\rho^2\frac{\partial}{\partial\rho}\right) + \frac{1}{\rho^2}\frac{1}{\sin\theta}\frac{\partial}{\partial\theta}\left(\sin\theta\frac{\partial}{\partial\theta}\right) + \frac{1}{\rho^2\sin^2\theta}\frac{\partial^2}{\partial\varphi^2} - \rho^2\right]. \tag{3.2.59}$$

Using (3.2.59) we obtain the wave equation

$$\frac{1}{\rho^2}\frac{\partial}{\partial\rho}\left(\rho^2\frac{\partial\psi_{v_t}}{\partial\rho}\right) + \frac{1}{\rho^2\sin\theta}\frac{\partial}{\partial\theta}\left(\sin\theta\frac{\partial\psi_{v_t}}{\partial\theta}\right) + \frac{1}{\rho^2\sin^2\theta}\frac{\partial^2\psi_{v_t}}{\partial\varphi^2} + \left(\frac{2E}{\hbar\lambda_t^{\frac{1}{2}}} - \rho^2\right)\psi_{v_t} = 0 \tag{3.2.60}$$

that can be solved exactly after separation of the variables.

Complete separation of the variables is obtained by using for ψ_{v_t} the form

$$\psi_{v_t} = R(\rho)\Theta(\theta)\Phi(\varphi) \tag{3.2.61}$$

Insertion of (3.2.61) into (3.2.60) leads, by simple manipulation, to the three equations

$$\frac{d^2\Phi}{d^2\varphi^2} = -m^2\Phi \qquad \text{(a)}$$

$$\frac{1}{\sin\theta}\frac{d}{d\theta}\left(\sin\theta\frac{d\Theta}{d\theta}\right) + \left[l(l+1) - \frac{m^2}{\sin^2\theta}\right]\Theta = 0 \qquad \text{(b)} \quad (3.2.62)$$

$$\frac{1}{\rho^2}\frac{d}{d\rho}\left(\rho^2\frac{dR}{d\rho}\right) + \left[\frac{2E}{\hbar\lambda_t^{\frac{1}{2}}} - \rho^2 - \frac{l(l+1)}{\rho^2}\right]R = 0 \qquad \text{(c)}$$

which have solutions

$$\Phi = \mathcal{N}_\Phi e^{im\varphi} \qquad \text{(a)}$$

$$\Theta = \mathcal{N}_{l,|m|}P_l^{|m|}(\cos\theta) \qquad \text{(b)} \quad (3.2.63)$$

$$R = \mathcal{N}_{v_t,l}\, e^{-\rho^2/2}\rho^l L_\tau^{l+\frac{1}{2}}(\rho^2) \qquad \text{(c)}$$

where $\tau = \frac{1}{2}(v_t + l + 1)$, l and m being quantum numbers that can assume only the values

$$l = v_t, v_t - 2, v_t - 4, \ldots, 1 \text{ or } 0$$
$$m = 0, \pm 1, \pm 2, \ldots, \pm l \qquad (3.2.64)$$

The functions $P_l^{|m|}(\cos\theta)$ are the associated Legendre polynomials[2] of order $|m|$ and degree l which are described in detail in many books of quantum mechanics. The functions $L_\tau^{l+\frac{1}{2}}(\rho^2)$ are the associated Laguerre polynomials of degree $\frac{1}{2}(v_t - 1)$ in ρ^2 and thus of degree $(v_t - 1)$ in ρ. The normalization factors $\mathcal{N}_\Phi$, $\mathcal{N}_{l,|m|}$ and $\mathcal{N}_{v_t,l}$ are found[7] to be

$$\mathcal{N}_\Phi = \frac{1}{\sqrt{(2\pi)}} \qquad \text{(a)}$$

$$\mathcal{N}_{l,|m|} = \frac{(-1)^l}{2^l l!}\left[\frac{(2l+1)}{2}\frac{(l-|m|)!}{(l+|m|)!}\right]^{\frac{1}{2}} \qquad \text{(b)} \quad (3.2.65)$$

$$\mathcal{N}_{v_t,l} = \sqrt{2}\frac{\{[(v_t-1)/2]!\}^{\frac{1}{2}}}{\{[(v_t+l+1)/2]!\}^{\frac{1}{2}}} \qquad \text{(c)}$$

In the configuration space spanned by q_{t1}, q_{t2} and q_{t3} we can define a vibrational angular momentum M with components

$$M_1 = q_{t2}p_{t3} - q_{t3}p_{t2} \qquad \text{(a)}$$

$$M_2 = q_{t3}p_{t1} - q_{t1}p_{t3} \qquad \text{(b)} \quad (3.2.66)$$

$$M_3 = q_{t1}p_{t2} - q_{t2}p_{t1} \qquad \text{(c)}$$

with magnitude

$$M^2 = M_1^2 + M_2^2 + M_3^2 \qquad (3.2.67)$$

The quantum-mechanical operators associated with M^2 and M_3 have in polar coordinates the form[2,7]

$$\mathcal{M}^2 = -\hbar^2\left(\frac{\partial^2}{\partial\theta^2} + \frac{\cos\theta}{\sin\theta}\frac{\partial}{\partial\theta} + \frac{1}{\sin^2\theta}\frac{\partial^2}{\partial\varphi^2}\right) \qquad \text{(a)}$$

$$\mathcal{M}_3 = -i\hbar\frac{\partial}{\partial\varphi} \qquad \text{(b)} \quad (3.2.68)$$

and it can be easily seen that the wavefunctions (3.2.61) are also eigenfunctions of both $\mathcal{M}^2$ and $\mathcal{M}_3$

$$\begin{aligned} \mathcal{M}^2\psi_{v_t} &= \hbar^2 l(l+1)\psi_{v_t} && \text{(a)} \\ \mathcal{M}_3\psi_{v_t} &= \hbar m\psi_{v_t} && \text{(b)} \end{aligned} \qquad (3.2.69)$$

Therefore, the quantum numbers l and m quantize the vibrational angular momentum and its component M_3.

The calculation of the matrix elements of the operators $Q_{t\sigma}$ and $P_{t\sigma}$ ($\sigma = 1, 2$ and 3) is very difficult and will not be considered here. References 6 and 7 deal with this problem in some detail.

3.3 The energy levels of a polyatomic molecule

According to equation (3.2.9) the vibrational energy of a polyatomic molecule is

$$E_v = \sum_{i=1}^{3N-6} (v_i + \tfrac{1}{2})hc\nu_i \qquad (3.3.1)$$

where ν_i is the classical frequency of vibration, in units of cm^{-1}, associated with the normal coordinate Q_i, and v_i is the corresponding quantum number. In vibrational spectroscopy it is customary to use units of cm^{-1} for the frequencies of vibration ν, defined as $1/\lambda$, λ being the wavelength in cm. The frequency in cycles/s, $\tilde{\nu}$, is then equal to νc, where c is the velocity of light.

The lowest vibrational level of the molecule is called the *ground level* and corresponds to a vibrational situation in which all normal coordinates have zero quantum number. The energy of the ground level is not zero but has from (3.3.1) the value

$$E_0 = \tfrac{1}{2}hc\sum_i \nu_i \qquad (3.3.2)$$

which can be relatively high for a polyatomic molecule. This quantity takes the name of *zero-point* energy and is of great importance in many physical processes.

If all quantum numbers have value zero except one, which is equal to unity, the corresponding energy levels take the name of *fundamental levels*. From (3.3.1) we see that the number of fundamental levels is equal to $3N - 6$ and each corresponds to the excitation of a single normal mode. If the molecule possesses two-fold degenerate normal coordinates Q_l and Q_k ($\nu_l = \nu_k$), the energy levels for $v_l = 1$, $v_k = 0$ and $v_l = 0$, $v_k = 1$ are two-fold degenerate. If there are three-fold degenerate normal coordinates, Q_l, Q_k and Q_m, the levels corresponding to v_l or v_k or $v_m = 1$ (all other $v_i = 0$) are triply degenerate.

If all quantum numbers have value zero except one, which has value greater than one, the corresponding energy levels take the name of *overtone* levels. In this case, if the quantum number v_k takes the value 2, we speak of the first overtone level, if it takes the value 3 we speak of the second overtone level of the ν_k vibration, and so on.

If two (or more) quantum numbers are different from zero, the resulting levels are called *combination* levels. The number of combination levels becomes very large as the energy increases, and above a given limit they build almost a continuum of energies. Happily these levels with many quantum numbers excited are of little interest in vibrational spectroscopy. As an example, some of the vibrational levels of a bent triatomic molecule like SO_2, with fundamental frequencies $\nu_1 = 1151\ cm^{-1}$, $\nu_2 = 519\ cm^{-1}$, $\nu_3 = 1361\ cm^{-1}$, are

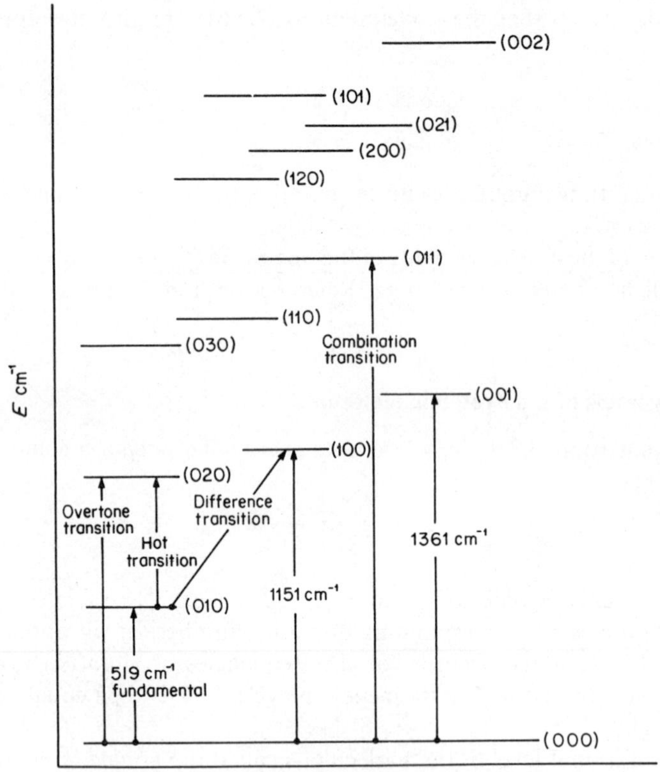

Figure 3.3.1 Schematic drawing of the harmonic energy levels, in units of cm^{-1}, of SO_2

shown schematically in Figure 3.3.1. Since the frequencies are measured in units of cm^{-1}, it is convenient to measure the vibrational energies in the same units, by dividing equation (3.3.1) by *hc*.

Transitions between two energy levels are governed by the famous Bohr rule

$$\nu_{v'v''} = \frac{E_{v''} - E_{v'}}{hc} \tag{3.3.3}$$

where $E_{v''}$ is the energy of the upper level and $E_{v'}$ the energy of the lower level. Among these, by far the most important in absorption spectra are the transitions between the ground and the fundamental levels, i.e. between the ground level and the levels corresponding to the excitation of a single quantum number. From equations (3.3.1) and (3.3.3) we see that the frequencies associated with these transitions are simply the classical normal frequencies. The name 'fundamental' is then used both for the level $v_k = 1$ and for the corresponding classical frequency ν_k. In the same way, transitions from the ground level to overtone levels are called overtones and transitions from the ground level to combination levels are called combinations.

In the harmonic approximation the fundamental and the overtone levels of a given mode ν_k are, according to (3.3.3), equally spaced. The transition from the ground to the fundamental level $v_k = 1$ thus has exactly the same frequency as the transition from the overtone level $v_k = m$ to the level $v_k = m + 1$. Such transitions between two excited levels of the same mode are called *hot transitions* since the lower excited level must be thermally populated in order that a sufficient number of molecules experience the transition.

Transitions can also occur between two excited levels corresponding to two different quantum numbers v_k and v_l. The frequency associated with this transition is called the *difference* frequency since, from (3.3.3), it is just the difference between the frequencies ν_k and ν_l of the transitions between the ground and the two excited levels, as shown graphically in Figure 3.3.1.

3.4 Generality on the light–matter interaction

The interaction of radiation with matter can be discussed in two different ways. The simpler approach is to use quantum mechanics to describe the material system and classical electromagnetic theory to describe the radiation field. Since it mixes classical and quantum-mechanical concepts, this kind of treatment is often called *semiclassical.* The second approach, more difficult but considerably more powerful, consists in treating both the molecules and the radiation field as quantized systems, and takes the name of *quantum field* theory.

Both treatments are based on time-dependent perturbation theory. In particular some of the interaction processes, such as absorption and emission of light, are satisfactorily handled by means of first-order perturbation theory. For this reason they are often classified as *first-order* optical phenomena. Some other processes, such as Raman scattering, require second-order perturbation theory and are therefore classified as *second-order* optical phenomena. A different kind of classification is based on the number of light quanta (photons) involved in each elementary process. In normal absorption and emission of light, one photon at a time is either absorbed or emitted by a molecule. For this reason absorption and emission are also called *one-photon* processes. With a much lower probability, however, a molecule can also absorb or emit two or more photons simultaneously. Such processes take the name of *two-photon* (or multi-photon) absorption and emission. Normal Raman scattering, in which at the same time one photon is absorbed and one photon is emitted is also a two-photon process.

The semiclassical theory can be used without difficulty to describe a one-photon absorption process and to reach final conclusions on the absorption selection rules. The semiclassical theory of emission is, however, much more complex and difficult and basically uses a different approach. Quantum field theory is much more general. It furnishes a unified treatment of absorption and emission and automatically leads to the concept of spontaneous emission. As far as second-order optical phenomena are concerned, the advantages of quantum field theory are even greater. The mathematical treatment of Raman scattering in the semiclassical theory is extremely complex, this complexity not being balanced by a greater generality. On the contrary, many second-order phenomena, which are readily interpreted in the quantum field treatment, are not easily covered by the semiclassical theory.

The only difficulty connected with the use of quantum field theory is that it employs a description of the radiation field which is not familiar to many spectroscopists. For this reason in the following three sections we shall present the time-dependent perturbation theory, the classical description of the radiation field and the basic aspects of quantum field theory, necessary to develop the absorption and scattering processes. Section 3.8 introduces the interaction Hamiltonian in its general form. The following Section, 3.9, presents the one-photon absorption and emission treatment. The reader not interested in the development of the theory but only in the final results which lead to the selection rules can jump directly to Section 3.10. Raman scattering is presented in Sections 3.13 and 3.14. Again the mathematical treatment of Section 3.13 can be ignored if only the final results are required.

For obvious reasons of space, the theory is presented in a very concise form which leaves very little space for the discussion of the physical problem often buried in the mathematical formalism. For this reason we strongly recommend the reader to consult one of the several excellent books available on the argument.[8,9,10]

3.5 Time-dependent perturbation theory

In problems in which the Hamiltonian operator is time-independent, wavefunctions of the type

$$\phi(x, t) = \psi(x)\,e^{-iEt/\hbar} \tag{3.5.1}$$

satisfy the time-dependent Schrödinger equation

$$\mathscr{H}\phi(x, t) = i\hbar\frac{\partial\phi(x, t)}{\partial t} \tag{3.5.2}$$

If, instead, $\mathscr{H}$ is time-dependent, solutions like (3.5.1) are not possible. In many cases, however, $\mathscr{H}$ can be written in the form

$$\mathscr{H} = \mathscr{H}_0 + \lambda\mathscr{H}_1 \tag{3.5.3}$$

where λ is a parameter ($0 \leqslant \lambda \leqslant 1$). The time-dependent part $\mathscr{H}_1$ is small compared to the time-independent part $\mathscr{H}_0$ and can thus be considered as a perturbation of the stationary states of the system, which are described by the eigenfunctions $\psi_k(x)$ and by the eigenvalues E_k of the unperturbed Hamiltonian $\mathscr{H}_0$.

Since the $\psi(x)$ are a complete set of functions, the wavefunction $\phi(x, t)$ of $\mathscr{H}$ can be expanded in terms of the wavefunctions $\psi(x)$ with appropriate coefficients

$$\phi(x, t) = \sum_k b_k(t)\psi_k(x)\exp\left[-iE_k t/\hbar\right] \tag{3.5.4}$$

By substitution into the time-dependent Schrödinger equation, we have

$$(\mathscr{H}_0 + \lambda\mathscr{H}_1)\sum_k b_k(t)\psi_k(x)\exp\left[-iE_k t/\hbar\right] = i\hbar\frac{\partial}{\partial t}\sum_k b_k(t)\psi_k(x)\exp\left[-iE_k t/\hbar\right] \tag{3.5.5}$$

and thus

$$\lambda\sum_k b_k(t)\mathscr{H}_1\psi_k(x)\exp\left[-iE_k t/\hbar\right] = i\hbar\sum_k\frac{db_k(t)}{dt}\psi_k(x)\exp\left[-iE_k t/\hbar\right] \tag{3.5.6}$$

Multiplication on the left by $\psi_m^*(x)$ and integration over the space coordinates yields a set of differential equations for the coefficients

$$\dot{b}_m(t) = \frac{db_m(t)}{dt} = -\frac{i}{\hbar}\sum_k \lambda b_k(t)H_{mk}\exp\left[-i(E_k - E_m)t/\hbar\right] \tag{3.5.7}$$

where

$$H_{mk} = \langle\psi_m|\mathscr{H}_1|\psi_k\rangle \tag{3.5.8}$$

Equation (3.5.7) has been obtained without introducing any approximation and is therefore quite general. It shows that the effect of the perturbation is to couple together all the stationary states for which a matrix element H_{mk} is different from zero.

The coefficients $b_k(t)$ satisfy the relation

$$\sum_k b_k^*(t)b_k(t) = 1 \tag{3.5.9}$$

which is easily obtained from (3.5.4) by left multiplication by ϕ^* and integration over the space coordinates. It is convenient, in order to use (3.5.7) at different orders of approximation, to develop each coefficient b_k in series of powers of the parameter λ

$$b_k = b_k^{(0)} + \lambda b_k^{(1)} + \lambda^2 b_k^{(2)} + \ldots \tag{3.5.10}$$

Introduction of (3.5.10) into (3.5.7) then yields

$$\sum_\nu \lambda^\nu \dot{b}_m^{(\nu)}(t) = -\frac{\mathrm{i}}{\hbar}\sum_\nu \sum_k \lambda^{\nu+1} b_k^{(\nu)}(t) H_{mk} \exp[\mathrm{i}(E_m - E_k)t/\hbar] \tag{3.5.11}$$

This equation is satisfied for all values of λ only if the coefficients of a given power of λ are the same on both sides. We thus obtain the equations

$$\text{zeroth-order} \qquad \dot{b}_m^{(0)}(t) = 0 \tag{3.5.12}$$

$$\text{first-order} \qquad \dot{b}_m^{(1)}(t) = -\frac{\mathrm{i}}{\hbar}\sum_k b_k^{(0)}(t) H_{mk} \exp[\mathrm{i}(E_m - E_k)t/\hbar] \tag{3.5.13}$$

$$\text{second-order} \qquad \dot{b}_m^{(2)}(t) = -\frac{\mathrm{i}}{\hbar}\sum_k b_k^{(1)}(t) H_{mk} \exp[\mathrm{i}(E_m - E_k)t/\hbar] \tag{3.5.14}$$

etc.

In order to obtain a unique solution for these equations it is necessary to supply some boundary conditions. Such conditions are normally chosen to reproduce correctly the physical situation being dealt with.

In the case of the interaction of the radiation field with a molecule, it is convenient to assume that, up to a given time, taken as $t = 0$, the system is in some stationary state of $\mathscr{H}_0$, described by a wavefunction ψ_0. It may be noticed that ψ_0 need not be the ground state wavefunction, although it is in most of the cases of interest. At the time $t = 0$ the perturbation $\mathscr{H}_1$ is switched on and there is a definite probability of transition to a different state.

We shall now calculate this probability for both first-order and second-order perturbations. The first-order treatment is required for the analysis of the absorption and emission processes. The second-order treatment is necessary for the development of the theory of scattering in general and of the Raman effect in particular.

(A) *First-order perturbation theory*

The kind of problems covered by the first-order perturbation theory, such as absorption and emission of light from a molecule, can be schematized as follows.

For $t \leqslant 0$ the system is in a stationary state defined by the wavefunction ψ_0. From equation (3.5.9) we obtain the initial conditions

$$|b_0(0)|^2 = 1$$

i.e.

$$b_0(0) = 1 \qquad b_k(0) = 0 \qquad k \neq 0 \tag{3.5.15}$$

In zeroth order we thus have

$$b_0^{(0)}(0) = 1 \qquad b_k^{(0)}(0) = 0 \qquad k \neq 0 \tag{3.5.16}$$

At $t = 0$ we switch on the perturbation described by $\mathscr{H}_1$ (in the case of interest to us, absorption and emission of light, this is the perturbation of the electromagnetic field). For small time intervals we can reasonably assume that all coefficients $b_k(t)$, different from $b_0(t)$ are very small, whereas $b_0(t)$ is still close to unity. In other words we assume that the zeroth-order situation remains unchanged over small time intervals and that the variation in the coefficients $b_k(t)$ occurs only at the first order. Higher-order variations will be ignored at this level. We thus have

$$b_0^{(0)}(t) = 1 \qquad b_k^{(0)}(t) = 0 \qquad k \neq 0 \tag{3.5.17}$$

Equation (3.5.13) then yields

$$\dot{b}_m^{(1)}(t) = -\frac{i}{\hbar} H_{m0} \exp\,[i(E_m - E_0)t/\hbar] \tag{3.5.18}$$

For our purposes it is sufficient to consider only the simpler case in which $\mathscr{H}_1$ is time independent. We notice that even if $\mathscr{H}_1$ is a constant perturbation for $t > 0$, the total perturbation from $t < 0$ to $t > 0$ is time-dependent since $\mathscr{H}$ behaves as a step function at $t = 0$. In the case of a constant perturbation $\mathscr{H}_1$, equation (3.5.18) can be easily integrated and gives

$$b_m^{(1)}(t) = H_{m0} \frac{1 - \exp\,[i(E_m - E_0)t/\hbar]}{E_m - E_0} \tag{3.5.19}$$

the probability of finding the system in the state m at time t, is then given by

$$b_m(t)b_m^*(t) = [b_m^{(0)}(t) + b_m^{(1)}(t)][b_m^{(0)}(t) + b_m^{(1)}(t)]^* = 2H_{m0}H_{m0}^* \left\{\frac{1 - \cos\,[(E_m - E_0)t/\hbar]}{(E_m - E_0)^2}\right\} \tag{3.5.20}$$

and for large values of t is very small unless $E_m \approx E_0$. We point out to the reader that E_m and E_0 are not molecular states here, but rather states of the whole system (for instance molecule plus field). The case in which E_m coincides with E_0 is thus of the greatest interest to us since it reflects the physical situation we shall deal with in the discussion of the interaction of the electromagnetic field with a molecule.

We assume therefore that E_m belongs to a continuum of states in which the discrete level E_0 is embedded. This corresponds to the assumption that $\mathscr{H}_0$ has both a continuum and a discrete spectrum of eigenvalues. The probability per unit of time of finding the system in one of the states with energies between E_m and $E_m + dE_m$, is given then, if $\rho(E)$ is the density of such states, by

$$\frac{b_m(t)b_m^*(t)\rho(E)}{t}\, dE_m = 2H_{m0}H_{m0}^* \left\{\frac{1 - \cos\,[(E_m - E_0)t/\hbar]}{(E_m - E_0)^2 t}\right\} \rho(E)\, dE_m \tag{3.5.21}$$

This expression can be easily integrated over the entire energy spectrum, by introducing the variable $x = (E_m - E_0)t/\hbar$ so that $dx = (t/\hbar)\, dE_m$. We obtain in this way the so-called 'transition probability'

$$w_{m0} = \frac{2H_{m0}H_{m0}^*}{\hbar} \rho(E) \int_{-\infty}^{+\infty} \frac{1 - \cos x}{x^2}\, dx = \frac{2\pi}{\hbar} H_{m0}H_{m0}^* \rho(E) \tag{3.5.22}$$

It is very convenient to rewrite equation (3.5.20) in a different form using a formalism due to Dirac, which utilizes the so-called 'δ-function'. The $\delta(y)$ function is an operational quantity that has a meaning only under an integral sign. It is an integral representation of unity and can be considered as a function that is zero everywhere except around $y = 0$ where it has a strong singularity and tends to infinity in such a way that it satisfies the condition

$$\int_{-\infty}^{+\infty} \delta(y)\, \mathrm{d}y = 1 \tag{3.5.23}$$

The function $\delta(y)$ can be represented in different ways since there are several functions that satisfy this condition. For our purposes it is convenient to use the following representation

$$\delta(y) = \frac{1}{\pi} \lim_{\alpha \to \infty} \frac{1 - \cos \alpha y}{\alpha y^2} \tag{3.5.24}$$

The δ-function properties of (3.5.24) are easily verified. When $\alpha \to \infty$, the expression

$$\frac{1 - \cos \alpha y}{\alpha y^2} \tag{3.5.25}$$

vanishes unless $y = 0$. At $y = 0$ it goes to infinity and furthermore, since

$$\int_{-\infty}^{+\infty} \frac{1 - \cos \alpha y}{\alpha y^2}\, \mathrm{d}y = \pi \tag{3.5.26}$$

condition (3.5.23) is satisfied. A more extensive discussion of the δ-function properties and representations can be found in References 8 and 9.

Using (3.5.24) with $\alpha = t/\hbar$ and $y = (E_m - E_0)$, equation (3.5.20) becomes for $t \to \infty$

$$b_m(t)b_m^*(t) = \frac{2\pi}{\hbar} t H_{m0} H_{m0}^* \delta(E_m - E_0) \tag{3.5.27}$$

and it is easily verified that the transition probability w_{m0} is obtained by integration of the previous relation using (3.5.23)

$$w_{m0} = \frac{2\pi}{\hbar} H_{m0} H_{m0}^* \rho(E) \int_{-\infty}^{+\infty} \delta(E_m - E_0)\, \mathrm{d}E_m = \frac{2\pi}{\hbar} H_{m0} H_{m0}^* \rho(E) \tag{3.5.28}$$

In deriving equation (3.5.20) we have explicitly stated that the validity of this equation is confined to short time intervals for which the coefficient $b_0^{(0)}$ could still be considered equal to unity and all other coefficients $b_k^{(0)}$ equal to zero. In equation (3.5.27) we have instead introduced the condition $t \to \infty$, which is apparently contradictory with the previous statement. It is therefore necessary to verify that the time intervals for which the coefficients $b_k^{(1)}$ remain very small are large enough to justify the extrapolation to infinity.

The condition that the coefficients $b_k^{(1)}$ are very small can be written

$$b_k^{(1)}(t) \ll 1$$

The upper limit for t can be obtained, therefore, by using this condition and expanding the exponential in (3.5.19). In this way we obtain

$$t \ll \frac{\hbar}{|H_{m0}|} \tag{3.5.29}$$

which determines a relatively large time interval, since $|H_{m0}|$ is normally very small.

(B) *Second-order perturbation theory*

We shall develop the second-order perturbation theory in its simplest form which is limited to infinitesimally sharp energy levels of the system and to short time intervals.

This form of the theory permits a relatively simple understanding of the scattering of light from a molecule, but offers serious difficulties if one wishes to analyse the form of the bands and to inquire into the lifetime of the excited states. In this case it is necessary to use a more general form of perturbation theory that has been devised to overcome these difficulties.[9,10] Since for the determination of the Raman selection rules it is unnecessary to go into this more advanced mathematical treatment, we shall limit ourselves to the simpler approach.

The kind of problems for which the second-order perturbation treatment is necessary can be reduced to the following scheme. For $t < 0$ the system is in a stationary state defined by the wavefunction ψ_0 so that the initial conditions (3.5.16)

$$b_0^{(0)}(0) = 1 \qquad b_k^{(0)}(0) = 0 \qquad k \neq 0 \tag{3.5.30}$$

given for the first-order treatment, are still valid.

At the time $t = 0$ we switch on the perturbation $\mathscr{H}_1$ and we select, among the excited states of $\mathscr{H}_0$, a stationary state m which is not directly connected to the state 0 by a first-order term, i.e. such that $H_{m0} = 0$. We now ask for the probability of transition from the state 0 to the state m at the second order since, as $H_{m0} = 0$, the first-order probability (3.5.22) is equal to zero. We assume the existence of series of states, denoted by the label k, which are connected by a first-order matrix element both to 0 and to m, such that H_{k0}, $H_{km} \neq 0$. We shall call these states intermediate states and for the moment we shall not impose any restriction on them.

For small positive time intervals, when the perturbation Hamiltonian is on, we shall again assume that no variations occur in the zero-order coefficients. This gives, by analogy with (3.5.17)

$$b_0^{(0)}(t) = 1 \qquad b_m^{(0)}(t) = 0 \qquad b_k^{(0)}(t) = 0 \tag{3.5.31}$$

From equation (3.5.13) we then obtain

$$\dot{b}_0^{(1)}(t) = 0 \qquad \dot{b}_m^{(1)}(t) = 0 \tag{3.5.32}$$

$$\dot{b}_k^{(1)}(t) = -\frac{\mathrm{i}}{\hbar} H_{k0} \exp\left[\mathrm{i}(E_k - E_0)t/\hbar\right] \tag{3.5.33}$$

and by integration over t we have, apart from a constant (that we assume equal to zero)

$$b_k^{(1)}(t) = H_{k0} \frac{\exp\left[\mathrm{i}(E_k - E_0)t/\hbar\right]}{E_0 - E_k} \tag{3.5.34}$$

The main difference between (3.5.34) and the corresponding equation (3.5.19) is that, in the first-order treatment, the value of $b_k^{(1)}(0)$ was taken to be zero to fulfil the initial conditions of the type (3.5.17) also at the first order.

In equation (3.5.34) we do not require the first-order coefficients $b_k^{(1)}(t)$ to fulfil such conditions. As a consequence instead of $b_k^{(1)}(0) = 0$, as from (3.5.19), we obtain from (3.5.34)

$$b_k^{(1)}(0) = \frac{H_{k0}}{E_0 - E_k} \tag{3.5.35}$$

For the two coefficients $b_0^{(1)}(t)$ and $b_m^{(1)}(t)$ we have from (3.5.32)

$$b_0^{(1)}(0) = 0 \qquad b_m^{(1)}(0) = 0 \tag{3.5.36}$$

We can now calculate from (3.5.14) the second-order coefficient $b_m^{(2)}(t)$. Using the previous results we obtain

$$\dot{b}_m^{(2)}(t) = -\frac{\mathrm{i}}{\hbar} \sum_k \frac{H_{k0} H_{mk}}{E_0 - E_k} \exp\left[\mathrm{i}(E_m - E_0)t/\hbar\right] \tag{3.5.37}$$

and, by integration of this equation, we can obtain the coefficient $b_m^{(2)}(t)$ and calculate the probability of a second-order transition $0 \rightarrow m$. If we put

$$K_{m0} = \sum_k \frac{H_{k0}H_{mk}}{E_0 - E_k} \tag{3.5.38}$$

equation (3.5.37) becomes formally identical to equation (3.5.18). The second-order transition probability is thus, by analogy with (3.5.22)

$$w_{m0} = \frac{2\pi}{\hbar} K_{m0} K_{m0}^* \rho(E) \tag{3.5.39}$$

3.6 The classical radiation field

The electromagnetic field is described in classical electromagnetic theory by means of the electric field strength **E** and the magnetic field strength **H**, related to the charge density ρ and the current **i** by the well known Maxwell–Lorentz equations

$$\begin{aligned}
&\nabla \times \mathbf{E} + \frac{1}{c}\frac{\partial \mathbf{H}}{\partial t} = \mathbf{0} && \text{(a)}\\
&\nabla \times \mathbf{H} - \frac{1}{c}\frac{\partial \mathbf{E}}{\partial t} = \frac{4\pi}{c}\mathbf{i} && \text{(b)}\\
&\nabla \cdot \mathbf{E} = 4\pi\rho && \text{(c)}\\
&\nabla \cdot \mathbf{H} = 0 && \text{(d)}
\end{aligned} \tag{3.6.1}$$

These field equations may be simplified by introducing a scalar potential ϕ and a vector potential **A**, related to **E** and **H** by the equations

$$\begin{aligned}
&\mathbf{H} = \nabla \times \mathbf{A} && \text{(a)}\\
&\mathbf{E} = -\nabla\phi - \frac{1}{c}\frac{\partial \mathbf{A}}{\partial t} && \text{(b)}
\end{aligned} \tag{3.6.2}$$

The vector **A** and the scalar ϕ are not uniquely determined by these equations and different choices are possible. For instance it is easily seen that, if $\mathbf{A}_0$ and ϕ_0 represent a given choice, we can replace them by a new choice **A** and ϕ, related to the first by equations of the type

$$\begin{aligned}
&\mathbf{A} = \mathbf{A}_0 - \nabla\chi && \text{(a)}\\
&\phi = \phi_0 + \frac{1}{c}\frac{\partial \chi}{\partial t} && \text{(b)}
\end{aligned} \tag{3.6.3}$$

where χ is an arbitrary scalar function, without affecting† **E** and **H** of equation (3.6.2). If we substitute (3.6.2) into (3.6.1) the first and the fourth are automatically satisfied‡ while the second and the third equations become

$$\begin{aligned}
&\nabla \times (\nabla \times \mathbf{A}) + \frac{1}{c}\nabla\left(\frac{\partial \phi}{\partial t}\right) + \frac{1}{c^2}\frac{\partial^2 \mathbf{A}}{\partial t^2} = \frac{4\pi}{c}\mathbf{i} && \text{(a)}\\
&-\nabla \cdot (\nabla\phi) - \frac{1}{c}\nabla \cdot \left(\frac{\partial \mathbf{A}}{\partial t}\right) = 4\pi\rho && \text{(b)}
\end{aligned} \tag{3.6.4}$$

† Notice that if **u** is an arbitrary vector, $\nabla \cdot (\nabla \times \mathbf{u}) = 0$ and that if ϕ is an arbitrary scalar function $\nabla \times (\nabla\phi) = \mathbf{0}$.
‡ Equation (3.6.2b) can be actually obtained by insertion of equation (3.6.2a) in equation (3.6.1a).

and since

$$\nabla \times (\nabla \times \mathbf{A}) = \nabla(\nabla \cdot \mathbf{A}) - \nabla^2 \mathbf{A} \tag{3.6.5}$$

they assume the form

$$\begin{aligned} \frac{1}{c^2}\frac{\partial^2 \mathbf{A}}{\partial t^2} - \nabla^2 \mathbf{A} + \nabla\left(\nabla \cdot \mathbf{A} + \frac{1}{c}\frac{\partial \phi}{\partial t}\right) &= \frac{4\pi \mathbf{i}}{c} \\ \nabla^2 \phi + \frac{1}{c}\nabla \cdot \left(\frac{\partial \mathbf{A}}{\partial t}\right) &= -4\pi\rho \end{aligned} \tag{3.6.6}$$

The different choices of **A** and ϕ described by equation (3.6.3), that leave **E** and **H** unchanged, are called *gauges* and the freedom we have in the choice of the gauge can be utilized to simplify equations (3.6.6). The two important gauges which can be utilized for this simplification are called the Lorentz and Coulomb gauges.[8,9]

The Lorentz gauge is defined by the condition

$$\nabla \cdot \mathbf{A} + \frac{1}{c}\frac{\partial \phi}{\partial t} = 0 \tag{3.6.7}$$

which, inserted into (3.6.6), yields

$$\begin{aligned} \frac{1}{c^2}\frac{\partial^2 \mathbf{A}}{\partial t^2} - \nabla^2 \mathbf{A} &= \frac{4\pi \mathbf{i}}{c} \\ \frac{1}{c^2}\frac{\partial^2 \phi}{\partial t^2} - \nabla^2 \phi &= 4\pi\rho \end{aligned} \tag{3.6.8}$$

The Coulomb gauge is defined by the condition

$$\nabla \cdot \mathbf{A} = 0 \tag{3.6.9}$$

which, inserted into (3.6.6), yields

$$\begin{aligned} \frac{1}{c^2}\frac{\partial^2 \mathbf{A}}{\partial t^2} - \nabla^2 \mathbf{A} + \frac{1}{c}\nabla\left(\frac{\partial \phi}{\partial t}\right) &= \frac{4\pi \mathbf{i}}{c} \\ \nabla^2 \phi &= -4\pi\rho \end{aligned} \tag{3.6.10}$$

The Coulomb gauge has the advantage of being simpler and of defining the scalar potential ϕ only from the charge density ρ. The disadvantages of the Coulomb gauge are due to the fact that equations (3.6.10) are not invariant under a relativistic transformation. The Lorentz gauge is slightly more complicated but offers the advantage that equations (3.6.8) are relativistically invariant. For this reason, and also because it allows a simpler description of the radiation field, the Lorentz gauge is customarily adopted for such a description.

Since the radiation field is defined as the electromagnetic field *in vacuo*, the field equations in the Lorentz gauge are simply obtained from (3.6.8) by assuming $\mathbf{i} = 0$ and $\rho = 0$. They are therefore

$$\nabla^2 \mathbf{A} = \frac{1}{c^2}\frac{\partial^2 \mathbf{A}}{\partial t^2} \tag{3.6.11}$$

$$\nabla^2 \phi = \frac{1}{c^2}\frac{\partial^2 \phi}{\partial t^2} \tag{3.6.12}$$

As discussed before, the potentials **A** and ϕ are not uniquely determined and may be replaced by a different choice **A**$'$ and ϕ' defined through equation (3.6.3), as long as the

Lorentz condition (3.6.7) is satisfied. By substitution of (3.6.3) into (3.6.7), this condition takes the form

$$\nabla^2\chi = \frac{1}{c^2}\frac{\partial^2\chi}{\partial t^2} \tag{3.6.13}$$

It is thus possible to choose χ in such a way that $\phi' = 0$ and that (3.6.13) is fulfilled.

The radiation field is then described only by means of the vector potential $\mathbf{A}'$, hereafter called simply $\mathbf{A}$, through the equations

$$\phi = 0 \tag{3.6.14}$$

$$\nabla . \mathbf{A} = 0$$
$$\nabla^2\mathbf{A} = \frac{1}{c^2}\frac{\partial^2\mathbf{A}}{\partial t^2} \tag{3.6.15}$$

A possible solution of (3.6.15) can be found by putting

$$\mathbf{A}(\mathbf{r}, t) = q_\lambda(t)\mathbf{A}_\lambda(\mathbf{r}) \tag{3.6.16}$$

where the index λ specifies the particular solution found.

By substitution into (3.6.15) we obtain

$$\nabla . \mathbf{A}_\lambda(\mathbf{r}) = 0 \tag{3.6.17}$$

$$\frac{1}{q_\lambda(t)}\frac{\partial^2 q_\lambda(t)}{\partial t^2} = \frac{c^2}{\mathbf{A}_\lambda(\mathbf{r})}\nabla^2\mathbf{A}_\lambda(\mathbf{r}) \tag{3.6.18}$$

In this last equation the left-hand side depends only on t and the right-hand side only on $\mathbf{r}$. By introducing a constant $-\omega_\lambda^2$, we can thus separate (3.6.18) into the two equations

$$\frac{\partial^2 q_\lambda(t)}{\partial t^2} + \omega_\lambda^2 q_\lambda(t) = 0 \tag{3.6.19}$$

$$\nabla^2\mathbf{A}_\lambda(\mathbf{r}) + \frac{\omega_\lambda^2}{c^2}\mathbf{A}_\lambda(\mathbf{r}) = \mathbf{0} \tag{3.6.20}$$

with solutions

$$q_\lambda(t) = |q_\lambda| \, e^{-i\omega_\lambda t} \tag{3.6.21}$$

$$\mathbf{A}_\lambda(\mathbf{r}) = \mathbf{v}_\lambda \, e^{i(\mathbf{k}_\lambda \cdot \mathbf{r})} \tag{3.6.22}$$

where $\mathbf{v}_\lambda$ is an arbitrary vector perpendicular to $\mathbf{k}_\lambda$

$$(\mathbf{v}_\lambda . \mathbf{k}_\lambda) = 0$$

and $\mathbf{k}_\lambda$ is a vector which satisfies the condition

$$|\mathbf{k}_\lambda| = \frac{\omega_\lambda}{c} \tag{3.6.23}$$

Equation (3.6.16) therefore describes a monocromatic wave of frequency† $\omega_\lambda/2\pi$ with a direction of propagation defined by the wavevector $\mathbf{k}_\lambda$ and a direction of oscillation of the field $\mathbf{A}$ defined by the polarization vector $\mathbf{v}_\lambda$.

The general solution of (3.6.15) is then a Fourier sum of the type

$$\mathbf{A}(\mathbf{r}, t) = \sum_\lambda (q_\lambda(t)\mathbf{A}_\lambda(\mathbf{r}) + q_\lambda^*(t)\mathbf{A}_\lambda^*(\mathbf{r})) \tag{3.6.24}$$

† In cycles/s.

In the present treatment we shall limit ourselves to linearly polarized radiation waves. Under these conditions it is possible to show[8,9] that a convenient representation of the field can be made by choosing orthogonal and normalized solutions of (3.6.20) in the form

$$\int (\mathbf{A}_\lambda(\mathbf{r}) \cdot \mathbf{A}^*_\mu(\mathbf{r}))\, \delta\tau = 4\pi c^2\, \delta_{\lambda\mu}$$

from which follows, using (3.6.22),

$$\mathbf{A}_\lambda = (4\pi c^2)^{\frac{1}{2}} \mathbf{u}_\lambda\, \mathrm{e}^{\mathrm{i}(\mathbf{k}_\lambda \cdot \mathbf{r})} \tag{3.6.25}$$

where $\mathbf{u}_\lambda$ is a unit vector in a direction perpendicular to the direction of propagation of the field. In order to switch to the quantum-mechanical description of the radiation field it is convenient to introduce the real canonical variables

$$Q_\lambda = q_\lambda(t) + q^*_\lambda(t) \tag{3.6.26}$$

$$P_\lambda = \dot{Q}_\lambda = -\mathrm{i}\omega_\lambda[q_\lambda(t) - q^*_\lambda(t)] \tag{3.6.27}$$

and the Hamiltonian function

$$H_\lambda = \tfrac{1}{2}(P^2_\lambda + \omega^2_\lambda Q^2_\lambda) \tag{3.6.28}$$

It is easily verified that the Hamiltonian function given above satisfies the Hamiltonian equations of motion

$$\frac{\partial H_\lambda}{\partial Q_\lambda} = -\dot{P}_\lambda \qquad \frac{\partial H_\lambda}{\partial P_\lambda} = \dot{Q}_\lambda \tag{3.6.29}$$

The complete Hamiltonian for the radiation field can be obtained from (3.6.28) by summing over all values of λ

$$H = \sum_\lambda H_\lambda = \tfrac{1}{2}\sum_\lambda (P^2_\lambda + \omega^2_\lambda Q^2_\lambda) \tag{3.6.30}$$

It can be shown[8,9] that (3.6.30) represents the total energy of the radiation field, thus proving that (3.6.28) is a correctly defined Hamiltonian.

By comparison with equation (3.2.1) it is easily realized that the energy of the radiation field is the sum, for all values of λ, of the energies of harmonic oscillators, each described by a normal coordinate Q_λ and a conjugated momentum P_λ. For future reference we notice also that the Hamiltonian function (3.6.30) can be written, using (3.6.26) and (3.6.27), in the form

$$H = \sum_\lambda 2\omega^2_\lambda q_\lambda q^*_\lambda \tag{3.6.31}$$

3.7 The quantum radiation field

A relatively simple quantum-mechanical description of the radiation field can be obtained from the classical description of the previous section by substitution of the classical Hamiltonian function (3.6.30) with a quantum-mechanical Hamiltonian operator $\mathscr{H}_\mathrm{f}$, according to standard quantum-mechanical procedures.

We shall thus define a wavefunction ϕ, an eigenfunction of $\mathscr{H}_\mathrm{f}$, which contains all the information about the field. Consequently the energy E of a stationary state of the radiation field will be obtained as eigenvalue of the Schrödinger equation

$$\mathscr{H}_\mathrm{f}\phi = E\phi \tag{3.7.1}$$

Furthermore, since from (3.6.30) the Hamiltonian $\mathscr{H}_f$ is the sum of Hamiltonians $\mathscr{H}_\lambda$, each a function of one coordinate and of the conjugated momentum, it is legitimate to write ϕ as a product

$$\phi = \prod_\lambda \psi_{n\lambda} \tag{3.7.2}$$

of wavefunctions $\psi_{n\lambda}$, and E as a sum

$$E = \sum_\lambda E_{n\lambda} \tag{3.7.3}$$

where $\psi_{n\lambda}$ and $E_{n\lambda}$ are eigenfunctions and eigenvalues of the Hamiltonians $\mathscr{H}_\lambda$ and n is a quantum number.

This description of the radiation field runs parallel to the quantum-mechanical treatment of a multidimensional harmonic oscillator for which a set of normal coordinates Q_λ and conjugate momenta P_λ can be defined. The parallelism is, however, purely formal since the physical meaning of the coordinates Q_λ is rather obscure and the nature of the wavefunction is even less understandable. For this reason we shall limit ourselves to developing the essential formalism for the understanding of the absorption and scattering processes. The reader interested in a less concise treatment can consult one of the several books dealing specifically with the argument. In particular see References 8 and 10.

If we replace the coordinates Q_λ and the conjugated momenta P_λ with the corresponding quantum-mechanical operators Q_λ and $(\hbar/\mathrm{i})\partial/\partial Q_\lambda$, equation (3.6.28) yields the Schrödinger equation

$$-\frac{\hbar^2}{2}\frac{\mathrm{d}^2\psi_{n\lambda}}{\mathrm{d}Q_\lambda^2} + \frac{\omega_\lambda^2}{2}Q_\lambda^2\psi_{n\lambda} = E_{n\lambda}\psi_{n\lambda} \tag{3.7.4}$$

which is the well-known wave equation for the harmonic oscillator. The eigenvalues $E_{n\lambda}$ are thus given by

$$E_{n\lambda} = (n_\lambda + \tfrac{1}{2})\hbar\omega_\lambda \tag{3.7.5}$$

and the energy of the radiation field becomes from (3.7.3)

$$E = \sum_\lambda (n_\lambda + \tfrac{1}{2})\hbar\omega_\lambda \tag{3.7.6}$$

Equation (3.7.6) offers the difficulty that, since there are an infinite number of possibilities for λ, the zero-point energy of the radiation field is infinite. Although one may argue on the possibility of an infinite zero-point energy for a radiation field defined as above, it is simpler to avoid this difficulty by using, instead of the Hamiltonian (3.6.28), the Hamiltonian

$$H_\lambda = \tfrac{1}{2}(P_\lambda^2 + \omega_\lambda^2 Q_\lambda^2 - \hbar\omega_\lambda) \tag{3.7.7}$$

where a constant $-\hbar\omega_\lambda$ has been added to (3.6.27), leaving equation (3.6.29) unchanged. Starting from the above given Hamiltonian the energy of the radiation field becomes

$$E = \sum_\lambda n_\lambda \omega_\lambda \hbar \tag{3.7.8}$$

and has now a zero-point energy equal to zero. Most of the information on the radiation field can be obtained by investigating the properties of the various operators involved in the quantum description. To do this, let us evaluate the matrix elements of Q_λ and P_λ. Since the wavefunctions $\psi_{n\lambda}$ are harmonic oscillator wavefunctions of the type (3.2.10), the matrix elements of Q_λ and P_λ are those given by equations (3.2.16) to (3.2.21).

We have therefore

$$
\begin{aligned}
(Q_\lambda)_{n,n+1} = (Q_\lambda)_{n+1,n} &= \left(\frac{\hbar}{\omega_\lambda}\right)^{\frac{1}{2}}\left(\frac{n+1}{2}\right)^{\frac{1}{2}} \\
(Q_\lambda)_{n,m} &= 0 \qquad m \neq n \pm 1 \\
(P_\lambda)_{n,n+1} &= -\mathrm{i}(\hbar\omega_\lambda)^{\frac{1}{2}}\left(\frac{n+1}{2}\right)^{\frac{1}{2}} \\
(P_\lambda)_{n+1,n} &= \mathrm{i}(\hbar\omega_\lambda)^{\frac{1}{2}}\left(\frac{n+1}{2}\right)^{\frac{1}{2}} \\
(P_\lambda)_{n,m} &= 0 \qquad m \neq n \pm 1
\end{aligned}
\tag{3.7.9}
$$

From (3.6.26) and (3.6.27) it follows that

$$q_\lambda = \frac{1}{2}\left(Q_\lambda + \frac{\mathrm{i}}{\omega_\lambda}P_\lambda\right) \tag{3.7.10}$$

$$q_\lambda^* = \frac{1}{2}\left(Q_\lambda - \frac{\mathrm{i}}{\omega_\lambda}P_\lambda\right) \tag{3.7.11}$$

and therefore

$$
\begin{aligned}
(q_\lambda)_{n,n+1} &= \left(\frac{\hbar}{\omega_\lambda}\right)^{\frac{1}{2}}\left(\frac{n+1}{2}\right)^{\frac{1}{2}} \\
(q_\lambda^*)_{n+1,n} &= \left(\frac{\hbar}{\omega_\lambda}\right)^{\frac{1}{2}}\left(\frac{n+1}{2}\right)^{\frac{1}{2}} \\
(q_\lambda)_{n+1,n} = (q_\lambda^*)_{n,n+1} &= 0
\end{aligned}
\tag{3.7.12}
$$

From these equations it follows that

$$q_\lambda\phi_n = (\hbar/2\omega_\lambda)^{\frac{1}{2}}\sqrt{n}\,\phi_{n-1} \tag{3.7.13}$$

$$q_\lambda^*\phi_n = (\hbar/2\omega_\lambda)^{\frac{1}{2}}\sqrt{(n+1)}\,\phi_{n+1} \tag{3.7.14}$$

It must be noticed that q_λ and q_λ^* are now time-independent operators whereas in the classical treatment they are introduced as time-dependent coefficients (see equation (3.6.16)). Their commutation properties are easily obtained from (3.7.13) and (3.7.14) and are

$$[q_\lambda, q_\lambda^*] = (q_\lambda q_\lambda^* - q_\lambda^* q_\lambda) = \hbar/2\omega_\lambda$$

On the basis of the properties of the operators defined above, we can complete the quantum-mechanical description of the radiation field. As shown by equation (3.7.2) the wavefunction ϕ which describes the field depends upon a set of quantum numbers n_λ, each related to a field oscillator with frequency $\omega_\lambda/2\pi$. A given quantum number n_λ therefore measures, in units of $\hbar\omega_\lambda$, the energy separation between an excited state of the field oscillator with frequency $\omega_\lambda/2\pi$ and the ground state. In other words n_λ counts the number of quanta of energy $\hbar\omega_\lambda$ for each oscillation mode λ of the field, i.e. the number of photons of energy $\hbar\omega_\lambda$. The wavefunction $\phi(n_{\lambda_1}, n_{\lambda_2}, \ldots, n_{\lambda_i}, \ldots)$ therefore describes a radiation field in which there are n_{λ_1} photons of species λ_1, n_{λ_2} photons of species λ_2 and so on.

Since the operator q_{λ_i} transforms, according to (3.7.13), the wavefunction $\phi(n_{\lambda_1}, n_{\lambda_2}, \ldots, n_{\lambda_i}, \ldots)$ into the function $\phi(n_{\lambda_1}, n_{\lambda_2}, \ldots, n_{\lambda_i} - 1, \ldots)$, its effect is the annihilation of a photon of energy $\hbar\omega_{\lambda_i}$. In the same way we see that the operator $q_{\lambda_i}^*$ creates a photon of energy $\hbar\omega_{\lambda_i}$. For this reason q_{λ_i} and $q_{\lambda_i}^*$ are called annihilation and creation operators respectively.

It is important to notice that for a satisfactory description of the radiation field we do not need to known in detail the number of photons of each field mode, but rather their variation when the field interacts with a molecule.

3.8 The interaction between the radiation field and a molecule

In Section 3.1 we have described the quantum-mechanical treatment of a polyatomic molecule in the absence of an external perturbation, through a molecule Hamiltonian $\mathscr{H}_{mol}$, a set of molecular wavefunctions ψ_{mol} and a set of molecular stationary states E_{mol}, related by the time-independent Schrödinger equation. Similarly in Section 3.7 we have described the quantum-mechanical treatment of the radiation field in the absence of atoms or molecules by a field Hamiltonian $\mathscr{H}_{f}$, a set of field wavefunctions ϕ_f and a set of field stationary states E_f. In the present section we shall consider a molecule in the presence of a radiation field and let them interact.

Consider first a system formed by a molecule and a radiation field which do not interact. Clearly in this case the system will be described by a Hamiltonian $\mathscr{H}$ which is the sum of $\mathscr{H}_{mol}$ and $\mathscr{H}_f$, by a set of wavefunctions ψ which are the product of the wavefunctions ψ_{mol} and ϕ_f and by stationary energy states which are the sum of E_{mol} and E_f.

Consider now the same system but let the molecule be able to interact with the field. In this case the Hamiltonian will be

$$\mathscr{H} = \mathscr{H}_{mol} + \mathscr{H}_f + \mathscr{H}_{int} \tag{3.8.1}$$

where $\mathscr{H}_{int}$ is the part of the Hamiltonian which describes the interaction. The form of $\mathscr{H}_{int}$ can be obtained in the usual way by writing first the classical Hamiltonian function for the complete system and then transforming it into the quantum-mechanical operator. Classically, a particle of mass m and charge e which moves in an electromagnetic field with velocity $\mathbf{v}$, is subject to a force

$$\mathbf{F} = e\left(\mathbf{E} + \frac{1}{c}[\mathbf{v} \times \mathbf{H}]\right) \tag{3.8.2}$$

where $\mathbf{E}$ and $\mathbf{H}$ are the electric and the magnetic field vectors respectively. Using equation (3.6.2) we can write this equation in terms of the scalar potential ϕ and of the vector potential $\mathbf{A}$ in the form

$$\mathbf{F} = e\left\{-\frac{1}{c}\frac{\partial \mathbf{A}}{\partial t} - \nabla\phi + \frac{1}{c}[\mathbf{v} \times (\nabla \times \mathbf{A})]\right\} \tag{3.8.3}$$

This equation can be easily derived from a Hamiltonian of the type

$$H = \frac{1}{2m}\left(\mathbf{p} - \frac{e}{c}\mathbf{A}\right)^2 + e\phi \tag{3.8.4}$$

by using the Hamiltonian equations of motion

$$\frac{\partial H}{\partial q_i} = -\dot{p}_i \qquad \frac{\partial H}{\partial p_i} = \dot{q}_i$$

We have, in fact, for the x-coordinate

$$\frac{\partial H}{\partial p_x} = v_x = \frac{1}{m}\left(p_x - \frac{e}{c}A_x\right) \tag{3.8.5}$$

$$-\frac{\partial H}{\partial x} = \dot{p}_x = -e\frac{\partial \phi}{\partial x} + \frac{e}{mc}\left[\left(p_x - \frac{e}{c}A_x\right)\frac{\partial A_x}{\partial x} + \left(p_y - \frac{e}{c}A_y\right)\frac{\partial A_y}{\partial x} + \left(p_z - \frac{e}{c}A_z\right)\frac{\partial A_z}{\partial x}\right] \tag{3.8.6}$$

From (3.8.5) we obtain

$$p_x = mv_x + \frac{e}{c}A_x \tag{3.8.7}$$

and, by substitution of the time derivative of p_x into (3.8.6)

$$m\dot{v}_x = -\frac{e}{c}\frac{\partial A_x}{\partial t} - e\frac{\partial \phi}{\partial x} + \frac{e}{c}\left[v_y\left(\frac{\partial A_y}{\partial x} - \frac{\partial A_x}{\partial y}\right) + v_z\left(\frac{\partial A_z}{\partial x} - \frac{\partial A_x}{\partial z}\right)\right] \tag{3.8.8}$$

which is the x-component of (3.8.3) and where use has been made of the relation

$$\frac{\mathrm{d}A_x}{\mathrm{d}t} = \frac{\partial A_x}{\partial t} + \frac{\partial A_x}{\partial x}\dot{x} + \frac{\partial A_x}{\partial y}\dot{y} + \frac{\partial A_x}{\partial z}\dot{z}$$

Similar equations hold for the y-, and z-components.

For a molecule consisting of n particles, the Hamiltonian in an external field is thus

$$H = \sum_{j=1}^{n} \frac{1}{2m_j}\left(\mathbf{p}_j - \frac{e_j}{c}\mathbf{A}_j\right)^2 + \sum_{j=1}^{n} e_j\phi_j + V \tag{3.8.9}$$

where V is the intramolecular potential.

The quantum-mechanical Hamiltonian operator $\mathscr{H}$ is simply obtained from H in the standard way[2] and is

$$\mathscr{H} = \sum_j \frac{1}{2m_j}\left(\frac{\hbar}{\mathrm{i}}\nabla_j - \frac{e_j}{c}\mathbf{A}_j\right)^2 + \sum_j e_j\phi_j + V \tag{3.8.10}$$

$$\mathscr{H} = \sum_j \frac{1}{2m_j}\left[-\hbar^2\nabla_j^2 - \frac{\hbar e_j}{\mathrm{i}c}(\nabla_j \cdot \mathbf{A}_j + \mathbf{A}_j \cdot \nabla_j) + \frac{e_j^2}{c^2}\mathbf{A}_j^2\right] + \sum_j e_j\phi_j + V \tag{3.8.11}$$

since

$$\nabla \cdot (\mathbf{A}\psi) = (\mathbf{A} \cdot \nabla)\psi + \psi(\nabla \cdot \mathbf{A})$$

We may write equation (3.8.11) in the form

$$\mathscr{H} = \left(-\sum_j \frac{\hbar^2}{2m_j}\nabla_j^2 + V\right) + \sum_j \frac{1}{2m_j}\left\{-\frac{\hbar e_j}{\mathrm{i}c}[(\nabla_j \cdot \mathbf{A}_j) + 2(\mathbf{A}_j \cdot \nabla_j)] + \frac{e_j^2}{c^2}\mathbf{A}_j^2\right\} + \sum_j e_j\phi_j \tag{3.8.12}$$

The first term of (3.8.12) is clearly the molecular Hamiltonian $\mathscr{H}_{\mathrm{mol}}$, while the last two terms describe the interaction between the molecule and the field and form, therefore, the interaction Hamiltonian $\mathscr{H}_{\mathrm{int}}$.

In the Lorentz gauge the above given expression for $\mathscr{H}_{\mathrm{int}}$ can be simplified since it is always possible to choose $\phi = 0$ and to describe the radiation field only by means of the vector potential $\mathbf{A}$ subject to the condition (3.6.15) $(\nabla \cdot \mathbf{A}) = 0$. We have therefore

$$\mathscr{H}_{\mathrm{int}} = \sum_j \left[-\frac{e_j\hbar}{\mathrm{i}cm_j}(\mathbf{A}_j \cdot \nabla) + \frac{e_j^2}{2m_jc^2}\mathbf{A}_j^2\right] \tag{3.8.13}$$

In order to insert the appropriate expression of $\mathbf{A}_j$ into (3.8.13), it is necessary to redefine the wavevector $\mathbf{k}_\lambda$ of equation (3.6.25). In the quantum-mechanical description the wavevector $\mathbf{k}_\lambda$ is actually defined as a vector whose direction gives the direction of propagation of the photon and whose magnitude is equal to the energy of the photon. Equation (3.6.24) using (3.6.25) and adopting the new definition of $\mathbf{k}_\lambda$ becomes

$$\mathbf{A}_j = (4\pi c^2)^{\frac{1}{2}} \sum_\lambda \mathbf{u}_\lambda \left\{q_\lambda \exp\left[\frac{\mathrm{i}}{\hbar c}(\mathbf{k}_\lambda \cdot \mathbf{r}_j)\right] + q_\lambda^* \exp\left[-\frac{\mathrm{i}}{\hbar c}(\mathbf{k}_\lambda \cdot \mathbf{r}_j)\right]\right\} \tag{3.8.14}$$

and thus

$$\mathscr{H}_{\text{int}} = -\frac{2\hbar\pi^{\frac{1}{2}}}{\text{i}}\sum_j\sum_\lambda \frac{e_j}{m_j}\left\{q_\lambda \exp\left[\frac{\text{i}}{\hbar c}(\mathbf{k}_\lambda\cdot\mathbf{r}_j)\right] + q_\lambda^* \exp\left[-\frac{\text{i}}{\hbar c}(\mathbf{k}_\lambda\cdot\mathbf{r}_j)\right]\right\}(\mathbf{u}_\lambda\cdot\nabla_j) + 2\pi\sum_j\sum_\lambda\sum_\mu \frac{e_j^2}{m_j}(\mathbf{u}_\lambda\cdot\mathbf{u}_\mu)$$

$$\times\left\{q_\lambda q_\mu \exp\left[\frac{\text{i}}{\hbar c}(\mathbf{k}_\lambda+\mathbf{k}_\mu)\cdot\mathbf{r}_j\right] + q_\lambda q_\mu^* \exp\left[\frac{\text{i}}{\hbar c}(\mathbf{k}_\lambda-\mathbf{k}_\mu)\cdot\mathbf{r}_j\right] + q_\lambda^* q_\mu \exp\left[\frac{\text{i}}{\hbar c}(\mathbf{k}_\mu-\mathbf{k}_\lambda)\cdot\mathbf{r}_j\right]\right.$$

$$\left. + q_\lambda^* q_\mu^* \exp\left[-\frac{\text{i}}{\hbar c}(\mathbf{k}_\lambda+\mathbf{k}_\mu)\cdot\mathbf{r}_j\right]\right\} \tag{3.8.15}$$

According to equations (3.7.13) and (3.7.14) and to the following discussion, the operator q_λ annihilates a photon of energy $\hbar\omega_\lambda$ and the operator q_λ^* creates a photon of the same energy. Therefore the first term of the interaction Hamiltonian $\mathscr{H}_{\text{int}}$ describes an absorption and the second an emission process. Since in this case only one photon is involved in each elementary process we shall call them one-photon processes.

The third term contains the operators q_λ and q_μ. It therefore describes a two-photon absorption process in which two photons, one with energy $\hbar\omega_\lambda$ and one with energy $\hbar\omega_\mu$, are simultaneously absorbed.

Similarly the last term, which contains the operators q_λ^* and q_μ^*, describes a two-photon emission process in which one photon of energy $\hbar\omega_\lambda$ and one of energy $\hbar\omega_\mu$ are simultaneously emitted.

The probability of a two-photon process is much smaller than that of a one-photon process. In the visible and ultraviolet region, using laser sources, such processes can be detected. In the infrared region however they are practically negligible and therefore will not be considered further.

The fourth and fifth terms of $\mathscr{H}_{\text{int}}$ describe a process in which one photon is absorbed and one is emitted at the same time. Among such two-photon processes the case of great importance for us is that in which the energy difference between the two photons is equal to the energy of a vibrational level, since in this case we have the Raman effect. We shall consider this part of the interaction Hamiltonian in Section 3.13.

3.9 One-photon processes. Absorption and emission

In this section we shall consider in more detail the one-photon processes occurring in a system formed by a radiation field and a molecule. As we have seen in the last section, the only part of the interaction Hamiltonian involved is

$$\mathscr{H}_{\text{int}}^{(1)} = -\frac{2\hbar\pi^{\frac{1}{2}}}{\text{i}}\sum_j\sum_\lambda \frac{e_j}{m_j}\left\{q_\lambda \exp\left[\frac{\text{i}}{\hbar c}(\mathbf{k}_\lambda\cdot\mathbf{r}_j)\right] + q_\lambda^* \exp\left[-\frac{\text{i}}{\hbar c}(\mathbf{k}_\lambda\cdot\mathbf{r}_j)\right]\right\}(\mathbf{u}_\lambda\cdot\nabla_j) \tag{3.9.1}$$

where all operators are time-independent, including q_λ and q_λ^* as pointed out in Section 3.7. Furthermore, the total Hamiltonian of the system is a typical example of a Hamiltonian with a discrete eigenvalue spectrum (the molecular part) embedded in a continuum (the field part). It is therefore legitimate to apply the first-order perturbation theory to evaluate the transition probability between stationary states of the system, in the form given by equation (3.5.21).

If the system is in the stationary state A of the total Hamiltonian there is, according to equation (3.5.20), a definite probability of finding the system in a different stationary state B, as consequence of the interaction between the field and the molecule, only if (a) the matrix element of $\mathscr{H}_{\text{int}}^{(1)}$ connecting the two states is different from zero and (b) if the energy of the two states is degenerate or nearly degenerate. If the two states differ because the energy of a photon has been transferred from the field to the molecule or vice versa, the total energy of the two states is the same and therefore the second condition is satisfied. As far as the first is concerned, the only restriction imposed by the harmonic-oscillator-type wavefunctions of the field (see Section 3.7) is that the number of photons of state B differs by unity from that of state A.

Hereafter we shall use the symbols ψ_{ij} and $\mathscr{E}_{ij}$ for the wavefunction and the energy of a stationary state of the system. Since the wavefunction for a stationary state is the product of the molecular wavefunction and of the field wavefunction while the energy is the sum of the corresponding energies, the subscript i specifies the molecular situation and the subscript j specifies the structure of the field.

It is worthwhile to remember that the field wavefunction is determined by a set of quantum numbers $n_{\lambda_1}, n_{\lambda_2}, \ldots, n_{\lambda_i}, \ldots$ each specifying the number of photons of a given species λ. The exact knowledge of these numbers is, however, unnecessary and all we have to do is to keep track of the variation of one of them, n_λ, between two stationary states. Therefore, given an initial state $(0, n_\lambda)$ with eigenfunction and eigenvalue

$$\psi_{0,n_\lambda} = \psi_0 \phi_{n_\lambda} \qquad \mathscr{E}_{0,n_\lambda} = E_0 + n_\lambda \hbar\omega_\lambda \tag{3.9.2}$$

the only states connected to $(0, n_\lambda)$ are the states $(m, n_\lambda + 1)$ and $(m, n_\lambda - 1)$ with eigenfunction and eigenvalues

$$\begin{aligned} \psi_{m,n_\lambda+1} &= \psi_m \phi_{n_\lambda+1} \qquad & \mathscr{E}_{m,n_\lambda+1} &= E_m + (n_\lambda + 1)\hbar\omega_\lambda \\ \psi_{m,n_\lambda-1} &= \psi_m \phi_{n_\lambda-1} \qquad & \mathscr{E}_{m,n_\lambda-1} &= E_m + (n_\lambda - 1)\hbar\omega_\lambda \end{aligned} \tag{3.9.3}$$

In Section 3.5 we have seen that the wavefunction of the total Hamiltonian of the system, including the perturbation term $\mathscr{H}^{(1)}_{\text{int}}$, can be expanded in terms of the wavefunctions of the unperturbed Hamiltonian with coefficients $b_m(t)$ which involve the matrix elements of $\mathscr{H}^{(1)}_{\text{int}}$ between the initial and other stationary states.

In our case, since only states of the type $(m, n_\lambda + 1)$ and $(m, n_\lambda - 1)$ are connected to the initial state $(0, n_\lambda)$, we have, according to (3.5.4)

$$\begin{aligned} \psi(t) = {} & b_{0,n_\lambda}(t)\psi_{0,n_\lambda} \exp[-\mathrm{i}\mathscr{E}_{0,n_\lambda}t/\hbar] + \sum_m \sum_\lambda \{b_{m,n_\lambda+1}(t)\psi_{m,n_\lambda+1} \exp[-\mathrm{i}\mathscr{E}_{m,n_\lambda+1}t/\hbar] \\ & + b_{m,n_\lambda-1}(t)\psi_{m,n_\lambda-1} \exp[-\mathrm{i}\mathscr{E}_{m,n_\lambda-1}t/\hbar]\} \end{aligned} \tag{3.9.4}$$

From equation (3.5.20) we obtain then that the probability of finding the system in the state $(m, n_\lambda + 1)$ or $(m, n_\lambda - 1)$ at the time t, is given by

$$b_{m,n_\lambda\pm1} b^*_{m,n_\lambda\pm1} = 2H_{m,n_\lambda\pm1;0,n_\lambda} H^*_{m,n_\lambda\pm1;0,n_\lambda} \left\{ \frac{1 - \cos[(E_m - E_0 \pm \hbar\omega_\lambda)t/\hbar]}{(E_m - E_0 \pm \hbar\omega_\lambda)^2} \right\} \tag{3.9.5}$$

where

$$H_{m,n_\lambda\pm1;0,n_\lambda} = \langle \psi_{m,n_\lambda\pm1} | \mathscr{H}^{(1)}_{\text{int}} | \psi_{0,n_\lambda} \rangle = \langle \psi_m \phi_{n_\lambda\pm1} | \mathscr{H}^{(1)}_{\text{int}} | \psi_0 \phi_{n_\lambda} \rangle \tag{3.9.6}$$

are the matrix elements connecting the states. By substitution of (3.9.1) into (3.9.6) we obtain

$$\begin{aligned} H_{m,n_\lambda\pm1;0,n_\lambda} = {} & -\frac{2\hbar\pi^{\frac{1}{2}}}{\mathrm{i}} \left\{ \left\langle \psi_m \phi_{n_\lambda\pm1} \left| \sum_j \sum_\lambda \frac{e_j}{m_j} q_\lambda \exp\left[\frac{\mathrm{i}}{\hbar c}(\mathbf{k}_\lambda \cdot \mathbf{r}_j)\right] (\mathbf{u}_\lambda \cdot \nabla_j) \right| \psi_0 \phi_{n_\lambda} \right\rangle \right. \\ & \left. + \left\langle \psi_m \phi_{n_\lambda\pm1} \left| \sum_j \sum_\lambda \frac{e_j}{m_j} q^*_\lambda \exp\left[-\frac{\mathrm{i}}{\hbar c}(\mathbf{k}_\lambda \cdot \mathbf{r}_j)\right] (\mathbf{u}_\lambda \cdot \nabla_j) \right| \psi_0 \phi_{n_\lambda} \right\rangle \right\} \end{aligned} \tag{3.9.7}$$

and using equations (3.7.12)

$$H_{m,n_\lambda+1;0,n_\lambda} = \mathrm{i}\left(\frac{2\pi\hbar^3}{\omega_\lambda}\right)^{\frac{1}{2}}\sqrt{(n_\lambda+1)}\left\langle\psi_m\left|\sum_j\frac{e_j}{m_j}\exp\left[-\frac{\mathrm{i}}{\hbar c}(\mathbf{k}_\lambda\cdot\mathbf{r}_j)\right](\mathbf{u}_\lambda\cdot\nabla_j)\right|\psi_0\right\rangle$$

$$= \mathrm{i}\left(\frac{2\pi\hbar^3}{\omega_\lambda}\right)^{\frac{1}{2}}\sqrt{(n_\lambda+1)}H^*_{m0} \tag{3.9.8}$$

$$H_{m,n_\lambda-1;0,n_\lambda} = \mathrm{i}\left(\frac{2\pi\hbar^3}{\omega_\lambda}\right)^{\frac{1}{2}}\sqrt{n_\lambda}\left\langle\psi_m\left|\sum_j\frac{e_j}{m_j}\exp\left[\frac{\mathrm{i}}{\hbar c}(\mathbf{k}_\lambda\cdot\mathbf{r}_j)\right](\mathbf{u}_\lambda\cdot\nabla_j)\right|\psi_0\right\rangle = \mathrm{i}\left(\frac{2\pi\hbar^3}{\omega_\lambda}\right)^{\frac{1}{2}}\sqrt{n_\lambda}H_{m0}$$

where

$$H_{m0} = \left\langle\psi_m\left|\sum_j\frac{e_j}{m_j}\exp\left[\frac{\mathrm{i}}{\hbar c}(\mathbf{k}_\lambda\cdot\mathbf{r}_j)\right](\mathbf{u}_\lambda\cdot\nabla_j)\right|\psi_0\right\rangle$$

$$H^*_{m0} = \left\langle\psi_m\left|\sum_j\frac{e_j}{m_j}\exp\left[-\frac{\mathrm{i}}{\hbar c}(\mathbf{k}_\lambda\cdot\mathbf{r}_j)\right](\mathbf{u}_\lambda\cdot\nabla_j)\right|\psi_0\right\rangle \tag{3.9.9}$$

are the matrix elements connecting together the molecular states only. We shall consider them in greater detail in Section 3.10.

The probabilities (3.9.5) can thus be written

$$b_{m,n_\lambda+1}b^*_{m,n_\lambda+1} = \frac{4\pi\hbar^3}{\omega_\lambda}(n_\lambda+1)H_{m0}H^*_{m0}\left\{\frac{1-\cos\left[(E_m-E_0+\hbar\omega_\lambda)t/\hbar\right]}{(E_m-E_0+\hbar\omega_\lambda)^2}\right\}$$

$$b_{m,n_\lambda-1}b^*_{m,n_\lambda\mp 1} = \frac{4\pi\hbar^3}{\omega_\lambda}n_\lambda H_{m0}H^*_{m0}\left\{\frac{1-\cos\left[(E_m-E_0-\hbar\omega_\lambda)t/\hbar\right]}{(E_m-E_0-\hbar\omega_\lambda)^2}\right\} \tag{3.9.10}$$

and using Dirac's δ-function, as in section 3.5, equation (3.5.27),

$$b_{m,n_\lambda+1}b^*_{m,n_\lambda+1} = \frac{4\pi^2\hbar^2 t}{\omega_\lambda}H_{m0}H^*_{m0}(n_\lambda+1)\delta(k_m+k_\lambda) \tag{3.9.11}$$

$$b_{m,n_\lambda-1}b^*_{m,n_\lambda-1} = \frac{4\pi^2\hbar^2 t}{\omega_\lambda}H_{m0}H^*_{m0}n_\lambda\delta(k_m-k_\lambda) \tag{3.9.12}$$

where

$$k_m = E_m - E_0$$

$$k_\lambda = \hbar\omega_\lambda$$

Owing to the properties of the δ-function (see equation 3.5.25) k_m must be negative in equation (3.9.11) and positive in equation (3.9.12). In the first case therefore E_m must be smaller than E_0 and equation (3.9.11) yields the probability of emission of a photon of energy $\hbar\omega_\lambda$, i.e. the probability of transition from an upper level E_0 to a lower level E_m with an energy separation equal to $\hbar\omega_\lambda$. In the second case E_m must be larger than E_0 and thus equation (3.9.12) yields the probability of absorption of a photon of energy $\hbar\omega_\lambda$, i.e. the probability of transition from a lower level E_0 to an upper level E_m.

It is important to observe that the probability of absorption is proportional to the number of photons n_λ of the field for the species λ, i.e. to the energy density of the species λ. The probability of emission is made up of two terms, one proportional to n_λ and equal to the probability of absorption and one independent of n_λ. We shall call the first contribution 'stimulated emission', since it corresponds to a process in which the emission of the photon from a molecule is caused by the radiation field, and the second contribution 'spontaneous emission', since it is independent of the field.

3.10 The electric dipole transitions

In the previous section we have seen that the probability of absorption of a photon of energy $\hbar\omega_\lambda$ by a molecule at time t is given by

$$b_{m,n_\lambda-1}b^*_{m,n_\lambda-1} = \frac{4\pi^2\hbar^2 t}{\omega_\lambda} H_{m0}H^*_{m0}n_\lambda\delta(k_m - k_\lambda) \tag{3.10.1}$$

and is appreciably different from zero only when the Bohr condition

$$\begin{aligned} k_m &= k_\lambda \\ \hbar\omega_\lambda &= E_m - E_0 \end{aligned} \tag{3.10.2}$$

is satisfied and when the matrix element H_{m0} is different from zero. According to equation (3.9.9) H_{m0} and H^*_{m0} are given by

$$H_{m0} = \left\langle \psi_m \middle| \sum_j \frac{e_j}{m_j} \exp\left[\frac{\mathrm{i}}{\hbar c}(\mathbf{k}_\lambda \cdot \mathbf{r}_j)\right](\mathbf{u}_\lambda \cdot \nabla_j) \middle| \psi_0 \right\rangle \tag{3.10.3}$$

$$H^*_{m0} = \left\langle \psi_m \middle| \sum_j \frac{e_j}{m_j} \exp\left[-\frac{\mathrm{i}}{\hbar c}(\mathbf{k}_\lambda \cdot \mathbf{r}_j)\right](\mathbf{u}_\lambda \cdot \nabla_j) \middle| \psi_0 \right\rangle \tag{3.10.4}$$

where the summation is over all atoms of the molecule and $\mathbf{r}_j$ is the position vector of the jth atom referred to an arbitrary origin.

Since the wavelength of the light absorbed by a molecule is some orders of magnitude greater than the molecular dimensions, it is convenient to locate the origin of $\mathbf{r}_j$ at the centre of mass of the molecule and to expand the exponential in a power series which is known to converge very rapidly. Using the expansion[8,9]

$$\exp\left[\frac{\mathrm{i}}{\hbar c}(\mathbf{k}_\lambda \cdot \mathbf{r}_j)\right] = 1 + \frac{\mathrm{i}}{\hbar c}(\mathbf{k}_\lambda \cdot \mathbf{r}_j) + \dots \tag{3.10.5}$$

we have

$$H_{m0} = \overline{M}_{m0} + \frac{\mathrm{i}}{\hbar c}\Omega_{m0} + \dots$$

$$H_{m0} = \left\langle \psi_m \middle| \sum_j \frac{e_j}{m_j}(\mathbf{u}_\lambda \cdot \nabla_j) \middle| \psi_0 \right\rangle + \frac{\mathrm{i}}{\hbar c}\left\langle \psi_m \middle| \sum_j \frac{e_j}{m_j}(\mathbf{k}_\lambda \cdot \mathbf{r}_j)(\mathbf{u}_\lambda \cdot \nabla_j) \middle| \psi_0 \right\rangle \tag{3.10.6}$$

The first term of (3.10.6) is normally much larger than all the others, so that it alone controls the absorption process. Only if this term is zero does the second term need to be considered. In such cases, however, the intensity of absorption is many orders of magnitude smaller.

Since the expansion (3.10.5) corresponds in the classical theory to the multipole expansion of a system of charges interacting with the electromagnetic field, the same nomenclature is used. The first term of (3.10.6) is then called the *electric dipole* matrix element, the second is called the *electric quadrupole* and the *magnetic dipole* matrix element, etc. In this section we shall consider only electric dipole transitions, which are by far the most important ones. The electric quadrupole and magnetic dipole term will be briefly discussed in Section 3.12.

For our purposes it is convenient to write the electric dipole matrix element

$$\overline{M}_{m0} = \left\langle \psi_m \middle| \sum_j \frac{e_j}{m_j}(\mathbf{u}_\lambda \cdot \nabla_j) \middle| \psi_0 \right\rangle \tag{3.10.7}$$

in a different form which contains a more familiar expression for the dipole operator. In order to do that, we need to prove the following relation

$$\left\langle \psi_m \left| \frac{1}{m_j} \frac{\partial}{\partial x_j} \right| \psi_0 \right\rangle = -\frac{(E_m - E_0)}{\hbar^2} \langle \psi_m | x_j | \psi_0 \rangle \tag{3.10.8}$$

For this we recall that ψ_m and ψ_0 are molecular wavefunctions which satisfy the Schrödinger equations

$$\begin{aligned} \left(\frac{\hbar^2}{2} \sum_j \frac{1}{m_j} \nabla_j^2 + V \right) \psi_m^* &= E_m \psi_m^* \\ \left(-\frac{\hbar^2}{2} \sum_j \frac{1}{m_j} \nabla_j^2 + V \right) \psi_0 &= E_0 \psi_0 \end{aligned} \tag{3.10.9}$$

Let us multiply the first equation by $x_i \psi_0$ on the right and the second by $\psi_m^* x_i$ on the left, where x_i is an arbitrary coordinate, integrate both equations and subtract one from the other.

Since all operators in (3.10.9) are Hermitian, we obtain

$$\left\langle \psi_m \left| x_i \sum_j \frac{1}{m_j} \nabla_j^2 - \sum_j \frac{1}{m_j} \nabla_j^2 x_i \right| \psi_0 \right\rangle = \frac{2}{\hbar^2} (E_m - E_0) \langle \psi_m | x_i | \psi_0 \rangle \tag{3.10.10}$$

The operator on the left-hand side of (3.10.10) is the commutator of the operators

$$x_i \quad \text{and} \quad \sum_j \frac{1}{m_j} \nabla_j^2$$

which can be easily found using the commutation rule

$$[R, S] = i\hbar \{R_c, S_c\} \tag{3.10.11}$$

where R_c and S_c are two classical quantities, R and S the corresponding quantum operators and $\{R_c, S_c\}$ the so-called 'Poisson bracket'.† The result is

$$\left[x_i, \sum_j \frac{1}{m_j} \nabla_j^2 \right] = -\frac{2}{m_i} \frac{\partial}{\partial x_i} \tag{3.10.12}$$

Substitution of (3.10.12) into (3.10.10) proves the validity of equation (3.10.8) and allows us to rewrite (3.10.7) in the form

$$\overline{M}_{m0} = -\frac{(E_m - E_0)}{\hbar^2} \left\langle \psi_m \left| \sum_j e_j (\mathbf{u}_\lambda \cdot \mathbf{r}_j) \right| \psi_0 \right\rangle = -\frac{(E_m - E_0)}{\hbar^2} M_{m0} \tag{3.10.13}$$

Using (3.10.13) we obtain for the probability (3.10.1) the expression

$$b_{m,n_\lambda - 1} b_{m,n_\lambda - 1}^* = \frac{4\pi^2}{\hbar} (E_m - E_0) t M_{m0} M_{m0}^* n_\lambda \, \delta(k_m - k_\lambda) \tag{3.10.14}$$

Since we never deal with a pure monochromatic radiation it is convenient to define the transition probability per unit of time over the entire frequency spectrum

$$w_{m0} = \int_{-\infty}^{+\infty} \frac{b_{m,n_\lambda - 1} b_{m,n_\lambda - 1}^*}{t} c \, d\nu = \frac{1}{2\pi\hbar} \int_{-\infty}^{+\infty} \frac{b_{m,n_\lambda - 1} b_{m,n_\lambda - 1}^*}{t} dk_\lambda = \frac{2\pi}{\hbar^2} (E_m - E_0) M_{m0} M_{m0}^* n_\lambda \tag{3.10.15}$$

† The Poisson bracket is defined as $\{R_c, S_c\} = \sum_\alpha [(\partial R_c/\partial \alpha)(\partial S_c/\partial p_\alpha) - (\partial R_c/\partial p_\alpha)(\partial S_c/\partial \alpha)]$ ($\alpha = x, y, z$).

When the Bohr conditions (3.10.2) are satisfied the quantity $n_\lambda(E_m - E_0) = n_\lambda \hbar\omega_\lambda$ represents the energy of the radiation field for each type of field oscillator λ that we shall call energy density of the field $\rho(E)$. We thus have

$$w_{m0} = \frac{2\pi}{\hbar^2} M_{m0} M^*_{m0} \rho(E) \qquad (3.10.16)$$

From equation (3.10.13), the matrix element $\overline{M}_{m0}$ can be written in the form

$$\overline{M}_{m0} = -\frac{\omega_\lambda}{\hbar} \mathbf{u} . \mathbf{P}_{m0} \qquad (3.10.17)$$

where

$$\mathbf{P}_{m0} = \left\langle \psi_m \middle| \sum_j e_j \mathbf{r}_j \middle| \psi_0 \right\rangle = \langle \psi_m | \boldsymbol{\mu} | \psi_0 \rangle \qquad (3.10.18)$$

$\mathbf{P}_{m0}$ is called the *electric transition dipole* since it defines, according to the principles of quantum mechanics, the average value of the electric dipole moment

$$\boldsymbol{\mu} = \sum_j e_j \mathbf{r}_j$$

in the transition between the states 0 and m.

Using (3.10.18) we can rewrite (3.10.16) in the form

$$w_{m0} = \left(\frac{2\pi}{\hbar^2}\right) (\mathbf{u}_\lambda . \mathbf{P}_{m0})(\mathbf{u}_\lambda . \mathbf{P}_{m0})^* \rho(E) \qquad (3.10.19)$$

Since we have defined the position vectors $\mathbf{r}_j$ with respect to a reference system attached to the centre of mass of the molecule, the electric moment operator is referred to the same system, whereas the polarization vector $\mathbf{u}_\lambda$ of the photons, is referred to as a space-fixed system. If we call θ the angle formed by $\mathbf{u}_\lambda$ and $\mathbf{P}_{m0}$, equation (3.10.19) becomes

$$w_{m0} = \frac{2\pi}{\hbar^2} P_{m0} P^*_{m0} \cos^2 \theta \rho(E) \qquad (3.10.20)$$

In order to relate this transition probability to observable quantities, we need to average it over all possible orientations of the molecule with respect to the field. Since

$$\overline{\cos^2 \theta} = \frac{1}{4\pi} \int_0^\pi \int_0^{2\pi} \cos^2 \theta \sin \theta \, d\theta \, d\varphi = \tfrac{1}{3}$$

we have

$$w_{m0} = \frac{2\pi}{3\hbar^2} (P_{m0} P^*_{m0}) \rho(E) = B_{m0} \rho(E) \qquad (3.10.21)$$

where

$$B_{m0} = \frac{2\pi}{3\hbar^2} (P_{m0} P^*_{m0}) \qquad (3.10.22)$$

is called the '*Einstein coefficient of absorption*'.

In an absorption experiment what we actually measure is the intensity of an absorption band. If we disregard the complications arising from the natural linewidth and from the further broadening due to other effects (Doppler effect, molecular collisions, intermolecular forces, etc.) the correlation with the theoretical quantities derived above is rather simple.

If we define as absorption intensity the energy absorbed by 1 cm^2 cross-section of the incident beam by a sample of thickness Δx, we have

$$I_{m0} = c\rho(E)(\hbar\omega_\lambda)B_{m0}N_0\,\Delta x = I_0(\hbar\omega_\lambda)B_{m0}N_0\,\Delta x$$

where N_0 is the number of molecules per cm^3 in the state 0 and $I_0 = c\rho(E)$ is the intensity of the incident beam.

3.11 Vibrational selection rules

In the previous section we have seen that the absorption intensity of a band is proportional to the transition dipole $\mathbf{P}_{m0}$. It is therefore important to find some rules which permit one to establish for which transitions $\mathbf{P}_{m0}$ is different from zero. Only in such cases can the transition between two levels take place with absorption of radiation. It would actually be more profitable to evaluate $\mathbf{P}_{m0}$ in each single case to obtain a quantitative determination of the 'theoretical intensity' of a band. This, however, is very difficult and does not help very much in the interpretation of the absorption spectra.

A sharp division between 'allowed' and 'forbidden' transitions, where by 'allowed' we mean simply that '*nihil obstat*' to the occurrence of an absorption band and by 'forbidden' that '*esse non potest*', is much more useful for the spectroscopist especially if it is given in terms of general rules easily applicable to specific cases.

Such rules are called selection rules and will be discussed in two separate sections. In the present section we shall investigate, in the case of vibrational transitions, the restrictions imposed by the harmonic oscillator wavefunctions on the transition dipole $\mathbf{P}_{m0}$. In Chapter 7, by using group theory, we shall further discuss the symmetry restrictions on $\mathbf{P}_{m0}$ which lead to the final form of the selection rules.

We recall from the previous section, that the transition dipole $\mathbf{P}_{m0}$ is given by a relation of the type

$$\mathbf{P}_{m0} = \langle\psi_m|\boldsymbol{\mu}|\psi_0\rangle \tag{3.11.1}$$

where the wavefunctions ψ_m and ψ_0 are the complete molecular wavefunctions and $\boldsymbol{\mu}$ is the molecular dipole moment. Under these conditions a wavefunction ψ_n can be written approximately in the form

$$\psi_n = \psi_{ne}\psi_{nv}\psi_{nr} \tag{3.11.2}$$

where ψ_{ne}, ψ_{nv} and ψ_{nr} indicate the electronic, the vibrational and the rotational wavefunctions respectively.

Since in this book we are concerned only with vibrational states, we shall consider only transitions in which no changes of the electronic wavefunctions occur. However, we still need to separate the rotational from the vibrational contribution in (3.11.1) and for this we shall use the technique described in Chapter 2 for the separation of vibrational and rotational motions. Accordingly we shall introduce another reference system, located at the centre of mass and rotating with the molecule so as to satisfy the second Sayvetz condition. If we call X_i, Y_i and Z_i the Cartesian coordinates of the ith atom with respect to the reference system which translates with the molecule with a fixed space orientation and x_i, y_i and z_i the Cartesian coordinates with respect to the rotating system, we have a relation of the type

$$\begin{aligned} X_i &= \phi_{Xx}x_i + \phi_{Xy}y_i + \phi_{Xz}z_i \\ Y_i &= \phi_{Yx}x_i + \phi_{Yy}y_i + \phi_{Yz}z_i \\ Z_i &= \phi_{Zx}x_i + \phi_{Zy}y_i + \phi_{Zz}z_i \end{aligned} \tag{3.11.3}$$

where the ϕ_{Rr} are the direction cosines of the rotating system with respect to the system with a fixed space orientation.

Since the dipole moment $\boldsymbol{\mu}$ is a vector, a component μ_R in the fixed orientation system is related to the components μ_r in the rotating system by the same relationship since

$$\mu_X = \sum_i e_i X_i = \sum_r \phi_{Xr} \sum_i e_i r_i = \phi_{Xx}\mu_x + \phi_{Xy}\mu_y + \phi_{Xz}\mu_z \tag{3.11.4}$$

From (3.11.1) we thus have, considering for simplicity only the X-component

$$\begin{aligned}(P_{m0})_X &= \langle \psi_{mv}\psi_{mr}|\mu_X|\psi_{0v}\psi_{0r}\rangle = \langle \psi_{mv}|\mu_x|\psi_{0v}\rangle\langle \psi_{mr}|\phi_{Xx}|\psi_{0r}\rangle \\ &+ \langle \psi_{mv}|\mu_y|\psi_{0v}\rangle\langle \psi_{mr}|\phi_{Xy}|\psi_{0r}\rangle + \langle \psi_{mv}|\mu_z|\psi_{0v}\rangle\langle \psi_{mr}|\phi_{Xz}|\psi_{0r}\rangle\end{aligned} \tag{3.11.5}$$

We therefore conclude that for a vibrational transition to occur one of the matrix elements

$$\langle \psi_{mv}|\mu_x|\psi_{0v}\rangle \qquad \langle \psi_{mv}|\mu_y|\psi_{0v}\rangle \qquad \langle \psi_{mv}|\mu_z|\psi_{0v}\rangle \tag{3.11.6}$$

must be different from zero. In addition there may be a simultaneous set of rotational selection rules which control the fine structure of the absorption bands.

The rotational selection rules are discussed in great detail in the book *Molecular Vib-rotors* by Allen and Cross.[3] In our treatment we shall ignore the rotational structure of vibrational transitions and treat them as pure vibrational transitions.

In dealing with vibrational transitions we need to specify what is meant by the electric moment

$$\boldsymbol{\mu} = \bar{\mathbf{i}}\mu_x + \bar{\mathbf{j}}\mu_y + \bar{\mathbf{k}}\mu_z = \bar{\mathbf{i}}\sum_j e_j x_j + \bar{\mathbf{j}}\sum_j e_j y_j + \bar{\mathbf{k}}\sum_j e_j z_j \tag{3.11.7}$$

for a vibrating molecule. In order to apply the conclusion of this chapter to the ball-and-spring model introduced in Chapter 2, we may think of the charges e_j of (3.11.7) as localized on the atoms, i.e. as 'effective atomic charges'. When the atoms move, the dipole moment changes because the distances and the 'effective atomic charges' change so that in general we can consider the electric moment as a non-linear function of the atomic coordinates.

Since the atomic coordinates are related to the normal coordinates by a linear transformation, it is more convenient to expand $\boldsymbol{\mu}$ in a power series of the normal coordinates. In this way we obtain

$$\boldsymbol{\mu} = \boldsymbol{\mu}_0 + \sum_i \boldsymbol{\mu}_1^{(i)} Q_i + \frac{1}{2}\sum_{ij} \boldsymbol{\mu}_2^{(ij)} Q_i Q_j + \ldots \tag{3.11.8}$$

where $\boldsymbol{\mu}_0$ is the 'static' dipole moment, i.e. the dipole moment for the equilibrium configuration, $\boldsymbol{\mu}_1^{(i)} = (\partial\boldsymbol{\mu}/\partial Q_i)_0$ and $\boldsymbol{\mu}_2^{(ij)} = (\partial^2\boldsymbol{\mu}/\partial Q_i\,\partial Q_j)_0$. Since the coefficients $\boldsymbol{\mu}_2^{(ij)}$ are normally very small, it is customary to consider only the first two terms of (3.11.8). This corresponds to the assumption that the dipole moment varies linearly with the normal coordinates.

In this case we say that the vibrations are electrically harmonic. If the third term of (3.11.8) is found to be relevant, we speak of electrical anharmonicity. For infinitesimal displacements of the atoms we can disregard the electrical anharmonicity and use the first two terms of (3.11.8) to calculate the vibrational selection rules.

According to the previous discussion, we need only consider the vibrational part of the wavefunctions ψ_m and ψ_0 which, from equation (3.2.3), is of the type

$$\prod_{i=1}^{3N-6} \psi_{v_i}(Q_i) \tag{3.11.9}$$

The vibrational transition dipole is thus, from (3.11.1)

$$\mathbf{P}_{m0} = \left\langle \psi_m \middle| \boldsymbol{\mu}_0 + \sum_i \boldsymbol{\mu}_1^{(i)} Q_i \middle| \psi_0 \right\rangle \quad \boldsymbol{\mu}_0 \langle \psi_m | \psi_0 \rangle + \sum_i \boldsymbol{\mu}_1^{(i)} \langle \psi_m | Q_i | \psi_0 \rangle \tag{3.11.10}$$

Since the harmonic oscillator wavefunctions are an orthonormal set, the first term of (3.11.10) vanishes unless $m = 0$. The permanent dipole moment therefore plays no role in the absorption of radiation for a vibrational transition.

The generic kth term of the summation occurring in (3.11.10) is, from (3.11.9), of the type

$$\begin{aligned} &\langle \psi_{v_1''}(Q_1)\psi_{v_2''}(Q_2)\cdots\psi_{v_k''}(Q_k)\cdots\psi_{v_{3N-6}''}(Q_{3N-6})|Q_k|\psi_{v_1'}(Q_1)\psi_{v_2'}(Q_2)\cdots\psi_{v_k'}(Q_k)\cdots\psi_{v_{3N-6}'}(Q_{3N-6})\rangle \\ &= \langle \psi_{v_1''}(Q_1)|\psi_{v_1'}(Q_1)\rangle\langle \psi_{v_2''}(Q_2)|\psi_{v_2'}(Q_2)\rangle\cdots\langle \psi_{v_k''}(Q_k)|Q_k|\psi_{v_k'}(Q_k)\rangle\cdots\langle \psi_{v_{3N-6}''}|\psi_{v_{3N-6}'}\rangle \end{aligned} \tag{3.11.11}$$

Again, because of the orthogonality of the wavefunctions, the integrals occurring in this expression are different from zero only if $v_1'' = v_1', v_2'' = v_2', \ldots, v_{3N-6}'' = v_{3N-6}'$ except for the quantum numbers v_k'' and v_k'. The matrix element relative to these two quantum numbers

$$\langle \psi_{v_k''} | Q_k | \psi_{v_k'} \rangle$$

has been evaluated in Section 3.2 and has been found to be different from zero only if $v_k'' = v_k' \pm 1$.

This result can be summarized as follows. In the harmonic approximation (harmonic oscillator wavefunction), for an electrically harmonic molecule, absorption and emission of radiation in a vibrational transition between two states occurs if the two states differ by one quantum number only and this quantum number changes by one unit in the transition. In addition, for such transitions $\boldsymbol{\mu}_1^{(i)}$ must be different from zero. From this statement it follows directly that only fundamental transitions and hot bands (between two consecutive levels) are allowed in absorption and emission. Overtone and combination transitions are forbidden. Furthermore, not all fundamental transitions originate absorption or emission bands but only those for which $\boldsymbol{\mu}_1^{(i)}$ is different from zero.

Experimentally it is found that the predictions of the harmonic model are only approximately fulfilled. Fundamental vibrations are normally much stronger than overtones and combination bands but these latter, even in conditions of minimum molecular perturbation, can still have appreciable intensities and in particular cases can be stronger than some fundamentals. The failure of the harmonic model in accounting for the spectral intensities of overtones and combination bands has not hindered its use nor reduced its popularity for many reasons. First of all, the model is simple whereas the more sophisticated model of the anharmonic oscillator is very difficult to treat mathematically. Secondly, slight deviations of the molecular system from harmonicity can have a strong effect on the intensity of the overtones but a much smaller effect on the frequencies of the fundamental modes. Thirdly, it is difficult to separate the effect of the mechanical anharmonicity (cubic and quartic terms in the potential function) from the effect of the electrical anharmonicity (non-linear variation of the electric moment with the normal coordinates). Finally, the conclusion, easily reached in the harmonic approximation, that the permanent moment has no effect and that the intensity of a fundamental band is zero unless $\boldsymbol{\mu}_1^{(i)}$ is different from zero, is quite general and remains true for the anharmonic model also.

We shall see in Chapter 7 that it is unnecessary to evaluate the matrix elements to calculate the variation of the dipole moment in a molecular vibration. Group theory furnishes powerful and easy rules to decide, from the symmetry of the vibrations, on the existence or the absence of a dipole variation.

3.12 Electric quadrupole and magnetic dipole transitions.

In Section 3.10 we have defined the matrix element of the electric quadrupole and magnetic dipole moments as

$$\Omega_{m0} = \left\langle \psi_m \left| \sum_j \frac{e_j}{m_j} (\mathbf{k}_\lambda \cdot \mathbf{r}_j)(\mathbf{u}_\lambda \cdot \nabla_j) \right| \psi_0 \right\rangle \tag{3.12.1}$$

and we have pointed out that this term is much smaller than the dipole moment matrix element and can normally be ignored.

If the transition dipole is zero, the quadrupole term can, however, supply a small contribution to the intensity of an absorption or emission band. Typically the ratio of the square of a quadrupole to a dipole matrix element is of the order of 10^{-6}, this being also roughly the ratio of the relative contribution to an absorption or emission band. For this reason very long path lengths, sometimes of the order of several kilometres, are required for the detection of quadrupole transitions. These path lengths are not easily available in a laboratory but are normally encountered in astrophysics and space spectroscopy. Quadrupole transitions have often been invoked to explain weak bands in the spectra of homonuclear diatomic molecules such as O_2, N_2, H_2 etc, contained in the atmosphere of planets.

In order to evaluate the matrix elements (3.12.1) it is convenient, as done for the dipole term, to rewrite it in a form which contains more familiar expressions for the electric quadrupole and for the magnetic dipole operators and allows us to separate their contributions. Using the well known vector product relation

$$(\mathbf{u} \cdot \mathbf{w})(\mathbf{v} \cdot \mathbf{s}) = (\mathbf{u} \times \mathbf{v}) \cdot (\mathbf{w} \times \mathbf{s}) + (\mathbf{v} \cdot \mathbf{w})(\mathbf{u} \cdot \mathbf{s}) \tag{3.12.2}$$

we can write (3.12.1), after some manipulation, in the form

$$\Omega_{m0} = \Omega'_{m0} + \Omega''_{m0} \tag{3.12.3}$$

where

$$\Omega'_{m0} = \frac{1}{2} \left\langle \psi_m \left| (\mathbf{k}_\lambda \times \mathbf{u}_\lambda) \cdot \left(\sum_j \frac{e_j}{m_j} [\mathbf{r}_j \times \nabla_j] \right) \right| \psi_0 \right\rangle \tag{3.12.4}$$

and

$$\Omega''_{m0} = \left\langle \psi_m \left| \sum_j \frac{e_j}{m_j} \left[\sum_\alpha (k_\alpha r_{j\alpha} u_\alpha \nabla_{j\alpha}) + \frac{1}{4} \sum_\alpha \sum_{\beta \neq \alpha} (k_\alpha u_\beta + k_\beta u_\alpha)(r_{j\alpha} \nabla_{j\beta} + r_{j\beta} \nabla_{j\alpha}) \right] \right| \psi_0 \right\rangle \tag{3.12.5}$$

with

$$\alpha, \beta = x, y, z$$

Following the procedure used in Section 3.10 to prove equation (3.10.8), it is possible to show that

$$\left\langle \psi_m \left| \frac{1}{m_i} x_i \frac{\partial}{\partial x_i} \right| \psi_0 \right\rangle = -\frac{\pi \nu_{m0}}{\hbar} \langle \psi_m | x_i^2 | \psi_0 \rangle \tag{3.12.6}$$

and that

$$\left\langle \psi_m \left| \frac{1}{m_i} \left(x_i \frac{\partial}{\partial y_i} + y_i \frac{\partial}{\partial x_i} \right) \right| \psi_0 \right\rangle = -\frac{2\pi \nu_{m0}}{\hbar} \langle \psi_m | x_i y_i | \psi_0 \rangle \tag{3.12.7}$$

Insertion of (3.12.6) and (3.12.7) into (3.12.5) yields

$$\Omega''_{m0} = \frac{1}{2} \left\langle \psi_m \left| \sum_\alpha \sum_\beta k_\alpha u_\beta T_{\alpha\beta} \right| \psi_0 \right\rangle$$

where

$$T_{\alpha\beta} = \sum_j e_j r_{j\alpha} r_{j\beta}$$

The reader acquainted with classical electromagnetism will easily recognize that Ω'_{m0} describes the interaction with the magnetic dipole and Ω''_{m0} the interaction with the electric quadrupole of the system.

3.13 Two-photon processes. Scattering

The scattering of electromagnetic radiation from a molecule is a typical two-photon process in which one photon is absorbed and one emitted at the same time. There are two fundamental features of a scattering phenomenon which show that it cannot be reduced simply to a sequence in time of two distinct one-photon processes. The first is that the energies of the incident and of the scattered photons bear no relationship to the energy of the stationary molecular states, in contrast to one-photon processes which are completely controlled by Bohr's rule. The second is that the annihilation of the incident photon and the creation of the scattered photon are physically non-separable and truly 'simultaneous' events, in the sense that it is physically impossible to establish which occurs first.

Scattering processes are classified essentially according to the energy difference between incident and scattered photons. If this difference is zero, i.e. if the frequency of the scattered radiation is the same as that of the exciting radiation, the process is called Rayleigh scattering. If instead the frequency of the scattered radiation is either greater or smaller than that of the exciting radiation, the process is called the Raman effect.

Let us consider again, as for one-photon processes, a molecule embedded in a radiation field. At $t < 0$ the total system is in a stationary state $A = (n, n_\lambda, n_\mu)$ defined, according to Section 3.9 by a wavefunction

$$\psi_A = \psi_n \phi_{n_\lambda n_\mu} \tag{3.13.1}$$

where ψ_n is a molecular wavefunction and $\phi_{n_\lambda n_\mu}$ is a field wavefunction, and by an energy

$$E_A = \mathscr{E}_{n,n_\lambda n_\mu} = E_n + n_\lambda \hbar\omega_\lambda + n_\mu \hbar\omega_\mu \tag{3.13.2}$$

In the case of one-photon processes the field wavefunction and energy were labelled with only one quantum number n_λ since the variations in the field structure were limited to only one photon species. In a two-photon process we have to follow the variations of two photon species and thus two labels are used.

Let us now define a final state $B = (m, n_\lambda - 1, n_\mu + 1)$ which differs from A because one photon of species λ has been annihilated and one photon of species μ has been created while the molecule is in a different molecular state m. The state B is thus described by a wavefunction

$$\psi_B = \psi_m \phi_{n_\lambda - 1, n_\mu + 1} \tag{3.13.3}$$

and is associated with an energy

$$E_B = \mathscr{E}_{m,n_\lambda - 1, n_\mu + 1} = E_m + (n_\lambda - 1)\hbar\omega_\lambda + (n_\mu + 1)\hbar\omega_\mu \tag{3.13.4}$$

We also assume, to avoid the concurrence of a direct one-photon transition from A to B, that the first-order matrix element of the type (3.9.6), connecting A and B, is equal to zero. At the time $t = 0$ we switch on the perturbation $\mathscr{H}_{\text{int}}$ and we ask for the probability that at the time t the system is found in state B. From the general expression (3.8.15) of the perturbation Hamiltonian $\mathscr{H}_{\text{int}}$, we know that the states A and B are connected by a two-photon matrix element of the type

$$H_{\text{BA}}^{(2)} = \langle\psi_{\text{B}}|\mathscr{H}_{\text{int}}^{(2)}|\psi_{\text{A}}\rangle \tag{3.13.5}$$

where

$$\mathscr{H}_{\text{in}}^{(2)} = 2\pi \sum_j \sum_{\lambda\mu} \frac{e_j^2}{m_j}(\mathbf{u}_\lambda \cdot \mathbf{u}_\mu)\left\{q_\lambda q_\mu^* \exp\left[\frac{\text{i}}{\hbar c}(\mathbf{k}_\lambda - \mathbf{k}_\mu)\cdot \mathbf{r}_j\right]\right\} \tag{3.13.6}$$

We notice that only one term of (3.8.15) has been used in (3.13.6), namely the term which creates a photon of species μ and destroys a photon of species λ. The other three terms of the two-photon part of the interaction Hamiltonian have different effects on the system and play no role in the transition A → B.

In addition to the matrix element (3.13.5) which couples the states A and B, we know from second-order perturbation theory that there is another path for the transition, which involves a set of 'virtual' intermediate states which also play the role of binding together the initial state A and the final state B. These virtual states can be thought of as a model representation of the mixing of states necessary to represent non-stationary states of the molecule under the effect of the time-dependent perturbation $\mathscr{H}_{\text{int}}$.

As discussed before, it is physically impossible to separate a two-photon process into a time-ordered sequence of two one-photon processes. This means that virtual states corresponding both to the absorption of one photon of species λ and to the emission of one photon of species μ have to be considered. We can therefore define two types of virtual states, indicated by the symbols k_1 and k_2, corresponding respectively to the absorption of a photon $k_\lambda = \hbar\omega_\lambda$ and to the emission of a photon $k_\mu = \hbar\omega_\mu$ with simultaneous transition of the molecule into a different molecular state l. These states are described by wavefunctions and energies

$$\begin{aligned} \psi_{k_1} &= \psi_l \phi_{n_\lambda - 1, n_\mu} \qquad E_{k_1} = \mathscr{E}_{l, n_\lambda - 1, n_\mu} = E_l + (n_\lambda - 1)\hbar\omega_\lambda + n_\mu \hbar\omega_\mu \\ \psi_{k_2} &= \psi_l \phi_{n_\lambda, n_\mu + 1} \qquad E_{k_2} = \mathscr{E}_{l, n_\lambda, n_\mu + 1} = E_l + n_\lambda \hbar\omega_\lambda + (n_\mu + 1)\hbar\omega_\mu \end{aligned} \tag{3.13.7}$$

It is important to notice that the energy is not preserved in the transition to the virtual states k_1 and k_2 but only in the total transition $A \rightarrow B$.

According to second-order perturbation theory, the states A and B are connected by the second-order matrix element (3.5.38)

$$K_{BA} = \sum_k \frac{H_{kA} H_{Bk}}{E_A - E_k} \tag{3.13.8}$$

which, using the division introduced before between states k_1 and k_2, can be formally rewritten in the form

$$K_{BA} = \sum_{k_1} \frac{H_{k_1 A} H_{Bk_1}}{E_A - E_{k_1}} + \sum_{k_2} \frac{H_{k_2 A} H_{Bk_2}}{E_A - E_{k_2}} \tag{3.13.9}$$

Using (3.13.2), (3.13.4) and (3.13.7) this expression becomes

$$K_{BA} = \sum_{k_1} \frac{H_{k_1 A} H_{Bk_1}}{E_n - E_l + k_\lambda} + \sum_{k_2} \frac{H_{k_2 A} H_{Bk_2}}{E_n - E_l - k_\mu} \tag{3.13.10}$$

and therefore the complete matrix element which connects the states A and B can be written from (3.13.5) and (3.13.10)

$$R_{BA} = H_{BA}^{(2)} + \sum_{k_1} \frac{H_{k_1 A} H_{Bk_1}}{E_n - E_l + k_\lambda} + \sum_{k_2} \frac{H_{k_2 A} H_{Bk_2}}{E_n - E_l - k_\mu} \tag{3.13.11}$$

where

$$
\begin{aligned}
H^{(2)}_{BA} &= \langle \psi_B | \mathscr{H}^{(2)}_{\text{int}} | \psi_A \rangle \\
H_{k_1 A} &= \langle \psi_{k_1} | \mathscr{H}^{(1)'}_{\text{int}} | \psi_A \rangle ; \; H_{k_2 A} = \langle \psi_{k_2} | \mathscr{H}^{(1)''}_{\text{int}} | \psi_A \rangle \\
H_{Bk_1} &= \langle \psi_B | \mathscr{H}^{(1)''}_{\text{int}} | \psi_{k_1} \rangle ; \; H_{Bk_2} = \langle \psi_B | \mathscr{H}^{(1)'}_{\text{int}} | \psi_{k_2} \rangle
\end{aligned}
\tag{3.13.12}
$$

with (see 3.9.1)

$$
\mathscr{H}^{(1)}_{\text{int}} = \mathscr{H}^{(1)'}_{\text{int}} + \mathscr{H}^{(1)''}_{\text{int}} \tag{3.13.13}
$$

We can now use the results of Section 3.7 to simplify (3.13.11). Using (3.7.12) we have

$$
H^{(2)}_{BA} = 2\pi\hbar^2 \left(\frac{n_\lambda (n_\mu + 1)}{k_\lambda k_\mu} \right)^{\frac{1}{2}} (\mathbf{u}_\lambda \cdot \mathbf{u}_\mu) \left\langle \psi_m \middle| \sum_j \frac{e_j^2}{m_j} \exp\left[\frac{\mathrm{i}}{\hbar c} (\mathbf{k}_\lambda - \mathbf{k}_\mu) \cdot \mathbf{r}_j \right] \middle| \psi_n \right\rangle \tag{3.13.14}
$$

$$
\sum_{k_1} \frac{H_{k_1 A} H_{Bk_1}}{E_n - E_l + \hbar\omega_\lambda} = 2\pi\hbar^2 \left[\frac{n_\lambda (n_\mu + 1)}{k_\lambda k_\mu} \right]^{\frac{1}{2}} \sum_l \frac{\langle \psi_l | \mathscr{L}_\lambda | \psi_n \rangle \langle \psi_m | \mathscr{L}^*_\mu | \psi_l \rangle}{-k_l + k_\lambda} \tag{3.13.15}
$$

$$
\sum_{k_2} \frac{H_{k_2 A} H_{Bk_2}}{E_n - E_l - \hbar\omega_\mu} = 2\pi\hbar^2 \left[\frac{n_\lambda (n_\mu + 1)}{k_\lambda k_\mu} \right]^{\frac{1}{2}} \sum_l \frac{\langle \psi_l | \mathscr{L}^*_\mu | \psi_n \rangle \langle \psi_m | \mathscr{L}_\lambda | \psi_l \rangle}{-k_l - k_\mu} \tag{3.13.16}
$$

where

$$
\begin{aligned}
\mathscr{L}_\lambda &= \sum_j \frac{e_j}{m_j} (\mathbf{u}_\lambda \cdot \mathbf{P}_j) \exp\left[\frac{\mathrm{i}}{\hbar c} (\mathbf{k}_\lambda \cdot \mathbf{r}_j) \right] \\
\mathscr{L}^*_\mu &= \sum_j \frac{e_j}{m_j} (\mathbf{u}_\lambda \cdot \mathbf{P}_j) \exp\left[-\frac{\mathrm{i}}{\hbar c} (\mathbf{k}_\mu \cdot \mathbf{r}_j) \right]
\end{aligned}
\tag{3.13.17}
$$

$\mathbf{P}_j$ being equal to $(\hbar/\mathrm{i})\nabla_j$. It is worthwhile pointing out that in (3.13.14) the factor 2 originates from the double sum on λ and μ.

Using the above given matrix elements we obtain for equation (3.13.11) the expression

$$
\begin{aligned}
R_{BA} = 2\pi\hbar^2 \left[\frac{n_\lambda (n_\mu + 1)}{k_\lambda k_\mu} \right]^{\frac{1}{2}} \Bigg\{ & (\mathbf{u}_\lambda \cdot \mathbf{u}_\mu) \left\langle \psi_m \middle| \sum_j \frac{e_j^2}{m_j} \exp\left[\frac{\mathrm{i}}{\hbar c} (\mathbf{k}_\lambda - \mathbf{k}_\mu) \cdot \mathbf{r}_j \right] \middle| \psi_n \right\rangle \\
& + \sum_l \left[\frac{\langle \psi_l | \mathscr{L}_\lambda | \psi_n \rangle \langle \psi_m | \mathscr{L}^*_\mu | \psi_l \rangle}{k_\lambda - k_l} + \frac{\langle \psi_l | \mathscr{L}^*_\mu | \psi_n \rangle \langle \psi_m | \mathscr{L}_\lambda | \psi_l \rangle}{-k_\mu - k_l} \right] \Bigg\}
\end{aligned}
\tag{3.13.18}
$$

This general expression for the matrix element connecting two states A and B of the total system (molecule plus field) can be considerably simplified using the dipole approximation discussed in Section 3.10. Quadrupole terms can be of interest in special cases of two-photon processes, but play a very secondary role in the normal Raman scattering. Since for vibrational molecular states we are essentially interested in the Raman effect, we shall ignore quadrupole terms in (3.13.18).

In the dipole approximation, exponentials of the type

$$
\exp\left[\frac{\mathrm{i}}{\hbar c} (\mathbf{k}_\lambda \cdot \mathbf{r}_j) \right]
$$

can be set equal to unity and the matrix elements (3.13.14, 16) assume the form

$$
\begin{aligned}
H^{(2)}_{BA} &= C_{\lambda\mu} (\mathbf{u}_\lambda \cdot \mathbf{u}_\mu) G^{(0)}_{mn} \\
\sum_{k_1} \frac{H_{k_1 A} H_{Bk_1}}{-k_l + k_\lambda} &= C_{\lambda\mu} \sum_l \frac{F^{(0)}_{ln,\lambda} F^{(0)}_{ml,\mu}}{-k_l + k_\lambda} \\
\sum_{k_2} \frac{H_{k_2 A} H_{Bk_2}}{-k_l - k_\mu} &= C_{\lambda\mu} \sum_l \frac{F^{(0)}_{ln,\mu} F^{(0)}_{ml,\lambda}}{-k_l - k_\mu}
\end{aligned}
\tag{3.13.19}
$$

where

$$
\begin{aligned}
k_l &= E_l - E_n \\
C_{\lambda\mu} &= 2\pi\hbar^2\left[\frac{n_\lambda(n_\mu + 1)}{k_\lambda k_\mu}\right]^{\frac{1}{2}} \\
G_{mn}^{(0)} &= \left(\sum_j \frac{e_j^2}{m_j}\right)\langle\psi_m|\psi_n\rangle = \left(\sum_j \frac{e_j^2}{m_j}\right)\delta_{mn} \\
F_{pq,\alpha}^{(0)} &= \left\langle \psi_p \left| \sum_j \frac{e_j}{m_j}(\mathbf{u}_\alpha \cdot \mathbf{P}_j) \right| \psi_q \right\rangle \qquad \alpha = \lambda, \mu; \quad p = l, m; \quad q = n, l
\end{aligned}
\tag{3.13.20}
$$

Using (3.13.19, 20) equation (3.13.18) becomes

$$
R_{BA} = C_{\lambda\mu}\left[(\mathbf{u}_\lambda \cdot \mathbf{u}_\mu)G_{mn}^{(0)} + \sum_l \frac{F_{ln,\lambda}^{(0)}F_{ml,\mu}^{(0)}}{-k_l + k_\lambda} + \sum_l \frac{F_{ln,\mu}^{(0)}F_{ml,\lambda}^{(0)}}{-k_l - k_\mu}\right] \tag{3.13.21}
$$

This equation can be further transformed by using equation (3.10.8) which yields for $F_{pq,\alpha}^{(0)}$

$$
F_{pq,\alpha}^{(0)} = \frac{\mathrm{i}}{\hbar}(E_p - E_q)(\mathbf{u}_\alpha \cdot \mathbf{P}_{pq}) \tag{3.13.22}
$$

where the dipole matrix element $\mathbf{P}_{pq}$ has been introduced in Section 3.10 and is given by

$$
\mathbf{P}_{pq} = \left\langle \psi_p \left| \sum_j e_j \mathbf{r}_j \right| \psi_q \right\rangle
$$

By substitution of (3.13.22), equation (3.13.21) assumes the form

$$
R_{BA} = 2\pi[n_\lambda k_\lambda (n_\mu + 1)k_\mu]^{\frac{1}{2}}\phi_{BA} \tag{3.13.23}
$$

where

$$
\phi_{BA} = (\mathbf{u}_\lambda \cdot \mathbf{u}_\mu)\frac{\hbar^2}{k_\lambda k_\mu}G_{mn}^{(0)} + \sum_l \frac{k_l(k_l - k_m)}{k_\lambda k_\mu}\left[\frac{(\mathbf{u}_\lambda \cdot \mathbf{P}_{ln})(\mathbf{u}_\mu \cdot \mathbf{P}_{ml})}{-k_l + k_\lambda} + \frac{(\mathbf{u}_\lambda \cdot \mathbf{P}_{ml})(\mathbf{u}_\mu \cdot \mathbf{P}_{ln})}{-k_l - k_\mu}\right] \tag{3.13.24}
$$

The first term in this equation is different from zero, according to (3.13.20), only if $m = n$. It therefore contributes only to the Rayleigh scattering. This term is anyway very small and can be neglected. Furthermore, since $k_\lambda \sim k_l$ and $k_\mu \sim k_l - k_m$, the coefficient of the second term can be set equal to unity. With these simplifications we have

$$
\phi_{BA} = \sum_l \left[\frac{(\mathbf{u}_\lambda \cdot \mathbf{P}_{ln})(\mathbf{u}_\mu \cdot \mathbf{P}_{ml})}{-k_l + k_\lambda} + \frac{(\mathbf{u}_\lambda \cdot \mathbf{P}_{ml})(\mathbf{u}_\mu \cdot \mathbf{P}_{ln})}{-k_l - k_\mu}\right] \tag{3.13.25}
$$

which is the well known Kramers–Heisenberg dispersion formula, used normally as the basis of the discussion of the Raman effect.

According to equations (3.5.39) and (3.10.15), the probability of finding the system in the state B at the time t, is given by

$$
w_{BA} = \frac{(2\pi)^3}{\hbar} n_\lambda k_\lambda (n_\mu + 1)k_\mu |\phi_{BA}|^2 \tag{3.13.26}
$$

Let us now calculate the intensity of the scattered radiation. For this we consider a purely monochromatic exciting radiation of frequency $\tilde{\nu}_\lambda = k_\lambda/h$ ($\tilde{\nu} = \nu c$; units of $\tilde{\nu}$: cycles/s), with an intensity†

$$
I_0 = c n_\lambda k_\lambda \tag{3.13.27}
$$

† The intensity of the exciting light is defined as the energy falling on 1 cm^2 of the sample in 1 s. Here n_λ is taken as the number of quanta k_λ per cm^3.

The intensity of the radiation of frequency $\tilde{\nu}_\mu = k_\mu/h$ and polarization vector $\mathbf{u}_\mu$, scattered per unit time by one molecule in the solid angle element $d\Omega$, is then given by

$$dI = k_\mu w_{BA}\rho(\mu)\, d\Omega \tag{3.13.28}$$

where $\rho(\mu)$ is the density of radiation of frequency $\tilde{\nu}_\mu$, defined as the number of photons in the unit of volume, with energy in the interval dk_μ and direction of propagation in the element of solid angle $d\Omega$, by the relation[8]

$$\rho(\mu)\, dk_\mu\, d\Omega = \frac{k_\mu^2}{(2\pi\hbar c)^3}\, dk_\mu\, d\Omega \tag{3.13.29}$$

Using (3.13.26), equation (3.13.28) becomes

$$dI = \frac{(2\pi)^3}{\hbar}\frac{I_0}{c}\frac{(n_\mu + 1)k_\mu^4}{(2\pi\hbar c)^3}|\phi_{BA}|^2\, d\Omega \tag{3.13.30}$$

i.e.

$$dI = I_0\left(\frac{k_\mu}{\hbar c}\right)^4 (n_\mu + 1)|\phi_{BA}|^2\, d\Omega \tag{3.13.31}$$

If we have Z molecules per unit volume, the incoherent scattering is Z times greater. In order to obtain the intensity of the light scattered in the complete solid angle 4π we have to integrate (3.13.31) over $d\Omega$, multiply by Z and sum over two mutually perpendicular directions of polarization $\mathbf{u}_\mu$. In the normal scattering processes with monochromatic exciting light $n_\mu = 0$ and thus

$$I = ZI_0\left(\frac{k_\mu}{\hbar c}\right)^4 \sum_{\mathbf{u}_\mu}\int |\phi_{BA}|^2\, d\Omega \tag{3.13.32}$$

In order to evaluate the integral in (3.13.32) we can proceed as follows.[11] From (3.13.25) we can write ϕ_{BA} as $(\mathbf{u}_\mu \cdot \mathbf{V}_{BA})$ where $\mathbf{V}_{BA}$ is a vector. If we call θ the angle formed by the vector $\mathbf{V}_{BA}$ and by the direction of propagation of a photon of μ species, and if we choose

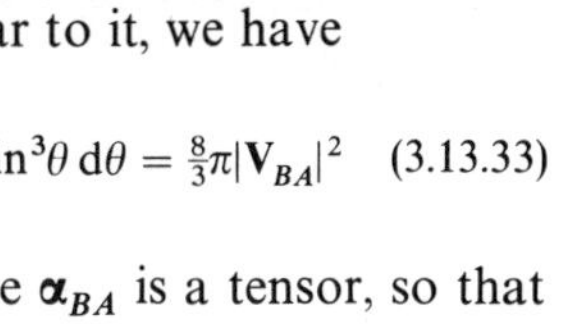

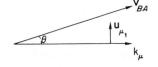

Figure 3.13.1 The angle θ formed by the vector $\mathbf{V}_{BA}$ with the direction of propagation of the photon μ

the two arbitrary mutually perpendicular polarization vectors of (3.13.32) so that one is in the plane defined by $\mathbf{V}_{BA}$ and $\mathbf{k}_\mu$ and the other is perpendicular to it, we have

$$\sum_{\mathbf{u}_\mu}\int |\phi_{BA}|^2\, d\Omega = \int_0^{2\pi} d\phi \int_0^{\pi} |\mathbf{u}_\mu \cdot \mathbf{V}_{BA}|^2 \sin\theta\, d\theta = 2\pi\int_0^{\pi} |\mathbf{V}_{BA}|^2 \sin^3\theta\, d\theta = \tfrac{8}{3}\pi|\mathbf{V}_{BA}|^2 \tag{3.13.33}$$

The vector $\mathbf{V}_{BA}$ can then be written in the form $-\boldsymbol{\alpha}_{BA} \cdot \boldsymbol{\mu}_\lambda$ where $\boldsymbol{\alpha}_{BA}$ is a tensor, so that (3.13.32) becomes

$$I = ZI_0\left(\frac{k_\mu}{\hbar c}\right)^4 \frac{8\pi}{3}|\boldsymbol{\alpha}_{BA} \cdot \mathbf{u}_\lambda|^2 = Z\frac{2^7\pi^5}{3c^4}I_0\tilde{\nu}_\mu^4|\boldsymbol{\alpha}_{BA} \cdot \mathbf{u}_\lambda|^2 \tag{3.13.34}$$

We conclude, therefore, that the scattered intensity is proportional to the fourth power of the frequency. If we average the above expression for all possible relative orientations of the molecules to the space-fixed polarization vector $\mathbf{u}_\lambda$ of the exciting light, as done in

Section 3.10, we obtain the final expression

$$I = Z\frac{2^7\pi^5}{3^2c^4}I_0\tilde{\nu}_\mu^4 \sum_{\rho\sigma} |(\alpha_{\rho\sigma})_{BA}|^2 \tag{3.13.35}$$

where ρ and σ indicate the Cartesian coordinates of a space-fixed Cartesian system and $(\alpha_{\rho\sigma})_{BA}$ indicates the element of the ρth row and the σth column of the $\boldsymbol{\alpha}_{BA}$ tensor.

Let us now go back to the form of the $\boldsymbol{\alpha}_{BA}$ tensor. According to the previous discussion we can write the matrix element ϕ_{BA} in the form

$$\phi_{BA} = -\mathbf{u}_\mu \cdot \boldsymbol{\alpha}_{BA} \cdot \mathbf{u}_\lambda \tag{3.13.36}$$

and thus the 'scattering' tensor $\boldsymbol{\alpha}_{BA}$ is, from (3.13.25),

$$\boldsymbol{\alpha}_{BA} = \boldsymbol{\alpha}_{mn} = \sum_l \left[\frac{\mathbf{P}_{ml}\mathbf{P}_{ln}}{k_l - k_\lambda} + \frac{\mathbf{P}_{ln}\mathbf{P}_{ml}}{k_l + k_\mu}\right] \tag{3.13.37}$$

It is important to discuss briefly the coherence problems associated with equations (3.13.35) and (3.13.37). We recall that a wavefunction ψ_r, which describes a stationary molecular state r, is associated with an arbitrary, undetermined phase factor $\exp(i\delta_r)$ such that $\psi_r \exp(i\delta_r)$ is also a solution of the Schrödinger equation. From this it follows that each matrix element $\mathbf{P}_{pq}$ has a phase factor $\exp[i(\delta_p - \delta_q)]$ which disappears only if $p = q$.

Let us then consider the radiation scattered by an assembly of molecules. If the initial and the final molecular states n and m coincide, we have Rayleigh scattering. In this case there is no phase difference between the photons scattered by the various centres, so that the scattered light is 'coherent' and interference can occur. The total scattered intensity depends strongly upon the relative orientation of the scattering centres. If, instead, n and m are different molecular states, we have the Raman effect. In this case the phase factors have random values and interference cannot occur. The total intensity is in this case simply proportional to the number of scattering centres per unit volume.

3.14 The Raman effect

In the normal Raman effect, light of intensity I and of frequency $\tilde{\nu}_0$, different from any absorption frequency of the molecules considered, is shone onto a sample and the light scattered is observed and analysed. In addition to the exciting frequency $\tilde{\nu}_0$, one observes that the scattered light also contains frequencies $\tilde{\nu}_0 \pm \tilde{\nu}_{mn}$, where $\tilde{\nu}_{mn}$ are vibrational frequencies of the molecules.

The intensity of the radiation scattered, at frequency $\tilde{\nu}_0 \pm \tilde{\nu}_{mn}$ and over the entire solid angle, by an assembly of Z molecules per unit volume, is given, according to the previous section, by

$$I = Z\frac{2^7\pi^5}{3^2c^4}I_0(\tilde{\nu}_0 \pm \tilde{\nu}_{mn})^4 \sum_{\rho\sigma} |(\alpha_{\rho\sigma})_{mn}|^2 \tag{3.14.1}$$

$\boldsymbol{\alpha}_{mn}$ is the so-called 'scattering tensor' for the vibrational transition $n \to m$ and is given by

$$\boldsymbol{\alpha}_{mn} = \sum_l \left[\frac{\mathbf{P}_{ml}\mathbf{P}_{ln}}{k_l - k_0} + \frac{\mathbf{P}_{ln}\mathbf{P}_{ml}}{k_l + (k_0 \pm k_{mn})}\right] \tag{3.14.2}$$

where the quantities $\mathbf{P}_{pq}$ are defined from the previous section as the transition dipoles between the molecular states p and q

$$\mathbf{P}_{pq} = \left\langle \psi_p \middle| \sum_j e_j\mathbf{r}_j \middle| \psi_q \right\rangle \tag{3.14.3}$$

The products $\mathbf{P}_{pq}\mathbf{P}_{rs}$ in (3.14.2) indicate the direct (dyadic) product of the vectors $\mathbf{P}_{pq}$ and $\mathbf{P}_{rs}$. Consequently $\boldsymbol{\alpha}_{mn}$ is a tensor with elements $(\alpha_{\rho\sigma})_{mn}$ where $\rho, \sigma = x, y, z$. The coordinates x, y and z are defined as a molecule-fixed, non-rotating, Cartesian system. The wavefunctions ψ_p and ψ_q are complete molecular wavefunctions and the summation in (3.14.3) runs over all charged particles of the molecule, nuclei and electrons.

In a normal Raman experiment the exciting frequency is in the visible region and is therefore much greater than any vibrational frequency of the molecule. We can then reasonably assume that only the electrons are perturbed by the incident radiation field, the nuclei being too heavy to follow the variation of the electromagnetic field. Using the Born–Oppenheimer approximation we can write each molecular wavefunction appearing in (3.14.3) as the product of an electronic wavefunction $\phi_e(\mathbf{r}, \mathbf{R})$ and a nuclear wavefunction $\chi(\mathbf{R})$ (see Section 3.1). Strictly, the nuclear wavefunction should be the product of a vibrational and a rotational wavefunction. As discussed in Section 3.1, however, the vibrational and rotational wavefunctions can be handled, to a good approximation, separately and since we are only interested in vibrational transitions, we shall ignore the rotational part of ψ. Hereafter we shall label the electronic wavefunctions with a quantum number e which specifies the electronic state and the vibrational wavefunctions with two quantum numbers e and v which indicate the electronic and the vibrational states respectively. In equation (3.14.2) we have three types of states to specify:

(1) The initial state n with wavefunction $\psi_n = \phi_{e_n}\chi_{v_n}^{(e_n)}$.
(2) The intermediate states l with wavefunctions $\psi_l = \phi_{e_l}\chi_{v_l}^{(e_l)}$.
(3) The final state m with wavefunction $\psi_m = \phi_{e_m}\chi_{v_m}^{(e_m)}$.

In Raman spectroscopy the initial and the final electronic states are the same and correspond to the ground electronic level. We can therefore write $e_m = e_n = e_0$.

The transition dipoles (3.14.3) then assume the form

$$\mathbf{P}_{pq} = \langle \phi_{e_p}\chi_{v_p}^{(e_p)} | \mathbf{M} | \phi_{e_q}\chi_{v_q}^{(e_q)} \rangle \tag{3.14.4}$$

where

$$\mathbf{M} = \sum_j e_j \mathbf{r}_j$$

Since the summation extends over all charged particles in the molecule, nuclei and electrons, we can consider $\mathbf{M}$ as made up of two parts, one, $\mathbf{M}_e$, due to the displacement of the electrons with respect to the nuclei and one, $\mathbf{M}_N$, due to the displacement of the nuclei in the charge distribution of the electrons. The matrix elements of the latter term play no role in the polarizability tensor (3.14.2), since in the expression

$$\mathbf{P}_{pq} = \langle \chi_{v_p}^{(e_p)} | \mathbf{M}_N | \chi_{v_q}^{(e_q)} \rangle \langle \phi_{e_p} | \phi_{e_q} \rangle + \langle \phi_{e_p}\chi_{v_p}^{(e_p)} | \mathbf{M}_e | \phi_{e_q}\chi_{v_q}^{(e_q)} \rangle$$

the first term is always zero, owing to the orthogonality of the electronic wavefunctions, unless $e_p = e_q$. For equation (3.14.2) we therefore need to consider only the second term. This we shall rewrite, by a formal integration over the coordinates of the electrons, as

$$\mathbf{P}_{e_p e_q}^{v_p v_q} = \langle \chi_{v_p}^{(e_p)} | \mathbf{P}_{e_p e_q} | \chi_{v_q}^{(e_q)} \rangle \tag{3.14.5}$$

where

$$\mathbf{P}_{e_p e_q} = \langle \phi_{e_p} | \mathbf{M}_e | \phi_{e_q} \rangle \tag{3.14.6}$$

is the electronic transition moment for the transition $e_q \to e_p$. Using (3.14.5) and remembering that $e_m = e_n = e_0$, we obtain for the scattering tensor (3.14.2) the expression

$$\boldsymbol{\alpha}_{mn} = \sum_e \sum_v \left[\frac{\mathbf{P}_{e_0 e}^{v_m v} \mathbf{P}_{e e_0}^{v v_n}}{k_{v v_n}^{e e_0} - k_0} + \frac{\mathbf{P}_{e e_0}^{v v_n} \mathbf{P}_{e_0 e}^{v_m v}}{k_{v v_n}^{e e_0} + k_0 \pm k_{v_m v_n}^{e_0 e_0}} \right] \tag{3.14.7}$$

In order to evaluate the matrix elements (3.14.5), we need to specify the dependence of the electronic transition moment upon the nuclear positions. For small displacements of the nuclei, we can expand the electronic transition moments in a power series of the normal coordinates of the initial level. Without serious loss of generality we can assume that $\mathbf{P}_{e_p e_q} = \mathbf{P}_{e_q e_p}$ and use for the expansion the normal coordinates of the ground electronic level. In this way we can write, cutting the series off after the linear terms,

$$\mathbf{P}_{e_p e_q} = \mathbf{P}^0_{e_p e_q} + \sum_l \mathbf{P}^l_{e_p e_q} Q_l \tag{3.14.8}$$

where $\mathbf{P}^l_{e_p e_q} = (\partial \mathbf{P}_{e_p e_q}/\partial Q_l)_0$.

Furthermore, since $k^{eeo}_{vv_n} = k^{eeo}_{v_0v_0} - k^{eoeo}_{v_nv_0} + k^{ee}_{vv_0}$ and since $k^{eeo}_{v_0v_0} \gg k^{eoeo}_{v_nv_0}$, $k^{eeo}_{v_0v_0} \gg k^{ee}_{vv_0}$ and $k^{eeo}_{v_0v_0} \gg k_0$, we can replace, without serious error, $k^{eeo}_{vv_n}$ with $k^{eeo}_{v_0v_0}$. In this way the denominators in (3.14.7) are independent of v and we can therefore write

$$\alpha_{mn} = \sum_e \left[\frac{1}{k^{eeo}_{v_0v_0} - k_0} \sum_v \mathbf{P}^{v_m v}_{eoe} \mathbf{P}^{vv_n}_{eeo} + \frac{1}{k^{eeo}_{v_0v_0} \pm k^{eoeo}_{v_mv_n} + k_0} \sum_v \mathbf{P}^{vv_n}_{eeo} \mathbf{P}^{v_m v}_{eoe} \right] \tag{3.14.9}$$

Using (3.14.8) we obtain

$$\begin{aligned} \alpha_{mn} = \sum_e \Bigg\{ & (\mathbf{P}^0_{eoe}\mathbf{P}^0_{eeo}) \sum_v [a_e \langle \chi^{(eo)}_{v_m} | \chi^{(e)}_v \rangle \langle \chi^{(e)}_v | \chi^{(eo)}_{v_n} \rangle + b_e \langle \chi^{(eo)}_{v_n} | \chi^{(e)}_v \rangle \langle \chi^{(e)}_v | \chi^{(eo)}_{v_m} \rangle] \\ & + \sum_l \Bigg[(\mathbf{P}^0_{eoe}\mathbf{P}^l_{eeo}) \sum_v (a_e \langle \chi^{(eo)}_{v_m} | \chi^{(e)}_v \rangle \langle \chi^{(e)}_v | Q_l | \chi^{(eo)}_{v_n} \rangle + b_e \langle \chi^{(eo)}_{v_n} | \chi^{(e)}_v \rangle \langle \chi^{(e)}_v | Q_l | \chi^{(eo)}_{v_m} \rangle) \\ & + (\mathbf{P}^l_{eoe}\mathbf{P}^0_{eeo}) \sum_v (a_e \langle \chi^{(eo)}_{v_m} | Q_l | \chi^{(e)}_v \rangle \langle \chi^{(e)}_v | \chi^{(eo)}_{v_n} \rangle + b_e \langle \chi^{(eo)}_{v_n} | Q_l | \chi^{(e)}_v \rangle \langle \chi^{(e)}_v | \chi^{(eo)}_{v_m} \rangle) \Bigg] \Bigg\} \end{aligned} \tag{3.14.10}$$

where

$$a_e = \frac{1}{k^{eeo}_{v_0v_0} - k_0} \qquad b_e = \frac{1}{k^{eeo}_{v_0v_0} \pm k^{eoeo}_{v_mv_n} + k_0}$$

and $(\mathbf{P}^0_{pq}\mathbf{P}^l_{qp})$ represents, as before, the direct product of the two vectors. This expression can be considerably simplified using some approximate sum rules due to Van Vleck[12]

$$\begin{aligned} &\sum_v \langle \chi^{(eo)}_{v_m} | \chi^{(e)}_v \rangle \langle \chi^{(e)}_v | \chi^{(eo)}_{v_n} \rangle = \langle \chi^{(eo)}_{v_m} | \chi^{(eo)}_{v_n} \rangle = \delta_{mn} \\ &\sum_v \langle \chi^{(eo)}_{v_m} | \chi^{(e)}_v \rangle \langle \chi^{(e)}_v | Q_l | \chi^{(eo)}_{v_n} \rangle = \sum_v \langle \chi^{(eo)}_{v_n} | \chi^{(e)}_v \rangle \langle \chi^{(e)}_v | Q_l | \chi^{(eo)}_{v_m} \rangle = \langle \chi^{(eo)}_{v_m} | Q_l | \chi^{(eo)}_{v_n} \rangle \\ &\sum_v \langle \chi^{(eo)}_{v_m} | Q_l | \chi^{(e)}_v \rangle \langle \chi^{(e)}_v | \chi^{(eo)}_{v_n} \rangle = \sum_v \langle \chi^{(eo)}_{v_n} | Q_l | \chi^{(e)}_v \rangle \langle \chi^{(e)}_v | \chi^{(eo)}_{v_m} \rangle = \langle \chi^{(eo)}_{v_m} | Q_l | \chi^{(eo)}_{v_n} \rangle \end{aligned} \tag{3.14.11}$$

In this way we obtain

$$\alpha_{mn} = \sum_e (a_e + b_e) \left\{ (\mathbf{P}^0_{eoe}\mathbf{P}^0_{eeo}) \langle \chi^{(eo)}_{v_m} | \chi^{(eo)}_{v_n} \rangle + \sum_l [(\mathbf{P}^0_{eoe}\mathbf{P}^l_{eeo}) + (\mathbf{P}^l_{eoe}\mathbf{P}^0_{eeo})] \langle \chi^{(eo)}_{v_m} | Q_l | \chi^{(eo)}_{v_n} \rangle \right\} \tag{3.14.12}$$

and we can use this equation to find the Raman selection rules with a procedure similar to that used in Section 3.11 for the absorption selection rules in the infrared.

First of all we observe that the first term of (3.14.12) vanishes unless $v_m = v_n$, i.e. unless the initial and final states coincide. This term is therefore responsible for the Rayleigh scattering. We recall that there is another small contribution to the undisplaced Rayleigh scattering that has been neglected from equation (3.14.25) on.

The second term of (3.14.12) is responsible for the Raman effect. From Section 3.11 we know that the matrix elements of the type

$$\langle \chi^{(eo)}_{v_m} | Q_i | \chi^{(eo)}_{v_n} \rangle \tag{3.14.13}$$

have a simple form in the harmonic approximation. In order to follow the same nomenclature as Section 3.11 we shall write the wavefunctions χ_{v_n} and χ_{v_m} for the two vibrational states v_n and v_m in the form

$$\begin{aligned}\chi_{v_n} &= \psi_{v'_1}(Q_1)\psi_{v'_2}(Q_2)\ldots\psi_{v'_i}(Q_i)\ldots\psi_{v'_{3N-6}}(Q_{3N-6})\\ \chi_{v_m} &= \psi_{v''_1}(Q_1)\psi_{v''_2}(Q_2)\ldots\psi_{v''_i}(Q_i)\ldots\psi_{v''_{3N-6}}(Q_{3N-6})\end{aligned} \tag{3.14.14}$$

where the ψ_v are harmonic wavefunctions. From Section 3.11 we known that the matrix elements (3.14.13) are different from zero only if $v'_1 = v''_1$, $v'_2 = v''_2, \ldots, v'_{3N-6} = v''_{3N-6}$ except v'_i and v''_i for which we have the condition $v''_i = v'_i \pm 1$.

This permits us to conclude that, in the dipole approximation, only fundamental transitions are allowed. Of course the condition $v''_i = v'_i \pm 1$ is necessary but not sufficient for Raman activity. In addition we have from (3.14.12) that at least one component of the direct product

$$\mathbf{P}^0_{e_0e}\mathbf{P}^i_{ee_0} + \mathbf{P}^i_{e_0e}\mathbf{P}^0_{ee_0}$$

must be different from zero.

Overtones and combination bands are forbidden within the framework of the approximations used. Higher-order terms in (3.14.12) are responsible for their activity.

Let us now briefly discuss the implications of the general equation (3.14.12) for Raman scattering. In a Stokes process, which involves the annihilation of a photon k_0 and the creation of a photon $k_0 - k_{v_m v_0}$ (for simplicity superscripts are omitted), the initial state is the ground vibrational state and the final state is a fundamental vibrational state. In an anti-Stokes process, a photon k_0 is annihilated and a photon $k_0 + k_{v_m v_0}$ is created. In this case the initial state is a fundamental vibrational state and the final state is the ground vibrational state. Application of the above discussed harmonic oscillator selection rules therefore yields for the generic element of the scattering tensor (3.14.12)

$$(\alpha_{\rho\sigma})_{v_i\pm 1, v_i} = \sum_e \frac{2k_{ee_0} \mp k_{m0}}{(k^2_{ee_0} - k^2_0) \mp k_{m0}(k_{ee_0} - k_0)}[(P^0_\rho)_{e_0e}(P^i_\sigma)_{ee_0} + (P^i_\rho)_{e_0e}(P^0_\sigma)_{ee_0}]\langle\psi_{v_i\pm 1}(Q_i)|Q_i|\psi_{v_i}(Q_i)\rangle \tag{3.14.15}$$

where

$$\rho, \sigma = x, y, z$$

and the upper sign holds for Stokes, the lower sign for anti-Stokes lines.

Since the population of the ground vibrational level is much greater, for not too high temperatures, than the population of excited vibrational levels, the intensity of Stokes lines is, according to (3.14.1), much greater than the intensity of anti-Stokes lines. Since the ratio of the two populations, according to the Boltzmann statistics, is proportional to the factor $\exp[-k_{m0}/KT]$, we have that the ratio of the intensities of an anti-Stokes to the corresponding Stokes line is given, from (3.14.1), by

$$\frac{I_{as}}{I_s} = \left(\frac{\tilde{\nu}_0 + \tilde{\nu}_{m0}}{\tilde{\nu}_0 - \tilde{\nu}_{m0}}\right)^4 \frac{\sum_{\rho\sigma}|(\alpha_{\rho\sigma})_{v_i-1,v_i}|^2}{\sum_{\rho\sigma}|(\alpha_{\rho\sigma})_{v_i+1,v_i}|^2}\exp[-k_{m0}/KT] \tag{3.14.16}$$

For this reason only the Stokes lines are normally used in Raman spectroscopy for the determination of vibrational frequencies.

The calculation of the elements of the scattering tensor $(\alpha_{\rho\sigma})_{m0}$ is very difficult, since it requires the evaluation of the electronic transition moments and of their dependence upon the normal coordinates. The discussion of this problem is beyond the scope of this book. The treatment of Raman intensities in terms of electronic transition moments and the calculation of these quantities are dealt with in References 13–21.

For the determination of the Raman selection rules it is, however, unnecessary to know the components of the scattering tensor exactly. We shall see in Chapter 7 that knowledge of their symmetry behaviour is sufficient to determine the Raman activity of vibrational transitions. For this reason it is convenient to rewrite (3.14.12) in the more concise and less detailed form

$$(\alpha_{\rho\sigma})_{mn} = \langle \psi_{v_i''} | \alpha_{\rho\sigma} | \psi_{v_i'} \rangle \qquad (3.14.17)$$

where

$$\alpha_{\rho\sigma} = \sum_{el} \frac{2k_{ee_0} - k_{mn}}{(k_{mn}^2 - k_0^2) - k_{mn}(k_{ee_0} - k_0)} [(P_\rho^0)_{e_0e}(P_\sigma^l)_{ee_0} + (P_\rho^l)_{e_0e}(P_\sigma^0)_{ee_0}] Q_l \qquad (3.14.18)$$

By comparison of equation (3.14.1) with the corresponding classical expression Placzek was able to correlate the components of the scattering tensor with the components of the classical polarizability tensor, defined by the relationship

$$\boldsymbol{\mu} = \boldsymbol{\alpha} \mathbf{E}$$

where $\boldsymbol{\mu}$ is the dipole moment induced by the perturbing electromagnetic field. A detailed discussion of the polarizability theory can be found in the book *Einführung in die Raman-spektroskopie* by Brandmüller and Moser.[20] It is worthwhile to observe that equation (3.4.17) can be made more general than it is in the present derivation. If the expansion of the electronic transition moments given in equation (3.14.8) is not cut after linear terms but is extended to higher terms, equation (3.14.17) can be written again in the same form. $\boldsymbol{\alpha}$ now contains additional terms in (3.14.18). We conclude, therefore, from equation (3.14.17) that a vibrational transition from an initial vibrational state n to a final vibrational state m is active in the Raman spectrum only if at least one component of $\boldsymbol{\alpha}_{mn}$ is different from zero.

A particular type of Raman effect is obtained when the exciting frequency ν_0 approaches one of the transition frequencies to a vibronic excited level. In this case we have the so-called 'Resonance Raman effect'. Equation (3.14.12) is no longer valid for the Resonance Raman effect and a different approach, based on a more general perturbation theory[9] should be used. The final expression is, however, not much different from (3.14.12). Briefly what is required to extend (3.14.12) to the Resonance Raman effect is to introduce a damping factor in the denominators of (3.14.7) to perform the summation over the vibrational states without the help of the sum rules and to limit the sum over the electronic states to the lowest excited level. The treatment of the Resonance Raman effect can be found in References 15, 16 and 21.

References

1. M. Born and R. Oppenheimer, *Ann. Physik*, **84**, 457 (1927).
2. H. Eyring, J. Walter and G. E. Kimball, *Quantum Chemistry*, Wiley, 1944.
3. H. C. Allen and P. C. Cross, *Molecular Vib-rotors*, Wiley, 1963.
4. J. T. Hougen, in *Physical Chemistry. An Advanced Treatise*, Academic Press, 1970, Vol. IV, p. 308.
5. G. Herzberg, *Molecular Spectra and Molecular Structure. Vol. III. Electronic Spectra*, Van Nostrand 1966, pp. 23–65.
6. H. H. Nielsen, in *Handbuch der Physik* (Ed. S. Flugge), Vol. XXXVIII, Springer Verlag, 1959.
7. M. P. Barchewitz, *Spectroscopie Infrarouge. Vol. I*, Gauthier-Villars, 1961.
8. W. Heitler, *The Quantum Theory of Radiation*, Oxford Clarendon Press, 1954.
9. H. F. Hameka, *Advanced Quantum Chemistry*, Addison Wesley, 1965.
10. A. Messiah, *Quantum Mechanics*, North Holland and J. Wiley, 1966.

11. G. Placzek, in *Handbuch der Radiologie* (Ed. E. Marx), Akademische Verlagsgesellschaft, 1934, Vol. VI.
12. J. H. Van Vleck, *Proc. Natl. Acad. Sci. U.S.*, **15**, 754 (1929).
13. P. Schorygin, *Ber. Akad. Wiss. UdSSR*, **81**, 201 (1952).
14. J. Behringer and J. Brandmüller, *Z. Elektrochem.*, **60**, 643 (1956).
15. J. Behringer, *Z. Elektrochem.*, **62**, 906 (1958).
16. A. C. Albrecht, *J. Chem. Phys.*, **34**, 1476 (1961).
17. F. A. Savin, *Optics and Spectrosc.*, **19**, 308 (1965); **19**, 412 (1965).
18. J. Tang and A. C. Albrecht, in *Raman Spectroscopy* (Ed. H. A. Szymanski), Plenum Press, 1970, Vol. II.
19. W. L. Peticolas, L. Nafie, P. Stein and B. Fanconi, *J. Chem. Phys.*, **52**, 1576 (1970).
20. J. Brandmüller and H. Moser, *Einführung in die Ramanspektroskopie*, D. Steinkopff Verlag, 1962.
21. J. Behringer, *Theorie der molekularen Lichtstreuung*, Sektion Physik der Universitat, Munchen Internal Report, 1967.

Chapter 4
The Vibrational Problem in Internal Coordinates

4.1 Internal coordinates

In Chapter 2 we have outlined the classical treatment of molecular vibrations and introduced, in addition to the Cartesian coordinates, a new set of 'normal coordinates', each associated with a normal vibration of the molecule.

In normal coordinates the equations of motion take a very simple form, as no cross-terms occur in the kinetic and potential energies. Unfortunately this is of little assistance to us. The normal coordinates are actually unknown until the secular equation has been solved. They are coordinates only in the generalized sense that they describe the molecular motion, but are of no use in the process of setting up the vibrational problem and of calculating the vibrational frequencies.

The Cartesian coordinates remain, for the moment, the only working set of coordinates we have at our disposal. This would not be a great limitation *per se* if our objective were just the calculation of vibrational frequencies. Let us consider, however, the process we have described in Chapter 2. It amounts to a method of calculation by which, given a set of force constants, defined in Cartesian coordinates, it is possible to obtain a set of vibrational frequencies for a molecule.

This process is actually purely theoretical in most cases. In practical situations the vibrational frequencies can be determined experimentally from the analysis of the infrared and Raman spectra whereas the force constants, i.e. the elements of the **f**-matrix, are unknown. The procedure in which we are interested is thus exactly the reverse of that described previously, that is, given the vibration frequencies for a molecule, to calculate the force constants.

Let us assume for a moment that we know how to do such a calculation. The result would be a set of Cartesian force constants which would mean practically nothing to us. As physicists or chemists we are in fact interested in using the observed frequencies to get information about the molecular structure and to derive parameters directly related to that structure. Such parameters, therefore, must be based on a choice of coordinates which is more characteristic of the molecular structure than the Cartesian coordinates of the atoms.

Changes in the bond length and in the angles between chemical bonds provide the most significant and physically meaningful set of coordinates for the description of the potential energy. We shall call these coordinates 'internal coordinates' since they describe (being unaffected by translations and rotations of the molecule as a whole) just the internal motions of the molecule, i.e. the molecular vibrations.

The types of internal coordinates which are generally used in vibrational problems are the following:

(1) *Bond stretching coordinate*—variation of the length of a chemical bond.

(2) *In-plane bending coordinate*—variation of the angle between two chemical bonds having one atom in common.

(3) *Our-of-plane bending coordinate*—variation of the angle between the plane defined by two bonds with one atom in common and a third bond connected to the common atom.

(4) *Torsion coordinate*—variation in the dihedral angle between the planes determined by three consecutive bonds connecting four atoms.

Since the variations in bond lengths and angles in a molecule can be written in terms of the variation in the Cartesian displacement coordinates of the atoms, the problem we have to face is that of writing the explicit form of the transformation from Cartesian coordinates to this new set of coordinates.

Let us call s_k a generic internal coordinate. The most general relation between s_k and the Cartesian displacement coordinates can be written in the form

$$s_k = \sum_i B_i^k x_i + \tfrac{1}{2}\sum_{ij} B_{ij}^k x_i x_j + \text{higher terms} \tag{4.1.1}$$

where the coefficients B_i^k, B_{ij}^k, etc., are determined by the molecular geometry.

A drastic simplification of equation (4.1.1) can be achieved if we restrict our treatment to the case of infinitesimal amplitudes of vibration. In this approximation, already adopted in Chapter 2, terms higher than the first can confidently be neglected in equation (4.1.1), which then becomes a linear relation between the internal and the Cartesian coordinates. If we call **s** and **x** the vectors whose components are the internal and the Cartesian coordinates, equation (4.1.1) reads in matrix notation

$$\mathbf{s} = \mathbf{B}\mathbf{x} \tag{4.1.2}$$

where **B** is a matrix whose elements are the coefficients B_i^k of equation (4.1.1).

The matrix **B** is not, in general, a square matrix. Let us consider, for example, a bent triatomic molecule (Figure 4.1.1) with $3n - 6 = 3$ vibrational degrees of freedom. If we

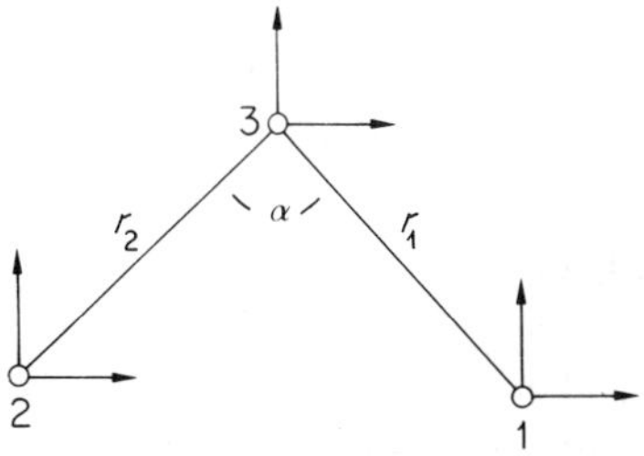

Figure 4.1.1 Internal coordinates for a bent triatomic molecule

choose as internal coordinates the variation in length of the two bonds r_1 and r_2, and the variation in the angle α between them, the **B**-matrix will have 3 rows and 9 columns (since there are nine Cartesian displacement coordinates for three atoms). We notice that in this case the number of internal coordinates is equal to the number of vibrational degrees of freedom. The translations and rotations are not included in the set of internal coordinates which describe, by definition, a molecular motion in which the relative positions of the atoms are changed. Since **B** is not a square matrix, it cannot be inverted.

We can always include in the **s**-vector six additional coordinates to describe the three translations and the three rotations. These six coordinates are most conveniently chosen in the form of equation (2.6.5). In this way **B** is a square and non-singular matrix (i.e. $|B| \neq 0$) and therefore can be inverted. We can thus define the inverse transformation of

(4.1.2) from internal to Cartesian coordinates in the form

$$\mathbf{x} = \mathbf{B}^{-1}\mathbf{s} \tag{4.1.3}$$

where $\mathbf{B}^{-1}$ is the inverse matrix of $\mathbf{B}$.

It must be pointed out that, as far as the vibrational problem is concerned, the six translations and rotations were introduced in the **s**-vector only to define the inverse of the **B**-matrix and therefore they need not be retained in the calculations.

4.2 The secular equation in internal coordinates

As discussed in the previous section, the use of internal coordinates gives a direct physical meaning to the elements of the potential energy matrix, in terms of molecular structure. The expressions of the potential and kinetic energies in internal coordinates are easily obtained by introduction of equation (4.1.3) in equations (2.4.1) and (2.4.2). We have

$$2T = \tilde{\dot{\mathbf{q}}}\dot{\mathbf{q}} = \tilde{\dot{\mathbf{x}}}\mathbf{M}\dot{\mathbf{x}} = \tilde{\dot{\mathbf{s}}}\tilde{\mathbf{B}}^{-1}\mathbf{M}\mathbf{B}^{-1}\dot{\mathbf{s}} = \tilde{\dot{\mathbf{s}}}\mathbf{G}^{-1}\dot{\mathbf{s}} \tag{4.2.1}$$

$$2V = \tilde{\mathbf{q}}\mathbf{f}\,\mathbf{q} = \tilde{\mathbf{x}}\,\mathbf{M}^{\frac{1}{2}}\mathbf{f}\,\mathbf{M}^{\frac{1}{2}}\mathbf{x} = \tilde{\mathbf{x}}\mathbf{f}_x\mathbf{x} = \tilde{\mathbf{s}}\tilde{\mathbf{B}}^{-1}\mathbf{f}_x\mathbf{B}^{-1}\mathbf{s} = \tilde{\mathbf{s}}\mathbf{F}\mathbf{s} \tag{4.2.2}$$

where

$$\mathbf{f}_x = \mathbf{M}^{\frac{1}{2}}\,\mathbf{f}\,\mathbf{M}^{\frac{1}{2}} \tag{4.2.3}$$

is the force constant matrix in Cartesian coordinates,

$$\mathbf{F} = \tilde{\mathbf{B}}^{-1}\mathbf{f}_x\mathbf{B}^{-1} \tag{4.2.4}$$

is the force constant matrix in terms of the internal coordinates and

$$\mathbf{G}^{-1} = \tilde{\mathbf{B}}^{-1}\,\mathbf{M}\,\mathbf{B}^{-1} \tag{4.2.5}$$

is the kinetic energy matrix in terms of the same coordinate basis. In internal coordinates therefore the kinetic energy matrix is no longer diagonal as in Cartesian coordinates.

Once we have introduced the internal coordinates we can combine equation (4.1.2) with equation (2.4.5) to define the transformation from internal to normal coordinates.

We then obtain

$$\mathbf{s} = \mathbf{B}\mathbf{x} = \mathbf{B}\mathbf{M}^{-\frac{1}{2}}\mathbf{q} = \mathbf{B}\mathbf{M}^{-\frac{1}{2}}\mathscr{L}\mathbf{Q} = \mathbf{L}\mathbf{Q} \tag{4.2.6}$$

where

$$\mathbf{L} = \mathbf{B}\mathbf{M}^{-\frac{1}{2}}\mathscr{L}$$

Using equation (4.2.6) we obtain with the same procedure of Section 2.4

$$2V = \tilde{\mathbf{Q}}\tilde{\mathbf{L}}\mathbf{F}\mathbf{L}\mathbf{Q} \tag{4.2.7}$$

$$2T = \tilde{\dot{\mathbf{Q}}}\tilde{\mathbf{L}}\mathbf{G}^{-1}\mathbf{L}\dot{\mathbf{Q}} \tag{4.2.8}$$

from these equations it follows, using (2.4.6) and (2.4.7), that

$$\tilde{\mathbf{L}}\mathbf{F}\mathbf{L} = \mathbf{\Lambda} \tag{4.2.9}$$

and

$$\tilde{\mathbf{L}}\mathbf{G}^{-1}\mathbf{L} = \mathbf{E} \tag{4.2.10}$$

By use of simple matrix algebra we obtain from equation (4.2.10)

$$\mathbf{G}^{-1} = \tilde{\mathbf{L}}^{-1}\mathbf{L}^{-1} \tag{4.2.11}$$

and by inversion

$$\mathbf{G} = \mathbf{L}\tilde{\mathbf{L}} \tag{4.2.12}$$

Equation (4.2.12) for the internal coordinate basis is the equivalent of equation (2.4.9). It shows that, when the relation between the coordinate basis adopted and the normal coordinates is a linear relation of the type (4.2.6), the inverse kinetic energy matrix can always be written in the form of equation (4.2.12). Equation (2.4.9) is, therefore, a special case of equation (4.2.12) the inverse kinetic-energy matrix being just the unit matrix.

By left multiplication of equation (4.2.9) by **L** we have

$$\mathbf{L}\ \tilde{\mathbf{L}}\ \mathbf{F}\ \mathbf{L} = \mathbf{L} \tag{4.2.13}$$

from which it follows, by use of (4.2.12), that

$$\mathbf{G}\ \mathbf{F}\ \mathbf{L} = \mathbf{L}\boldsymbol{\Lambda} \tag{4.2.14}$$

This is the most general form, in the harmonic approximation, of the secular equation for the vibrational problem. The matrix **L** is the matrix of the eigenvectors of $\mathbf{H} = \mathbf{G}\ \mathbf{F}$ and $\boldsymbol{\Lambda}$ is the eigenvalue matrix. Equation (2.4.13) is again a special case of this equation, the **G**-matrix being, in mass-weighted Cartesian displacement coordinates, the unit matrix.

Using (4.2.12), equation (4.2.14) can be rewritten in the form

$$\mathbf{F}\ \mathbf{G}\ \tilde{\mathbf{L}}^{-1} = \tilde{\mathbf{L}}^{-1}\boldsymbol{\Lambda} \tag{4.2.15}$$

The eigenvalues of **FG** and **GF** are thus the same although the matrix of the eigenvectors of **FG** is the transposed inverse of that of **GF**. It is easily seen that, as discussed in Chapter 2, for Cartesian coordinates, the eigenvalues of **FG** or **GF** satisfy the compatibility relations

$$|\mathbf{G}\ \mathbf{F} - \mathbf{E}\lambda| = 0 \qquad \text{(a)}$$
$$|\mathbf{F}\ \mathbf{G} - \mathbf{E}\lambda| = 0 \qquad \text{(b)} \tag{4.2.16}$$

The solution of the secular equation (4.2.14) is the central problem in vibrational spectroscopy. It can be solved by one of the several methods available in numerical calculus for the diagonalization of non-symmetric matrices, or as shown in Section 4.6, it can first be reduced to symmetric form and then diagonalized. The construction of the **G**-matrix is obviously the preliminary step in setting up the vibrational problem. By inversion of equation (4.2.5) the **G**-matrix is given by

$$\mathbf{G} = \mathbf{B}\mathbf{M}^{-1}\tilde{\mathbf{B}} \tag{4.2.17}$$

4.3 Construction of the B-matrix

The construction of the elements of the **B**-matrix is a straightforward process once the molecular geometry is known. It amounts to the expression of the internal parameters (bond lengths, angles between adjacent bonds, torsion angles, etc.) in terms of the Cartesian coordinates of the atoms. A small variation in these parameters is then obtained, to the first order, by differentiation of these expressions and, in the approximation of infinitesimal amplitudes of vibration, it is assumed to be an internal coordinate.

The numerical computations represent a tedious and time consuming process for a polyatomic molecule. However, computer programs are available to perform the calculations in a very short time.

In the following part of this section we shall illustrate the construction of the **B**-matrix elements for the types of internal coordinates discussed in Section 4.1. A compact method, more suitable for hand calculations, will be discussed in Section 4.4.

(A) *Stretching coordinate*

Consider two atoms bonded together as in Figure 4.3.1. The vectorial distance from

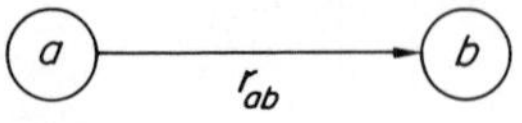

Figure 4.3.1 Vectorial distance between two atoms

atom a to atom b can be written in terms of any Cartesian system as

$$\mathbf{r}_{ab} = (x_b - x_a)\bar{\mathbf{i}} + (y_b - y_a)\bar{\mathbf{j}} + (z_b - z_a)\bar{\mathbf{k}} \qquad (4.3.1)$$

where $\bar{\mathbf{i}}$, $\bar{\mathbf{j}}$ and $\bar{\mathbf{k}}$ are unit vectors along the x-, y- and z-axes of the reference system. The modulus r_{ab} of the $\mathbf{r}_{ab}$ vector is then given by

$$r_{ab}^2 = \mathbf{r}_{ab} \cdot \mathbf{r}_{ab} = (x_b - x_a)^2 + (y_b - y_a)^2 + (z_b - z_a)^2 \qquad (4.3.2)$$

A small variation in the bond length is then obtained by forming the differential of (4.3.2)

$$r_{ab}\,\Delta r_{ab} = (x_b - x_a)(\Delta x_b - \Delta x_a) + (y_b - y_a)(\Delta y_b - \Delta y_a) + (z_b - z_a)(\Delta z_b - \Delta z_a) \qquad (4.3.3)$$

The non-zero elements of the row of the **B**-matrix corresponding to the internal coordinate Δr_{ab}, the variation of the length of the bond between atoms a and b, are then

$$\begin{aligned} B_{x_a} &= -A_{ab} \qquad & B_{y_a} &= -B_{ab} \qquad & B_{z_a} &= -C_{ab} \\ B_{x_b} &= A_{ab} & B_{y_b} &= B_{ab} & B_{z_b} &= C_{ab} \end{aligned} \qquad (4.3.4)$$

where

$$\begin{aligned} A_{ab} &= \frac{1}{r_{ab}}(x_b - x_a) \\ B_{ab} &= \frac{1}{r_{ab}}(y_b - y_a) \\ C_{ab} &= \frac{1}{r_{ab}}(z_b - z_a). \end{aligned} \qquad (4.3.5)$$

(B) *Valence angle bending coordinate*

Consider two bonds with one atom in common, as in Figure 4.3.2. The scalar product

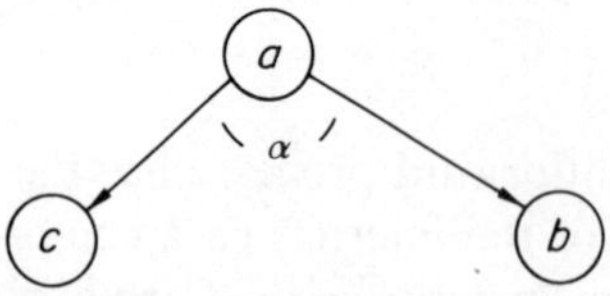

Figure 4.3.2 Angle α between two bonds with a common atom

between the vectors $\mathbf{r}_{ab}$ and $\mathbf{r}_{ac}$ is

$$\mathbf{r}_{ab} \cdot \mathbf{r}_{ac} = r_{ab} r_{ac} \cos\alpha = (x_b - x_a)(x_c - x_a) + (y_b - y_a)(y_c - y_a) + (z_b - z_a)(z_c - z_a) \qquad (4.3.6)$$

and therefore a small variation in the angle α, corresponding to an independent small variation in each Cartesian coordinate of the atoms is obtained by differentiating equation (4.3.6)

$$r_{ab}\cos\alpha\,\Delta r_{ac} + r_{ac}\cos\alpha\,\Delta r_{ab} - r_{ab}r_{ac}\sin\alpha\,\Delta\alpha$$
$$= (x_b - x_a)(\Delta x_c - \Delta x_a) + (x_c - x_a)(\Delta x_b - \Delta x_a) + (y_b - y_a)(\Delta y_c - \Delta y_a)$$
$$+ (y_c - y_a)(\Delta y_b - \Delta y_a) + (z_b - z_a)(\Delta z_c - \Delta z_a) + (z_c - z_a)(\Delta z_b - \Delta z_a) \qquad (4.3.7)$$

By use of (4.3.3) we then obtain

$$\Delta\alpha = \frac{(A_{ab}\cos\alpha - A_{ac})}{r_{ab}\sin\alpha}(\Delta x_b - \Delta x_a) + \frac{(A_{ac}\cos\alpha - A_{ab})}{r_{ac}\sin\alpha}(\Delta x_c - \Delta x_a)$$
$$+ \frac{(B_{ab}\cos\alpha - B_{ac})}{r_{ab}\sin\alpha}(\Delta y_b - \Delta y_a) + \frac{(B_{ac}\cos\alpha - B_{ab})}{r_{ac}\sin\alpha}(\Delta y_c - \Delta y_a)$$
$$+ \frac{(C_{ab}\cos\alpha - C_{ac})}{r_{ab}\sin\alpha}(\Delta z_b - \Delta z_a) + \frac{(C_{ac}\cos\alpha - C_{ab})}{r_{ac}\sin\alpha}(\Delta z_c - \Delta z_a) \qquad (4.3.8)$$

The row of **B**, corresponding to the internal coordinate $\Delta\alpha_{bac}$, the variation of the angle between the bonds *ab* and *ac*, therefore contains nine non-zero elements given by

$$B^{\alpha}_{x_b} = (A_{ab}\cos\alpha - A_{ac})/r_{ab}\sin\alpha \qquad B^{\alpha}_{x_c} = (A_{ac}\cos\alpha - A_{ab})/r_{ac}\sin\alpha$$
$$B^{\alpha}_{y_b} = (B_{ab}\cos\alpha - B_{ac})/r_{ab}\sin\alpha \qquad B^{\alpha}_{y_c} = (B_{ac}\cos\alpha - B_{ab})/r_{ac}\sin\alpha$$
$$B^{\alpha}_{z_b} = (C_{ab}\cos\alpha - C_{ac})/r_{ab}\sin\alpha \qquad B^{\alpha}_{z_c} = (C_{ac}\cos\alpha - C_{ab})/r_{ac}\sin\alpha \qquad (4.3.9)$$
$$B^{\alpha}_{x_a} = -(B^{\alpha}_{x_b} + B^{\alpha}_{x_c}) \qquad B^{\alpha}_{y_a} = -(B^{\alpha}_{y_b} + B^{\alpha}_{y_c}) \qquad B^{\alpha}_{z_a} = -(B^{\alpha}_{z_b} + B^{\alpha}_{z_c})$$

(C) *Out-of-plane bending coordinate*

Consider three bonds having one atom in common as shown in Figure 4.3.3 and define as internal coordinate the variation in the angle between bond *ab* and the plane defined by the other two bonds.

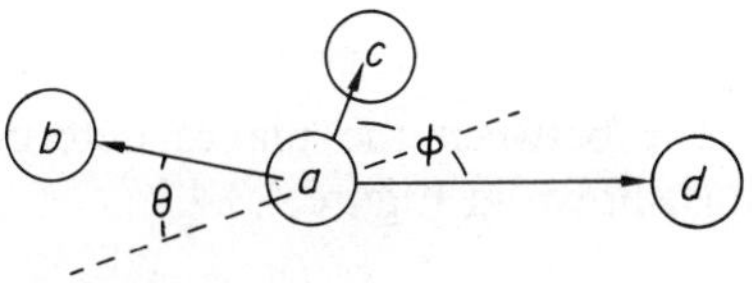

Figure 4.3.3 Out-of-plane displacement of an atom

In this case we have

$$\mathbf{r}_{ab}\cdot(\mathbf{r}_{ac}\times\mathbf{r}_{ad}) = r_{ab}r_{ac}r_{ad}\sin\phi\sin\theta = (x_b - x_a)(y_d - y_a)(z_c - z_a)$$
$$-(x_b - x_a)(z_d - z_a)(y_c - y_a) + (y_b - y_a)(z_d - z_a)(x_c - x_a)$$
$$-(y_b - y_a)(x_d - x_a)(z_c - z_a) + (z_b - z_a)(x_d - x_a)(y_c - y_a)$$
$$-(z_b - z_a)(y_d - y_a)(x_c - x_a) \qquad (4.3.10)$$

A small variation $\Delta\theta$ of the angle θ, due to arbitrary and independent small variations in the Cartesian coordinates of the atoms, is obtained by differentiation of (4.3.10) and by use of (4.3.8) and (4.3.3). We obtain in this way

$$\begin{aligned}\Delta\theta = {} & P_{ab}(\Delta x_b - \Delta x_a) + Q_{ab}(\Delta y_b - \Delta y_a) + R_{ab}(\Delta z_b - \Delta z_a)\\ & + S_{ad}(\Delta x_d - \Delta x_a) + T_{ad}(\Delta y_d - \Delta y_a) + U_{ad}(\Delta z_d - \Delta z_a)\\ & + S_{ac}(\Delta x_c - \Delta x_a) + T_{ac}(\Delta y_c - \Delta y_a) + U_{ac}(\Delta z_c - \Delta z_a)\end{aligned} \tag{4.3.11}$$

where

$$\begin{aligned}P_{ab} &= \frac{1}{r_{ab}}\left[\frac{B_{ad}C_{ac} - C_{ad}B_{ac}}{\cos\theta\sin\phi} - A_{ab}\tan\theta\right]\\ Q_{ab} &= \frac{1}{r_{ab}}\left[\frac{C_{ad}A_{ac} - A_{ad}C_{ac}}{\cos\theta\sin\phi} - B_{ab}\tan\theta\right]\\ R_{ab} &= \frac{1}{r_{ab}}\left[\frac{A_{ad}B_{ac} - B_{ad}A_{ac}}{\cos\theta\sin\phi} - C_{ab}\tan\theta\right]\\ S_{ax} &= \frac{1}{r_{ax}}\left[\frac{B_{ay}C_{ab} - C_{ay}B_{ab}}{\cos\theta\sin\phi} - \frac{\tan\theta}{\sin^2\phi}(A_{ax} - A_{ay}\cos\phi)\right]\\ T_{ax} &= \frac{1}{r_{ax}}\left[\frac{C_{ay}A_{ab} - A_{ay}C_{ab}}{\cos\theta\sin\phi} - \frac{\tan\theta}{\sin^2\phi}(B_{ax} - B_{ay}\cos\phi)\right] \qquad x, y = c, d\\ U_{ax} &= \frac{1}{r_{ax}}\left[\frac{A_{ay}B_{ab} - B_{ay}A_{ab}}{\cos\theta\sin\phi} - \frac{\tan\theta}{\sin^2\phi}(C_{ax} - C_{ay}\cos\phi)\right]\end{aligned} \tag{4.3.12}$$

The row of the **B**-matrix corresponding to the internal coordinate $\Delta\theta_{bacd}$, the variation in the angle between the bond ab and the plane defined by bonds ac and ad, therefore contains twelve non-zero elements given by

$$\begin{aligned}& B_{x_b} = P_{ab} \quad B_{y_b} = Q_{ab} \quad B_{z_b} = R_{ab} \quad B_{x_c} = S_{ac} \quad B_{y_c} = T_{ac} \quad B_{z_c} = U_{ac}\\ & B_{x_d} = S_{ad} \quad B_{y_d} = T_{ad} \quad B_{z_d} = U_{ad}\\ & B_{x_a} = -(P_{ab} + S_{ac} + S_{ad}) \quad B_{y_a} = -(Q_{ab} + T_{ac} + T_{ad}) \quad B_{z_a} = -(R_{ab} + U_{ac} + U_{ad})\end{aligned} \tag{4.3.13}$$

(D) *Torsion coordinate*

Consider the dihedral angle τ between the planes determined by three consecutive bonds connecting four atoms as shown in Figure 4.3.4.

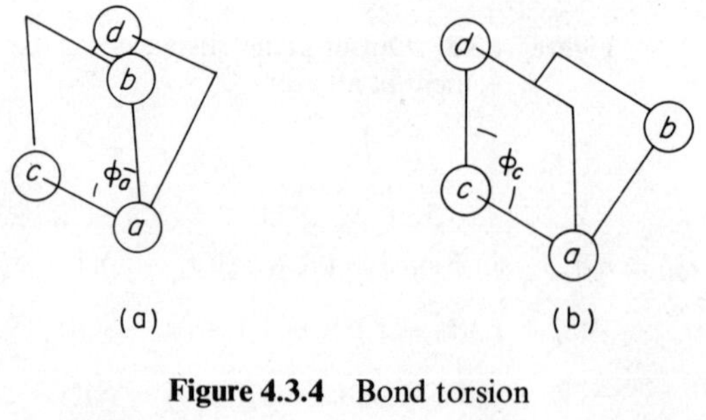

Figure 4.3.4 Bond torsion

By convention we shall restrict τ to the interval $-\pi \leqslant \tau \leqslant \pi$ and consider τ positive if, viewing the atoms in the direction of bond ac, the angle from the plane bac to the plane acd, with a nearer to the observer, is traced in the clockwise direction. Figure 4.3.4(a) shows an example of a positive τ, whereas Figure 4.3.4(b) shows an example of a negative τ-angle.

The analytical relation connecting τ with the Cartesian coordinates of the atoms can be found by forming the scalar product between a vector perpendicular to the plane *bac* and a vector perpendicular to the plane *acd*. We have in this case

$$
\begin{aligned}
(\mathbf{r}_{ab} \times \mathbf{r}_{ac}) \cdot (\mathbf{r}_{ac} \times \mathbf{r}_{cd}) &= r_{ab} r_{ac}^2 r_{cd} \sin\phi_a \sin\phi_c \cos\tau \\
&= [(x_c - x_a)(y_d - y_c) - (y_c - y_a)(x_d - x_c)][(x_a - x_b)(y_c - y_a) - (y_a - y_b)(x_c - x_a)] \\
&\quad + [(z_c - z_a)(x_d - x_c) - (x_c - x_a)(z_d - z_c)][(z_a - z_b)(x_c - x_a) - (x_a - x_b)(z_c - z_a)] \\
&\quad + [(y_c - y_a)(z_d - z_c) - (z_c - z_a)(y_d - y_c)][(y_a - y_b)(z_c - z_a) - (z_a - z_b)(y_c - y_a)]
\end{aligned}
\tag{4.3.14}
$$

A small variation $\Delta\tau$ in the torsion angle τ, due to arbitrary and independent small variations in the Cartesian coordinates of the atoms, is then obtained by differentiation of (4.3.14) and by use of (4.3.3) and (4.3.8)

$$
\begin{aligned}
\Delta\tau = {} & \frac{(C_{ba}B_{ac} - B_{ba}C_{ac})}{r_{ab}\sin^2\phi_a}\Delta x_b + \frac{(A_{ba}C_{ac} - C_{ba}A_{ac})}{r_{ab}\sin^2\phi_a}\Delta y_b + \frac{(B_{ba}A_{ac} - A_{ba}B_{ac})}{r_{ab}\sin^2\phi_a}\Delta z_b \\
& + \left[\frac{r_{ac} - r_{ab}\cos\phi_a(B_{ba}C_{ac} - C_{ba}B_{ac})}{r_{ab}r_{ac}\sin^2\phi_a} - \frac{\cos\phi_c}{r_{ac}\sin^2\phi_c}(B_{dc}C_{ca} - C_{dc}B_{ca})\right]\Delta x_a \\
& + \left[\frac{r_{ac} - r_{ab}\cos\phi_a(C_{ba}A_{ac} - A_{ba}C_{ac})}{r_{ab}r_{ac}\sin^2\phi_a} - \frac{\cos\phi_c}{r_{ac}\sin^2\phi_c}(C_{dc}A_{ca} - A_{dc}C_{ca})\right]\Delta y_a \\
& + \left[\frac{r_{ac} - r_{ab}\cos\phi_a(A_{ba}B_{ac} - B_{ba}A_{ac})}{r_{ab}r_{ac}\sin^2\phi_a} - \frac{\cos\phi_c}{r_{ac}\sin^2\phi_c}(A_{dc}B_{ca} - B_{dc}A_{ca})\right]\Delta z_a
\end{aligned}
$$

+equivalent terms in Δx_d, Δy_d, Δz_d, Δx_c, Δy_c and Δz_c with a permutation of the indices *a* and *b* with *c* and *d* respectively. (4.3.15)

The row of the **B**-matrix, corresponding to the internal coordinate $\Delta\tau_{bacd}$, the variation in the dihedral angle between the planes defined by the bonds *ab–ac* and *ac–cd*, therefore contains twelve non-zero elements given by

$$
\begin{aligned}
B_{x_i} &= \frac{(C_{ik}B_{kl} - B_{ik}C_{kl})}{r_{ki}\sin^2\phi_k} & B_{x_k} &= \frac{r_{kl} - r_{ki}\cos\phi_k(B_{ik}C_{kl} - C_{ik}B_{kl})}{r_{ki}r_{kl}\sin^2\phi_k} - \frac{\cos\phi_l}{r_{kl}\sin^2\phi_l}(B_{jl}C_{lk} - C_{jl}B_{lk}) \\
B_{y_i} &= \frac{(A_{ik}C_{kl} - C_{ik}A_{kl})}{r_{ki}\sin^2\phi_k} & B_{y_k} &= \frac{r_{kl} - r_{ki}\cos\phi_k(C_{ik}A_{kl} - A_{ik}C_{kl})}{r_{ki}r_{kl}\sin^2\phi_k} - \frac{\cos\phi_l}{r_{kl}\sin^2\phi_l}(C_{jl}A_{lk} - A_{jl}C_{lk}) \\
B_{z_i} &= \frac{(B_{ik}A_{kl} - A_{ik}B_{kl})}{r_{ki}\sin^2\phi_k} & B_{z_k} &= \frac{r_{kl} - r_{ki}\cos\phi_k(A_{ik}B_{kl} - B_{ik}A_{kl})}{r_{ki}r_{kl}\sin^2\phi_k} - \frac{\cos\phi_l}{r_{kl}\sin^2\phi_l}(A_{jl}B_{lk} - B_{jl}A_{lk})
\end{aligned}
\tag{4.3.16}
$$

$$i = b, d \qquad j = d, b \qquad k = a, c \qquad l = c, a$$

(E) *Linear angle bending coordinate*

For linear molecules (or parts of molecules) the definition (4.3.8) of the valence angle bending coordinate cannot be used since at equilibrium $\alpha = 180°$ and thus the **B**-matrix elements given by (4.3.9) become infinite. In addition, for a linear arrangement of three atoms, one needs to consider two bending coordinates in two perpendicular planes owing to the extra degree of vibrational freedom that occurs for linear molecules.

These difficulties can be circumvented by using a different definition of the bending coordinate for linear molecules. We shall define, following Reference 26, two linear bending coordinates in two perpendicular planes, choosen as the xz and yz planes in Figure 4.3.5,

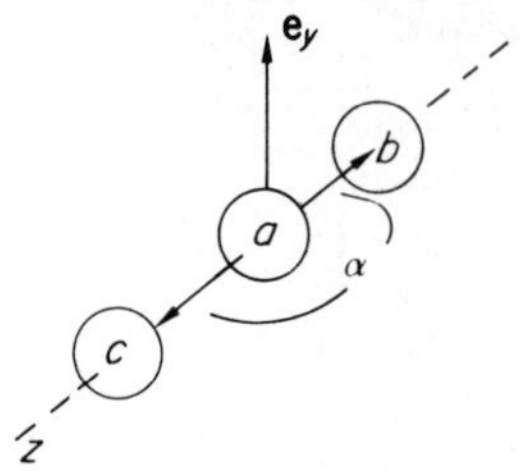

Figure 4.3.5 Linear angle bending

as

$$R_y = +\mathbf{e}_y \cdot (\mathbf{r}_{ab} \times \mathbf{r}_{ac})/r_{ab}r_{ac} = \frac{(z_b - z_a)(x_c - x_a) - (x_b - x_a)(z_c - z_a)}{r_{ab}r_{ac}}$$

$$R_x = -\mathbf{e}_x \cdot (\mathbf{r}_{ab} \times \mathbf{r}_{ac})/r_{ab}r_{ac} = \frac{(z_b - z_a)(y_c - y_a) - (y_b - y_a)(z_c - z_a)}{r_{ab}r_{ac}} \qquad (4.3.17)$$

In these expressions $-\mathbf{e}_x$ and $+\mathbf{e}_y$ are unit vectors oriented along the x- and y-axes of a reference system with z along the molecular axis of the equilibrium configuration. The unit vectors are chosen in this way to ensure that the coordinates transform in the correct way under a positive rotation about the z-axis.

By forming the differential of R_y and R_x and by using the fact that only the z-coordinates are different from zero at equilibrium, we obtain the result that the only non-vanishing **B**-elements are those relative to the x-coordinates for R_y and to the y-coordinates for R_x:

$$B_{x_a} = 1/r_{ab} \qquad B_{y_a} = 1/r_{ac}$$
$$B_{x_b} = -1/r_{ab} \quad B_{y_c} = -1/r_{ac} \qquad (4.3.18)$$

We notice that many authors have used a different definition of the linear bending coordinate, related to the one discussed above by the equations

$$\gamma_y = \arcsin R_y$$
$$\gamma_x = \arcsin R_x \qquad (4.3.19)$$

The **B**-matrix elements corresponding to definition (4.3.19) are easily obtained from (4.3.18) by changing the sign.

4.4 The s-vector method

A very useful method for the construction of the **B**-matrix elements was proposed independently by Eliashevich[f] and by Wilson[2] and is generally known in the western literature as the 's-vector method' of Wilson. According to equation (4.1.2) a generic internal coordinate s_t is given by

$$s_t = \sum_{i=1}^{3n} B_{ti}\xi_i \qquad (4.4.1)$$

where ξ_i represents the generic Cartesian displacement coordinate and B_{ti} the corresponding element of the tth row of the **B**-matrix. The $3n$ terms occurring in the summation above can be conveniently divided into n groups of three elements, each group involving only one atom. If we introduce for the atom α a vector $\boldsymbol{\rho}_\alpha$, whose components are three

Cartesian displacements of the atom, and a vector $\mathbf{s}_{t\alpha}$, whose components are the corresponding three B_{ti} elements, equation (4.4.1) takes the simple vectorial form

$$s_t = \sum_{\alpha=1}^{n} \mathbf{s}_{t\alpha} \cdot \boldsymbol{\rho}_\alpha \tag{4.4.2}$$

This way of writing the internal coordinates offers the advantage that it is independent of the choice of the Cartesian axes and that simple recipes can be worked out for finding the $\mathbf{s}_{t\alpha}$ vectors.

It is easily seen in fact that if all atoms but atom α are in their equilibrium positions, $\mathbf{s}_{t\alpha}$ is just the gradient† of the scalar internal coordinate $s_t(x_\alpha, y_\alpha, z_\alpha)$. The direction of $\mathbf{s}_{t\alpha}$ is therefore the direction in which a given displacement $\boldsymbol{\rho}_\alpha$ of atom α produces the greatest increase is $\mathbf{s}_{t\alpha}$.

Furthermore the magnitude $|s_{t\alpha}|$ of $\mathbf{s}_{t\alpha}$ is equal to the increase in $\mathbf{s}_{t\alpha}$ produced by a unit displacement of atom α.

It has been shown by Malhiot and Ferigle[3] that, in order to fulfil the Sayvetz conditions, the $\mathbf{s}$-vectors must obey the conditions

$$\sum_\alpha \mathbf{s}_{t\alpha} = \mathbf{0} \tag{4.4.3}$$

$$\sum_\alpha \mathbf{R}_\alpha \times \mathbf{s}_{t\alpha} = \mathbf{0} \tag{4.4.4}$$

where $\mathbf{R}_\alpha$ is a vector which defines the equilibrium position of atom α for an arbitrary origin.

The $\mathbf{s}$-vectors for the four fundamental types of internal coordinates discussed before have been worked out by a number of authors.[4–7] We shall derive them presently and show that, when the $\mathbf{s}$-vectors are expressed in terms of unit vectors along the bonds, their components are just the **B**-matrix elements derived in Section 4.3.

(A) *Bond stretching*

Since a bond-stretching coordinate involves only two atoms, we have two $\mathbf{s}_{t\alpha}$ vectors, called $\mathbf{s}_{ta}$ and $\mathbf{s}_{tb}$ in Figure 4.4.1.

Figure 4.4.1 Definition of the s-vectors in a stretching coordinate

The most efficient direction to displace the atoms is obviously along the bond away from each other. The magnitude of $\mathbf{s}_{ta}$ is equal, according to equation (4.4.2), to the increase in the bond length produced by a unit displacement of atom *a*. The same is true for $\mathbf{s}_{tb}$ and atom *b*. Thus

$$\begin{aligned} \mathbf{s}_{ta} &= -\mathbf{e}_{ab} = \mathbf{e}_{ba} \\ \mathbf{s}_{tb} &= -\mathbf{e}_{ba} = \mathbf{e}_{ab} \end{aligned} \tag{4.4.5}$$

where $\mathbf{e}_{ij}$ is a unit vector along the bond, oriented from atom *i* to atom *j*.

It is clear that the Cartesian components of $\mathbf{s}_{ta}$ and $\mathbf{s}_{tb}$ are the elements of the **B**-matrix for a bond-stretching coordinate, i.e. are the quantities A_{ab}, B_{ab} and C_{ab} of equation (4.3.4).

† See Margenau and Murphy, *The Mathematics for Physics and Chemistry*, D. Van Nostrand, p. 145.

(B) *Valence angle bending*

A valence angle bending coordinate involves three atoms (Figure 4.4.2) and thus requires three $\mathbf{s}_{t\alpha}$ vectors, $\mathbf{s}_{ta}$, $\mathbf{s}_{tb}$ and $\mathbf{s}_{tc}$. Let us consider the displacement of atom *b*. Since the most efficient way to displace atom *b* is in the direction perpendicular to $\mathbf{r}_{ab}$, $\mathbf{s}_{tb}$ will be oriented in this direction. The magnitude of $\mathbf{s}_{tb}$ is equal to the increase in the angle α produced by a unit displacement of atom *b* in this direction. If we call $\mathbf{f}_{ab}^{c}$ a unit vector perpendicular to $\mathbf{r}_{ab}$ and oriented toward atom *c*, the increase in the angle α, produced by a displacement $\mathbf{f}_{ab}^{c}$, will be equal to $-\mathbf{f}_{ab}^{c}/r_{ab}$.

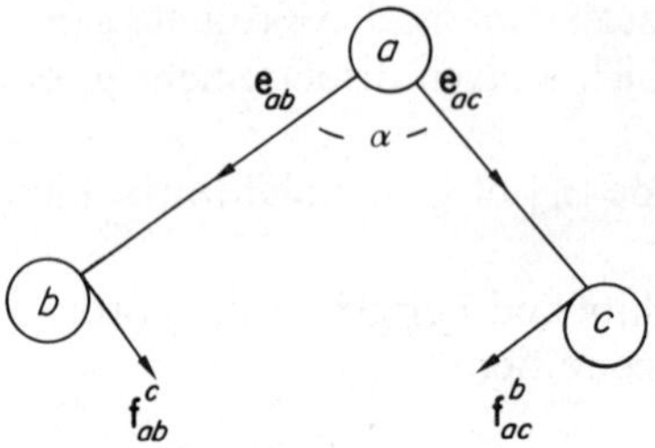

Figure 4.4.2 Definition of the s-vectors in the bending coordinate

Let us express $\mathbf{f}_{ab}^{c}$ in terms of two unit vectors $\mathbf{e}_{ab}$ and $\mathbf{e}_{ac}$ oriented along the bonds as shown in Figure 4.4.2. If we call *m* and *n* the components of $\mathbf{f}_{ab}^{c}$ we have

$$\mathbf{f}_{ab}^{c} = m\,\mathbf{e}_{ab} + n\mathbf{e}_{ac}$$

$$\mathbf{f}_{ab}^{c} \cdot \mathbf{e}_{ab} = m + n\cos\alpha = 0 \tag{4.4.6}$$

$$\mathbf{f}_{ab}^{c} \cdot \mathbf{e}_{ac} = m\cos\alpha + n = \sin\alpha$$

hence

$$\mathbf{f}_{ab}^{c} = (\mathbf{e}_{ac} - \mathbf{e}_{ab}\cos\alpha)/\sin\alpha \tag{4.4.7}$$

and

$$\mathbf{s}_{tb} = -\mathbf{f}_{ab}^{c}/r_{ab} = (\mathbf{e}_{ab}\cos\alpha - \mathbf{e}_{ac})/r_{ab}\sin\alpha \tag{4.4.8}$$

In a similar way we obtain

$$\mathbf{s}_{tc} = (\mathbf{e}_{ac}\cos\alpha - \mathbf{e}_{ab})/r_{ac}\sin\alpha \tag{4.4.9}$$

and from (4.4.3)

$$\mathbf{s}_{ta} = -\mathbf{s}_{tb} - \mathbf{s}_{tc} \tag{4.4.10}$$

We wish to point out once more that the s-vector method furnishes the same results as Section 4.3 in vectorial form. The components of $\mathbf{s}_{ta}$, $\mathbf{s}_{tb}$ and $\mathbf{s}_{tc}$ are in fact the coefficients of equation (4.3.7).

(C) *Out-of-plane bending*

In this case, since four atoms are involved in the coordinate, there are four $\mathbf{s}_{t\alpha}$ vectors. The most efficient way to displace atom *b*, is in the direction perpendicular to bond *ab*, as shown in Figure 4.4.3. $\mathbf{s}_{tb}$ is then given by

$$\mathbf{s}_{tb} = \mathbf{u}/r_{ab} \tag{4.4.11}$$

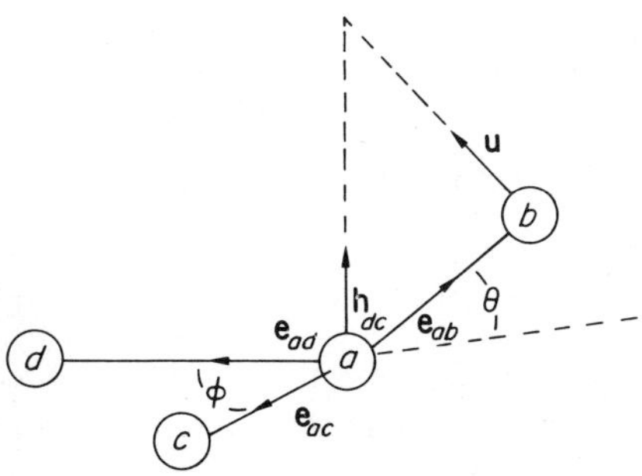

Figure 4.4.3 Construction of the s-vectors in the out-of-plane coordinate

where **u** is a unit vector perpendicular to bond *ab*.

The following vector equations will then hold

$$\mathbf{u} = m\mathbf{h}_{dc} + n\mathbf{e}_{ab} \tag{4.4.12}$$

$$\mathbf{u} \cdot \mathbf{e}_{ab} = m \sin\theta + n = 0 \tag{4.4.13}$$

$$\mathbf{u} \cdot \mathbf{h}_{dc} = m + n \sin\theta = \cos\theta \tag{4.4.14}$$

where $\mathbf{h}_{dc}$ is a unit vector perpendicular to the plane *cad* and given, in terms of the unit vectors $\mathbf{e}_{ad}$ and $\mathbf{e}_{ac}$ by

$$\mathbf{h}_{dc} = (\mathbf{e}_{ad} \times \mathbf{e}_{ac})/\sin\phi \tag{4.4.15}$$

From equations (4.4.13) and (4.4.14) we obtain

$$m = 1/\cos\theta \qquad n = -\tan\theta \tag{4.4.16}$$

and therefore

$$\mathbf{s}_{tb} = \frac{\mathbf{e}_{ad} \times \mathbf{e}_{ac}}{r_{ab} \sin\phi \cos\theta} - \frac{\mathbf{e}_{ab} \tan\theta}{r_{ab}} \tag{4.4.17}$$

The most efficient way to displace atom *d* is in the direction of $\mathbf{h}_{dc}$. In order to obtain the **s**-vector for this atom, we use equation (4.4.4), putting the origin at atom *a*. We then have

$$(\mathbf{e}_{ad} \times \mathbf{s}_{td})r_{ad} + (\mathbf{e}_{ac} \times \mathbf{s}_{tc})r_{ac} + (\mathbf{e}_{ab} \times \mathbf{s}_{tb})r_{ab} = \mathbf{0} \tag{4.4.18}$$

Dotting $\mathbf{e}_{ac}$ into equation (4.4.18), the second term drops out and we obtain, after substitution of equation (4.4.17),

$$\mathbf{s}_{td} \cdot \mathbf{h}_{dc} = \frac{1}{r_{ad}} \frac{(\mathbf{e}_{ac} \times \mathbf{e}_{ab})}{\sin\phi \cos\theta} \cdot \mathbf{h}_{dc} \tag{4.4.19}$$

The vector $\mathbf{s}_{td}$, as pointed out before, is parallel to the vector $\mathbf{h}_{dc}$ whereas the unit vector $(\mathbf{e}_{ac} \times \mathbf{e}_{ab})/\sin\phi$ is not. To obtain $\mathbf{s}_{td}$ we have to subtract from the previous vector its component perpendicular to $\mathbf{h}_{dc}$ which is easily expressed in terms of the unit vector $\mathbf{f}_{ac}$. The vector $\mathbf{s}_{td}$ then becomes

$$\mathbf{s}_{td} = (1/r_{ad})\left[\frac{\mathbf{e}_{ac} \times \mathbf{e}_{ab}}{\sin\phi\cos\theta} - \frac{\tan\theta}{\sin^2\phi}(\mathbf{e}_{ad} - \mathbf{e}_{ac}\cos\phi)\right] \tag{4.4.20}$$

By analogy we have

$$\mathbf{s}_{tc} = (1/r_{ac})\left[\frac{\mathbf{e}_{ab} \times \mathbf{e}_{ad}}{\sin\phi\cos\theta} - \frac{\tan\theta}{\sin^2\phi}(\mathbf{e}_{ac} - \mathbf{e}_{ad}\cos\phi)\right] \tag{4.4.21}$$

and finally from (4.4.3)

$$\mathbf{s}_{ta} = -(\mathbf{s}_{tb} + \mathbf{s}_{tc} + \mathbf{s}_{td}) \tag{4.4.22}$$

The components of the four $\mathbf{s}_{t\alpha}$ vectors are given in equation (4.3.13).

(D) *Torsion coordinate*

Again in this case we have to compute four $\mathbf{s}_{t\alpha}$ vectors. According to Figure 4.4.4, the most efficient way to displace atom b is in the direction perpendicular to the plane bac.

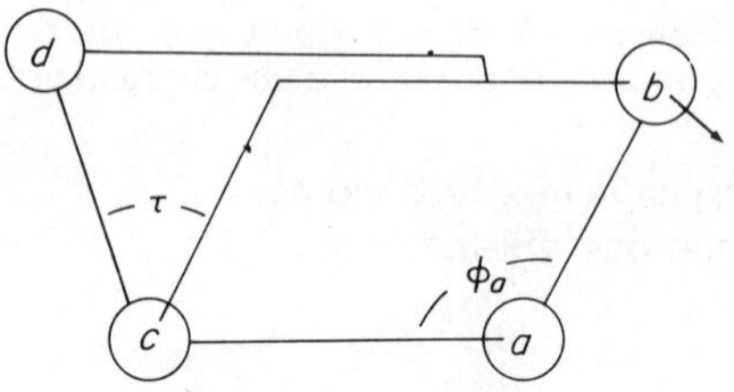

Figure 4.4.4 Torsion angle τ. Displacement of the b-atom

The magnitude of $\mathbf{s}_{tb}$ is equal to the increase in the angle τ produced by a unit displacement of atom b in this direction. We have then

$$\mathbf{s}_{tb} = -(1/r_{ba} \sin \phi_a)\mathbf{h}_{bc} = -\frac{\mathbf{e}_{ba} \times \mathbf{e}_{ac}}{r_{ba} \sin 2\phi_a} \tag{4.4.23}$$

In the same way we obtain

$$\mathbf{s}_{td} = -(1/r_{dc} \sin \phi_c)\mathbf{h}_{da} = -\frac{\mathbf{e}_{dc} \times \mathbf{e}_{ca}}{r_{dc} \sin 2\phi_c} \tag{4.4.24}$$

In order to obtain the $\mathbf{s}_{tc}$ vector, we again use equation (4.4.4), with the origin at atom a. We have

$$(\mathbf{e}_{ac} \times \mathbf{s}_{tc}) r_{ac} + (\mathbf{e}_{ab} \times \mathbf{s}_{tb}) r_{ab} + (\mathbf{e}_{ac} r_{ac} + \mathbf{e}_{cd} r_{cd}) \times \mathbf{s}_{td} = \mathbf{0} \tag{4.4.25}$$

Crossing $\mathbf{e}_{ac}$ into this equation we obtain, after substitution of $\mathbf{s}_{tb}$ and $\mathbf{s}_{td}$ from equations (4.4.23) and (4.4.24),

$$\mathbf{e}_{ac}(\mathbf{e}_{ac} \cdot \mathbf{s}_{tc}) - \mathbf{s}_{tc} = (1/r_{ac})\left[-\frac{\mathbf{e}_{ac} \times (\mathbf{e}_{ba} \times \mathbf{h}_{bc})}{\sin \phi_a} + \frac{\mathbf{e}_{ac} \times (\mathbf{e}_{ac} \times \mathbf{h}_{da}) r_{ac}}{\sin \phi_c r_{dc}} - \frac{\mathbf{e}_{ac} \times (\mathbf{e}_{dc} \times \mathbf{h}_{da})}{\sin \phi_c}\right] \tag{4.4.26}$$

Since $\mathbf{s}_{tc}$ is the gradient of τ, it must be normal to $\mathbf{e}_{ac}$ and thus, making use of the vector relation

$$\mathbf{f}_{ij}^{k} = -(\mathbf{e}_{ij} \times \mathbf{h}_{ki})/\sin \alpha \tag{4.4.27}$$

where α is the angle between the two vectors, we have

$$\mathbf{s}_{tc} = (1/r_{ac})\left[\frac{\cos \phi_a}{\sin \phi_a}\mathbf{h}_{bc} + \frac{r_{ac}}{r_{dc}} \frac{\mathbf{h}_{da}}{\sin \phi_c} - \frac{\cos \phi_c}{\sin \phi_c}\mathbf{h}_{da}\right] \tag{4.4.28}$$

or, by use of equation (4.4.15),

$$\mathbf{s}_{tc} = \frac{(r_{ac} - r_{dc} \cos \phi_c)}{r_{ac} r_{dc} \sin \phi_c} \frac{\mathbf{e}_{dc} \times \mathbf{e}_{ca}}{\sin \phi_c} - \frac{\cos \phi_a}{r_{ac} \sin \phi_a} \frac{\mathbf{e}_{ba} \times \mathbf{e}_{ac}}{\sin \phi_a} \tag{4.4.29}$$

The vector $\mathbf{s}_{ta}$ is given by the same expression, with a permutation of the indices b and d, and a and c.

The Cartesian components of these vectors are given in equation (4.3.16).

4.5 Construction of the G-matrix

As discussed in Section 4.2 the **G**-matrix is given by

$$\mathbf{G} = \mathbf{B}\ \mathbf{M}^{-1}\tilde{\mathbf{B}}$$

and is therefore simply obtained once the **B**-matrix has been constructed according to the methods outlined in the previous sections.

As an example let us consider the bent triatomic YX_2 molecule of Figure 4.1.1, Section 4.1. The **B**-matrix has the form

	Δx_1	Δy_1	Δz_1	Δx_2	Δy_2	Δz_2	Δx_3	Δy_3	0
Δr_{31}	A_{31}	B_{31}	0	0	0	0	$-A_{31}$	$-B_{31}$	0
Δr_{32}	0	0	0	A_{32}	B_{32}	0	$-A_{32}$	$-B_{32}$	0
$\Delta\alpha$	$\dfrac{A_{31}\cos\alpha - A_{32}}{r\sin\alpha}$	$\dfrac{B_{31}\cos\alpha - B_{32}}{r\sin\alpha}$	0	$\dfrac{A_{32}\cos\alpha - A_{31}}{r\sin\alpha}$	$\dfrac{B_{32}\cos\alpha - B_{31}}{r\sin\alpha}$	0	$\dfrac{(A_{31}+A_{32})(1-\cos\alpha)}{r\sin\alpha}$	$\dfrac{(B_{31}+B_{32})(1-\cos\alpha)}{r\sin\alpha}$	0

From equations (4.3.4) and (4.3.9) it follows that

$$A_{31}^2 + B_{31}^2 = A_{32}^2 + B_{32}^2 = 1$$

$$A_{31}A_{32} + B_{31}B_{32} = \cos\alpha$$

and thus, performing the matrix multiplication we have for **G**

	Δr_{31}	Δr_{32}	$\Delta\alpha$
Δr_{31}	$\mu_x + \mu_y$	$\mu_y\cos\alpha$	$-\mu_y\sin\alpha/r$
Δr_{32}	$\mu_y\cos\alpha$	$\mu_x + \mu_y$	$-\mu_y\sin\alpha/r$
$\Delta\alpha$	$-\mu_y\sin\alpha/r$	$-\mu_y\sin\alpha/r$	$[2\mu_x + 2\mu_y(1 - \cos\alpha)]/r^2$

where $\mu_x = 1/m_x$ and $\mu_y = 1/m_y$ are the reciprocal of the masses of the atoms. Since each column of $\tilde{\mathbf{B}}$ is equal to a row of **B**, instead of multiplying the three matrices use can be made of the formula

$$G_{tr} = \sum_{i=1}^{3n} \mu_i B_{ti} B_{ri} \tag{4.5.1}$$

which gives the generic element G_{tr} of **G** in terms of the rows of **B**. If the **s**-vector method is used, the previous equation becomes, using equation (4.4.2)

$$G_{tr} = \sum_{\alpha=1}^{n} \mu_\alpha \mathbf{s}_{t\alpha} \cdot \mathbf{s}_{r\alpha} \tag{4.5.2}$$

where the dot represents the scalar product and μ_α is the reciprocal of the mass of atom α.

For the YX_2 molecule discussed above we have

$$\mathbf{s}_{r_1 1} = \mathbf{e}_{31} \qquad \mathbf{s}_{r_1 2} = \mathbf{0} \qquad \mathbf{s}_{r_1 3} = -\mathbf{e}_{31}$$

$$\mathbf{s}_{r_2 1} = \mathbf{0} \qquad \mathbf{s}_{r_2 2} = \mathbf{e}_{32} \qquad \mathbf{s}_{r_2 3} = -\mathbf{e}_{32}$$

$$\mathbf{s}_{\alpha 1} = \frac{\mathbf{e}_{31}\cos\alpha - \mathbf{e}_{32}}{r\sin\alpha} \qquad \mathbf{s}_{\alpha 2} = \frac{\mathbf{e}_{32}\cos\alpha - \mathbf{e}_{31}}{r\sin\alpha} \qquad \mathbf{s}_{\alpha 3} = (\mathbf{e}_{31} + \mathbf{e}_{32})\frac{(1 - \cos\alpha)}{r\sin\alpha}$$

Use of equation (4.5.2) gives the same results as before. In the past the construction of the **G**-matrix elements has been the most tedious and time-consuming step in building up the secular equation for polyatomic molecules. Today computer programs are available for this process and there is therefore little need for hand calculations. If needed, however, **G**-matrix elements of normal occurrence in polyatomic molecules have been tabulated by Decius.[4] Some of them are also listed in Reference (8).

4.6 The secular equation in symmetric form

The secular equation in terms of internal coordinates

$$\mathbf{GFL} = \mathbf{L\Lambda} \tag{4.6.1}$$

is most conveniently reduced to a symmetric form by a method due to Miyazawa.[9]

Consider the equation

$$\mathbf{GA} = \mathbf{AT} \tag{4.6.2}$$

where **A** is the matrix of the eigenvectors and **T** is the matrix (diagonal) of the eigenvalues of **G**. Since **G** is real symmetric, **A** is orthogonal and thus

$$\mathbf{G} = \mathbf{AT\tilde{A}} = \mathbf{AT^{\frac{1}{2}}T^{\frac{1}{2}}\tilde{A}} = \mathbf{W\tilde{W}} \tag{4.6.3}$$

where $\mathbf{W} = \mathbf{AT^{\frac{1}{2}}}$. Using equation (4.6.3) the secular equation (4.2.14) can be written in the form

$$\mathbf{\tilde{W}FW(W^{-1}L)} = \mathbf{(W^{-1}L)\Lambda} \tag{4.6.4}$$

or

$$\mathbf{\tilde{W}FWC} = \mathbf{C\Lambda} \tag{4.6.5}$$

Comparison of equation (4.6.5) with equation (4.2.14) shows that the real symmetric matrix

$$\mathbf{H} = \mathbf{\tilde{W}FW} \tag{4.6.6}$$

has the same eigenvalues as the unsymmetrical **GF** matrix and eigenvectors

$$\mathbf{C} = \mathbf{W^{-1}L} \tag{4.6.7}$$

Once the secular equation (4.6.5) in symmetric form has been solved, the **L**-matrix can be obtained from equation (4.6.7), being given by

$$\mathbf{L} = \mathbf{WC} \tag{4.6.8}$$

The Miyazawa method, owing to its simplicity, is currently used today in force-constant calculations.

The solution of the secular equation (4.6.5) corresponds to finding a matrix **C** which diagonalizes the **H**-matrix since from (4.6.5) it follows that

$$\mathbf{\tilde{C}HC} = \mathbf{\Lambda} \tag{4.6.9}$$

Diagonalization methods for symmetrical matrices, based on iterative procedures, are very suited for computer programs. The iterative method of Jacobi[28] is one of the simplest and it is therefore often used.

4.7 The calculation of force constants

The methods outlined in the previous sections furnish the necessary mathematical machinery for the calculation of the normal frequencies of vibration of a polyatomic molecule. The whole process amounts to the computation of the elements of the **G**-matrix, to the assignment of numerical values to the elements of the **F**-matrix and to the solution of the secular equation (4.2.14).

The elements of **G** are easily evaluated once the molecular geometry is known. The molecular structure can be determined experimentally by means of several physico-chemical methods such as X-ray or electron diffraction, microwave and infrared spectroscopy, etc. For the discussion which follows we shall therefore assume that the molecular geometry is known. The elements of **F**, on the other hand, are completely unknown. We therefore need to devise a method by which these elements can be determined using the available experimental data, i.e. the molecular structure and the vibrational frequencies. In the language of the previous sections the vibrational problem assumes the following form: given the elements of the **G**-matrix and of the **Λ**-matrix, we wish to calculate the elements of the **F**-matrix by means of equation (4.2.14).

In principle the force constants could be expressed as explicit functions of the vibrational frequencies and of the elements of **G** by expansion in minors of the secular determinant. In practice this method is applicable only for very small sizes of the secular equation. Iterative methods by which a trial set of force constants is refined to fit the experimental frequencies and any other additional data (Coriolis constants, centrifugal distortion constants, etc.) are much more powerful and convenient.[10–21]

Suppose $\mathbf{F}_0$ is an initial trial **F**-matrix. Let us compute the eigenvalues $\mathbf{\Lambda}_0$ and the eigenvectors $\mathbf{L}_0$ of the secular equation

$$\mathbf{G}\mathbf{F}_0\mathbf{L}_0 = \mathbf{L}_0\mathbf{\Lambda}_0 \tag{4.7.1}$$

Since $\mathbf{F}_0$ is not the true **F**-matrix, the eigenvalues $\mathbf{\Lambda}_0$ will in general be different from those determined experimentally. We can then use first-order perturbation theory to refine the elements of $\mathbf{F}_0$ so that the difference between observed and calculated frequencies is minimized. Following I. A. Mills,[15] we consider a small change in the **F**-matrix

$$\mathbf{F} = \mathbf{F}_0 + \Delta\mathbf{F} \tag{4.7.2}$$

which produces a corresponding change in the eigenvalues and eigenvectors of the secular equation

$$\mathbf{\Lambda} = \mathbf{\Lambda}_0 + \Delta\mathbf{\Lambda} \tag{4.7.3}$$

$$\mathbf{L} = \mathbf{L}_0 + \Delta\mathbf{L} = \mathbf{L}_0(\mathbf{E} + \mathbf{L}_0^{-1}\,\Delta\mathbf{L}) = \mathbf{L}_0(\mathbf{E} + \mathbf{B}) = \mathbf{L}_0\mathbf{A} \tag{4.7.4}$$

and we try to determine both $\Delta\mathbf{\Lambda}$ and $\Delta\mathbf{L}$.

First of all we notice that the matrix **A** of equation (4.7.4) is orthogonal since it connects the unperturbed normal coordinates Q_0 with the normal coordinates Q after the perturbation, which are both normalized. We have therefore

$$\mathbf{L}^{-1} = \tilde{\mathbf{A}}\ \mathbf{L}_0^{-1} \tag{4.7.5}$$

The secular equation after the change of the **F**-matrix is given by

$$\begin{aligned} \mathbf{GFL} &= \mathbf{L\Lambda} \\ \mathbf{GFL}_0\mathbf{A} &= \mathbf{L}_0\mathbf{A\Lambda} \\ \tilde{\mathbf{L}}_0\mathbf{FL}_0\mathbf{A} &= \mathbf{A\Lambda} \end{aligned} \tag{4.7.6}$$

Using (4.7.2) and (4.7.3) this last equation can be rewritten, neglecting terms of higher order than the first

$$\tilde{\mathbf{L}}_0\mathbf{F}_0\mathbf{L}_0\mathbf{A} + \tilde{\mathbf{L}}_0\,\Delta\mathbf{F}\mathbf{L}_0 = \mathbf{A}\boldsymbol{\Lambda}_0 + \Delta\boldsymbol{\Lambda} \tag{4.7.7}$$

or, since $\tilde{\mathbf{L}}_0\mathbf{F}_0\mathbf{L} = \boldsymbol{\Lambda}_0$ (see equation 4.2.9),

$$\boldsymbol{\Lambda}_0\mathbf{A} + \tilde{\mathbf{L}}_0\,\Delta\mathbf{F}\mathbf{L}_0 = \mathbf{A}\boldsymbol{\Lambda}_0 + \Delta\boldsymbol{\Lambda} \tag{4.7.8}$$

By writing the diagonal elements of (4.7.8) in expanded form, we obtain a system of n equations of the type

$$\Delta\lambda_i = \sum_{k,l} (L_0)_{ki}(L_0)_{li}\,\Delta F_{kl} \qquad i, k, l = 1, 2, \ldots, n \tag{4.7.9}$$

from which it is possible to determine n corrections for the matrix $\mathbf{F}_0$. In matrix form the system of equations (4.7.9) can be written as

$$\mathbf{J}\,\Delta\mathbf{F} = \Delta\boldsymbol{\lambda} \tag{4.7.10}$$

where $\Delta\boldsymbol{\lambda}$ is a column vector with elements $(\lambda^{\text{obs}} - \lambda^{\text{calc}})$ arranged in a convenient order, $\Delta\mathbf{F}$ is a column vector of the corrections to $\mathbf{F}_0$ and $\mathbf{J}$ is the so-called Jacobian matrix of $\boldsymbol{\Lambda}$ with respect to $\mathbf{F}$ with elements

$$\begin{aligned} \frac{\partial\lambda_i}{\partial F_{kk}} &= (L_0)_{ki}(L_0)_{ki} \\ \frac{\partial\lambda_i}{\partial F_{kl}} &= 2(L_0)_{ki}(L_0)_{li} \end{aligned} \tag{4.7.11}$$

which are obtained from (4.7.9) by letting $\Delta\boldsymbol{\Lambda} \to 0$ and $\Delta\mathbf{F} \to 0$. In many cases the number of force constants refined is different from the number of frequencies. It is then more convenient to solve equation (4.7.10) by least squares, introducing a diagonal weighting matrix $\mathbf{P}$ and forming the matrix equation

$$(\tilde{\mathbf{J}}\ \mathbf{P}\ \mathbf{J})\,\Delta\mathbf{F} = (\tilde{\mathbf{J}}\ \mathbf{P})\,\Delta\boldsymbol{\lambda} \tag{4.7.12}$$

Since the normal coordinates used were approximate, the corrections obtained from equation (4.7.12) are insufficient to reduce the errors at once. The perturbation must be repeated several times using new eigenvectors each time to set up a new Jacobian matrix until the error is below a given limit.

The elements of the weighting matrix $\mathbf{P}$ are normally taken as equal to $1/\lambda_{\text{obs}}$ although different choices are possible. In particular if a given frequency of vibration λ_k is unknown it is convenient to take the corresponding P_k element equal to zero.

Equations (4.7.8) can be used also for the calculation of the first-order correction $\Delta\mathbf{L}$ to the eigenvectors of the secular equation. By writing the off-diagonal terms of (4.7.8) in expanded form we have

$$\lambda_i^0 A_{ij} + \sum_{k,l} (L_0)_{ki}(L_0)_{lj}\,\Delta F_{kl} = \lambda_j^0 A_{ij} \tag{4.7.13}$$

and thus we obtain

$$A_{ij} = \frac{\sum_{k,l} (L_0)_{ki}(L_0)_{lj}\,\Delta F_{kl}}{(\lambda_j^0 - \lambda_i^0)} \qquad i \neq j \tag{4.7.14}$$

Equation (4.7.14) leaves the diagonal elements A_{ii} undetermined. These can be found by the condition that the matrix $\mathbf{A}$ is normalized, the result being

$$A_{ii} = 1 \tag{4.7.15}$$

For future reference it is convenient to specify, according to equation (4.7.4), the elements of the matrix **B**

$$\mathbf{B} = \mathbf{A} - \mathbf{E} = \mathbf{L}_0^{-1}\,\Delta\mathbf{L} \tag{4.7.16}$$

From equations (4.7.14) and (4.7.15) we obtain directly

$$B_{ij} = A_{ij} \tag{4.7.17}$$

$$B_{ii} = 0 \tag{4.7.18}$$

4.8 Coriolis coupling constants

In Section 2.7 we have pointed out that the Coriolis coupling coefficients, obtained from the analysis of vibrational bands, can be used for the refinement of force constants together with the vibrational frequencies. In this section we shall discuss this problem in more detail and present the method[16] for the inclusion of Coriolis coefficients in the refinement procedure.

We recall from Section 2.7 that the Coriolis energy is given by

$$T_{\text{Cor}} = \Omega_x\omega_x + \Omega_y\omega_y + \Omega_z\omega_z \tag{4.8.1}$$

where ω_σ ($\sigma = x, y, z$) is a component of the angular velocity of the rotating system and Ω_σ are coefficients that can be expressed in a normal coordinate basis in the matrix form

$$\Omega_\sigma = \tilde{\mathbf{Q}}_v \boldsymbol{\zeta}_v^\sigma \dot{\mathbf{Q}}_v \tag{4.8.2}$$

where $\mathbf{Q}_v$ represents the vector of the $3N - 6$ internal normal coordinates. If instead of a normal coordinate we use a mass-weighted Cartesian displacement basis, the coefficients Ω_σ are given by the relationship

$$\Omega_\sigma = \tilde{\mathbf{q}}\,\mathscr{M}^\sigma \dot{\mathbf{q}} \tag{4.8.3}$$

where the matrices $\mathscr{M}^\sigma$ are matrices of dimension $3N \times 3N$ formed by N identical 3×3 blocks along the diagonal. These blocks, that we shall indicate with the symbol $(\mathscr{M}^\sigma)_\alpha$, to specify that each block refers to a given atom α, have the form

$$(\mathscr{M}^x)_\alpha = \begin{pmatrix} 0 & 0 & 0 \\ 0 & 0 & 1 \\ 0 & -1 & 0 \end{pmatrix} \quad (\mathscr{M}^y)_\alpha = \begin{pmatrix} 0 & 0 & -1 \\ 0 & 0 & 0 \\ 1 & 0 & 0 \end{pmatrix} \quad (\mathscr{M}^z)_\alpha = \begin{pmatrix} 0 & 1 & 0 \\ -1 & 0 & 0 \\ 0 & 0 & 0 \end{pmatrix} \tag{4.8.4}$$

Since the normal and the mass-weighted Cartesian displacement coordinates are related by equation (2.4.4),

$$\mathbf{Q}_v = \tilde{\mathscr{L}}_v \mathbf{q} \tag{4.8.5}$$

we obtain from (4.8.2) and (4.8.3) that

$$\boldsymbol{\zeta}_v^\sigma = \tilde{\mathscr{L}}_v \mathscr{M}^\sigma \mathscr{L}_v \tag{4.8.6}$$

From this it follows that the element† ζ_{ij}^σ is given by

$$\zeta_{ij}^\sigma = \sum_\alpha (\mathscr{L}_{\alpha i}^\tau \mathscr{L}_{\alpha j}^\rho - \mathscr{L}_{\alpha i}^\rho \mathscr{L}_{\alpha j}^\tau) \qquad \sigma, \tau, \rho = x, y, z \tag{4.8.7}$$

where the sum runs over all atoms in the molecule. Equation (4.8.7) means that the Coriolis

† We shall omit the subscript v from now on unless otherwise specified.

coupling constant between two normal coordinates is given by the sum over all atoms of the vector products of atomic displacements in the two vibrations.

Since the vibrational problem is normally solved in internal coordinates, it is convenient to reformulate the above expressions in this basis. We recall that equation (4.2.6) furnishes the relationship between the $\mathscr{L}$ and $\mathbf{L}$ matrices. Using (4.8.5) and setting $\mathbf{D} = \mathbf{BM}^{-\frac{1}{2}}$, we obtain

$$\mathbf{s} = \mathbf{L}\ \mathbf{Q} = \mathbf{L}\tilde{\mathscr{L}}\mathbf{q} = \mathbf{BM}^{-\frac{1}{2}}\mathbf{q} = \mathbf{Dq}$$

$$\mathbf{D} = \mathbf{L}\tilde{\mathscr{L}}$$

$$\tilde{\mathscr{L}} = \mathbf{L}^{-1}\mathbf{D} \tag{4.8.8}$$

Use of (4.8.8) in connection with (4.8.6) yields

$$\boldsymbol{\zeta}^{\sigma} = \mathbf{L}^{-1}\ \mathbf{D}\mathscr{M}^{\sigma}\ \tilde{\mathbf{D}}\ \tilde{\mathbf{L}}^{-1} = \mathbf{L}^{-1}\ \mathbf{C}^{\sigma}\ \tilde{\mathbf{L}}^{-1} \tag{4.8.9}$$

where

$$\mathbf{C}^{\sigma} = \mathbf{D}\mathscr{M}^{\sigma}\tilde{\mathbf{D}} \tag{4.8.10}$$

Since the matrix $\mathbf{D}$ depends only on the masses and geometry of the molecule, the matrix $\mathbf{C}^{\sigma}$ is easily obtained. From equation (4.8.9) we see that an element ζ^{σ}_{ij} is given by

$$\zeta^{\sigma}_{ij} = \sum_{u}\sum_{w} L^{-1}_{ui}\ L^{-1}_{wj}\ C^{\sigma}_{uw} \tag{4.8.11}$$

where the indices u and w label the internal coordinates. Meal and Polo have derived some useful relationships involving the $\boldsymbol{\zeta}^{\sigma}$ matrices. By combining equation (4.8.9) with equations (4.2.9) and (4.2.10), we obtain

$$\tilde{\mathbf{L}}\ \mathbf{F}\ \mathbf{C}^{\sigma}\ \tilde{\mathbf{L}}^{-1} = \boldsymbol{\Lambda}\boldsymbol{\zeta}^{\sigma} \tag{4.8.12}$$

$$\tilde{\mathbf{L}}\ \mathbf{G}^{-1}\ \mathbf{C}^{\sigma}\ \tilde{\mathbf{L}}^{-1} = \boldsymbol{\zeta}^{\sigma} \tag{4.8.13}$$

Equation (4.8.13) is very important since it shows that the matrix $\boldsymbol{\zeta}^{\sigma}$ and the matrix $(\mathbf{G}^{-1}\ \mathbf{C}^{\sigma})$ are related by a similarity transformation and therefore have the same eigenvalues. Because both $\mathbf{G}^{-1}$ and $\mathbf{C}^{\sigma}$ depend only on the masses and geometry of the molecule, the eigenvalues of $(\mathbf{G}^{-1}\ \mathbf{C}^{\sigma})$ are independent of the force field and the same is then true for $\boldsymbol{\zeta}^{\sigma}$. In order to use the Coriolis constants for the refinement of the force field, it is necessary to compute, from equation (4.8.9), the dependence of the ζ^{σ}_{ij} elements upon small variations in the $\mathbf{L}$-matrix elements which are themselves produced by small changes in the force constants. This problem has been discussed by several authors and requires a straightforward extension of the perturbation treatment given in the previous section.

Let $\Delta\mathbf{F}$ be a small change in the $\mathbf{F}$-matrix. This produces a corresponding change in the $\boldsymbol{\Lambda}$- and $\mathbf{L}$-matrices given in equations (4.7.3) and (4.7.4) and a change in the $\boldsymbol{\zeta}^{\sigma}$ matrices given by

$$\boldsymbol{\zeta}^{\sigma} = \boldsymbol{\zeta}^{\sigma}_{0} + \Delta\boldsymbol{\zeta}^{\sigma} \tag{4.8.14}$$

From equations (4.8.9) and (4.7.4) we have

$$\boldsymbol{\zeta}^{\sigma} = \tilde{\mathbf{A}}\ \mathbf{L}_{0}^{-1}\ \mathbf{C}^{\sigma}\ \tilde{\mathbf{L}}_{0}^{-1}\ \mathbf{A} = (\mathbf{E} + \tilde{\mathbf{B}})\ \mathbf{L}_{0}^{-1}\mathbf{C}^{\sigma}\ \tilde{\mathbf{L}}_{0}^{-1}(\mathbf{E} + \mathbf{B}) \tag{4.8.15}$$

and neglecting, as in Section 4.7, terms higher than those of first order

$$\boldsymbol{\zeta}^{\sigma} = \mathbf{L}_{0}^{-1}\ \mathbf{C}^{\sigma}\ \tilde{\mathbf{L}}_{0}^{-1} + \tilde{\mathbf{B}}\ \mathbf{L}_{0}^{-1}\ \mathbf{C}^{\sigma}\ \tilde{\mathbf{L}}_{0}^{-1} + \mathbf{L}_{0}^{-1}\ \mathbf{C}^{\sigma}\ \tilde{\mathbf{L}}_{0}^{-1}\ \mathbf{B} \tag{4.8.16}$$

or

$$\Delta\zeta^{\sigma} = \tilde{\mathbf{B}}\zeta_0^{\sigma} + \zeta_0^{\sigma}\mathbf{B} \tag{4.8.17}$$

Using equation (4.7.17) for the elements of **B** we obtain

$$\Delta\zeta_{ij}^{\sigma} = \sum_{t \neq i} (\zeta_0^{\sigma})_{jt}(\lambda_t - \lambda_i)^{-1} \sum_{k,l} (L_0)_{ki}(L_0)_{lt}\,\Delta F_{kl} + \sum_{t \neq j} (\zeta_0^{\sigma})_{it}(\lambda_j - \lambda_t)^{-1} \sum_{k,l} (L_0)_{kj}(L_0)_{lt}\,\Delta F_{kl} \tag{4.8.18}$$

Finally, letting $\Delta\zeta^{\sigma} \to \mathbf{0}$ and $\Delta\mathbf{F} \to \mathbf{0}$, we obtain the elements of the Jacobian of ζ^{σ} with respect to **F** as the coefficients of ΔF_{kl} in (4.8.18)

$$\begin{aligned} \frac{\partial \zeta_{ij}^{\sigma}}{\partial F_{kl}} &= \sum_{t \neq i} (\zeta_0^{\sigma})_{jt} \frac{(L_{ki}L_{lt} + L_{li}L_{kt})}{(\lambda_t - \lambda_i)} + \sum_{t \neq j} (\zeta^{\sigma})_{it} \frac{(L_{kj}L_{lt} + L_{lj}L_{kt})}{(\lambda_j - \lambda_t)} \\ \frac{\partial \zeta_{ij}^{\sigma}}{\partial F_{kk}} &= \sum_{t \neq i} (\zeta_0^{\sigma})_{jt} \frac{L_{ki}L_{kt}}{(\lambda_t - \lambda_i)} + \sum_{t \neq j} (\zeta_0^{\sigma})_{it} \frac{L_{kj}L_{kt}}{(\lambda_j - \lambda_t)} \end{aligned} \tag{4.8.19}$$

Equations (4.8.19) can then be used, together with the corresponding equations (4.7.11), for the refinement of the force constants.

4.9 Centrifugal distortion constants

Other useful information on the force field of polyatomic molecules can be obtained from the analysis of rotational transitions. Since molecules are actually non-rigid, they are increasingly distorted in higher rotational levels due to centrifugal forces. This fact produces slight changes of rotational energy that can be related to the molecular force field in terms of the 'centrifugal stretching constants'. The theory of centrifugal distortion constants is discussed in detail in the book *Molecular Vib-rotors* by H. C. Allen and P. C. Cross[27] and the reader is referred to this book and to the references contained therein. Most of the recent developments, in particular those due to Watson, can be found in the list of references given in the paper by W. H. Kirchhoff.[22] A matrix formulation of the theory of centrifugal distortion has been discussed by P. Pulay and W. Savodny[23] and by S. J. Cyvin.[24] Since this book is concerned only with vibrational spectra, we shall not describe the theory of centrifugal distortion but only discuss the relationship of centrifugal stretching constants with the molecular force field and their use in the refinement of force constants. We shall follow the treatment given by Kivelson and Wilson in their classic paper[25] on the subject. These authors express the centrifugal stretching constants as a linear combination of a different set of constants τ defined by the relationship

$$\tau_{\alpha\beta\gamma\delta} = -\tfrac{1}{2} \sum_{k,l} \mu_{\alpha\beta}^{k}(F^{-1})_{kl}\mu_{\gamma\delta}^{l} \qquad \alpha, \beta, \gamma, \delta = x, y, z \tag{4.9.1}$$

where

$$\mu_{\alpha\beta}^{k} = \frac{\partial \mu_{\alpha\beta}}{\partial s_k} \tag{4.9.2}$$

In these equations $\mu_{\alpha\beta}$ is an element of the inverse moment of inertia tensor and s_k is one of the internal or symmetry coordinates of the molecule. It is often convenient to express the τ-constants in terms of the inertia tensor **I**. According to Kivelson and Wilson

$$\mu_{\alpha\beta}^{k} = -\frac{J_{\alpha\beta}^{k}}{I_{\alpha\alpha}^{0} I_{\beta\beta}^{0}} \tag{4.9.3}$$

where

$$J_{\alpha\beta}^{k} = \left(\frac{\partial I_{\alpha\beta}}{\partial s_k}\right) \tag{4.9.4}$$

and therefore

$$\tau_{\alpha\beta\gamma\delta} = -\frac{1}{2I^0_{\alpha\alpha}I^0_{\beta\beta}I^0_{\gamma\gamma}I^0_{\delta\delta}}\sum_{k,l} J^k_{\alpha\beta}(F^{-1})_{kl}J^l_{\gamma\delta} \tag{4.9.5}$$

The elements of the inertia tensor $I_{\alpha\alpha}$, $I_{\alpha\beta}$, etc., are given in Cartesian coordinates by the well known relationships

$$I_{\alpha\alpha} = \sum_i m_i(\beta^2 + \gamma^2) \tag{4.9.6}$$

$$I_{\alpha\beta} = -\sum_i m_i\alpha_i\beta_i \tag{4.9.7}$$

where α, β and γ designate Cartesian coordinates x, y, z taken in cyclic order.

It is important to notice that the constants τ of equations (4.9.1) or (4.9.5) are related to the inverse force constant matrix $\mathbf{F}^{-1}$ rather than to $\mathbf{F}$. We need therefore, in order to be able to utilize the τ-constants for the refinement of the force field, to obtain an expression which relates a variation $\Delta\mathbf{F}$ in the $\mathbf{F}$-matrix with the corresponding variation in the $\mathbf{F}^{-1}$ matrix. Again following Mills' treatment,[16] we write a small change $\Delta\mathbf{F}^{-1}$, produced by a small change $\Delta\mathbf{F}$ in $\mathbf{F}$, in the form

$$\mathbf{F}^{-1} = \mathbf{F}_0^{-1} + \Delta\mathbf{F}^{-1} \tag{4.9.8}$$

Since $\mathbf{F}^{-1}\mathbf{F} = \mathbf{E}$, we have from (4.9.8) and (4.7.2), neglecting terms of order higher than the first,

$$(\mathbf{F}_0 + \Delta\mathbf{F})(\mathbf{F}_0^{-1} + \Delta\mathbf{F}^{-1}) = \mathbf{E}$$

$$\mathbf{F}_0\,\Delta\mathbf{F}^{-1} + \Delta\mathbf{F}\mathbf{F}_0^{-1} = 0$$

and thus

$$\Delta\mathbf{F}^{-1} = -\mathbf{F}_0^{-1}\,\Delta\mathbf{F}\mathbf{F}_0^{-1} \tag{4.9.9}$$

The element F_{kl}^{-1} is then, according to (4.9.8) and (4.9.9),

$$F_{kl}^{-1} = (F_0^{-1})_{kl} - \sum_{rt}(F_0^{-1})_{kr}(F_0^{-1})_{tl}\,\Delta F_{rt} \tag{4.9.10}$$

and thus, by substitution into (4.9.5), we obtain

$$\Delta\tau_{\alpha\beta\gamma\delta} = \frac{1}{2I^0_{\alpha\alpha}I^0_{\beta\beta}I^0_{\gamma\gamma}I^0_{\delta\delta}}\sum_{kl}\sum_{rt} J^k_{\alpha\beta}J^l_{\gamma\delta}(F_0^{-1})_{kr}(F_0^{-1})_{tl}\,\Delta F_{rt}$$

Finally, letting $\Delta\tau \to 0$ and $\Delta\mathbf{F} \to 0$, the coefficients of $(\Delta F)_{rt}$ give the Jacobian elements of τ with respect to $\mathbf{F}$

$$\begin{aligned}\left(\frac{\partial\tau_{\alpha\beta\gamma\delta}}{\partial F_{rt}}\right) &= \frac{1}{2I^0_{\alpha\alpha}I^0_{\beta\beta}I^0_{\gamma\gamma}I^0_{\delta\delta}}\sum_{kl} J^k_{\alpha\beta}J^l_{\gamma\delta}[(F_0^{-1})_{kr}(F_0^{-1})_{tl} + (F_0^{-1})_{kt}(F_0^{-1})_{rl}]\\ \left(\frac{\partial\tau_{\alpha\beta\gamma\delta}}{\partial F_{rr}}\right) &= \frac{1}{2I^0_{\alpha\alpha}I^0_{\beta\beta}I^0_{\gamma\gamma}I^0_{\delta\delta}}\sum_{kl} J^k_{\alpha\beta}J^l_{\gamma\delta}(F_0^{-1})_{kr}(F_0^{-1})_{rl}\end{aligned} \tag{4.9.11}$$

Equations (4.9.11) can be then used together with (4.8.19) and (4.7.11) for the refinement of the force constants.

References

1. M. Eliashevich, *Compt. Rend. Acad. Sci. U.R.S.S.*, **28**, 605 (1940).
2. E. B. Wilson jr, *J. Chem. Phys.*, **9**, 76 (1941).

3. R. J. Malhiot and S. M. Ferigle, *J. Chem. Phys.*, **22**, 717 (1954).
4. J. C. Decius, *J. Chem. Phys.*, **16**, 1025 (1948).
5. S. M. Ferigle and A. G. Meister, *J. Chem. Phys.*, **19**, 982 (1951).
6. J. B. Lohman, *Office of Naval Research—Tech. Report*, **17** (1951).
7. R. J. Malhiot and S. M. Ferigle, *J. Chem. Phys.*, **23**, 30 (1955).
8. E. B. Wilson jr, J. C. Decius and P. C. Cross, *Molecular Vibrations*, McGraw-Hill, 1955.
9. T. Miyazawa, *J. Chem. Phys.*, **29**, 246 (1958).
10. T. Miyazawa, *J. Chem. Soc. Japan*, **76**, 1132 (1955).
11. W. T. King, *Dissertation*, University of Minnesota (1956).
12. D. E. Mann, T. Shimanouchi, J. H. Meal and L. Fano, *J. Chem. Phys.*, **27**, 43 (1957).
13. A. Curtis, *Dissertation*, University of Minnesota (1959).
14. J. Overend and J. R. Sherer, *J. Chem. Phys.*, **32**, 1289 (1960).
15. I. M. Mills, *Spectrochim. Acta*, **16**, 35 (1960).
16. I. M. Mills, *J. Mol. Spectrosc.*, **5**, 334 (1960); erratum, **17**, 164 (1965).
17. R. C. Lord and I. Nakagawa, *J. Chem. Phys.*, **39**, 2951 (1963).
18. I. Nakagawa and T. Shimanouchi, *J. Chem. Soc. Japan*, **80**, 128 (1959).
19. N. Neto, *Gazz. Chim. It.*, **96**, 1094 (1966).
20. R. G. Sneyder and J. H. Schachtschneider, *Spectrochim. Acta*, **19**, 117 (1963).
21. J. N. Gayles, W. T. King and J. H. Schachtschneider, *Spectrochim. Acta*, **23A**, 703 (1967).
22. W. H. Kirchhoff, *J. Mol. Spectrosc.*, **41**, 333 (1972).
23. P. Pulay and W. Savodny, *J. Mol. Spectrosc.*, **26**, 150 (1968).
24. S. J. Cyvin, B. N. Cyvin and G. Hagen, *Z. Naturforschg.*, **A23**, 1649 (1968).
25. D. Kivelson and E. B. Wilson jr, *J. Chem. Phys.*, **20**, 1575 (1952); **21**, 1229 (1953).
26. A. R. Hoy, I. M. Mills and G. Strey, *Mol. Phys.*, **24**, 1265 (1972).
27. H. C. Allen and P. C. Cross, *Molecular Vib-rotors*, Wiley, 1963.
28. J. H. Wilkinson, *The Algebraic Eigenvalue Problem*, Oxford University Press, 1965, p. 266.

Chapter 5
Group Theory

5.1 Symmmetry considerations

The fact that many molecules possess some symmetry is of enormous help in the interpretation of their vibrational spectrum. The occurrence of molecular symmetry enables the spectroscopist to make full use of group theory, a formal mathematical tool by means of which symmetry can be treated systematically and can be used to derive most of the information on the vibrational behaviour of the molecules.

In this chapter we shall present the basic principles of group theory necessary to understand the following chapters. We wish, however, to emphasize that such a concise presentation is by no means intended to furnish a real understanding of the theory but only a survey of the fundamental rules and theorems as well as an introduction to the language of group theory. Many excellent books, at different levels of mathematical complexity, are available and are strongly recommended for a complete handling of the theory.[1–5]

Before entering the subject it is convenient to discuss briefly the problem of molecular symmetry. By *symmetry* of a molecule we mean the symmetry of the nuclear arrangement in the equilibrium configuration. The symmetry of the molecule is determined by the collection of all *symmetry operations* which send the equilibrium configuration into an equivalent indistinguishable configuration, simply exchanging equivalent atoms. A *symmetry element* is thus defined through the symmetry operations associated with it.

The possible elements of symmetry of a physical object such as a molecule are:

(1) The identity E.
(2) The plane of symmetry σ.
(3) The centre of inversion i.
(4) The p-fold axis of rotation C_p.
(5) The p-fold rotation–reflection axis S_p.

The *identity* E is the simplest symmetry element and the operation associated with it is merely to leave the molecule unchanged. There are mathematical reasons for the introduction of such a trivial operation, which are explained in the next section. A *plane of symmetry* σ is represented in Figure 5.1.1. The operation associated with it is the reflection through the plane. In the case of Figure 5.1.1 the left-hand side of the molecule is exchanged with the right-hand side giving rise to a configuration indistinguishable from the previous one. If x, y, z are the coordinates of one atom referred to a Cartesian system XYZ with the origin on the plane and the axes Y and Z lying on it, the coordinates of a symmetrically equivalent atom are $-x$, y, z.

A *centre of symmetry* i is shown in Figure 5.1.2. If a line is drawn from a given atom through the centre of symmetry and continued, it will meet an equivalent atom at the same distance as the first atom from the centre. If x, y, z are the coordinates of one atom referred to a coordinate system with the origin at the centre of inversion, the coordinates of a symmetrically equivalent atom are $-x$, $-y$, $-z$.

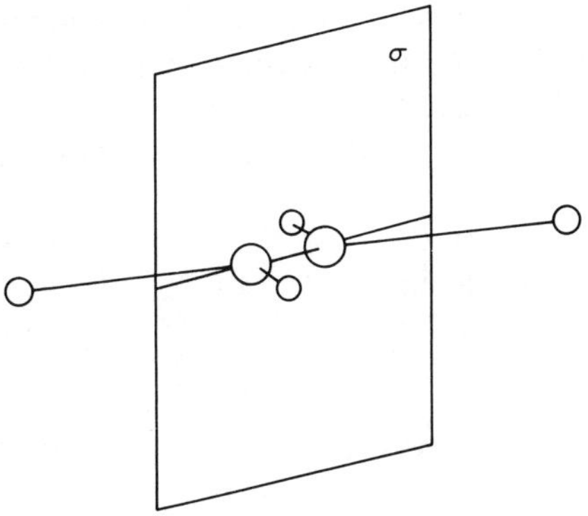

Figure 5.1.1. Example of a symmetry plane for the molecule $CH_2{=}CCl_2$

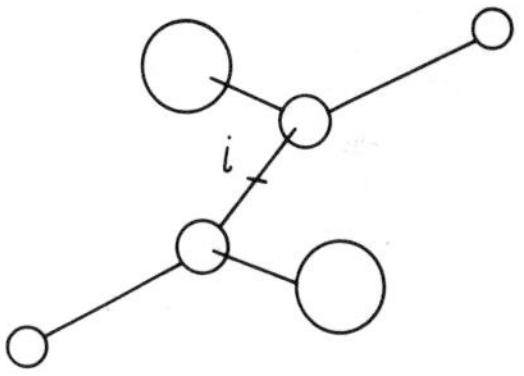

Figure 5.1.2 Example of a centre of symmetry for the molecule CHCl=CHCl

A p-fold *axis of rotation* C_p is shown in Figure 5.1.3 for the case $p = 3$. The symmetry operation associated with a p-fold axis is the rotation around the axis of $360°/p$. By convention we shall use the counterclockwise direction as the positive direction of rotation. When $p = 3$ the symmetry operation consists of a rotation of 120° around the axis. If a line is drawn for a given atom perpendicularly to the axis, there must be p such lines all crossing at the same point at angles of $360°/p$.

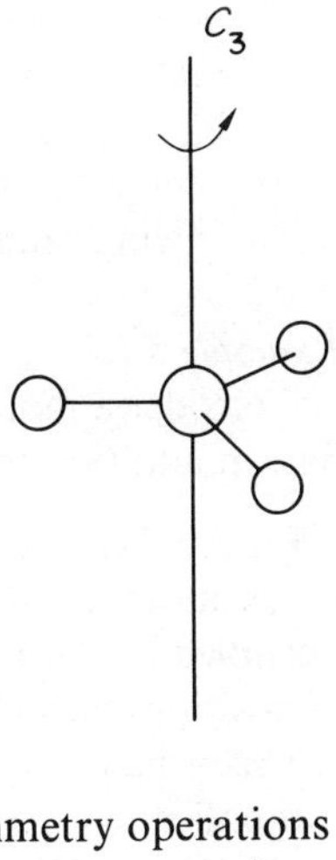

Figure 5.1.3 Example of a 3-fold axis of rotation for the molecule CO_3^{--}

If $p = 1$ the operation becomes the identity. For $p > 1$ there are p symmetry operations associated with a p-fold axis of rotation. For example, if $p = 3$ the symmetry operations are the rotations of 120°, 240° and 360°. Obviously if a p-fold rotation is performed p times in the same direction the molecule is brought into the original situation. A p-fold rotation performed p times is therefore equivalent to the identity.

We shall use the symbol C_p^m to indicate that a p-fold rotation is performed m times in the counterclockwise direction. We shall also use the symbol C_p^{-1} to indicate a p-fold rotation in the clockwise direction. Obviously C_p^{-1} is equivalent to C_p^{p-1} and C_p^p is equivalent to E.

In principle all values of p are permitted for a molecule, although values of p greater than 6 are of no interest in practical systems. The only exception is the ∞ axis of rotation (C_∞) encountered in linear molecules. In this case every infinitesimal rotation around the axis brings the molecules into self-coincidence.

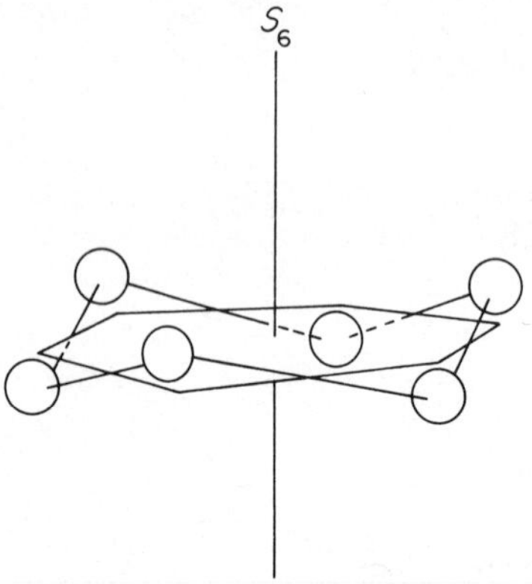

Figure 5.1.4 Example of a six-fold axis of rotation–reflection S_6

A p-fold axis of rotation–reflection S_p is shown in Figure 5.1.4, for the case $p = 6$. The symmetry operation associated with a p-fold rotation–reflection axis is a counterclockwise rotation of $360°/p$ followed by a reflection in a plane perpendicular to the axis. We shall use the symbol S_p^m to indicate that a p-fold rotation–reflection is performed m times. We shall also use the symbol S_p^{-1} to indicate a p-fold rotation–reflection in the clockwise direction. If p is even, S_p^{-1} is equivalent to S_p^{p-1} and S_p^p is equivalent to E, whereas if p is odd S_p^{2p} is equivalent to E. It is clear that S_2 is equivalent to the operation i.

For a molecule the occurrence of a p-fold rotation axis C_p and a symmetry plane perpendicular to it implies the occurrence of an axis S_p. The inverse is true only if p is odd. If p is even an axis S_p implies the occurrence of an axis $C_{p/2}$.

In a molecule several elements of symmetry can occur together. This is illustrated in the following example.

Example 5.1.1

Consider a molecule of ethane in the configuration shown in Figure 5.1.5. The molecule has, in addition to the trivial element E, the following elements of symmetry:

(1) An axis C_3 in the direction of the C–C bond. A rotation of 120° around this axis sends atom 1 into the position occupied before by atom 2, atom 2 into the position occupied by 3 and atom 3 into the position occupied by 1. A corresponding permutation of positions occurs for atoms 4, 5 and 6. Atoms 7 and 8 are left unchanged. If this operation is performed twice (C_3^2), it is equivalent to a clockwise rotation of 120°. We shall write symbolically $C_3^2 = C_3^{-1}$ to indicate this equivalence. The same operation performed three times is equivalent to the identity E. Symbolically $C_3^3 = E$.

(2) An axis S_6 coincident with C_3. A rotation of 60° followed by a reflection in a plane perpendicular to the C–C bond and going through the centre i of the bond sends $1 \to 6$, $2 \to 4$, $3 \to 5$, $4 \to 3$, $5 \to 1$, $6 \to 2$, $7 \to 8$ and $8 \to 7$. This operation performed twice (S_6^2) is

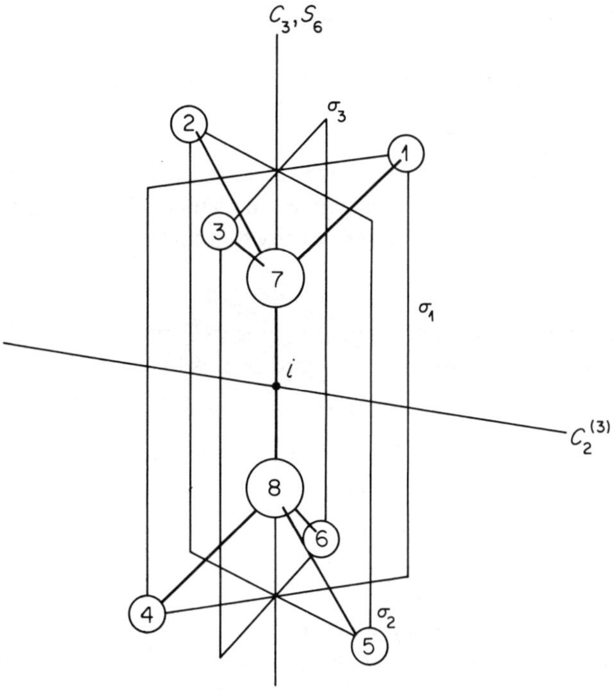

Figure 5.1.5 Ethane in the staggered configuration

equivalent to C_3. The same operation, if performed three times, is equivalent to the operation of inversion. We shall indicate this symbolically by writing $S_6^3 = i$. In the same way we find $S_6^4 = C_3^{-1}$.

(3) A centre of inversion i at the centre of the C–C bond. The operation of inversion at the centre sends $1 \to 4$, $2 \to 5$, $3 \to 6$, $4 \to 1$, $5 \to 2$, $6 \to 3$, $7 \to 8$ and $8 \to 7$.

(4) Three planes of symmetry indicated in the figure by σ_1, σ_2 and σ_3. Each plane contains the atoms 7 and 8 and two hydrogen atoms in trans-position. The reflection at the plane σ_1, for instance, leaves the atoms 1, 4, 7 and 8 unchanged and sends $2 \to 3$, $3 \to 2$, $5 \to 6$ and $6 \to 5$. In the same way the reflections at σ_2 and σ_3 leave atoms 2, 5, 7, 8 and 3, 6, 7, 8 unchanged respectively while $1 \leftrightarrow 3$, $4 \leftrightarrow 6$ and $1 \leftrightarrow 2$, $5 \leftrightarrow 4$ respectively.

(5) Three binary axes, perpendicular in turn to the three symmetry planes and going through the centre i. For simplicity only the one perpendicular to σ_3 and indicated with the symbol $C_2^{(3)}$ is shown in the figure. A rotation of 180° around $C_2^{(3)}$ sends $1 \to 5$, $2 \to 4$, $3 \to 6$, $5 \to 1$, $4 \to 2$, $6 \to 3$, $7 \to 8$, $8 \to 7$. The rotation around the axis $C_2^{(2)}$ (perpendicular to σ_2) sends $1 \to 6$, $2 \to 5$, $3 \to 4$, $4 \to 3$, $5 \to 2$, $6 \to 1$, $7 \to 8$, $8 \to 7$. The rotation around $C_2^{(1)}$ (perpendicular to σ_1) sends $1 \to 4$, $2 \to 6$, $3 \to 5$, $4 \to 1$, $5 \to 3$, $6 \to 2$, $7 \to 8$, $8 \to 7$. The classification of the possible molecular symmetries in terms of symmetry elements or of symmetry operations is discussed in the next chapter.

5.2 Definitions and group properties

A *group* G is a collection of distinct *elements* $\{A, B, \ldots, R, \ldots\}$ together with a binary combination law (multiplication) such that:

(a) Multiplication of two elements R, S of the group produces a single-valued product RS which is equivalent to some other element T of the collection (closure property).

(b) Multiplication is associative, i.e. $A(BC) = (AB)C$.
(c) There exists an *identity* element E in the group such that $RE = ER = R$.
(d) Every element R has an inverse R^{-1}, relative to the identity E, such that $RR^{-1} = R^{-1}R = E$.

In the definition of groups given above we have used the term *multiplication* in the most general sense of a combination law which gives an element T as a product of two elements R and S of the group. A few examples will clarify this point.

Example 5.2.1
The collection of all positive and negative integers forms a group if simple arithmetic addition is chosen as the multiplication law and zero is taken as the identity, since

$$3 + 4 = 7 \qquad \text{condition (a)}$$
$$(5 + 2) + 1 = 5 + (2 + 1) \qquad \text{condition (b)}$$
$$6 + 0 = 6 \qquad \text{condition (c)}$$
$$3 + (-3) = 0 \qquad \text{condition (d)}$$

Example 5.2.2
The collection of all positive rational numbers forms a group if simple arithmetic multiplication is chosen as the multiplication law and unity is taken as the identity, since

$$3 \times 4 = 12$$
$$(5 \times 2) \times 3 = 5 \times (2 \times 3)$$
$$4 \times 1 = 1 \times 4 = 4$$
$$5 \times (\tfrac{1}{5}) = 1$$

Example 5.2.3
Let us consider the symmetry operations which bring the equilateral pyramid of Figure 5.2.1 into a position indistinguishable from the original one. These are: (a) the identity operation E which consists in leaving the original position unchanged; (b) the anticlockwise rotation C_3 of 120° around an axis going through the apex and perpendicular to the

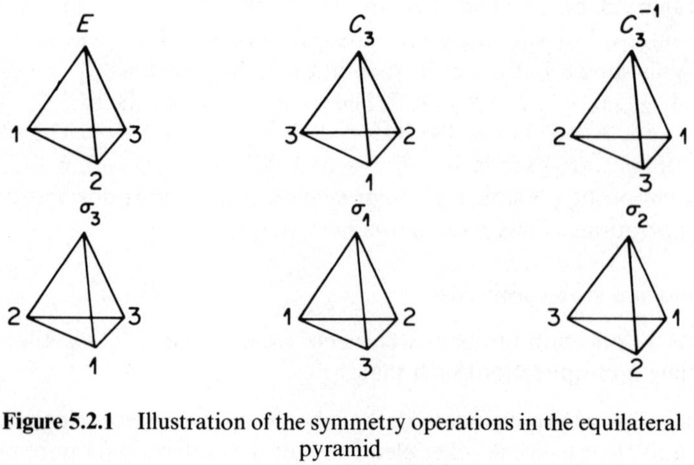

Figure 5.2.1 Illustration of the symmetry operations in the equilateral pyramid

base; (c) the clockwise rotation C_3^{-1} of 120° around the same axis; (d, e, f) the three reflections σ_1, σ_2 and σ_3 in one of the three planes going through one of the sides of the pyramid and the opposite face.

We can now define as multiplication of two symmetry operations defined above, their sequential performance. For instance, the product $\sigma_1 C_3$ means for us operation C_3 followed by operation σ_1. Since the sequential application of two symmetry operations must bring the pyramid into self-coincidence, it must correspond to one of the symmetry operations defined above. We shall indicate this correspondence with the symbol =. The reader will easily verify that the symmetry operations of the equilateral pyramid form a group (hereafter called C_{3v}), since

$RS = T$	for example, $\sigma_1 C_3 = \sigma_2$, $C_3\sigma_1 = \sigma_3$, $\sigma_1\sigma_3 = C_3^{-1}$, $\sigma_1\sigma_2 = C_3$, etc.
$R(TS) = (RT)S$	for example, $\sigma_2(C_3\sigma_1) = \sigma_2\sigma_3 = C_3$, $(\sigma_2 C_3)\sigma_1 = \sigma_3\sigma_1 = C_3$, etc.
$RE = R$	for example, $\sigma_1 E = \sigma_1$, etc.
$RR^{-1} = E$	for example, $C_3 C_3^{-1} = E$, $\sigma_1\sigma_1 = E$, $\sigma_2\sigma_2 = E$, $\sigma_3\sigma_3 = E$, etc.

Example 5.2.4

The numbers 1, −1, i and −i form a group if arithmetic multiplication is chosen as the combination law and if the number 1 is taken as the identity.

Example 5.2.5

Let us consider the symmetry operations which brings the object of Figure 5.2.2 into self-coincidence. These are: the identity operation E which leaves the object unchanged; the anticlockwise rotation C_4 of 90° around the quaternary axis shown in the figure; the

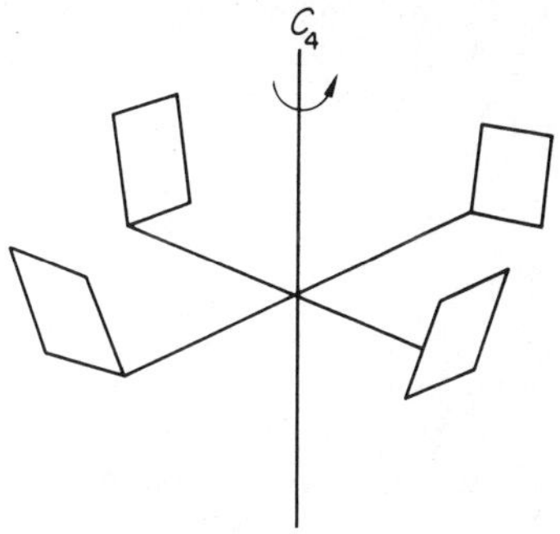

Figure 5.2.2 Example of C_4 symmetry

rotation C_2 of 180° around the same axis; the clockwise rotation C_4^{-1} of 90° around the same axis.

Since the rotation C_2 is equivalent to a rotation C_4 followed by a second rotation C_4 we can write

$$C_4 \times C_4 = C_4^2 = C_2$$

In the same way we obtain

$$C_4 \times C_4 \times C_4 = C_4^3 = C_4^{-1} \qquad C_4 \times C_4 \times C_4 \times C_4 = C_4^4 = E$$

where C_4^2 means that the operation C_4 is performed twice, C_4^3 that it is performed three times. These four symmetry operations form the group $\{E, C_2, C_4, C_4^{-1}\}$.

It is important to observe that the multiplication of two elements of a group is not necessarily commutative. For instance, we have seen in Example 5.2.3 that the product

$\sigma_1 C_3$ is different from the product $C_3\sigma_1$. In many groups, however, the product of two elements is commutative. These groups are called *Abelian.* The groups of Examples 5.2.1, 5.2.2, 5.2.4 and 5.2.5 are Abelian groups. The number g of elements contained in a group G is called the *order* of the group. The groups of the examples given above are of order ∞ (Examples 5.2.1 and 5.2.2), six (Example 5.2.3) and four (Examples 5.2.4 and 5.2.5) respectively. In what follows we shall be mainly concerned with groups of *finite* order. In this case a group is completely defined by specifying the products of all possible ordered pairs of elements. This is conveniently done by collecting systematically in a *multiplication table* the possible products.

Example 5.2.6

The multiplication table for the group of Example 5.2.3 is given below. The entries in the table represent the single element to which each product is equivalent. The product RS is placed at the intersection of the R-row and the S-column.

	E	C_3	C_3^{-1}	σ_1	σ_2	σ_3
E	E	C_3	C_3^{-1}	σ_1	σ_2	σ_3
C_3	C_3	C_3^{-1}	E	σ_3	σ_1	σ_2
C_3^{-1}	C_3^{-1}	E	C_3	σ_2	σ_3	σ_1
σ_1	σ_1	σ_2	σ_3	E	C_3	C_3^{-1}
σ_2	σ_2	σ_3	σ_1	C_3^{-1}	E	C_3
σ_3	σ_3	σ_1	σ_2	C_3	C_3^{-1}	E

Example 5.2.7

Let us consider the group of order four of Example 5.2.4. The multiplication table is

	1	−1	i	−i
1	1	−1	i	−i
−1	−1	1	−i	i
i	i	−i	−1	1
−i	−i	i	1	−1

If we assign to each element of this group a new symbol, *A*, *B*, *C*, *D* the table reads

	A	B	C	D
A	A	B	C	D
B	B	A	D	C
C	C	D	B	A
D	D	C	A	B

which defines an *abstract group* of order four with the same multiplication table as the group $\{1, -1, i, -i\}$.

The multiplication table of an abstract group defines all the properties of the group, irrespective of what meaning we assign to each symbol. Furthermore, it also defines the properties of all the groups that can be *realized* from the abstract group by specifying the meaning of each symbol. The group $\{1, -1, \mathrm{i}, -\mathrm{i}\}$ is thus a *realization* of this abstract group. The group $\{E, C_2, C_4, C_4^{-1}\}$ of Example 5.2.5 is another realization of the same abstract group.

It is often possible to define a group completely by using only a small number of its elements. These elements are then called the *generators* of the group. For instance, in the group $\{1, -1, \mathrm{i}, -\mathrm{i}\}$ of Example 5.2.4 the complete multiplication table can be built up by forming powers of the single element i through the equations

$$\mathrm{i} \times \mathrm{i} = -1 \qquad -1 \times \mathrm{i} = -\mathrm{i} \qquad -\mathrm{i} \times \mathrm{i} = 1$$

The same is true for the group $\{E, C_2, C_4, C_4^{-1}\}$ of Example 5.2.5 with the single generator C_4. When a group contains only the n distinct powers $R, R^2, \ldots, R^n = E$ of an element R, the group is called a *cyclic group*. The groups of Examples 5.2.4 and 5.2.5 are thus cyclic. Since in a cyclic group all products commute, a cyclic group is always Abelian. The elements of the C_{3v} group of Example 5.2.3 can be formed by two generators, C_3 and σ_1, through the equations

$$C_3 \times C_3 = C_3^{-1} \qquad C_3\sigma_1 = \sigma_3 \qquad \sigma_1 C_3 = \sigma_2 \qquad C_3 \times C_3 \times C_3 = E$$

The choice of the generators of a group is not always unique. For instance $-\mathrm{i}$ and C_4^{-1} are other possible choices of generators for the groups of Examples 5.2.4 and 5.2.5 and C_3, σ_2 or C_3, σ_3 are other possible choices for the group C_{3v}. If the elements of two groups, when properly arranged, satisfy the same multiplication table, i.e. they are different realizations of the same abstract group, the two groups are said to be *isomorphic* to each other. In this case there is a one-to-one correspondence between the elements of the two groups and this correspondence is called an *isomorphism*. For instance, the groups $\{1, -1, i, -i\}$ and $\{E, C_2, C_4, C_4^{-1}\}$ are isomorphic via the one-to-one correspondence

$$\begin{aligned} \mathrm{i} &\leftrightarrow C_4 \\ -1 &\leftrightarrow C_2 \\ -\mathrm{i} &\leftrightarrow C_4^{-1} \\ 1 &\leftrightarrow E \end{aligned}$$

since both obey the same abstract multiplication table as shown in Example 5.2.7.

The one-to-one correspondence between the elements of two groups is a special case of a more general correlation between groups, involving a many-to-one correspondence between elements of groups. This generalization of the concept of isomorphism is called *homomorphism*.

Consider, for instance, the group C_{3v} of Example 5.2.3 and associate with E and the rotations the element $\bar{E}$, and with the reflections the element σ, as shown below, with the conditions that $\bar{E}\sigma = \sigma$ and $\sigma\sigma = \bar{E}$

Under these conditions the elements of the C_{3v} group and the associated elements $\bar{E}$ and σ, which obviously also form a group (of order 2), obey the same multiplication table and are said to be homomorphic. In this case the elements E, C_3, C_3^{-1} are said to be *mapped* on to the element $\bar{E}$ and the elements $\sigma_1, \sigma_2, \sigma_3$ on to the element σ of the smaller group.

5.3 Partition of a group

If we consider the multiplication table of the C_{3v} group, we observe that the elements E, C_3, C_3^{-1} form a group by themselves, with a multiplication table given by the first three elements of the first three rows of the C_{3v} multiplication table. This leads to the definition of a subgroup H of a group G as the collection of some elements of G which form a group by themselves when used in conjunction with the law of multiplication of G. In the above example the subgroup is a cyclic group of order 3, with C_3 as generator. The C_{3v} group contains, in addition to $\{E, C_3, \quad C_3^{-1}\}$, three other subgroups $\{E, \sigma_1\}$, $\{E, \sigma_2\}$ and $\{E, \sigma_3\}$.

If G is any group and $H = \{H_1, H_2, \ldots, H_k\}$ any subgroup of G, the collection $HR = [H_1R, H_2R, \ldots, H_kR]$, R being any element of G not contained in H, is called a *right coset* of H with every respect to R. In the same way the collection RH is called a *left coset*. Every coset of H contains the same number of elements as H, but it is not a subgroup since it contains no elements in common with H and therefore cannot include the identity element. This is easily proved by assuming the contrary and showing that this leads to a contradiction.

Let $H_iR = H_j$, H_i and H_j being both members of H. Then $R = H_i^{-1}H_j$ in contradiction to the assumption that R is external to H.

In the same way it is easy to prove that two right (or left) cosets of H are either identical or have no element in common and that no element can appear more than once in a coset. It follows that the g elements of G can be partitioned into distinct sets, each containing h elements, by considering the h elements of any subgroup H and all its different cosets. Thus $g = nh$, where n is an integer. This proves that the order h of a subgroup H must be a divisor of the order g of the group G.

Example 5.3.1

Consider the subgroup $\{E, \sigma_1\}$ in the C_{3v} group. There are only two different left cosets of $\{E, \sigma_1\}$, i.e.

$$C_3\{E, \sigma_1\} = \sigma_3\{E, \sigma_1\} = [C_3, \sigma_3] \quad \text{and} \quad C_3^{-1}\{E, \sigma_1\} = \sigma_2\{E, \sigma_1\} = [C_3^{-1}, \sigma_2]$$

We have then $6 = 3 \times 2$.

A much more useful partitioning of the elements of a group can be obtained through the concept of classes. We define two conjugate elements A and B in a group through the relation

$$B = R\ A\ R^{-1} \tag{5.3.1}$$

and, by collecting all the distinct elements conjugate to A, as R varies in the group, we obtain the class of these elements.

Example 5.3.2

By use of (5.3.1) the C_{3v} group is partitioned into the three classes

$$(E) \qquad (C_3, C_3^{-1}) \qquad (\sigma_1, \sigma_2, \sigma_3)$$

Once we have defined conjugate elements, we can define in the same way conjugate subgroups through the relation

$$RHR^{-1} = \{RH_1R^{-1}, RH_2R^{-1}, \ldots, RH_hR^{-1}\}$$

The collection of conjugates defined in this way is a subgroup and has the same multiplication table as H since, if $H_iH_j = H_k$, then $(RH_iR^{-1})(RH_jR^{-1}) = RH_kR^{-1}$.

If $RHR^{-1} = H$, H is said to be an invariant subgroup of the group G. For instance, the subgroup $\{E, C_3, C_3^{-1}\}$ of C_{3v} is invariant, as can be verified by performing all operations of the type RER^{-1}, RC_3R^{-1} and $RC_3^{-1}R^{-1}$, with the aid of the multiplication table (Example 5.2.6). The concept of an invariant subgroup is of great importance and leads to that of a factor group.

If we consider, for a group G the collection formed by an invariant subgroup H and all its distinct cosets

$$H \qquad HR_1 \qquad HR_2, \ldots$$

this can be considered as a new group, called the *factor group* of G with respect to H, in which the subgroup H and its cosets play the role of single elements.

For example, in the C_{3v} group, there is an invariant subgroup $H = \{E, C_3, C_3^{-1}\}$ and one distinct coset $H\sigma_1 = [\sigma_1, \sigma_2, \sigma_3]$. If we perform the multiplication $H\{H\sigma_1\}$ we obtain as distinct results only σ_1, σ_2 and σ_3 repeated three times. H and $\{H\sigma_1\}$ form, therefore, a group of order two with the multiplication law

$$\{H\} \cdot \{H\sigma_1\} = \{H\sigma_1\}$$

5.4 Theory of representations

The theory of group representations provides the most significant relationship between the abstract group properties discussed in the previous section and physical problems. The main reason for this is that representation theory establishes a direct connection between group properties and operations in the vector space. Since operations in the vector space are easily represented with matrices, *matrix groups* and *homomorphism* constitute the key concepts in representation theory. Since homomorphism has been already discussed in the previous section, we shall briefly discuss matrix groups before going on to representation theory.

A matrix group is a group in which the elements are square matrices, the multiplication law being matrix multiplication.

Example 5.4.1

The set of four matrices

$$\mathbf{I} = \begin{pmatrix} 1 & 0 \\ 0 & 1 \end{pmatrix} \quad \mathbf{D}_1 = \begin{pmatrix} -1 & 0 \\ 0 & -1 \end{pmatrix} \quad \mathbf{D}_2 = \begin{pmatrix} -\mathrm{i} & 0 \\ 0 & \mathrm{i} \end{pmatrix} \quad \mathbf{D}_3 = \begin{pmatrix} \mathrm{i} & 0 \\ 0 & -\mathrm{i} \end{pmatrix}$$

forms a group of order four with the same multiplication table as the group of Example 5.2.7 (they are therefore isomorphic).

Example 5.4.2

The set of six matrices

$$\mathbf{I} = \begin{pmatrix} 1 & 0 \\ 0 & 1 \end{pmatrix} \quad \mathbf{D}_1 = \begin{pmatrix} 0 & -1 \\ 1 & -1 \end{pmatrix} \quad \mathbf{D}_2 = \begin{pmatrix} -1 & 1 \\ -1 & 0 \end{pmatrix} \quad \mathbf{D}_3 = \begin{pmatrix} 1 & -1 \\ 0 & -1 \end{pmatrix}$$

$$\mathbf{D}_4 = \begin{pmatrix} -1 & 0 \\ -1 & 1 \end{pmatrix} \quad \mathbf{D}_5 = \begin{pmatrix} 0 & 1 \\ 1 & 0 \end{pmatrix}$$

forms a group of order six with the same multiplication table as the group C_{3v} of Example 5.2.3.

Example 5.4.3

The set of six matrices

$$\mathbf{I} = \begin{pmatrix} 1 & 0 \\ 0 & 1 \end{pmatrix} \quad \mathbf{D}_1 = \begin{pmatrix} -\frac{1}{2} & -\frac{\sqrt{3}}{2} \\ \frac{\sqrt{3}}{2} & -\frac{1}{2} \end{pmatrix} \quad \mathbf{D}_2 = \begin{pmatrix} -\frac{1}{2} & \frac{\sqrt{3}}{2} \\ -\frac{\sqrt{3}}{2} & -\frac{1}{2} \end{pmatrix} \quad \mathbf{D}_3 = \begin{pmatrix} 1 & 0 \\ 0 & -1 \end{pmatrix}$$

$$\mathbf{D}_4 = \begin{pmatrix} -\frac{1}{2} & -\frac{\sqrt{3}}{2} \\ -\frac{\sqrt{3}}{2} & \frac{1}{2} \end{pmatrix} \quad \mathbf{D}_5 = \begin{pmatrix} -\frac{1}{2} & \frac{\sqrt{3}}{2} \\ \frac{\sqrt{3}}{2} & \frac{1}{2} \end{pmatrix}$$

forms another group of order six, isomorphic with the one of the preceding example.

The properties of matrix groups are of great importance for representation theory. A brief review of the most elementary ones will be given here. For a more complete discussion the reader is referred to one of the many excellent books available on the subject.[2,3,4,5]

(1) In a group, for every element R there must always exist the inverse element R^{-1}. It follows that only non-singular matrices can be elements of a matrix group. In addition the determinant of any element $\mathbf{D}_j$ of a matrix group must be a rational root of unity. Since some integral power $\mathbf{D}_j^d$ must be equal to the unity matrix $\mathbf{I}$, it follows that $\det \mathbf{D}_j^d = 1$.

(2) The character $\chi(\mathbf{D}_j)$ of the element $\mathbf{D}_j$ of a matrix group is defined as the trace of the matrix $\mathbf{D}_j$

$$\chi(\mathbf{D}_j) \equiv \operatorname{tr} \mathbf{D}_j \tag{5.4.4}$$

For future reference let us note that if $\mathbf{D}_1$ and $\mathbf{D}_2$ are square matrices of the same dimension, then

$$\operatorname{tr}(\mathbf{D}_1\mathbf{D}_2) = \operatorname{tr}(\mathbf{D}_2\mathbf{D}_1) \tag{5.4.5}$$

(3) Two matrix groups $\{\mathbf{I}, \mathbf{D}_1, \mathbf{D}_2, \ldots, \mathbf{D}_j, \ldots\}$ and $\{\mathbf{I}, \overline{\mathbf{D}}_1, \overline{\mathbf{D}}_2, \ldots, \overline{\mathbf{D}}_j\}$ are said to be equivalent if there is a square matrix $\mathbf{T}$ such that

$$\overline{\mathbf{D}}_j = \mathbf{T}^{-1}\mathbf{D}_j\mathbf{T} \qquad j = 1, 2, \ldots \tag{5.4.6}$$

It can be proved that such a similarity transformation exists if, and only if, the two matrix groups are isomorphic and corresponding matrices have the same character.

Example 5.4.4

The matrix groups of Examples 5.4.2 and 5.4.3 are isomorphic since they have the same multiplication table. In addition, corresponding matrices have the same character, i.e.

Element	$\mathbf{I}$	$\mathbf{D}_1$	$\mathbf{D}_2$	$\mathbf{D}_3$	$\mathbf{D}_4$	$\mathbf{D}_5$
χ	2	-1	-1	0	0	0

(5.4.7)

The two matrix groups are therefore equivalent.

(4) A *unitary* matrix group is defined as a matrix group consisting of unitary matrices. A fundamental theorem of representation theory, the Schur–Auerbach theorem, states that every matrix group is equivalent to a unitary matrix group. A proof of this theorem is given in Section 5.6.

These few properties of matrix groups are sufficient to understand the concept of representation of a group. A *representation* Γ of a group G is a matrix group homomorph of G, together with a specification of the homomorphism. The *dimension* of the representation is given by the dimension of the matrices of the group.

For example the matrix group of Example 5.4.3 is a representation of the C_{3v} group. In this case the homomorphism involved is actually an isomorphism since there is a one-to-one correspondence between the elements of the C_{3v} group and the matrices of Example 5.4.3. For this reason such a representation is called *faithful* (or true). This representation, shown in the third row of Table 5.4.1, is bidimensional since it involves 2×2 square matrices.

Table 5.4.1 Representations of the C_{3v} group

C_{3v}	E	C_3	C_3^{-1}	σ_1	σ_2	σ_3
Γ_1	1	1	1	1	1	1
Γ_2	1	1	1	-1	-1	-1
Γ_3	$\begin{pmatrix} 1 & 0 \\ 0 & 1 \end{pmatrix}$	$\begin{pmatrix} -\frac{1}{2} & -\frac{\sqrt{3}}{2} \\ \frac{\sqrt{3}}{2} & -\frac{1}{2} \end{pmatrix}$	$\begin{pmatrix} -\frac{1}{2} & \frac{\sqrt{3}}{2} \\ -\frac{\sqrt{3}}{2} & -\frac{1}{2} \end{pmatrix}$	$\begin{pmatrix} 1 & 0 \\ 0 & -1 \end{pmatrix}$	$\begin{pmatrix} -\frac{1}{2} & -\frac{\sqrt{3}}{2} \\ -\frac{\sqrt{3}}{2} & \frac{1}{2} \end{pmatrix}$	$\begin{pmatrix} -\frac{1}{2} & \frac{\sqrt{3}}{2} \\ \frac{\sqrt{3}}{2} & \frac{1}{2} \end{pmatrix}$

The matrix group of Example 5.4.2 is another bidimensional representation of the C_{3v} group, equivalent to the preceding one, since corresponding matrices in the two groups are equivalent (see Example 5.4.4). It will be shown later on that any other possible bidimensional representation of the C_{3v} group is equivalent to that of Example 5.4.3.

We can also find one-dimensional representations of the C_{3v} group, involving 1×1 matrices (i.e. numbers), by mapping the elements of this group onto a smaller homomorphic group. If, for instance, the six elements of C_{3v} are mapped onto the single quantity $+1$ we obtain the representation shown in the first row of Table 5.4.1.

Another possible one-dimensional representation of C_{3v} is given in the second row of Table 5.4.1. In this case the elements E, C_3 and C_3^{-1} are mapped onto the quantity $+1$ and the elements σ_1, σ_2 and σ_3 onto the quantity -1. These representations are called *unfaithful* since different elements of the larger group have the same image in the smaller group.

We wish to point out that the number $+1$ is a matrix group of order 1, homomorphic with all possible groups. It is therefore a representation of any group.

Obviously we can look for other representations of the C_{3v} group involving in general $n \times n$ matrices. We shall see, however, in Section 5.5 that any larger representation of C_{3v} can always be reduced to those given in Table 5.4.1. Before closing this section we notice that if $\Gamma_i(\mathbf{D}'_1, \mathbf{D}'_2, \ldots, \mathbf{D}'_n)$ and $\Gamma_j(\mathbf{D}''_1, \mathbf{D}''_2, \ldots, \mathbf{D}''_n)$ are two representations of a group of dimension d_i and d_j respectively, another representation of the group of dimension $d_i + d_j$ can be found by forming the *direct sum* of corresponding matrices of the two representations. The resulting representation Γ is

$$\mathbf{D}_1 = \begin{pmatrix} \mathbf{D}'_1 & 0 \\ 0 & \mathbf{D}''_1 \end{pmatrix} \quad \mathbf{D}_2 = \begin{pmatrix} \mathbf{D}'_2 & 0 \\ 0 & \mathbf{D}''_2 \end{pmatrix} \quad \mathbf{D}_n = \begin{pmatrix} \mathbf{D}'_n & 0 \\ 0 & \mathbf{D}''_n \end{pmatrix} \tag{5.4.8}$$

where the original matrices appear as diagonal blocks and all remaining elements are zeros. The representation Γ is called the direct sum of the representations Γ_i and Γ_j.

5.5 Vector spaces

As pointed out in the previous section, representation theory directly relates the operations in the vector space with group theory. We shall discuss this correlation presently.

Let us define an n-dimensional space by choosing n linearly independent basic vectors $\mathbf{e}_1, \mathbf{e}_2, \ldots, \mathbf{e}_n$. We shall refer to this set of vectors as a *basis* and we shall say that it spans the n-dimensional space. Any $(n+1)$th vector is related to the basis by a linear relation of the type

$$\mathbb{r} = \mathbf{e}_1 r_1 + \mathbf{e}_2 r_2 + \ldots + \mathbf{e}_n r_n = \sum_j \mathbf{e}_j r_j \tag{5.5.1}$$

where $r_1, r_2, \ldots, r_n$ are called the components of the vector relative to the basis $\mathbf{e}_1, \mathbf{e}_2, \ldots, \mathbf{e}_n$.

For simplicity let us use matrix notation. By writing the components of the vector $\mathbb{r}$ as a column matrix $\mathbf{r}$ and the basic vectors as a row matrix $\mathbf{e}$, equation (5.5.1) takes the much simpler form

$$\mathbb{r} = (\mathbf{e}_1, \mathbf{e}_2, \ldots, \mathbf{e}_n)\begin{pmatrix} r_1 \\ r_2 \\ \vdots \\ r_n \end{pmatrix} = \mathbf{er} \tag{5.5.2}$$

Let us now choose a second set of n vectors $\bar{\mathbf{e}}_1, \bar{\mathbf{e}}_2, \ldots, \bar{\mathbf{e}}_n$ to define a new basis $\bar{\mathbf{e}}$. Using (5.5.1) the vectors of the new basis $\bar{\mathbf{e}}$ can be written in terms of those of the basis $\mathbf{e}$

$$\bar{\mathbf{e}}_i = \sum_j \mathbf{e}_j R_{ij} \tag{5.5.3}$$

and in matrix notation, according to (5.5.2)

$$\bar{\mathbf{e}} = \mathbf{eR} \tag{5.5.4}$$

where $\mathbf{R}$ is a square $n \times n$ matrix with elements R_{ji}.

The components $\bar{\mathbf{r}}$ of the vector $\mathbb{r}$ of equation (5.5.2) in the new basis $\bar{\mathbf{e}}$ are now simply found from (5.5.2) and (5.5.4). We have

$$\mathbb{r} = \mathbf{er} = \bar{\mathbf{e}}\mathbf{R}^{-1}\mathbf{r} = \bar{\mathbf{e}}\bar{\mathbf{r}} \tag{5.5.5}$$

and therefore the components in the new basis are related to those in the old basis by the matrix relation

$$\bar{\mathbf{r}} = \mathbf{R}^{-1}\mathbf{r} \tag{5.5.6}$$

A change of basis, from $\mathbf{e}$ to $\bar{\mathbf{e}}$, can be accomplished in several ways. If for the moment we limit ourselves to a three-dimensional space, we can easily visualize any change of basis. In the most general case we can shift the origin, change the orientation, change the length of the basic vectors and change the angles between them. In special cases we can perform only one of these changes.

In particular we shall essentially be interested in changes of basis in which the lengths and the angles between basic vectors are preserved, whereas the origin is not shifted, as shown in Figure 5.5.1. Such changes of basis can be considered as rotations of the basis around a particular axis or reflection of the basis through a plane. For instance in Figure 5.5.1 is shown a change from a basis $\mathbf{e}$ to a new basis $\bar{\mathbf{e}}$ corresponding to a counterclockwise rotation θ of the basis $\mathbf{e}$ around an axis parallel to $\mathbf{e}_3$.

It is very convenient to consider the change of basis $\mathbf{e} \to \bar{\mathbf{e}}$ as the result of an operation $\mathscr{R}$ which when applied to the basis $\mathbf{e}$ transforms it into the basis $\bar{\mathbf{e}}$. The operator $\mathscr{R}$ can in general be made up of different operations and can involve a translation, a rotation, a reflection and a deformation of the basis. In the case of Figure 5.5.1 it simply consists of a rotation of an angle θ and henceforth, unless otherwise specified, the operator $\mathscr{R}$ will always represent a symmetry operation (called generically rotation) which preserves the

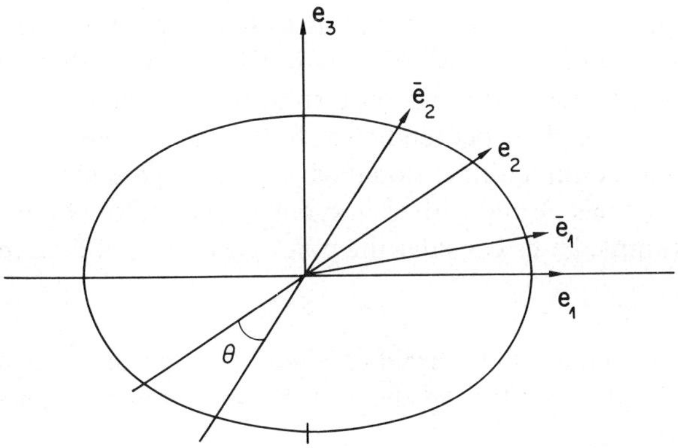

Figure 5.5.1 Rotation of a basis **e** around an axis parallel to $\mathbf{e}_3$

lengths and angles of the basis and leaves the origin of the basis invariant. Symbolically we can write

$$\mathscr{R}\mathbf{e} = \bar{\mathbf{e}} \tag{5.5.7}$$

and using (5.5.4)

$$\mathscr{R}\mathbf{e} = \mathbf{eR} \tag{5.5.8}$$

Equation (5.5.8) is of great importance since it establishes a one-to-one correspondence between a symmetry operation $\mathscr{R}$ and a matrix **R**. To every operation $\mathscr{R}$ there corresponds a matrix **R** for each choice of the basis **e**. An example will clarify to the reader the meaning of equation (5.5.8).

Example 5.5.1

Let us consider the rotation of the orthogonal basis **e** shown in Figure 5.5.1. In this case we have

$$\begin{aligned} \bar{\mathbf{e}}_1 &= \mathbf{e}_1 \cos\theta + \mathbf{e}_2 \sin\theta \\ \bar{\mathbf{e}}_2 &= -\mathbf{e}_1 \sin\theta + \mathbf{e}_2 \cos\theta \\ \bar{\mathbf{e}}_3 &= \mathbf{e}_3 \end{aligned} \tag{5.5.9}$$

and thus equations (5.5.7) and (5.5.8) give

$$(\bar{\mathbf{e}}_1, \bar{\mathbf{e}}_2, \bar{\mathbf{e}}_3) = (\mathbf{e}_1, \mathbf{e}_2, \mathbf{e}_3)\begin{pmatrix} \cos\theta & -\sin\theta & 0 \\ \sin\theta & \cos\theta & 0 \\ 0 & 0 & 1 \end{pmatrix} \tag{5.5.10}$$

The matrix **R** associated with the operator $\mathscr{R}$ is therefore

$$\mathbf{R} = \begin{pmatrix} \cos\theta & -\sin\theta & 0 \\ \sin\theta & \cos\theta & 0 \\ 0 & 0 & 1 \end{pmatrix}$$

The one-to-one correspondence between rotation operators and matrices furnishes a powerful method for finding matrix representations.

We have already seen in Examples 5.2.3 and 5.2.5 that the symmetry operations which bring an object into self-coincidence form a group. The symmetry operations of a physical object have the property of leaving invariant at least one point of the object and for this reason are said to form 'point groups'. We shall prove later on that any point symmetry operation can be considered either as a pure rotation (proper rotation) or as a rotation followed by a reflection in a plane perpendicular to the axis of rotation (improper rotation). If we embed in the object a three-dimensional basis **e**, each symmetry operation will change the basis **e** into a rotated basis $\bar{\mathbf{e}}$ and with each symmetry operator $\mathscr{R}$ there will be associated a matrix **R**. As an example let us consider the point group C_{3v} of Example 5.2.3.

Example 5.5.2

Let us embed in the equilateral pyramid of Figure 5.2.1 an orthogonal basis **e** as shown in Figure 5.5.2, and let us consider the effect of the six symmetry operations E, C_3, C_3^{-1},

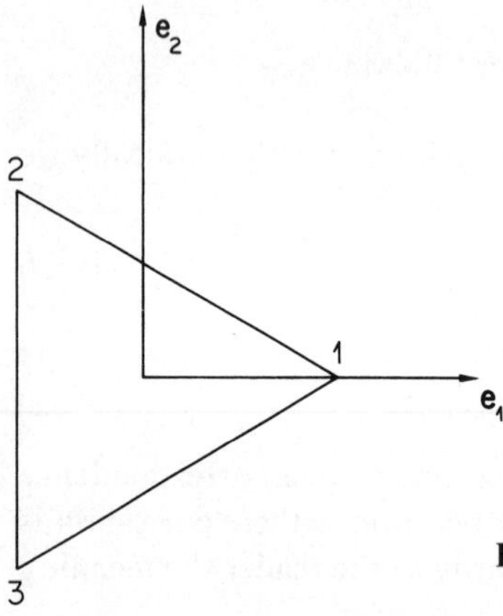

Figure 5.5.2 View along C_3 of the equilateral pyramid of Figure 5.2.1 with a basis **e** located with $\mathbf{e}_3$ parallel to C_3

σ_1, σ_2 and σ_3 on **e**. The identity operation E clearly leaves the basis **e** invariant. It corresponds therefore to a pure rotation of an angle $\theta = 0$. In this case equation (5.5.8) gives

$$(\bar{\mathbf{e}}_1, \bar{\mathbf{e}}_2, \bar{\mathbf{e}}_3) = (\mathbf{e}_1, \mathbf{e}_2, \mathbf{e}_3)\begin{pmatrix} 1 & 0 & 0 \\ 0 & 1 & 0 \\ 0 & 0 & 1 \end{pmatrix} \tag{5.5.12}$$

The effect of C_3 and C_3^{-1} on the basis **e** is simply obtained from Example 5.5.1 with $\theta = 120°$ and $240°$ respectively. Equation (5.5.8) gives

$$(\bar{\mathbf{e}}_1, \bar{\mathbf{e}}_2, \bar{\mathbf{e}}_3) = (\mathbf{e}_1, \mathbf{e}_2, \mathbf{e}_3)\begin{pmatrix} -\frac{1}{2} & -\frac{\sqrt{3}}{2} & 0 \\ \frac{\sqrt{3}}{2} & -\frac{1}{2} & 0 \\ 0 & 0 & 1 \end{pmatrix} \tag{5.5.13}$$

$$(\bar{\mathbf{e}}_1, \bar{\mathbf{e}}_2, \bar{\mathbf{e}}_3) = (\mathbf{e}_1, \mathbf{e}_2, \mathbf{e}_3)\begin{pmatrix} -\frac{1}{2} & \frac{\sqrt{3}}{2} & 0 \\ -\frac{\sqrt{3}}{2} & -\frac{1}{2} & 0 \\ 0 & 0 & 1 \end{pmatrix} \tag{5.5.14}$$

The effect of σ_1 on the basis **e** is to leave $\mathbf{e}_1$ and $\mathbf{e}_3$ unchanged and to send $\mathbf{e}_2$ into $-\mathbf{e}_2$. We have, therefore,

$$(\bar{\mathbf{e}}_1, \bar{\mathbf{e}}_2, \bar{\mathbf{e}}_3) = (\mathbf{e}_1, \mathbf{e}_2, \mathbf{e}_3)\begin{pmatrix} 1 & 0 & 0 \\ 0 & -1 & 0 \\ 0 & 0 & 1 \end{pmatrix} \tag{5.5.15}$$

The effect of σ_2 and σ_3 is easily determined in the same way from Figure 5.5.2 and is found to be

$$\text{(for } \sigma_2\text{)} \qquad (\bar{\mathbf{e}}_1, \bar{\mathbf{e}}_2, \bar{\mathbf{e}}_3) = (\mathbf{e}_1, \mathbf{e}_2, \mathbf{e}_3)\begin{pmatrix} -\frac{1}{2} & -\frac{\sqrt{3}}{2} & 0 \\ -\frac{\sqrt{3}}{2} & \frac{1}{2} & 0 \\ 0 & 0 & 1 \end{pmatrix} \tag{5.5.16}$$

$$\text{(for } \sigma_3\text{)} \qquad (\bar{\mathbf{e}}_1, \bar{\mathbf{e}}_2, \bar{\mathbf{e}}_3) = (\mathbf{e}_1, \mathbf{e}_2, \mathbf{e}_3)\begin{pmatrix} -\frac{1}{2} & \frac{\sqrt{3}}{2} & 0 \\ \frac{\sqrt{3}}{2} & \frac{1}{2} & 0 \\ 0 & 0 & 1 \end{pmatrix} \tag{5.5.17}$$

The six matrices obtained in this way clearly form a representation of the C_{3v} group. To prove it, consider three symmetry operations $\mathscr{R}, \mathscr{S}, \mathscr{T}$, such that $\mathscr{R}\mathscr{S} = \mathscr{T}$. Their effect on the basis **e** is

$$\mathscr{R}\mathbf{e} = \mathbf{eR} \qquad \mathscr{S}\mathbf{e} = \mathbf{eS} \qquad \mathscr{T}\mathbf{e} = \mathbf{eT} \tag{5.5.18}$$

where **R**, **S** and **T** are the three matrices associated with the operators $\mathscr{R}$, $\mathscr{S}$ and $\mathscr{T}$ in the basis **e**. We have then

$$\mathscr{R}\mathscr{S}\mathbf{e} = \mathscr{R}\mathbf{eS} = \mathbf{eRS} \tag{5.5.19}$$

from which it follows that if $\mathscr{R}\mathscr{S} = \mathscr{T}$ also $\mathbf{RS} = \mathbf{T}$. Inspection of the matrices (5.5.12) to (5.5.17) shows an interesting property of the representation found using the three-dimensional basis **e**. Since the off-diagonal terms in the third row and in the third column of all six matrices are always zero, the 3×3 matrices can be partitioned into 2×2 and 1×1 matrices which never mix upon multiplication. This property of the matrices reflects the fact that, with the choice made for the basis **e**, only the vectors $\mathbf{e}_1$ and $\mathbf{e}_2$ mix together upon application of a symmetry operation. The vector $\mathbf{e}_3$, for every symmetry operation, always goes into itself. This can be stated by saying that $\mathbf{e}_1$ and $\mathbf{e}_2$ *carry* together a bidimensional representation of the C_{3v} group and that $\mathbf{e}_3$ *carries* a monodimensional representation. The three-dimensional representation carried by the basis **e** can therefore be reduced to one bi- and one mono-dimensional representation, carried respectively by a bi- and a mono-dimensional space. In other words the three-dimensional representation carried by **e** is the direct sum of a mono- and a bi-dimensional representation (see Section 5.4). The reduction of a matrix representation is one of the most important concepts of group theory and we shall discuss it in more detail in the next section. For the moment we wish to point out that the representations carried by $\mathbf{e}_1$, $\mathbf{e}_2$ and by $\mathbf{e}_3$ are just those labelled Γ_1 and Γ_3 in Table 5.4.1.

As discussed before, the form of the matrix representation found is strictly bound to the basis **e** chosen. If we change the basis the representation also changes, even if the new representation can be correlated to the old. In order to prove this, let us consider another example.

Example 5.5.3

Let us embed in the equilateral pyramid of Figure 5.2.1 a non-orthogonal basis $\mathbf{e}'$ as shown in Figure 5.5.3 and consider the effect of the symmetry operations on $\mathbf{e}'$. Since $\mathbf{e}'_3$ is parallel to C_3 as $\mathbf{e}_3$ of Example 5.5.2, it will carry the same representation Γ_1 carried by $\mathbf{e}_3$. We need therefore consider only the bidimensional representation carried by $\mathbf{e}'_1$ and $\mathbf{e}'_2$. The identity E leaves $\mathbf{e}'_1$ and $\mathbf{e}'_2$ unchanged and therefore

$$(\bar{\mathbf{e}}'_1, \bar{\mathbf{e}}'_2) = (\mathbf{e}'_1, \mathbf{e}'_2)\begin{pmatrix} 1 & 0 \\ 0 & 1 \end{pmatrix} \tag{5.5.20}$$

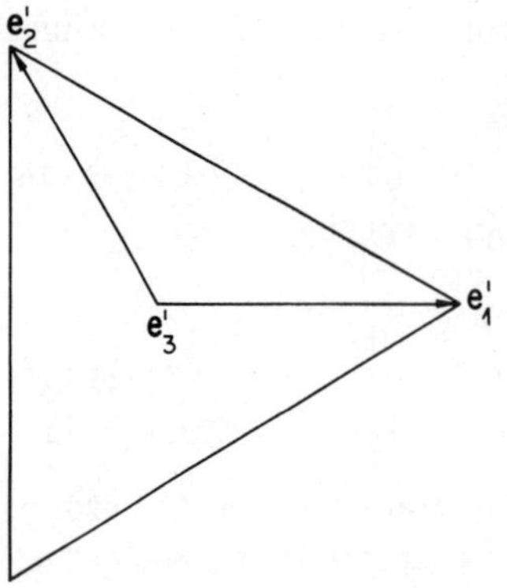

Figure 5.5.3 View along C_3 of the equilateral pyramid of Figure 5.2.1 with a basis $\mathbf{e}'$ with $\mathbf{e}'_3$ parallel to C_3

The effect of C_3 is clearly to send $\mathbf{e}'_1$ into $\mathbf{e}'_2$ and $\mathbf{e}'_2$ into $-\mathbf{e}'_1 - \mathbf{e}'_2$. Thus

$$(\bar{\mathbf{e}}'_1, \bar{\mathbf{e}}'_2) = (\mathbf{e}'_1, \mathbf{e}'_2)\begin{pmatrix} 0 & -1 \\ 1 & -1 \end{pmatrix} \tag{5.5.21}$$

The effect of C_3^{-1} is to send $\mathbf{e}'_1$ into $-\mathbf{e}'_1 - \mathbf{e}'_2$ and $\mathbf{e}'_2$ into $\mathbf{e}'_1$.

$$(\bar{\mathbf{e}}'_1, \bar{\mathbf{e}}'_2) = (\mathbf{e}'_1, \mathbf{e}'_2)\begin{pmatrix} -1 & 1 \\ -1 & 0 \end{pmatrix} \tag{5.5.22}$$

The effect of σ_1 is to send $\mathbf{e}'_1$ into itself and $\mathbf{e}'_2$ into $-\mathbf{e}'_1 - \mathbf{e}'_2$

$$(\bar{\mathbf{e}}'_1, \bar{\mathbf{e}}'_2) = (\mathbf{e}'_1, \mathbf{e}'_2)\begin{pmatrix} 1 & -1 \\ 0 & -1 \end{pmatrix} \tag{5.5.23}$$

The effect of σ_2 is to send $\mathbf{e}'_2$ into itself and $\mathbf{e}'_1$ into $-\mathbf{e}'_1 - \mathbf{e}'_2$

$$(\bar{\mathbf{e}}'_1, \bar{\mathbf{e}}'_2) = (\mathbf{e}'_1, \mathbf{e}'_2)\begin{pmatrix} -1 & 0 \\ -1 & 1 \end{pmatrix} \tag{5.5.24}$$

Finally, the effect of σ_3 is to send $\mathbf{e}'_1$ into $\mathbf{e}'_2$ and $\mathbf{e}'_2$ into $\mathbf{e}'_1$. Thus

$$(\bar{\mathbf{e}}'_1, \bar{\mathbf{e}}'_2) = (\mathbf{e}'_1, \mathbf{e}'_2)\begin{pmatrix} 0 & 1 \\ 1 & 0 \end{pmatrix} \tag{5.5.25}$$

The reader will realize at once that the six matrices obtained form another bidimensional representation of C_{3v} already given in Example 5.4.2, which is equivalent to the bidimensional representation carried by $\mathbf{e}_1$ and $\mathbf{e}_2$ in Example 5.5.2, as discussed in Example 5.4.4.

Let us investigate the equivalence of representations more thoroughly. It is obvious that, since the choice of the basis is arbitrary, we can in principle obtain an infinite number of representations. We shall prove, however, that these representations are all equivalent and that once one is found, all others can be derived by a similarity transformation.

Consider for instance two bases $\mathbf{e}$ and $\mathbf{e}'$, each carrying an n-dimensional representation of a group, and suppose that $\mathbf{e}' = \mathbf{eT}$. From equation (5.5.8) we have

$$\mathcal{R}\mathbf{e} = \mathbf{eR} \tag{5.5.26}$$

$$\mathcal{R}\mathbf{e}' = \mathbf{e}'\mathbf{R}' \tag{5.5.27}$$

$$\mathcal{R}\mathbf{e}' = \mathcal{R}\mathbf{eT} = \mathbf{eRT} = \mathbf{e}'\mathbf{T}^{-1}\mathbf{RT} \tag{5.5.28}$$

and comparing equations (5.5.27) and (5.5.28)

$$\mathbf{R}' = \mathbf{T}^{-1}\mathbf{R}\mathbf{T} \tag{5.5.29}$$

Since, according to equation (5.5.4) two bases are always related by a matrix,† we conclude that all representations found by a change of basis are equivalent and that they are related by the similarity transformation (5.5.29).

Once we know that all representations found by a change of basis are equivalent, we can select one which is convenient to deal with. In Section 5.4 we have seen that a fundamental theorem of group theory states that every matrix group is equivalent to a unitary matrix group. Since unitary matrices are the most convenient to handle, we shall always choose, as representative of every representation, a unitary matrix group.

We can now prove that the choice of a unitary matrix group representation corresponds to the choice of a unitary basis and, in real space, of an orthogonal basis. In order to prove it, let us introduce the definition of the metrical matrix **M** of a space spanned by a basis **e** as

$$\mathbf{M} = \mathbf{e}^{\dagger} \cdot \mathbf{e} = \begin{pmatrix} \mathbf{e}_1^* \\ \mathbf{e}_2^* \\ \vdots \\ \mathbf{e}_n^* \end{pmatrix} (\mathbf{e}_1, \mathbf{e}_2, \ldots, \mathbf{e}_n) = \begin{pmatrix} \mathbf{e}_1^* \cdot \mathbf{e}_1 & \mathbf{e}_1^* \cdot \mathbf{e}_2 & \ldots & \mathbf{e}_1^* \cdot \mathbf{e}_n \\ \mathbf{e}_2^* \cdot \mathbf{e}_1 & \mathbf{e}_2^* \cdot \mathbf{e}_2 & \ldots & \mathbf{e}_2^* \cdot \mathbf{e}_n \\ \vdots & \vdots & & \vdots \\ \mathbf{e}_n^* \cdot \mathbf{e}_1 & \mathbf{e}_n^* \cdot \mathbf{e}_2 & & \mathbf{e}_n^* \cdot \mathbf{e}_n \end{pmatrix} \tag{5.5.30}$$

The metrical matrix completely defines the metrical properties of the space, the diagonal elements $\mathbf{e}_i^* \cdot \mathbf{e}_i$ defining the lengths and the off-diagonal elements $\mathbf{e}_i^* \cdot \mathbf{e}_j$ the angles of the basis. A basis for which $\mathbf{e}_i^* \cdot \mathbf{e}_j = \delta_{ij}$, δ_{ij} being the Kronecker delta, is called an *orthonormal* or *unitary* basis. For such a basis obviously $\mathbf{M} = \mathbf{E}$.

Let us now apply to a unitary basis **e** a symmetry operator $\mathscr{R}$. According to equations (5.5.7) and (5.5.8) we have

$$\mathscr{R}\mathbf{e} = \bar{\mathbf{e}} = \mathbf{e}\mathbf{R} \tag{5.5.31}$$

and the metrical matrix $\overline{\mathbf{M}}$ in the 'rotated' basis $\bar{\mathbf{e}}$ is given, from (5.5.30), by

$$\overline{\mathbf{M}} = \bar{\mathbf{e}}^{\dagger} \cdot \bar{\mathbf{e}} = (\mathbf{e}\mathbf{R})^{\dagger} \cdot (\mathbf{e}\mathbf{R}) = \mathbf{R}^{\dagger}\mathbf{e}^{\dagger} \cdot \mathbf{e}\mathbf{R} = \mathbf{R}^{\dagger}\mathbf{M}\mathbf{R} \tag{5.5.32}$$

But, if **e** and $\bar{\mathbf{e}}$ are both unitary bases, as will happen if $\mathscr{R}$ does not change lengths and angles of the basis, then

$$\overline{\mathbf{M}} = \mathbf{M} = \mathbf{E}$$

and therefore

$$\mathbf{R}^{\dagger}\mathbf{R} = \mathbf{E} \tag{5.5.33}$$

This proves that the matrix **R** is a unitary matrix and that the representation carried by the unitary basis **e** is a unitary matrix representation.

The fact that with the choice of a unitary basis we obtain for point groups unitary representations has some interesting consequences.

Let us choose a three-dimensional Cartesian basis. This choice makes the unitary matrix **R** a real orthogonal matrix ($\mathbf{R}^{-1} = \tilde{\mathbf{R}}$), since all the vectors are real. From equation (5.5.33) it follows that the determinant $|\mathbf{R}|$ of **R** is equal to ± 1 since

$$|\mathbf{R}^{\dagger}||\mathbf{R}| = |\tilde{\mathbf{R}}||\mathbf{R}| = |\mathbf{R}|^2 = 1 \tag{5.5.34}$$

† It is necessary to point out that the matrix **T** which correlates two bases need not be generated by a symmetry operator and therefore need not preserve lengths and angles of the basis as do the matrices **R** and **R**′, and in general all matrices generated by equation (5.5.8).

and thus

$$|\mathbf{R}| = \pm 1 \tag{5.5.35}$$

This equation is very important and states that point group operations can be of two types: (a) those associated with matrices **R** whose determinant is $+1$ which we shall call *proper rotations*; (b) those associated with matrices whose determinant is -1 which we shall call *improper rotations.*

Inspection of the matrices of Example 5.5.2 shows that the operations E, C_3 and C_3^{-1} are proper rotations ($|\mathbf{R}| = +1$) and that the operations σ_1, σ_2 and σ_3 are improper rotations ($|\mathbf{R}| = -1$).

In Example 5.5.1 we have already seen the general form of the matrix **R** associated with a proper rotation of an angle θ. By orienting the rotation axis in the direction of $\mathbf{e}_3$, the matrix assumes the form

$$\mathbf{R} = \begin{pmatrix} \cos\theta & -\sin\theta & 0 \\ \sin\theta & \cos\theta & 0 \\ 0 & 0 & 1 \end{pmatrix} \tag{5.5.36}$$

which is an orthogonal matrix with determinant equal to $+1$. This covers all symmetry operations which are pure rotations. Particular cases are the symmetry operations $E(\theta = 0)$, $C_2(\theta = 180°)$, $C_3(\theta = 120°)$, $C_4(\theta = 90°)$ and $C_6(\theta = 60°)$. The corresponding matrices are listed below.

$$\begin{array}{ccccc} E & C_2 & C_3 & C_4 & C_6 \\ \theta = 0 & \theta = 180° & \theta = 120° & \theta = 90° & \theta = 60° \\ \begin{pmatrix} 1 & 0 & 0 \\ 0 & 1 & 0 \\ 0 & 0 & 1 \end{pmatrix} & \begin{pmatrix} -1 & 0 & 0 \\ 0 & -1 & 0 \\ 0 & 0 & 1 \end{pmatrix} & \begin{pmatrix} -\frac{1}{2} & -\frac{\sqrt{3}}{2} & 0 \\ \frac{\sqrt{3}}{2} & -\frac{1}{2} & 0 \\ 0 & 0 & 1 \end{pmatrix} & \begin{pmatrix} 0 & -1 & 0 \\ 1 & 0 & 0 \\ 0 & 0 & 1 \end{pmatrix} & \begin{pmatrix} \frac{1}{2} & -\frac{\sqrt{3}}{2} & 0 \\ \frac{\sqrt{3}}{2} & \frac{1}{2} & 0 \\ 0 & 0 & 1 \end{pmatrix} \end{array} \tag{5.5.37}$$

By orienting the rotation axis in the direction of $\mathbf{e}_3$, the general form[4] of the matrix associated with an improper rotation is

$$\mathbf{R} = \begin{pmatrix} \cos\theta & -\sin\theta & 0 \\ \sin\theta & \cos\theta & 0 \\ 0 & 0 & -1 \end{pmatrix} \tag{5.5.38}$$

which is an orthogonal matrix with determinant equal to -1. Particular cases are the symmetry operations $\sigma(\theta = 0)$, $\mathrm{i}(\theta = 180°)$, $S_3(\theta = 120°)$, $S_4(\theta = 90°)$ and $S_6(\theta = 60°)$. The corresponding matrices are

$$\begin{array}{ccccc} \sigma & \mathrm{i} & S_3 & S_4 & S_6 \\ \theta = 0 & \theta = 180° & \theta = 120° & \theta = 90° & \theta = 60° \\ \begin{pmatrix} 1 & 0 & 0 \\ 0 & 1 & 0 \\ 0 & 0 & -1 \end{pmatrix} & \begin{pmatrix} -1 & 0 & 0 \\ 0 & -1 & 0 \\ 0 & 0 & -1 \end{pmatrix} & \begin{pmatrix} -\frac{1}{2} & -\frac{\sqrt{3}}{2} & 0 \\ \frac{\sqrt{3}}{2} & -\frac{1}{2} & 0 \\ 0 & 0 & -1 \end{pmatrix} & \begin{pmatrix} 0 & -1 & 0 \\ 1 & 0 & 0 \\ 0 & 0 & -1 \end{pmatrix} & \begin{pmatrix} \frac{1}{2} & -\frac{\sqrt{3}}{2} & 0 \\ \frac{\sqrt{3}}{2} & \frac{1}{2} & 0 \\ 0 & 0 & -1 \end{pmatrix} \end{array} \tag{5.5.39}$$

An improper rotation can be considered as the product of a proper rotation and a reflection in a plane perpendicular to the axis of rotation, since

$$\begin{pmatrix} \cos\theta & -\sin\theta & 0 \\ \sin\theta & \cos\theta & 0 \\ 0 & 0 & 1 \end{pmatrix} \begin{pmatrix} 1 & 0 & 0 \\ 0 & 1 & 0 \\ 0 & 0 & -1 \end{pmatrix} = \begin{pmatrix} \cos\theta & -\sin\theta & 0 \\ \sin\theta & \cos\theta & 0 \\ 0 & 0 & -1 \end{pmatrix} \tag{5.5.40}$$

We can now show that the trace of a rotation matrix is independent of the choice of the basis. Consider the bases **e** and **e**′ of equations (5.5.26) and (5.5.27). We have from (5.4.5) and (5.5.29)

$$\operatorname{tr} \mathbf{R}' = \operatorname{tr} (\mathbf{T}^{-1}\mathbf{R}\mathbf{T}) = \operatorname{tr} (\mathbf{R}\ \mathbf{T}\ \mathbf{T}^{-1}) = \operatorname{tr} \mathbf{R} \tag{5.5.41}$$

This means that the trace of a matrix remains unchanged in a similarity transformation and thus that equivalent matrices have the same character.

From equations (5.5.36) and (5.5.38) it follows that the trace of a rotation matrix is

$$2 \cos \theta \pm 1 \tag{5.5.42}$$

As discussed before, equation (5.5.8) relates any symmetry operation $\mathscr{R}$ to an associated rotation matrix **R** through the rotation of a basis **e**. We can, however, use a different approach to define symmetry operations in the vector space. Instead of rotating the basis **e**, we can leave it fixed in space and rotate a vector $\mathbb{r}$ defined in terms of the components of the basis **e** by equation (5.5.1). Symbolically we write

$$\mathscr{R}\mathbb{r} = \mathbb{r}' \tag{5.5.43}$$

where $\mathbb{r}'$ represents the rotated vector. In order to find the components of $\mathbb{r}'$ in the fixed basis **e**, we can use another basis $\bar{\mathbf{e}}$ which rotates with $\mathbb{r}$ so that $\mathbb{r}'$ has in the rotating basis $\bar{\mathbf{e}}$ the same components that $\mathbb{r}$ had in the fixed basis **e**. We then have

$$\mathscr{R}\mathbb{r} = \mathbb{r}' = \bar{\mathbf{e}}\mathbf{r} = \mathbf{e}\mathbf{R}\mathbf{r} = \mathbf{e}\mathbf{r}' \tag{5.5.44}$$

and thus

$$\mathbf{r}' = \mathscr{R}\mathbf{r} = \mathbf{R}\mathbf{r} \tag{5.5.45}$$

If we compare equation (5.5.44) with (5.5.8) we realize that both equations associate a symmetry operator $\mathscr{R}$ with the same matrix **R**. Thus in principle both equations can be used to discuss the relation between the vector space and the matrix representations of point groups. In the general discussion of this chapter it is more convenient to use (5.5.8) by following the rotation of the basis **e**. In the discussion of the symmetry of molecular vibrations we shall, however, follow the other approach, leaving the basis **e** fixed and rotating the components of the displacement vectors of the atoms.

5.6 Irreducible representations

In the previous two sections we have discussed the representations of groups with matrices. In the process of finding matrix representations we have found that some representations of a group are equivalent, i.e. they can be transformed into one another by a similarity transformation and differ merely through a different choice of basis.

There are, however, representations of a group which are not equivalent in the sense that a similarity transformation among two of them cannot be found. For instance the representations Γ_1, Γ_2 and Γ_3 of the C_{3v} point group are inequivalent. The problem we have to face is to find the number of inequivalent representations of a group in their simplest form.

We recall from Section 5.4 that with two representations Γ_1 and Γ_2 we can always, by forming the direct sum of Γ_1 and Γ_2, construct a third representation with matrices

$$\mathbf{D}_1 = \begin{pmatrix} \mathbf{D}_1' & \mathbf{0} \\ \mathbf{0} & \mathbf{D}_1'' \end{pmatrix} \quad \mathbf{D}_2 = \begin{pmatrix} \mathbf{D}_2' & \mathbf{0} \\ \mathbf{0} & \mathbf{D}_2'' \end{pmatrix} \ldots \mathbf{D}_n = \begin{pmatrix} \mathbf{D}_n' & \mathbf{0} \\ \mathbf{0} & \mathbf{D}_n'' \end{pmatrix} \tag{5.6.1}$$

With a similarity transformation it is always possible to transform the representation Γ into an equivalent representation $\bar{\Gamma}$ which is not in block form as Γ. If $\mathbf{D}_i$ and $\bar{\mathbf{D}}_i$ are two corresponding matrices of Γ and $\bar{\Gamma}$ respectively, we have

$$\mathbf{T}^{-1}\mathbf{D}_i\mathbf{T} = \bar{\mathbf{D}}_i \tag{5.6.2}$$

$$\mathbf{T}^{-1}\begin{pmatrix} \mathbf{D}'_i & \mathbf{0} \\ \mathbf{0} & \mathbf{D}''_i \end{pmatrix}\mathbf{T} = \begin{pmatrix} \bar{\mathbf{D}}_{i1} & \bar{\mathbf{D}}_{i2} \\ \bar{\mathbf{D}}_{i3} & \bar{\mathbf{D}}_{i4} \end{pmatrix} \tag{5.6.3}$$

It is obvious that, in order to obtain simple representations of a group, we shall perform the inverse procedure. Given a representation $\bar{\Gamma}$, we wish to find a similarity transformation which transforms the representation $\bar{\Gamma}$ into an equivalent representation Γ which is in block form and is therefore the direct sum of two other representations Γ_i and Γ_j of smaller dimensions. We shall call this process *reduction* and we shall say that the representation $\bar{\Gamma}$ is reduced through the similarity transformation to the representations Γ_i and Γ_j.

Clearly we can proceed, through consecutive similarity transformations, in the process of reduction of a representation Γ to block form, until we reach a situation in which each block cannot be reduced further. This means that we can reduce the representation Γ to the direct sum of representations $\Gamma_1, \Gamma_2, \ldots, \Gamma_n$ of smaller dimension which cannot be reduced further and are therefore called *irreducible representations.*

All the basic properties of irreducible representations can be deduced from a very few theorems of group theory which establish the possibility of complete reduction of a representation to unitary form by unitary transformations. The first theorem has been already introduced in Section 5.4. This theorem, known as the Schur–Auerbach theorem, states that any representation of a finite group by non-singular matrices is equivalent to a unitary representation. The proof is outlined below.

We wish to show that, given a representation $\Gamma(\mathbf{D}_1, \mathbf{D}_2, \ldots, \mathbf{D}_n)$ there is an equivalent representation $\bar{\Gamma}(\bar{\mathbf{D}}_1, \bar{\mathbf{D}}_2, \ldots, \bar{\mathbf{D}}_n)$ which is unitary. There is therefore a similarity transformation such that

$$\bar{\mathbf{D}}_i = \mathbf{V}\mathbf{D}_i\mathbf{V}^{-1} \tag{5.6.4}$$

$$\bar{\mathbf{D}}_i^{\dagger}\bar{\mathbf{D}}_i = \mathbf{E} \tag{5.6.5}$$

Let us define a matrix $\mathbf{H}$ by the equation

$$\mathbf{H} = \sum_{i=1}^{n} \mathbf{D}_i^{\dagger}\mathbf{D}_i \tag{5.6.6}$$

The matrix $\mathbf{H}$ is Hermitian, since

$$\mathbf{H}^{\dagger} = \sum_{i=1}^{n} \mathbf{D}_i^{\dagger}(\mathbf{D}_i^{\dagger})^{\dagger} = \sum_{i=1}^{n} \mathbf{D}_i^{\dagger}\mathbf{D}_i = \mathbf{H} \tag{5.6.7}$$

and positive definite being the sum of positive definite terms $\mathbf{D}_i^{\dagger}\mathbf{D}_i$ (each $\mathbf{D}_i$ is non-singular). The matrix $\mathbf{H}$ has the property that

$$\mathbf{D}_j^{\dagger}\mathbf{H}\mathbf{D}_j = \mathbf{H} \tag{5.6.8}$$

since

$$\mathbf{D}_j^{\dagger}\mathbf{H}\mathbf{D}_j = \mathbf{D}_j^{\dagger}\left(\sum_{i=1}^{n} \mathbf{D}_i^{\dagger}\mathbf{D}_i\right)\mathbf{D}_j = \sum_{i=1}^{n} (\mathbf{D}_j^{\dagger}\mathbf{D}_i^{\dagger})(\mathbf{D}_i\mathbf{D}_j) = \sum_{k=1}^{n} \mathbf{D}_k^{\dagger}\mathbf{D}_k = \mathbf{H} \tag{5.6.9}$$

and, being Hermitian, can be written in the form

$$\mathbf{H} = \mathbf{V}^{\dagger}\mathbf{V} \tag{5.6.10}$$

as can be found by diagonalization of **H**, according to the equations.

$$\mathbf{HL} = \mathbf{L\Lambda} \tag{5.6.11}$$

in which $\mathbf{\Lambda}$ is diagonal and real.

$$\mathbf{H} = \mathbf{L\Lambda L}^\dagger = \mathbf{L\Lambda}^{\frac{1}{2}}\mathbf{\Lambda}^{\frac{1}{2}}\mathbf{L}^\dagger = (\mathbf{L\Lambda}^{\frac{1}{2}})(\mathbf{L\Lambda}^{\frac{1}{2}})^\dagger = \mathbf{V}^\dagger\mathbf{V} \tag{5.6.12}$$

We can now prove that

$$\overline{\mathbf{D}}_i = \mathbf{VD}_i\mathbf{V}^{-1} \tag{5.6.13}$$

is unitary since, according to (5.6.7), (5.6.8) and (5.6.10),

$$\overline{\mathbf{D}}_i^\dagger\overline{\mathbf{D}}_i = \mathbf{V}^{-1\dagger}\mathbf{D}_i^\dagger\mathbf{V}^\dagger\mathbf{VD}_i\mathbf{V}^{-1} = \mathbf{V}^{-1\dagger}\mathbf{D}_i^\dagger\mathbf{HD}_i\mathbf{V}^{-1} = \mathbf{V}^{-1\dagger}\mathbf{HV}^{-1} = \mathbf{V}^{-1\dagger}\mathbf{V}^\dagger\mathbf{VV}^{-1} = \mathbf{E} \tag{5.6.14}$$

The second theorem states that if two unitary representations $\Gamma(\mathbf{D}_1, \mathbf{D}_2, \ldots, \mathbf{D}_n)$ and $\overline{\Gamma}(\overline{\mathbf{D}}_1, \overline{\mathbf{D}}_2, \ldots, \overline{\mathbf{D}}_n)$ are equivalent, they can be related by a similarity transformation made with a unitary matrix.

We wish to prove that, if the representations Γ and $\overline{\Gamma}$ are both unitary and are related by the similarity transformation (5.6.13), we can also find a unitary transformation matrix **U** such that

$$\overline{\mathbf{D}}_j = \mathbf{U}^\dagger\mathbf{D}_j\mathbf{U} \tag{5.6.15}$$

Let us assume that we know a matrix **M** which commutes with any $\overline{\mathbf{D}}_j$

$$\overline{\mathbf{D}}_j\mathbf{M} = \mathbf{M}\overline{\mathbf{D}}_j \tag{5.6.16}$$

in this case equation (5.6.13) becomes, using (5.6.16),

$$\mathbf{D}_j = \mathbf{V}^{-1}\overline{\mathbf{D}}_j\mathbf{V} = \mathbf{V}^{-1}\mathbf{M}^{-1}\overline{\mathbf{D}}_j\mathbf{MV} = (\mathbf{MV})^{-1}\overline{\mathbf{D}}_j(\mathbf{MV}) \tag{5.6.17}$$

and we can choose **M** such that **MV** is unitary.

From equation (5.6.13) it follows that

$$\overline{\mathbf{D}}_j\mathbf{V} = \mathbf{VD}_j \tag{5.6.18}$$

and taking the Hermitian transpose

$$\mathbf{D}_j^{-1}\mathbf{V}^\dagger = \mathbf{V}^\dagger\overline{\mathbf{D}}_j^{-1} \tag{5.6.19}$$

Since equation (5.6.19) holds for any pair of matrices of Γ and $\overline{\Gamma}$ we can replace $\mathbf{D}_j^{-1}$ and $\overline{\mathbf{D}}_j^{-1}$ with $\mathbf{D}_j$ and $\overline{\mathbf{D}}_j$, obtaining

$$\mathbf{D}_j\mathbf{V}^\dagger = \mathbf{V}^\dagger\overline{\mathbf{D}}_j \tag{5.6.20}$$

or

$$\mathbf{D}_j = \mathbf{V}^\dagger\overline{\mathbf{D}}_j\mathbf{V}^{\dagger-1} \tag{5.6.21}$$

By substitution of (5.6.21) into (5.6.18) we have

$$\overline{\mathbf{D}}_j\mathbf{V} = \mathbf{VV}^\dagger\overline{\mathbf{D}}_j\mathbf{V}^{\dagger-1} \tag{5.6.22}$$

or

$$\overline{\mathbf{D}}_j\mathbf{VV}^\dagger = \mathbf{VV}^\dagger\overline{\mathbf{D}}_j \tag{5.6.23}$$

This shows that the matrix $\mathbf{VV}^\dagger$ commutes with $\overline{\mathbf{D}}_j$, and therefore also commutes with **M**. In order to make **MV** unitary we can use this property, thus obtaining

$$\mathbf{MVV}^\dagger\mathbf{M}^\dagger = \mathbf{E} \tag{5.6.24}$$

$$\mathbf{VV^\dagger MM^\dagger} = \mathbf{E} \tag{5.6.25}$$

$$\mathbf{MM^\dagger} = (\mathbf{VV^\dagger})^{-1} \tag{5.6.26}$$

From the commutation of $\mathbf{M}$ with $\mathbf{VV^\dagger}$ and from the fact that $(\mathbf{VV^\dagger})^\dagger = \mathbf{VV^\dagger}$ it follows that $\mathbf{M^\dagger} = \mathbf{M}$. We have, therefore, from (5.6.26)

$$\mathbf{M}^2 = (\mathbf{VV^\dagger})^{-1} \tag{5.6.27}$$

$$\mathbf{M} = (\mathbf{VV^\dagger})^{-\frac{1}{2}} \tag{5.6.28}$$

The unitary matrix $\mathbf{MV}$ therefore has the form

$$\mathbf{MV} = (\mathbf{VV^\dagger})^{-\frac{1}{2}}\mathbf{V} \tag{5.6.29}$$

The third theorem states that if a representation Γ is reducible to an equivalent representation $\bar{\Gamma}$ in which the matrices have the form

$$\bar{\mathbf{D}}_j = \begin{pmatrix} \mathbf{D}_{j1} & \mathbf{X}_j \\ \mathbf{0} & \mathbf{D}_{j2} \end{pmatrix}$$

then the representation is completely reducible, i.e. is equivalent to the representation obtained from $\bar{\Gamma}$ by putting all $\mathbf{X}_j = 0$. In order to prove the theorem it is enough to observe that if

$$\bar{\mathbf{D}}_j = \begin{pmatrix} \mathbf{D}_{j1} & \mathbf{X} \\ \mathbf{0} & \mathbf{D}_{j2} \end{pmatrix} \tag{5.6.30}$$

is one of the matrices of the representation $\bar{\Gamma}$, there must be, from the definition of representation, another matrix in the representation equal to $\bar{\mathbf{D}}_j^{-1}$. But, if $\bar{\mathbf{D}}_j$ is unitary, we have

$$\bar{\mathbf{D}}_j^{-1} = \bar{\mathbf{D}}_j^\dagger = \begin{pmatrix} \mathbf{D}_{j1}^\dagger & \mathbf{0} \\ \mathbf{X}^\dagger & \mathbf{D}_{j2}^\dagger \end{pmatrix} \tag{5.6.31}$$

which does not have the required form (5.6.30) unless $\mathbf{X}_j = 0$.

These three theorems ensure that any reducible representation of a finite group can be reduced, by unitary transformations, to the direct sum of reduced unitary representations. The process can be repeated until no further reduction is possible. The final result is the reduction of the reducible representation to the direct sum of irreducible unitary representations. Generally some of these irreducible representations will be equivalent, i.e. will differ only by a similarity transformation and therefore will not be considered as distinct. If n_α is the number of times that the Γ_α irreducible representation occurs in Γ we have

$$\Gamma \cdot\!\equiv\!\cdot\, n_1\Gamma_1 + n_2\Gamma_2 + \ldots = \sum_\alpha n_\alpha \Gamma_\alpha \tag{5.6.32}$$

where the symbol $\cdot\!\equiv\!\cdot$ means equivalent to and the symbol $+$ indicates the direct matrix sum.

5.7 Orthogonality relations

Once a representation Γ is reduced to the direct sum of irreducible representations $\Gamma_\alpha, \Gamma_\beta, \ldots$ we can analyse the structure of these representations and the relationships among them. These relationships between matrices of two irreducible representations of a group are governed by a famous theorem, known as Schur's lemma.

Let $\Gamma_\alpha(\mathbf{D}_1^\alpha, \mathbf{D}_2^\alpha, \ldots, \mathbf{D}_g^\alpha)$ and $\Gamma_\beta(\mathbf{D}_1^\beta, \mathbf{D}_2^\beta, \ldots, \mathbf{D}_g^\beta)$ be two irreducible representations of a group or, more generally, two collections of irreducible matrices, with dimensions g_α and g_β ($g_\alpha \geqslant g_\beta$) respectively. Then Schur's lemma states that if there exists a matrix $\mathbf{S}(g_\alpha \times g_\beta)$

such that

$$\mathbf{D}_j^\alpha \mathbf{S} = \mathbf{S}\mathbf{D}_j^\beta \tag{5.7.1}$$

then either (a) **S** contains only zeros or (b) **S** is square and non-singular (i.e. $g_\alpha = g_\beta$).

To prove the theorem let us write $\mathbf{S} = (|\mathbf{s}_1|, |\mathbf{s}_2|, \ldots, |\mathbf{s}_{g_\beta}|)$ where $|\mathbf{s}_j|$ represents the jth column of the **S**-matrix. Then, by the rules of matrix multiplication we have from (5.7.1)

$$\mathbf{D}_j^\alpha \mathbf{S} = (|\mathbf{D}_j^\alpha \mathbf{s}_1|\,|\mathbf{D}_j^\alpha \mathbf{s}_2|, \ldots, |\mathbf{D}_j^\alpha \mathbf{s}_{g_\beta}|) \tag{5.7.2}$$

$$\mathbf{S}\mathbf{D}_j^\beta = \left(\left|\sum_l (D_j^\beta)_{l1}\mathbf{s}_l\right|, \left|\sum_l (D_j^\beta)_{l2}\mathbf{s}_l\right|, \ldots, \left|\sum_l (D_j^\beta)_{lg_\beta}\mathbf{s}_l\right|\right) \tag{5.7.3}$$

hence

$$\mathbf{D}_j^\alpha \mathbf{s}_k = \sum_l (D_j^\beta)_{lk}\mathbf{s}_l \tag{5.7.4}$$

which states that the matrix $\mathbf{D}_j^\alpha$ sends each of the g_β vectors $\mathbf{s}_l$ into a linear combination of themselves. The g_β vectors $\mathbf{s}_l$ therefore span an invariant subspace of dimension $g_\beta \leqslant g_\alpha$. Since the matrices $\mathbf{D}_j^\alpha$ are members of an irreducible representation of dimension g_α, they represent the action of symmetry operations on a basis which spans an invariant space of dimension g_α which cannot possess invariant subspaces. It follows that either (a) $\mathbf{S} = 0$, i.e. the invariant subspace is the zero space, or (b) $g_\alpha = g_\beta$, i.e. **S** is a square matrix. Since in this case the g_α columns of **S** span a g_α dimensional space, **S** is also non-singular.

There is an important corollary of Schur's lemma which states[4] that if **S** is a matrix which commutes with an irreducible set of matrices $\mathbf{D}_j^\alpha$, then **S** must be a scalar matrix, i.e. must be of the form $\lambda\mathbf{E}$, where λ is a scalar and **E** the unit matrix.

We can now prove that if $\Gamma_\alpha(\mathbf{D}_1^\alpha, \mathbf{D}_2^\alpha, \ldots, \mathbf{D}_g^\alpha)$ and $\Gamma_\beta(\mathbf{D}_1^\beta, \mathbf{D}_2^\beta, \ldots, \mathbf{D}_g^\beta)$ are two irreducible representations of a finite group, a matrix **S** which satisfies equation (5.7.1) can always be written in the form

$$\mathbf{S} = \sum_r (\mathbf{D}_r^\alpha)^{-1}\mathbf{M}\mathbf{D}_r^\beta \tag{5.7.5}$$

where **M** is an arbitrary $g_\alpha \times g_\beta$ matrix, since

$$\mathbf{D}_k^\alpha \mathbf{S} = \mathbf{D}_k^\alpha \sum_r (\mathbf{D}_r^\alpha)^{-1}\mathbf{M}\mathbf{D}_r^\beta(\mathbf{D}_k^\beta)^{-1}\mathbf{D}_k^\beta = \sum_i (\mathbf{D}_i^\alpha)^{-1}\mathbf{M}\mathbf{D}_i^\beta\mathbf{D}_k^\beta = \mathbf{S}\mathbf{D}_k^\beta \tag{5.7.6}$$

where the product $\mathbf{D}_k^\alpha(\mathbf{D}_r^\alpha)^{-1} = (\mathbf{D}_i^\alpha)^{-1}$ and the product $\mathbf{D}_r^\beta(\mathbf{D}_k^\beta)^{-1} = \mathbf{D}_i^\beta$ are always corresponding matrices of the two representations and the index i runs through the group as does the index r.

Let us now apply Schur's lemma. We have two possibilities:

(a) $g_\alpha \neq g_\beta$. In this case $\mathbf{S} = 0$. From equation (5.7.5) we have that the element S_{il} is given by

$$S_{il} = \sum_r \sum_{pq} (D_r^\alpha)_{ip}^{-1} M_{pq}(D_r^\beta)_{ql} = 0 \tag{5.7.7}$$

where

$$p, i = 1, 2, \ldots, g_\alpha \qquad q, l = 1, 2, \ldots, g_\beta$$

Since **M** is completely arbitrary we can set $M_{pq} = 1$ if $p = j$ and $q = k$ and $M_{pq} = 0$ otherwise.

Equation (5.7.7) then reads

$$\sum_r (D_r^\alpha)_{ij}^{-1}(D_r^\beta)_{kl} = 0 \tag{5.7.8}$$

(b) $g_\alpha = g_\beta$. In this case if Γ_α and Γ_β are inequivalent, then there is no similarity transformation which transforms Γ_α into Γ_β. Therefore in equation (5.7.6) again $\mathbf{S} = 0$. If instead Γ_α and Γ_β are equivalent we can always make them identical by a similarity transformation. Thus we can assume $\Gamma_\alpha = \Gamma_\beta$ and equation (5.7.6) becomes

$$\mathbf{D}_k^\alpha \mathbf{S} = \mathbf{S}\mathbf{D}_k^\alpha \tag{5.7.9}$$

Since $\mathbf{S}$ commutes with any matrix $\mathbf{D}_k^\alpha$ ($k = 1, 2, \ldots, g$) of the Γ_αth representation, it follows from the corollary of Schur's lemma that $\mathbf{S}$ must be a scalar matrix, i.e. must be of the form $\lambda\mathbf{E}$. From equation (5.7.5) we thus have for the S_{il} element

$$S_{il} = \sum_r \sum_{pq} (D_r^\alpha)_{ip}^{-1} M_{pq} (D_r^\alpha)_{ql} = \lambda\delta_{il} \tag{5.7.10}$$

and again assuming that all elements of $\mathbf{M}$ equal zero except the element $M_{jk} = 1$ we have

$$\sum_r (D_r^\alpha)_{ij}^{-1} (D_r^\alpha)_{kl} = \lambda\delta_{il} \tag{5.7.11}$$

Since every diagonal element ($l = i$) of $\mathbf{S}$ is equal to λ, all others being zero, if we sum together all the g_α diagonal elements of $\mathbf{S}$ (by putting $i = l$ and summing over i) we obtain

$$\sum_r \sum_i (D_r^\alpha)_{ij}^{-1} (D_r^\alpha)_{ki} = \lambda g_\alpha \tag{5.7.12}$$

but, owing to the fact that $(\mathbf{D}_r^\alpha)^{-1}(\mathbf{D}_r^\alpha) = \mathbf{E}$, the left-side sum over i gives a factor δ_{kj}, whereas the summation over the g matrices of the group gives $g\delta_{kj}$. We have therefore

$$g\delta_{kj} = \lambda g_\alpha \tag{5.7.13}$$

or

$$\lambda = \frac{g}{g_\alpha}\delta_{kj} \tag{5.7.14}$$

If we insert this value of λ in equation (5.7.11) we obtain

$$\sum_r (D_r^\alpha)_{ij}^{-1} (D_r^\alpha)_{kl} = (g/g_\alpha)\delta_{il}\delta_{kj} \tag{5.7.15}$$

Combining together (5.7.8) and (5.7.15) we obtain

$$\sum_r (D_r^\alpha)_{ij}^{-1} (D_r^\beta)_{kl} = (g/g_\alpha)\delta_{\alpha\beta}\delta_{kj}\delta_{il} \tag{5.7.16}$$

If the representations are unitary, $(D_r^\alpha)_{ij}^{-1} = (D_r^\alpha)_{ji}^*$ and equation (5.7.16) becomes

$$\sum_r (D_r^\alpha)_{ij}^* (D_r^\beta)_{kl} = (g/g_\alpha)\delta_{\alpha\beta}\delta_{ki}\delta_{jl} \tag{5.7.17}$$

Equation (5.7.17) is the most important equation of the whole chapter and it is known as the orthogonality relation. It is very convenient to introduce a geometric interpretation of this equation by pointing out that it states the orthogonality of a set of vectors whose components are simply the corresponding elements $(D_r^\alpha)_{ij}$ of the matrices of the irreducible representations. Thus, if a representation Γ^α is monodimensional we can formally consider the g-number of Γ^α as the components of a vector in a g-dimensional space that we shall call the 'group element space'. If the representation is bidimensional, we can form four such vectors, collecting corresponding matrix elements. In general, for a g_α dimensional representation we have g_α^2 vectors collecting together, for each pair of indices i and j, the matrix elements $(D_r^\alpha)_{ij}$ for $r = 1, 2, \ldots, g$. These vectors are all orthogonal to each other and have length $(g/g_\alpha)^{\frac{1}{2}}$. This leads us to a very important conclusion. The total number of such vectors is clearly

$$\sum_\alpha g_\alpha^2$$

and since in a g-dimensional space the maximum number of orthogonal vectors is just equal to g, it follows that

$$\sum_\alpha g_\alpha^2 \leqslant g \tag{5.7.18}$$

We shall prove later that only the equality holds in (5.7.18). We obtain, therefore, the so-called 'dimensionality theorem'

$$\sum_\alpha g_\alpha^2 = g \tag{5.7.19}$$

which shows that the number of irreducible representations of a finite group is finite and such that the sum of the squares of the dimensions of the representations must be equal to the order of the group. This furnishes an important test in working out the irreducible representations of a group. For instance, in the case of the C_{3v} point group, of order 6, we have found three irreducible representations Γ_1, Γ_2 and Γ_3 of dimension $g_1 = 1$, $g_2 = 1$ and $g_3 = 2$ respectively. Equation (5.7.19) then yields

$$1^2 + 1^2 + 2^2 = 6$$

The dimensionality theorem tells us, therefore, that no other irreducible representation can exist for C_{3v}.

In Section 5.5 we have shown (equation 5.5.41) that the character of a matrix is independent of the choice of the basis and remains unchanged in a similarity transformation. The character is thus a specific property of any matrix $\mathbf{D}_r^\alpha$ belonging to the Γ^α representation, and can be used to identify an irreducible representation. The set of n characters of the n matrices of a representation is called the character system of the representation or simply the character of the representation. Equivalent representations have the same character system whereas inequivalent representations have different character systems. Furthermore, the characters of the matrices corresponding to group elements belonging to the same class must be the same since if two elements R and S belong to the same class they are related by a similarity transformation $R = TST^{-1}$ and the same must be true for the corresponding matrices. It is therefore unnecessary in writing the character systems of an irreducible representation, to specify the characters of all elements. Those belonging to the same class can be collected together and written only once. A table of the characters of the irreducible representations of a group, called a character table, will therefore specify the number of elements in each class and the character for each representative element of the class.

In appendix I are collected the tables of the characters of the irreducible representations of the point groups of interest in molecular spectroscopy.

Example 5.7.2

For the C_{3v} group the character systems of the irreducible representations are shown in the next table.

	E	C_3	C_3^{-1}	σ_1	σ_2	σ_3
Γ_1	1	1	1	1	1	1
Γ_2	1	1	1	-1	-1	-1
Γ_3	2	-1	-1	0	0	0

In the C_{3v} group there are three classes, (E), $(C_3C_3^{-1})$ and $(\sigma_1, \sigma_2, \sigma_3)$. From the table it is clear that the character in a representation is the same for elements belonging to the same class. The character table can therefore be written in the form

	E	$2C_3$	3σ
Γ_1	1	1	1
Γ_2	1	1	-1
Γ_3	2	-1	0

The orthogonality relation (5.7.17) is easily written in terms of characters, by considering only diagonal matrix elements. Putting $i = j$ and $k = l$, we have

$$\sum_r (D_r^\alpha)_{ii}^*(D_r^\beta)_{kk} = (g/g_\alpha)\delta_{\alpha\beta}\delta_{ik} \tag{5.7.20}$$

and summing over all the diagonal elements, i.e. over i and k, we have

$$\sum_r\left(\sum_i (D_r^\alpha)_{ii}^*\right)\left(\sum_k (D_r^\beta)_{kk}\right) = (g/g_\alpha)\delta_{\alpha\beta}\sum_{ik}\delta_{ik} \tag{5.7.21}$$

Since

$$\sum_i (D_r^\alpha)_{ii}^* = (\chi_r^\alpha)^*, \qquad \sum_k (D_r^\beta)_{kk} = (\chi_r^\beta) \quad \text{and} \quad \sum_{ik}\delta_{ik} = g_\alpha$$

equation (5.7.21) becomes

$$\sum_r (\chi_r^\alpha)^*\chi_r^\beta = g\delta_{\alpha\beta} \tag{5.7.22}$$

An important consequence of (5.7.22) can be seen by putting $\alpha = \beta$. We have then a convenient criterion for irreducibility in the form

$$\sum_r (\chi_r^\alpha)^*\chi_r^\alpha = g \tag{5.7.23}$$

For instance for the representation Γ_3 of C_{34}, we have

$$2 \times 2 + (-1) \times (-1) + 0 \times 0 = 6$$

and thus the representation is irreducible.

Let us now return to the problem of reduction of a representation. In Section 5.6 we have seen that through similarity transformations we can reduce a representation Γ to the direct sum of irreducible representations. If n_α is the number of times that the Γ_α irreducible representation occurs in Γ we have formally

$$\Gamma = \sum_\alpha n_\alpha \Gamma_\alpha$$

Since the trace of a matrix is invariant under a similarity transformation the trace of a matrix $\mathbf{D}$ of Γ must be equal to the sum of the traces of the matrices of the irreducible representations, as shown in the following schema

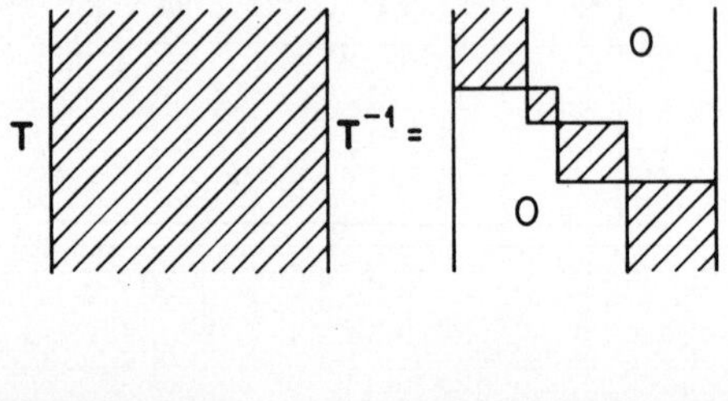

We thus have

$$\chi_r = \sum_\alpha n_\alpha \chi_r^\alpha \tag{5.7.24}$$

where χ_r is the character of the rth matrix of the reducible representation Γ and χ_r^α is the character of the rth matrix of the αth irreducible representation. If we multiply equation (5.7.24) by $(\chi_r^\gamma)^*$ and sum over r we have from (5.7.22)

$$\sum_r (\chi_r^\gamma)^* \chi_r = \sum_\alpha n_\alpha \sum_r (\chi_r^\gamma)^* \chi_r^\alpha = \sum_\alpha n_\alpha g \delta_{\alpha\gamma} = g n_\gamma \tag{5.7.25}$$

and thus

$$n_\gamma = \frac{1}{g} \sum_r (\chi_r^\gamma)^* \chi_r \tag{5.7.26}$$

This equation is of extreme importance for us since it allows the number of times an irreducible representation occurs in the reducible one to be calculated, once the characters of the irreducible representations are known. We can now prove that the number of irreducible representations of a group is equal to the number of classes of the group. We have already seen (Example 5.7.2) that the characters of the matrices corresponding to elements of a group belonging to the same class must be the same. We can thus rewrite equation (5.7.22), collecting together the characters of elements belonging to the same classes and summing over h, the number of classes, instead of r, the number of elements of the group,

$$\sum_{r=1}^{g} \chi_r^{\alpha *} \chi_r^\beta = \sum_{i=1}^{h} h_i \chi_i^{\alpha *} \chi_i^\beta = g \delta_{\alpha\beta} \tag{5.7.27}$$

where h_i is the number of elements in the ith class. Equation (5.7.27) can be rewritten in the form

$$\sum_h \left[\left(\frac{h_i}{g} \right)^{\frac{1}{2}} \chi_i^\alpha \right]^* \left[\left(\frac{h_i}{g} \right)^{\frac{1}{2}} \chi_i^\beta \right] = \delta_{\alpha\beta} \tag{5.7.28}$$

which can be interpreted as the scalar product of two orthonormal vectors with components

$$\left(\frac{h_1}{g} \right)^{\frac{1}{2}} \chi_1^\alpha \quad \left(\frac{h_2}{g} \right)^{\frac{1}{2}} \chi_2^\alpha \quad \cdots \quad \left(\frac{h_n}{g} \right)^{\frac{1}{2}} \chi_n^\alpha$$

$$\left(\frac{h_1}{g} \right)^{\frac{1}{2}} \chi_1^\beta \quad \left(\frac{h_2}{g} \right)^{\frac{1}{2}} \chi_2^\beta \quad \cdots \quad \left(\frac{h_n}{g} \right)^{\frac{1}{2}} \chi_n^\beta$$

spanning an h-dimensional space.

Since there is one vector of this kind for each irreducible representation the number of irreducible representations must be $\leqslant h$. By similar arguments, it is also possible to show (see Lomont,[2] appendix II) that the number of irreducible representations must be $\geqslant h$. We can therefore conclude that only the equality is possible.

We shall now introduce the concept of the regular representation of a group, that will enable us to prove the validity of equation (5.7.19). Consider a reducible representation of dimension g of a group G of order g, such that the matrix associated with the identity element E has the number 1 along the diagonal and zero otherwise and the matrices associated with other elements of the group have all zero elements except for one entry 1 in each row and column in such a way that the group multiplication table is satisfied. Such a representation is called the regular representation and has the interesting property that it contains each irreducible representation of the group as many times as the dimensionality of the irreducible representation. This is simply proved using equation (5.7.26). If the structure of the matrices of the representation is as described above, the character χ_E of the matrix associated with the unity element E is obviously equal to g whereas the character of any other matrix of the representation is zero. We then have from (5.7.26)

$$n_\gamma = \frac{1}{g}(\chi_E^\gamma)^*\chi_E = (\chi_E^\gamma)^* = g_\gamma \tag{5.7.29}$$

Thus one-dimensional irreducible representations will appear only once, two-dimensional representations twice and in general g_γ dimensional irreducible representations will appear g_γ times in the regular representation.

There is a simple method to construct the regular representation of a group. We rearrange the rows of the multiplication table so that they correspond to the inverses of the elements labelling the columns. Then we obtain the matrix associated with the generic element R of the group by replacing R in the multiplication table with the number 1 and putting zero otherwise. A simple example will clarify the procedure.

Example 5.7.3

Consider the multiplication table of the C_{3v} group given in Example 5.2.6. By rearranging the rows as described above we have

	E	C_3	C_3^{-1}	σ_1	σ_2	σ_3
E	E	C_3	C_3^{-1}	σ_1	σ_2	σ_3
C_3^{-1}	C_3^{-1}	E	C_3	σ_2	σ_3	σ_1
C_3	C_3	C_3^{-1}	E	σ_3	σ_1	σ_2
σ_1	σ_1	σ_2	σ_3	E	C_3	C_3^{-1}
σ_2	σ_2	σ_3	σ_1	C_3^{-1}	E	C_3
σ_3	σ_3	σ_1	σ_2	C_3	C_3^{-1}	E

The matrix associated, for instance, with the elements E, C_3 and σ_1 in the regular representation will therefore have the form

$$\overset{E}{\begin{vmatrix} 1&0&0&0&0&0\\ 0&1&0&0&0&0\\ 0&0&1&0&0&0\\ 0&0&0&1&0&0\\ 0&0&0&0&1&0\\ 0&0&0&0&0&1 \end{vmatrix}} \quad \overset{C_3}{\begin{vmatrix} 0&1&0&0&0&0\\ 0&0&1&0&0&0\\ 1&0&0&0&0&0\\ 0&0&0&0&1&0\\ 0&0&0&0&0&1\\ 0&0&0&1&0&0 \end{vmatrix}} \quad \overset{\sigma_1}{\begin{vmatrix} 0&0&0&1&0&0\\ 0&0&0&0&0&1\\ 0&0&0&0&1&0\\ 1&0&0&0&0&0\\ 0&0&1&0&0&0\\ 0&1&0&0&0&0 \end{vmatrix}}$$

The character of the matrix $\mathbf{D}(E)$ is thus equal to 6 whereas the character of all other matrices is equal to zero.

We shall now prove that the g matrices obtained in this way for a group of order g form a representation of the group. For this we need to show that if R_a, R_b and R_c are elements of the group such that $R_aR_b = R_c$, then also $\mathbf{D}_{R_a}^{\text{reg}} \cdot \mathbf{D}_{R_b}^{\text{reg}} = \mathbf{D}_{R_c}^{\text{reg}}$.

A generic element of the product is given by

$$(D_{R_c})_{R_k^{-1}R_i} = \sum_{R_j} (D_{R_a})_{R_k^{-1}R_j}(D_{R_b})_{R_j^{-1}R_i} \tag{5.7.30}$$

where the subscripts label the row and the columns of the multiplication table rearranged in the way described before. Since by definition of the table

$$(D_{R_a})_{R_k^{-1}R_j} = \begin{cases} 1 & \text{if } R_k^{-1}R_j = R_a \\ 0 & \text{otherwise} \end{cases} \quad \text{and} \quad (D_{R_b})_{R_j^{-1}R_i} = \begin{cases} 1 & \text{if } R_j^{-1}R_i = R_b \\ 0 & \text{otherwise} \end{cases} \tag{5.7.31}$$

the sum (5.7.30) vanishes except in the case in which the indices satisfy the relation

$$R_a R_b = (R_k^{-1}R_j)(R_j^{-1}R_i) = R_k^{-1}R_i \tag{5.7.32}$$

the element being in this case equal to unity. This is the combination of indices for which the element on the left-hand side of equation (5.7.30) must be equal to unity also. We conclude, therefore, that the matrices constructed using this procedure form a representation of the group. Since the character of the matrix associated with E is equal to g, and all other characters are zero, the representation is regular.

We can now show that only the equality holds in equation (5.7.18). In a regular representation the character of the matrix associated with the identity element is equal to the order g of the group, as stated before. On the other hand, it must also be equal to the sum of the characters of the irreducible matrices into which it can be decomposed. Since the character of an irreducible matrix of dimension g_α is equal to g_α, and since this representation occurs g_α times in the regular representation, we have

$$\sum_\alpha g_\alpha \cdot g_\alpha = g \tag{5.7.33}$$

which is just equation (5.7.19).

5.8 The direct product

In many applications of group theory it is important to know the representations carried by quantities which can be expressed as products of other quantities whose symmetry behaviour is already known. This is conveniently done by using a method of combination of matrices called the *direct product of matrices*. Since representations are actually collections of matrices, the direct product of matrices makes it possible to define a *direct product of representations* which simplifies the group-theoretical treatment of complicated functions.

The direct product of representations leads in turn to the concept of the *direct product of groups*, which is a way of combining together the elements of two separate groups to form a new group of larger dimension. For convenience we shall discuss these three different aspects of the direct product in three separate subsections.

(A) *The direct product of matrices*

The direct product (or Kronecker or outer product) of two square matrices **A** and **B** is a matrix **C** whose elements are defined by the relation

$$C_{ij,kl} = A_{ik}B_{jl} \tag{5.8.1}$$

where the elements of **C** have been labelled using double subscripts in order to indicate their origin from the elements of the matrices **A** and **B**. We shall use the symbol $\otimes$ to indicate the direct product and thus write

$$\mathbf{C} = \mathbf{A} \otimes \mathbf{B} \tag{5.8.2}$$

If the matrix **A** is of dimension g_α and the matrix **B** is of dimension g_β, the matrix **C** will be of dimension $g_\alpha g_\beta$. For example, if **A** is a 2×2 and **B** a 3×3 matrix, the matrix **C** will have the form

$$
\mathbf{A}\otimes\mathbf{B} = \left|\begin{array}{ccc|ccc}
A_{11}B_{11} & A_{11}B_{12} & A_{11}B_{13} & A_{12}B_{11} & A_{12}B_{12} & A_{12}B_{13} \\
A_{11}B_{21} & A_{11}B_{22} & A_{11}B_{23} & A_{12}B_{21} & A_{12}B_{22} & A_{12}B_{23} \\
A_{11}B_{31} & A_{11}B_{32} & A_{11}B_{33} & A_{12}B_{31} & A_{12}B_{32} & A_{12}B_{33} \\
\hline
A_{21}B_{11} & A_{21}B_{12} & A_{21}B_{13} & A_{22}B_{11} & A_{22}B_{12} & A_{22}B_{13} \\
A_{21}B_{21} & A_{21}B_{22} & A_{21}B_{23} & A_{22}B_{21} & A_{22}B_{22} & A_{22}B_{23} \\
A_{21}B_{31} & A_{21}B_{32} & A_{21}B_{33} & A_{22}B_{31} & A_{22}B_{32} & A_{22}B_{33}
\end{array}\right| = \left|\begin{array}{c|c} A_{11}\mathbf{B} & A_{12}\mathbf{B} \\ \hline A_{21}\mathbf{B} & A_{22}\mathbf{B} \end{array}\right|
$$

From the definition (5.8.1) of the direct product, it follows that

$$\mathrm{tr}\,(\mathbf{C}) = \mathrm{tr}\,(\mathbf{A}) \times \mathrm{tr}\,(\mathbf{B}) \tag{5.8.3}$$

$$(\mathbf{AB}) \otimes (\mathbf{CD}) = (\mathbf{A} \otimes \mathbf{C})(\mathbf{B} \otimes \mathbf{D}) \tag{5.8.4}$$

(B) *The direct product of representations*

Let $\Gamma^\alpha(\mathbf{D}_1^\alpha, \mathbf{D}_2^\alpha, \ldots, \mathbf{D}_g^\alpha)$ and $\Gamma^\beta(\mathbf{D}_1^\beta, \mathbf{D}_2^\beta, \ldots, \mathbf{D}_g^\beta)$ be two representations of a group, of dimension g_α and g_β respectively. The representation $\Gamma^\gamma(\mathbf{D}_1^\gamma, \mathbf{D}_2^\gamma, \ldots, \mathbf{D}_g^\gamma)$, where

$$\mathbf{D}_1^\gamma = \mathbf{D}_1^\alpha \otimes \mathbf{D}_1^\beta \qquad \mathbf{D}_2^\gamma = \mathbf{D}_2^\alpha \otimes \mathbf{D}_2^\beta \ldots \mathbf{D}_g^\gamma = \mathbf{D}_g^\alpha \otimes \mathbf{D}_g^\beta \tag{5.8.5}$$

is then another representation of dimension $g_\alpha \times g_\beta$ of the group since, according to (5.8.4), if $\mathbf{D}_i^\varepsilon \mathbf{D}_j^\varepsilon = \mathbf{D}_k^\varepsilon$ $(\varepsilon = \alpha, \beta)$, we have

$$(\mathbf{D}_i^\alpha \otimes \mathbf{D}_i^\beta)(\mathbf{D}_j^\alpha \otimes \mathbf{D}_j^\beta) = (\mathbf{D}_i^\alpha \mathbf{D}_j^\alpha) \otimes (\mathbf{D}_i^\beta \mathbf{D}_j^\beta) = \mathbf{D}_k^\alpha \otimes \mathbf{D}_k^\beta = \mathbf{D}_k^\gamma \tag{5.8.6}$$

The representation Γ^γ is called the direct product (inner product) of the representations Γ^α and Γ^β. Symbolically we write

$$\Gamma^\gamma = \Gamma^\alpha \otimes \Gamma^\beta \tag{5.8.7}$$

The inner product constitutes, therefore, a method of combining together the representations of a group to obtain other representations. In particular, if Γ^α and Γ^β are irreducible representations of a group, the representation Γ^γ will not necessarily be irreducible. It is easily seen that Γ^γ will be an irreducible representation only if at least one of the two representations Γ^α and Γ^β is monodimensional. It is important to notice that the inner product of representations refers to representations of the same group and yields g matrices if g is the order of the group. Later we shall discuss a different form of direct product between representations of different groups. In order to understand the importance of the direct product in many physical problems, it is convenient to discuss the relation existing between the spaces which carry the representations Γ^α and Γ^β and the space carrying the representation Γ^γ.

Let $\mathbf{e}^\alpha$ and $\mathbf{e}^\beta$ be the two irreducible spaces, with basis vectors

$$\mathbf{e}_1^\alpha, \mathbf{e}_2^\alpha, \ldots, \mathbf{e}_{g_\alpha}^\alpha \qquad \mathbf{e}_1^\beta, \mathbf{e}_2^\beta, \ldots, \mathbf{e}_{g_\beta}^\beta$$

which carry the irreducible representations Γ^α and Γ^β respectively, so that for a given symmetry operator $\mathscr{R}$ we have

$$\begin{aligned} \mathscr{R}\mathbf{e}^\alpha &= \mathbf{e}^\alpha \mathbf{D}^\alpha \\ \mathscr{R}\mathbf{e}^\beta &= \mathbf{e}^\beta \mathbf{D}^\beta \end{aligned} \tag{5.8.8}$$

For a given element $\mathbf{e}_i^\alpha$ of the vector $\mathbf{e}^\alpha$ and $\mathbf{e}_j^\beta$ of the vector $\mathbf{e}^\beta$, equations (5.8.8) read

$$\begin{aligned}\mathscr{R}\mathbf{e}_i^\alpha &= \sum_k \mathbf{e}_k^\alpha D_{ki}^\alpha \\ \mathscr{R}\mathbf{e}_j^\beta &= \sum_l \mathbf{e}_l^\beta D_{lj}^\beta\end{aligned} \tag{5.8.9}$$

Consider now the effect of the symmetry operator $\mathscr{R}$ on the product $\mathbf{e}_i^\alpha\mathbf{e}_j^\beta$. According to (5.8.9) we have

$$\mathscr{R}\mathbf{e}_i^\alpha\mathbf{e}_j^\beta = \sum_{k,l} \mathbf{e}_k^\alpha\mathbf{e}_l^\beta D_{ki}^\alpha D_{lj}^\beta = \sum_{k,l} \mathbf{e}_k^\alpha\mathbf{e}_l^\beta \mathbf{D}_{ki,lj}^{\alpha\beta} \tag{5.8.10}$$

From the definition of direct product of matrices given by equation (5.8.1) it follows then that the $g_\alpha \times g_\beta$ products of the type $\mathbf{e}_i^\alpha\mathbf{e}_j^\beta$ one can form by multiplication of each of the g_α vectors of the space $\mathbf{e}^\alpha$ by each of the g_β vectors of the space $\mathbf{e}^\beta$, can be considered as the components of a new space, of dimension $g_\alpha \times g_\beta$, which carries a representation which is the direct product of the representations carried by $\mathbf{e}^\alpha$ and $\mathbf{e}^\beta$.

In Chapter 7 the reader will find several examples of direct products of representations, as well as a set of general rules for finding the structure of the direct product representations. In Chapter 7 we shall also discuss the special case of the so-called 'symmetrized direct product' which is needed to describe the symmetry behaviour of a symmetric tensor.

The definition of the direct product of representations given by (5.8.7) is easily generalized. Thus if $\Gamma^\alpha, \Gamma^\beta, \Gamma^\gamma, \ldots$ are representations of a group G, the direct product

$$\Gamma^\mu = \Gamma^\alpha \otimes \Gamma^\beta \otimes \Gamma^\gamma \otimes \ldots \tag{5.8.11}$$

is another representation of the group. The character of the representation Γ^μ is, according to (5.8.3), equal to the product of the characters of the representations $\Gamma^\alpha, \Gamma^\beta, \Gamma^\gamma, \ldots$.

(C) *The direct product of groups*

In some groups it can happen that the symmetry elements can be divided into two or more sets such that all elements of one set commute with the elements of the others. This always happens when the two symmetry operators operate on different coordinates. A typical example is the commutation of a proper rotation axis C_n which acts on the xy coordinates of a Cartesian basis, with a symmetry plane σ_h (perpendicular to C_n) which acts on the z-coordinate. Another possibility of commutation is related to the special form of a given operator. For instance, the inversion operator commutes with all others since it is associated with a diagonal matrix which is equal to the unity matrix multiplied by -1.

The possibility of dividing the symmetry elements of a group into sets, such that the elements of one set commute with those of the other, offers a convenient method of reducing the structure of the group to those of smaller groups through the concept of direct product of groups discussed below. Let $G_a\{E, R_2^a, R_3^a, \ldots, R_{ga}^a\}$ and $G_b\{E, R_2^b, \ldots, R_{gb}^b\}$ be two groups such that all elements of G_a commute with those of G_b. We define as the direct product of G_a and G_b a new group G_{ab}, with $g_a g_b$ elements, by forming all possible products of the elements of G_a with those of G_b and assuming that only one element in the new group is the identity

$$G_{ab} = G_a \otimes G_b\{E, R_2^a, R_3^a, \ldots, R_{ga}^a, R_2^b, R_2^aR_2^b, R_3^aR_2^b, \ldots, R_{ga}^aR_2^b, R_3^b, R_2^aR_3^b, \ldots, R_{gb}^b, R_2^aR_{gb}^b, \ldots, R_{ga}^aR_{gb}^b$$

It is easily shown that $G_a \otimes G_b$ satisfies the closure property since

$$(R_i^aR_j^b)(R_k^aR_l^b) = (R_i^aR_k^a)(R_j^bR_l^b) = R_s^aR_t^b \tag{5.8.12}$$

The properties of the new group G_{ab} are completely determined by those of the groups G_a and G_b, once the direct product of matrices, discussed before, is used to form the products between the matrices associated with symmetry elements of the type $R_i^a R_j^b$. Using Schur's lemma it can be shown[3] that the direct product of an irreducible representation Γ^α of G_a by an irreducible representation Γ^β of G_b forms an irreducible representation $\Gamma^{\alpha\beta}$ of G_{ab}. Furthermore, it is easily proved that all the irreducible representations of G_{ab} are obtained from those of G_a and G_b. It is important to avoid confusion between the direct product of the representations of two commuting groups discussed here and the direct product of the representations of the same group discussed in Section 5.8 (B). In the former case the direct product of two irreducible representations always yields an irreducible representation of the new group, whereas in the latter case the irreducibility of the direct product representation is preserved only if at least one of the initial representations is monodimensional.

A further property of the direct product of groups is that the number of classes of G_{ab} is equal to the product of the number of classes of G_a and the number of classes in G_b. Furthermore the character system of an irreducible representation of the direct product group G_{ab} is the product of the character systems of the component representations of G_a and G_b since

$$\chi_{ab}^{\gamma}(\mathscr{R}) = \mathrm{tr}\,\mathbf{D}_{ab}^{\gamma}(\mathscr{R}) = \sum_{ij} (D_{ij,ij}^{\gamma}(\mathscr{R}))_{ab} = \sum_{ij} (D_{ii}^{\alpha}(\mathscr{R}))_a (D_{jj}^{\beta}(\mathscr{R}))_b = \chi_a^{\alpha}(\mathscr{R})\chi_b^{\beta}(\mathscr{R}) \tag{5.8.13}$$

Whenever a group can be written as a direct product of two smaller groups, the complete character table of the group can thus be constructed from the knowledge of the character tables of the component groups.

Example 5.8.1

The group D_{3h} can be obtained as the direct product of the group D_3 and the group C_s (see Appendix I),

$$D_{3h} = D_3(E, C_3, C_3^{-1}, C_2^{(1)}, C_2^{(2)}, C_2^{(3)}) \otimes C_s(E, \sigma_h)$$

if the plane of symmetry of the C_s point group is taken perpendicular to the C_3 axis of the D_3 group. In this way σ_h commutes with all the operations of D_3, since it makes no difference if we first reflect a point through σ_h and then rotate it or vice versa. This is shown schematically in Figure 5.8.1.

According to the definition of direct product group, we form the twelve products (see Figure 5.8.1)

$$D_3 \otimes C_s = D_{3h}(E, C_3, C_3^{-1}, C_2^{(1)}, C_2^{(2)}, C_2^{(3)}, \sigma_h, S_3, S_3^{-1}, \sigma_1, \sigma_2, \sigma_3)$$

where

$$S_3 = \sigma_h C_3 \qquad S_3^{-1} = \sigma_h C_3^{-1}$$

$$\sigma_h C_2^{(1)} = \sigma_1 \qquad \sigma_h C_2^{(2)} = \sigma_2 \qquad \sigma_h C_2^{(3)} = \sigma_3$$

The character tables of the D_3 and C_s groups are given below. The nomenclature used for the irreducible representations will be discussed in the next chapter.

D_3	E	$2C_3$	$3C_2$
A_1	1	1	1
A_2	1	1	-1
E	2	-1	0

C_s	E	σ_h
A'	1	1
A''	1	-1

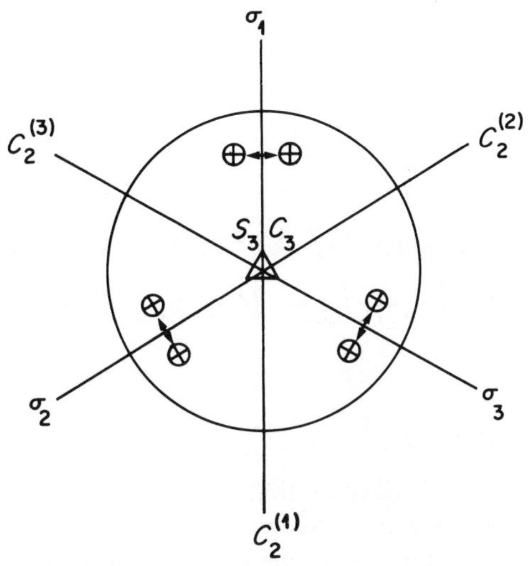

Figure 5.8.1 Symmetry operations of D_{3h}. Points above the symmetry plane σ_h are indicated by a cross, whereas points below the plane are indicated by circles. The operations of C_{3v} always send crosses into crosses and circles into circles. The operation σ_h sends each cross into the circle beneath. From the figure it is easily seen that σ_h commutes with all operations of the C_{3v} group and that $\sigma_h\sigma_i = C_2^{(i)}$, $\sigma_h C_3^n = S_3^n$

From these, the character table of the D_{3h} point group is easily obtained. In the original groups we have three and two classes and therefore three and two irreducible representations respectively. In the direct product group we then have six classes and thus six irreducible representations. The first three representations of D_{3h} are obtained by multiplying

D_{3h}	E	$2C_3$	$3C_2$	σ_h	$2S_3$	$3\sigma_v$
A_1'	1	1	1	1	1	1
A_2'	1	1	-1	1	1	-1
E'	2	-1	0	2	-1	0
A_1''	1	1	1	-1	-1	-1
A_2''	1	1	-1	-1	-1	1
E''	2	-1	0	-2	1	0

the three representations of D_3 by the representation A' of C_s. The next three representations are then obtained by multiplication by the representation A''. The nomenclature of the representations of D_{3h} follows directly from those of D_3 and C_s.

The D_{3h} group can be alternatively obtained by the direct product of the C_{3v} group with the C_s group.

References

1. A. P. Cracknell, *Applied Group Theory*, Pergamon Press, 1968.
2. J. S. Lomont, *Applications of Finite Groups*, Academic Press, 1959.
3. M. Tinkham, *Group Theory and Quantum Mechanics*, McGraw-Hill, 1964.
4. R. McWeeny, *Symmetry: An Introduction to Group Theory and its Applications*, Pergamon Press, 1963.
5. V. Heine, *Group Theory in Quantum Mechanics*, Pergamon Press, 1960.

Chapter 6
Applications of Group Theory to Molecular Vibrations

6.1 The classification of point groups

We have seen in the previous chapter that the symmetry operations which bring a physical object, like a molecule, into self-coincidence always form a group. At least one point of the object is left invariant by all symmetry operations and for this reason the groups of symmetry operations are called point groups.

Not all combinations of symmetry operations are possible, however, for a physical object. For instance, it is impossible to combine together a two-fold rotation and a reflection through a symmetry plane, unless the two-fold axis is contained in the plane or is perpendicular to it. This is due to the fact that each symmetry operation acts upon the symmetry elements, sending them into equivalent elements in much the same way as it acts upon the atoms of a molecule sending them into equivalent atoms. If we have, for instance, a two-fold axis C_2 and a symmetry plane σ at an arbitrary angle θ, the axis will produce from the plane σ a new plane σ', equivalent to σ, by rotation of 180°. In turn the two planes σ and σ' will produce from the C_2 axis two new axes C_2' and C_2'' by reflection at σ and σ' respectively, thus contradicting the hypothesis of a single rotation axis, unless $\theta = 0$ or $\theta = 90°$. On the other hand, the presence of some symmetry elements often implies the occurrence of certain others. In Figure 6.1.1, for instance, it is shown that a two-fold axis perpendicular to a symmetry plane introduces a centre of inversion at the point of intersection. The binary axis C_2 sends the point A into the symmetrically equivalent point A'. The plane σ_h then sends A' into the symmetrically equivalent point A''. The result of the two symmetry operations is therefore equivalent to an inversion at the centre i which also sends A into A''.

These considerations show that the number of possible combinations of symmetry elements is severely limited. Exhaustive combination of symmetry elements yields a relatively small number of point groups which describe the symmetry of physical objects.

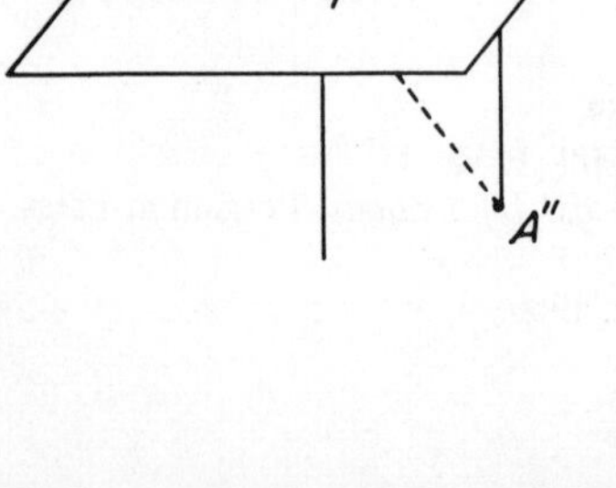

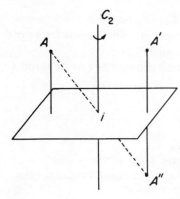

Figure 6.1.1 Generation of a centre of inversion i by the symmetry operations C_2 and σ_h

If we limit ourselves to the symmetry elements of interest in molecular systems, considering, in addition to symmetry planes and inversion centres, only rotation axes C_n and rotation–inversion axes S_n with $n = 1, 2, \ldots, 6, \infty$, we obtain the point groups collected in Table 6.1.1.

Table 6.1.1

		n						
Groups	Generators	1	2	3	4	5	6	∞
C_n	C_n	C_1	C_2	C_3	C_4	C_5	C_6	—
S_n	S_n	—	S_2	—	S_4	—	S_6	—
C_{nv}	C_n, σ_v	—	C_{2v}	C_{3v}	C_{4v}	C_{5v}	C_{v6}	$C_{\infty v}$
C_{nh}	C_n, σ_h	C_{1h}	C_{2h}	C_{3h}	C_{4h}	C_{5h}	C_{6h}	—
D_n	C_n, C_2	—	D_2	D_3	D_4	D_5	D_6	—
	C_n^z, C_3^{xyz}	—	T	—	O	—	—	—
	S_n^z, C_3^{xyz}	—	—	—	T_d	—	—	—
D_{nd}	C_n, C_2, σ_d	—	D_{2d}	D_{3d}	D_{4d}	D_{5d}	D_{6d}	—
D_{nh}	C_n, C_2, σ_h	—	D_{2h}	D_{3h}	D_{4h}	D_{5h}	D_{6h}	$D_{\infty h}$
	C_n^z, C_3^{xyz}, i	—	T_h	—	O_h	—	—	—

In their classification we have used the nomenclature introduced first by Schoenflies in the theory of crystal structure and commonly used in molecular spectroscopy and molecular structure problems.

In Table 6.1.1 we can distinguish three types of groups: (a) axial groups; (b) tetrahedral groups; (c) octahedral groups.

The axial point groups are characterized by a principal axis of order greater than (or at most equal to) any other axis of the group. The order of the principal axis is indicated by a subscript to the symbol of the axis (C or S). Additional subscripts specify the occurrence of other symmetry elements. For instance the subscript v indicates that there are symmetry planes containing the principal axis and the symbol h indicates a plane of symmetry perpendicular to the principal axis. If there are binary axes perpendicular to the principal axis, the symbol D is used for the group. The subscript d indicates symmetry planes, containing the principal axis and bisecting the angle between two adjacent twofold axes.

The tetrahedral and octahedral groups describe the symmetry of the regular tetrahedron and octahedron at various levels of completion. The fundamental feature of these symmetries is the occurrence of four three-fold axes oriented along the diagonal of a cube in which the tetrahedron and the octahedron can be inscribed. The nomenclature used follows that of the axial groups. The symbols T and O are used respectively for the pure rotation groups and subscripts are added to indicate other symmetry elements.

The simplest classification of the point groups of Table 6.1.1 is in terms of a set of generators, i.e. of a set of symmetry elements whose combination produces all the elements of the group.

The simplest type of point groups are the cyclic groups C_n of order n. There is only one generator and each element C_n^m ($m = 1, 2, \ldots, n$) of the group is a class by itself. C_1 is obviously a point group without symmetry elements except the identity.

If the single generator is an axis of rotation–reflection S_n, only even values of n are possible since, for n odd, S_n is equivalent to an axis C_n plus a symmetry plane σ_h and is therefore not a unique generator. For $n = 2$, S_2 is equivalent to the single generator i and the group S_2 is often indicated with the symbol C_i. If the single generator is a centre of inversion, only one group, the group $C_i \equiv S_2$ is produced. For this reason the row corresponding to the generator i is not included in the table. If the single generator is a symmetry plane again only one point group, the group C_s of order two is generated. For convenience the group C_s is classified in the table as formally generated by a symmetry plane and an axis of order one perpendicular to it. In this way the row corresponding to the generator σ is omitted in the table.

In combining together two generators, only the five types of point groups shown in the second part of the table are obtained.

The C_{nv} point groups have as generators an axis C_n and a plane of symmetry containing C_n. This combination automatically generates $n - 1$ other symmetry planes all containing C_n, as shown in Figure 6.1.2 for the case $n = 2$.

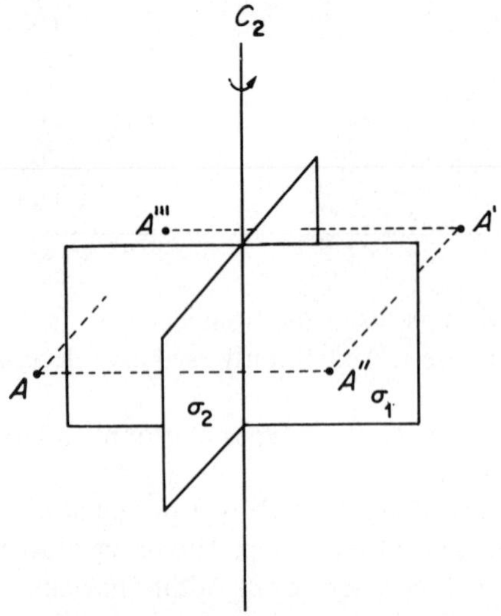

Figure 6.1.2 The symmetry elements of the C_{2v} point group. The generator C_2 sends the point A into the point A'. The generator σ_1 then sends A' into A''. Clearly the two operations applied one after the other are equivalent to a reflection in the new plane σ_2. Thus

$$\sigma_1 C_2 = C_2 \sigma_1 = \sigma_2$$

Another combination of two generators is that of an axis C_n with a symmetry plane perpendicular to it which gives rise to the C_{nh} point groups. An axis S_n coincident with C_n is automatically generated as shown in Figure 6.1.1 for the case $n = 2$.

The combination of two axes C_n and C_m not parallel to each other is possible only in few special cases. For the axial groups the only possibility is that of an axis C_n and of an axis C_2 perpendicular to C_n. These two elements generate the D_n point groups. The principal axis C_n generates from the C_2 axis $n - 1$ other C_2 axes, all going through the same point at angles of $360°/n$.

A little thought shows that in the axial groups any other combination of two generators always generates groups already discussed. For instance the combination of C_n with a centre of symmetry generates, if n is even, the groups C_{2h}, C_{4h} and C_{6h} already discussed and, if n is odd, the groups S_{2n} of order double that of the axis (for example, C_3 and i generate S_6; C_5 and i generate S_{10};† etc.). In the same way the combination of a C_n axis with an S_n axis parallel to C_n generates the C_{nh} groups.

The other combinations of two generators relevant for molecular systems are those connected with the symmetry of the tetrahedron and of the octahedron. Since both the tetrahedron and the octahedron can be inscribed in a cube, the tetrahedral and octahedral groups are often called cubic groups. A tetrahedron inscribed in a cube is shown in Figure 6.1.3. If we use as generators one of the C_3 axes and one of the C_2 axes of the tetrahedron, the pure rotation group T is obtained.

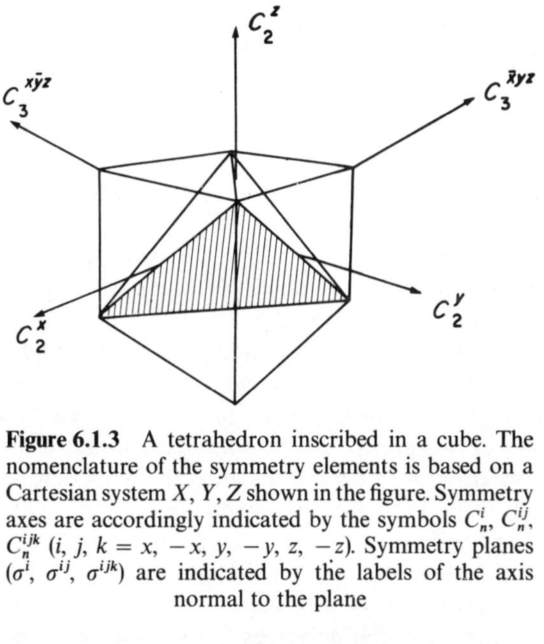

Figure 6.1.3 A tetrahedron inscribed in a cube. The nomenclature of the symmetry elements is based on a Cartesian system X, Y, Z shown in the figure. Symmetry axes are accordingly indicated by the symbols C_n^i, C_n^{ij}, C_n^{ijk} ($i, j, k = x, -x, y, -y, z, -z$). Symmetry planes ($\sigma^i$, σ^{ij}, σ^{ijk}) are indicated by the labels of the axis normal to the plane

The C_3^{xyz} axis, acting upon C_2^z generates the other two binary axes C_2^y and C_2^x. Each of these two-fold axes then generates a new three-fold axis from C_3^{xyz}.

An octahedron inscribed in a cube is shown in Figure 6.1.4. If we use as generators one of the C_4 and one of the C_3 axes, the pure rotation group O is generated. C_3^{xyz} acting upon C_4 generates the other two four-fold axes C_4^x and C_4^y. In turn C_4^z, acting upon C_3^{xyz} generates the other three three-fold axes. In addition six two-fold axes of type C_2^{xz}, passing through two opposite edges of the cube, are generated by operations of the type $C_3^{xyz}C_4^z$. The full symmetry of the tetrahedron is represented in the T_d group. In this case we have, in addition to the symmetry elements of the T point group, six symmetry planes and three four-fold rotation–reflection S_4 axes. Choosing as generators one C_3 and one S_4 axis, all other symmetry elements are generated. The C_3 axis generates from one S_4, for instance S_4^z, the other two S_4 axes S_4^x and S_4^y. The S_4^z axis generates from one C_3 the other three C_3 axes. Each S_4 operation, repeated twice, generates a binary axis ($S_4^z S_4^z = C_2^z$) parallel to it.

† Not included in Table 6.1.1.

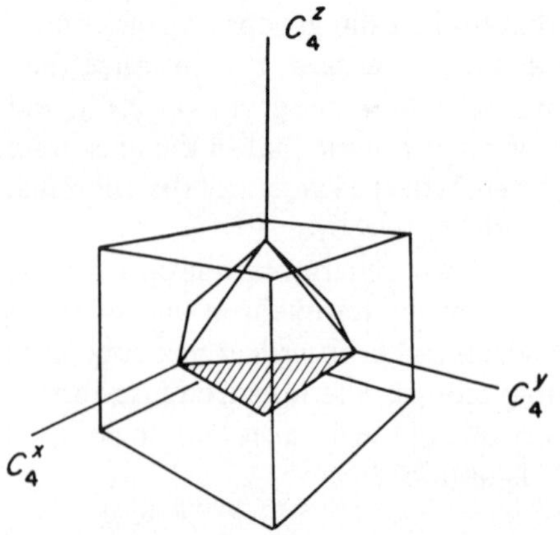

Figure 6.1.4 The symmetry of the octahedron

Finally operations of type $C_3^{xyz}S_4^z$ generate six symmetry planes, each containing two C_3 axes and one C_2 axis.

In combining together three generators, point groups shown in the last part of Table 6.1.1 are obtained.

The D_{nd} groups are obtained from the D_n by adding as a new generator a symmetry plane bisecting the angle between two successive binary axes and containing the n-fold principal axis. The n-fold axis then generates from this plane, $n - 1$ other planes. The D_{nh} groups are also obtained from the corresponding D_n groups by adding as a new generator a symmetry plane perpendicular to the principal axis.

The full symmetry of the octahedron (Figure 6.1.4) is represented in the O_h point group. All symmetry elements of this group can be obtained adding to the generators C_3^{xyz} and C_4^2 of the pure rotation group O, a centre of symmetry. This new generator produces the nine symmetry planes of the cube (the full symmetry of the octahedron is also the full symmetry of the cube) as well as three S_4 axes coinciding with the three C_4 axes and four S_6 axes coinciding with the four C_3 axes of the O group.

If we add, to the generators of the T group, a centre of inversion (a third generator), we obtain the point group T_h which describes the symmetry of the didodecahedron, i.e. the symmetry of two interlocking tetrahedra related by the centre of inversion.

Before closing this section we shall briefly discuss the character systems of the irreducible representations of point groups given in Appendix I.

(A) *Cyclic groups*

The character tables of the cyclic groups C_n are the simplest to derive. We have seen in Section 5.2 that all cyclic groups are Abelian, i.e. that the product of any two elements of the group is commutative. From this it follows directly that the number of classes is equal to the order of the group since

$$R^{-1}AR = AR^{-1}R = A \tag{6.1.1}$$

and thus each element is in a class by itself. A cyclic group of order n possesses n distinct classes and thus n irreducible representations which, according to the dimensionality theorem (5.7.19) must necessarily be monodimensional. Any monodimensional representation of a cyclic group is of the form

$$\frac{C_n \quad C_n^2 \ldots C_n^n = E}{\varepsilon \quad \varepsilon^2 \ldots \varepsilon^n = 1} \tag{6.1.2}$$

since, if we associate the number ε with C_n, we must associate ε^2 with C_n^2 and so on. Consequently ε must be an nth root of unity, of the form

$$\varepsilon = \mathrm{e}^{\mathrm{i}\theta} \qquad \theta = 2\pi l/n \tag{6.1.3}$$

where l is an integer which can assume n distinct values, one for each irreducible representation. Since

$$\mathrm{e}^{2\pi\mathrm{i}(n+l)/n} = \mathrm{e}^{2\pi\mathrm{i}l/n} \tag{6.1.4}$$

the numbers associated with the group elements remain unchanged when l is increased by n. A value of $l = n$ therefore gives the same result as $l = 0$, $l = n + 1$ as $l = 1$ and so on. The possible values of l are then

$$l = 0, 1, 2, \ldots, n - 1 \tag{6.1.5}$$

Furthermore, if $\varepsilon = e^{2\pi\mathrm{i}k/n}$ is the number associated with C_n in the Γ^k ($l = k$) representation, the number associated with C_n in the $\Gamma^{(n-k)}$ representation will be $\varepsilon^* = \mathrm{e}^{-2\pi\mathrm{i}k/n} = \mathrm{e}^{2\pi\mathrm{i}(n-k)/n}$. The representations of a cyclic group therefore occur in conjugate pairs.

Example 6.1.1

Consider the cyclic group of order 3 formed by C_3 and its powers. There are three elements in the group ($C_3, C_3^2 = C_3^{-1}, C_3^3 = E$), i.e. three classes and thus three irreducible representations corresponding to the values 0, 1 and 2 of l in equation (6.1.1).

Using 6.1.2 and 6.1.3 we have

	l	C_3	C_3^{-1}	E
Γ^0	0	1	1	1
Γ^1	1	ε	ε^*	1
Γ^2	2	ε^*	ε	1

(6.1.6)

where

$$\varepsilon = \mathrm{e}^{2\pi\mathrm{i}/3}$$
$$\varepsilon^* = \mathrm{e}^{4\pi\mathrm{i}/3} = \mathrm{e}^{-2\pi\mathrm{i}/3} \tag{6.1.7}$$

The conjugate representations Γ^1 and Γ^2 are carried by a complex basis. The direct sum Γ of Γ^1 and Γ^2

$$\begin{matrix} E & C_3 & C_3^{-1} \\ \begin{pmatrix} 1 & 0 \\ 0 & 1 \end{pmatrix} & \begin{pmatrix} \varepsilon & 0 \\ 0 & \varepsilon^* \end{pmatrix} & \begin{pmatrix} \varepsilon^* & 0 \\ 0 & \varepsilon \end{pmatrix} \end{matrix} \tag{6.1.8}$$

can, however, be transformed, by means of a similarity transformation $\mathbf{U}^\dagger\mathbf{R}\mathbf{U}$, with the unitary matrix

$$\mathbf{U} = \frac{1}{\sqrt{2}}\begin{pmatrix} 1 & \mathrm{i} \\ 1 & -\mathrm{i} \end{pmatrix} \tag{6.1.9}$$

into a two-dimensional representation in the real space, yielding

$$\mathbf{U}^{\dagger}\mathbf{R}(E)\mathbf{U} = \begin{pmatrix} 1 & 0 \\ 0 & 1 \end{pmatrix} \qquad \mathbf{U}^{\dagger}\mathbf{R}(C_3)\mathbf{U} = \begin{pmatrix} \cos\theta & -\sin\theta \\ \sin\theta & \cos\theta \end{pmatrix} \qquad \mathbf{U}^{\dagger}\mathbf{R}(C_3^{-1})\mathbf{U} = \begin{pmatrix} \cos\theta & \sin\theta \\ -\sin\theta & \cos\theta \end{pmatrix} \tag{6.1.10}$$

It is easily seen that this two-dimensional representation is just the one carried by the vectors $\mathbf{e}_1$ and $\mathbf{e}_2$ of a real orthogonal basis $\mathbf{e}$ with $\mathbf{e}_3$ oriented along the rotation axis C_3.

(B) *Dihedral groups*

The dihedral groups D_n can be obtained from the cyclic groups (see Table 6.1.1) by adding to the generator C_n a second generator C_2', which is orthogonal to C_n. Since in the group we always have n C_n^k elements plus n C_2' elements, the number of elements is $g = 2n$. The classes of the dihedral groups involve in general more than one symmetry element, since C_n^k and its inverse C_n^{n-k} always occur in the same class

$$C_2' C_n^{n-k} C_2'^{-1} = C_2' C_n^{n-k} C_2' = C_n^k \tag{6.1.11}$$

where C_2' is any one of the axes perpendicular to the generator C_n. Since:

(a) E always constitutes a class by itself;
(b) $C_n^{n/2}$ is a class by itself if n is even, being equal to C_n^{-n+k};
(c) the n C_2' elements are either in the same class if n is odd or fall into two classes (C_2' and C_2'') if n is even;

we have the following number of classes:

n even
the number of classes is

1	C_n	and	C_n^{n-1}
2	C_n^2	and	C_n^{n-2}
⋮	⋮	⋮	⋮
$\frac{1}{2}n - 1$	$C_n^{\frac{1}{2}n-1}$	and	$C_n^{\frac{1}{2}n+1}$
1	$C_n^{n/2}$		
1	$\frac{1}{2}nC_2'$		
1	$\frac{1}{2}nC_2''$		
1	E		

$$\left(\frac{n}{2} - 1\right) + 4 = \frac{n}{2} + 3$$

n odd
the number of classes is

1	C_n	and	C_n^{n-1}
2	C_n^2	and	C_n^{n-2}
⋮	⋮	⋮	⋮
$\frac{n-1}{2}$	$C_n^{(n-1)/2}$	and	$C_n^{(n+1)/2}$
1	nC_2'		
1	E		

$$\frac{n-1}{2} + 2 = \frac{n}{2} + \frac{3}{2}$$

The following considerations can now be made to find the characters of the monodimensional representations:

(i) Since $C_2' \times C_2' = E$, the character of C_2' in a monodimensional representation can only be ± 1.

(ii) The character associated with the generator C_n must be, as in the cyclic groups, an nth root of unity. Since C_n and $C_n^{n-1} = C_n^{-1}$ fall in the same class, we have

$$\chi(C_n) = \chi(C_n^{-1}) = \chi(C_n)^* \tag{6.1.12}$$

and thus, if n is even $\chi(C_n)$ can be ± 1 whereas if n is odd $\chi(C_n)$ can only be $+1$.

We conclude, therefore, that if n is even we have four monodimensional representations in D_n corresponding to the combinations

C_n	C'_2
+1	+1
+1	−1
−1	+1
−1	−1

for the two generators. If n is odd we have only two monodimensional representations in D_n corresponding to the combinations

C_n	C'_2
+1	+1
+1	−1

We can now show that all other representations of D_n are two-dimensional. For this we shall use the dimensionality theorem (5.7.19) and the fact that the number of irreducible representations must be equal to the number of classes. Thus, if s_1, s_2 and s_3 are the number of mono-, bi- and tri-dimensional representations, we have from (5.7.19)

$$s_1 + 4s_2 + 9s_3 = 2n \tag{6.1.13}$$

and from the relation

number of irreducible representations = number of classes

we have

$$s_1 + s_2 + s_3 = n/2 + 3 \quad \text{for } n \text{ even} \qquad \text{(a)}$$
$$s_1 + s_2 + s_3 = n/2 + \tfrac{3}{2} \quad \text{for } n \text{ odd} \qquad \text{(b)} \tag{6.1.14}$$

For n odd, by putting $s_1 = 2$ as discussed before and by using (6.1.13) and (6.1.14b) we obtain

$$5s_3 = 0$$

and thus

$$s_3 = 0 \qquad s_2 = (n-1)/2 \tag{6.1.15}$$

Likewise for n even, by putting $s_1 = 4$, we have

$$s_3 = 0 \qquad s_2 = (n-2)/2 \tag{6.1.16}$$

In order to obtain the characters of the two-dimensional representations it is clearly sufficient to deduce only those associated with the generators C_n and C'_2.

It can be shown that the characters associated with C_2' must always be zero in a two-dimensional representation of the dihedral groups. To prove this we shall make use of the orthonormality of the normalized characters $[(h_i g)^{\frac{1}{2}}\chi_i^\alpha]$ of equation (5.7.28), which can be interpreted as the components of h orthonormal vectors in an h-dimensional space and thus obey the relation

$$\sum_\alpha [(h_i/g)^{\frac{1}{2}}\chi_i^\alpha][(h_i/g)^{\frac{1}{2}}\chi_i^\alpha]^* = 1 \tag{6.1.17}$$

where α runs over the h irreducible representations of the group. From (6.1.17) we obtain

$$\sum_\alpha \chi_i^\alpha \chi_i^{\alpha *} = g/h_i \tag{6.1.18}$$

where i is the index of the class and h_i the number of elements in each class. We then have

$$\text{for } n \text{ even} \quad \sum_\alpha \chi_i^\alpha \chi_i^{\alpha *} = \frac{2n}{n/2} = 4 \qquad \text{(a)}$$

$$\text{for } n \text{ odd} \quad \sum_\alpha \chi_i^\alpha \chi_i^{\alpha *} = \frac{2n}{n} = 2 \qquad \text{(b)} \tag{6.1.19}$$

We have already seen that if n is even there are four monodimensional species with character ± 1 for C_2'. The summation (6.1.19a), when extended over these four species, yields a value of 4 and thus the summation over the two-dimensional species must be equal to zero. Since no term $\chi_i^\alpha \chi_i^{\alpha *}$ can be negative, each is necessarily zero. The same argument applies when n is odd.

As far as the generator C_n is concerned, we already know that in the real space it is associated with the matrix

$$\begin{pmatrix} \cos\phi & -\sin\phi \\ \sin\phi & \cos\phi \end{pmatrix}$$

and has character equal to $2\cos\phi$.

Since $C_n^n = E$, we have

$$\begin{pmatrix} \cos\phi & -\sin\phi \\ \sin\phi & \cos\phi \end{pmatrix}^n = \begin{pmatrix} \cos n\phi & -\sin n\phi \\ \sin n\phi & \cos n\phi \end{pmatrix} = \begin{pmatrix} 1 & 0 \\ 0 & 1 \end{pmatrix} \tag{6.1.20}$$

from which it follows that

$$\phi = \frac{2\pi l}{n} = \alpha l \qquad \alpha = 2\pi/n \tag{6.1.21}$$

The values of l which give rise to distinct representations are then (see 6.1.15 and 6.1.16)

$$\text{for } n \text{ even} \quad l = 1, 2, \ldots, \frac{n}{2} - 1$$

$$\text{for } n \text{ odd} \quad l = 1, 2, \ldots, \frac{n-1}{2}$$

Using the previous treatment and putting $n/2 = m$ and $(n-1)/2 = q$ we can write down the character table for the D_n groups with n even and odd respectively (Tables 6.1.2 and 6.1.3).

Table 6.1.2 Character table of D_n groups with n even

	E	$2C_n$	$2C_n^2$		$2C_n^{m-1}$	$2C_n^m$	mC_2'	mC_2''
Γ^1	1	1	1	...	1	1	1	1
Γ^2	1	1	1	...	1	1	-1	-1
Γ^3	1	-1	1	...	$(-1)^{m-1}$	$(-1)^m$	1	-1
Γ^4	1	-1	1	...	$(-1)^{m-1}$	$(-1)^m$	-1	1
Γ^5	2	$2\cos\alpha$	$2\cos 2\alpha$	...	$2\cos(m-1)\alpha$	$2\cos m\alpha$	0	0
Γ^6	2	$2\cos 2\alpha$	$2\cos 4\alpha$	...	$2\cos 2(m-1)\alpha$	$2\cos 2m\alpha$	0	0
	⋮	⋮	⋮	⋱	⋮	⋮	⋮	⋮
$\Gamma^{(n+6)/2}$	2	$2\cos(m-1)\alpha$	$2\cos(m-1)2\alpha$	...	$2\cos(m-1)^2\alpha$	$2\cos m(m-1)\alpha$	0	0

Table 6.1.3 Character table of D_n groups with n odd

	E	$2C_n$	$2C_n^2$		$2C_n^m$	nC_2'
Γ^1	1	1	1	...	1	1
Γ^2	1	1	1	...	1	-1
Γ^3	2	$2\cos\alpha$	$2\cos 2\alpha$	...	$2\cos q\alpha$	0
Γ^4	2	$2\cos 2\alpha$	$2\cos 4\alpha$	...	$2\cos 2q\alpha$	0
	⋮	⋮	⋮	⋱	⋮	⋮
$\Gamma^{(n+3)/2}$	2	$2\cos q\alpha$	$2\cos 2q\alpha$	...	$2\cos q^2\alpha$	0

(C) C_{nv} *Groups*

The character tables of the C_{nv} groups are identical to those of the D_n groups since the C_{nv} groups are isomorphic to the D_n groups.

(D) C_{nh} *Groups*

The C_{nh} groups can be written as the direct product

$$C_n \otimes C_s$$

if n is odd or as

$$C_n \otimes C_i$$

if n is even.

The character table of the two isomorphic groups $C_s\{E, \sigma_h\}$ and $C_i\{E, i\}$ is very simple since there are only two operations in the group. It has the form

	E	$\begin{cases} i \\ \sigma_h \end{cases}$
Γ^1	1	1
Γ^2	1	−1

The character tables of the C_{nh} group are thus obtained by forming the direct product of the representations of C_n with the representations of C_s (or C_i). The number of irreducible representations of the C_{nh} groups is thus double that of the corresponding C_n groups.

(E) *D_{nh} Groups*

The D_{nh} groups can be written as the direct product

$$D_n \otimes C_i$$

for n even or as the product

$$D_n \otimes C_s$$

for n odd. As in the previous case, the number of irreducible representations is doubled with respect to the corresponding D_n groups.

(F) *D_{nd} Groups*

The D_{nd} groups can be written as the direct product

$$D_n \otimes C_i$$

for n odd. For n even they are isomorphic to the corresponding $D_{(2n)}$ groups, i.e. to the dihedral groups of order $2n$.

(G) *S_n Groups*

The S_n groups are isomorphic with the C_n groups and thus have identical character tables. Alternatively the direct product

$$C_n \otimes C_i$$

with n odd generates the S_{2n} groups.

(H) *Tetrahedral and octahedral groups*

The derivation of the character tables of tetrahedral and octahedral groups is more complex and will not be discussed here. The interested reader is referred to the book[1] by Lomont *Applications of Finite Groups.*

(I) *Infinite groups $C_{\infty v}$ and $D_{\infty h}$*

The $C_{\infty v}$ and $D_{\infty h}$ groups have 2 and 4 monodimensional representations respectively. All other representations are two-dimensional and are obviously infinite in number.

No further discussion of these groups will be given here. A detailed discussion of infinitesimal rotation operators can be found in Reference 2.

A standard nomenclature is used to designate the irreducible representations of point groups. One-dimensional representations are indicated by the letters A or B. In particular A and B are used for representations symmetric or antisymmetric respectively with respect to the principal axis C_n (or S_n), i.e. for representations in which $+1$ or -1 respectively is associated with the operation C_n (or S_n). Additional labels to these letters are then used to specify the behaviour of the representation with respect to other generators of the group.

Subscripts 1 and 2 are used to indicate that $+1$ and -1 respectively are associated with the second generator σ_v of the C_{nv} groups or C_2 of the D_n groups. D_2 is a special case which has three equivalent C_2 axes for which the labels 1, 2 and 3 are used. The subscripts g and u are used to indicate that the representations are symmetric or antisymmetric respectively with respect to the inversion i. Finally the labels $'$ and $''$ are used to indicate symmetry or antisymmetry with respect to σ_h.

Two-dimensional representations are designated by the letter E and three-dimensional representations by the letter F (or T by some authors). Numerical subscripts are also used with the letter E for two-fold degenerate species to indicate the value of l in equation (6.1.3). Finally, subscripts 1 and 2 are used in the T_d group for F representations to indicate symmetry and antisymmetry respectively for the σ_d reflections.

When a group can be written as the direct product of two smaller groups the notations used in the two separate groups are combined (see Example 5.8.1).

6.2 The symmetry of molecular vibrations

In this section we begin to apply to molecular vibrations the machinery of group theory, discussed in the previous chapter. In studying molecular symmetry we have, until now, considered only molecules in their equilibrium configuration. A symmetry operation was defined in this case as an operation able to bring the equilibrium configuration into self-coincidence. In dealing with vibrating molecules we need, however, to consider distorted molecular conformations which in general do not preserve the symmetry of the equilibrium conformation and therefore are not brought into self-coincidence by symmetry operations.

If we submit a distorted molecule to one of the symmetry operations, which would bring the equilibrium configuration into self-coincidence, we obtain a new configuration, not coincident with the initial configuration, but still equivalent to it since the same interatomic distances occur in both cases.

This correlation of two conformations, connected through a symmetry operation of the undistorted molecule, suggests that a redefinition of symmetry operation is necessary for the vibrating molecule. In order to be able to make full use of the representation theory, it is most convenient to redefine a symmetry operation for a vibrating molecule, as a *virtual* operation which exchanges the displacements between equivalent atoms, without exchanging the atoms themselves. A rotation of the molecule (proper or improper), treated before in terms of the rotation of a suitable basis $\mathbf{e}$ associated with the molecule, becomes a virtual rotation in the displacement space, i.e. a rotation in the $3N$-dimensional space of a vector $\mathbf{x}$ whose components are the $3N$ Cartesian displacements of the atoms. It is clear that, as the 3-dimensional basis $\mathbf{e}$ carries a representation of the molecular point group, the $3N$-dimensional vector $\mathbf{x}$ carries a representation of the same group. There is a one-to-one correspondence (i.e. isomorphism) between the real rotation of the molecular point group and the virtual rotations acting on the displacement coordinates.

The displacement vectors $\mathbf{x}$ and $\mathbf{x}'$, before and after the symmetry operation, are related by an equation of the type

$$\mathbf{x}' = \mathbf{D}(\mathscr{R})\mathbf{x} \tag{6.2.1}$$

analogous to equation (5.5.45) of the previous chapter, where $\mathbf{D}(\mathscr{R})$ is a $3N \times 3N$ matrix, associated with the operation $\mathscr{R}$, which transforms the vector $\mathbf{x}$ into the rotated vector $\mathbf{x}'$. Obviously there is one matrix for each operation $\mathscr{R}$ of the molecular point group.

The potential and the kinetic energy have, in terms of the displacement vector $\mathbf{x}$, the form

$$2V = \tilde{\mathbf{x}}\mathbf{f}\mathbf{x} \tag{6.2.2}$$

$$2T = \tilde{\dot{\mathbf{x}}}\mathbf{M}\dot{\mathbf{x}} \tag{6.2.3}$$

Since V and T are functions only of the distances between the atoms and of the velocities of the atoms respectively, they are unchanged through virtual operations in which the relative positions of the atoms and their velocities remain unchanged.

Since equations (6.2.2) and (6.2.3) are true for arbitrary values of the displacements, we have

$$2V = \tilde{\mathbf{x}}'\mathbf{f}\mathbf{x}' = \tilde{\mathbf{x}}\tilde{\mathbf{D}}(\mathscr{R})\mathbf{f}\mathbf{D}(\mathscr{R})\mathbf{x} = \tilde{\mathbf{x}}\mathbf{f}\mathbf{x} \tag{6.2.4}$$

$$2T = \dot{\mathbf{x}}'\mathbf{M}\dot{\mathbf{x}}' = \dot{\mathbf{x}}\tilde{\mathbf{D}}(\mathscr{R})\mathbf{M}\mathbf{D}(\mathscr{R})\dot{\mathbf{x}} = \dot{\mathbf{x}}\mathbf{M}\dot{\mathbf{x}} \tag{6.2.5}$$

and therefore

$$\left.\begin{aligned}\tilde{\mathbf{D}}(\mathscr{R})\mathbf{f}\,\mathbf{D}(\mathscr{R}) &= \mathbf{f}\\ \tilde{\mathbf{D}}(\mathscr{R})\mathbf{M}\mathbf{D}(\mathscr{R}) &= \mathbf{M}\end{aligned}\right. \tag{6.2.6}$$

The last two equations describe the important property of the potential and kinetic energy matrices of being invariant forms under the symmetry operations of the molecular point group.

If internal coordinates are used instead of Cartesian displacements, equation (6.2.1) becomes

$$\mathbf{s}' = \mathbf{T}(\mathscr{R})\mathbf{s} \tag{6.2.7}$$

where $\mathbf{T}(\mathscr{R})$ is a matrix associated with the operation $\mathscr{R}$ which transforms the internal coordinate vector $\mathbf{s}$ into the rotated vector $\mathbf{s}'$.

Using the same arguments as before, equations (6.2.6) become

$$\tilde{\mathbf{T}}(\mathscr{R})\mathbf{F}_s\mathbf{T}(\mathscr{R}) = \mathbf{F}_s \tag{6.2.8}$$

$$\tilde{\mathbf{T}}(\mathscr{R})\mathbf{G}^{-1}\mathbf{T}(\mathscr{R}) = \mathbf{G}^{-1} \tag{6.2.9}$$

where $\mathbf{F}$ and $\mathbf{G}^{-1}$ are the potential and kinetic energy matrices in the $\mathbf{s}$-space (see Chapter 4, Section 4.2).

Since the internal coordinates are related to the Cartesian displacements, in the first approximation, by a linear transformation of the type (equation 4.1.2)

$$\mathbf{s} = \mathbf{B}\mathbf{x} \tag{6.2.10}$$

the representation carried by the $\mathbf{s}$-vector and the representation carried by the $\mathbf{x}$-vector, are equivalent. This can be seen, using equations (4.2.4) and (6.2.6), which give†

$$\tilde{\mathbf{D}}(\mathscr{R})(\tilde{\mathbf{B}}\mathbf{F}_s\mathbf{B})\mathbf{D}(\mathscr{R}) = \tilde{\mathbf{B}}\mathbf{F}_s\mathbf{B} \tag{6.2.11}$$

$$(\tilde{\mathbf{B}}^{-1}\tilde{\mathbf{D}}(\mathscr{R})\tilde{\mathbf{B}})\mathbf{F}_s(\mathbf{B}\mathbf{D}(\mathscr{R})\mathbf{B}^{-1}) = \mathbf{F}_s \tag{6.2.12}$$

† For simplicity we assume here that the **B**-matrix is square and non-singular. This corresponds to the use of $3N - 3$ internal coordinates and of 6 external coordinates in the form of Eckart's conditions. It must be noticed that these 6 coordinates are introduced here only as dummy coordinates and that they need not be considered further.

Comparing equations (6.2.12) and (6.2.8), we have

$$\mathbf{BD}(\mathscr{R})\mathbf{B}^{-1} = \mathbf{T}(\mathscr{R}) \tag{6.2.13}$$

which proves that the two representations are equivalent, being related to each other by a similarity transformation. Using internal coordinates there is the advantage that only $3N - 6$ ($3N - 5$ for linear molecules) coordinates are used instead of $3N$.

Obviously, any other set of coordinates, related to the Cartesian displacements by a linear transformation, will furnish a representation equivalent to that carried by the latter coordinates. We shall see that the representation carried by the normal coordinates is of great importance, being completely reduced. Before discussing the representation carried by the normal coordinates, we shall investigate in more detail the form of the matrices $\mathbf{D}(\mathscr{R})$ and illustrate their construction with some simple examples.

As discussed before, we perform a symmetry operation upon a distorted molecular conformation, by rotating the Cartesian displacements of the atoms, but leaving the atoms unshifted. This corresponds, if atoms i and j are exchanged by a symmetry operation of the undistorted molecule, to a rotation of the Cartesian displacements on atom i, according to the symmetry operation $\mathscr{R}$, and a shift of the rotated displacements to atom j. Simultaneously, the displacements at atom j undergo the same rotation followed by a shift to atom i. This is shown schematically in Figure 6.2.1 for two atoms related by a symmetry

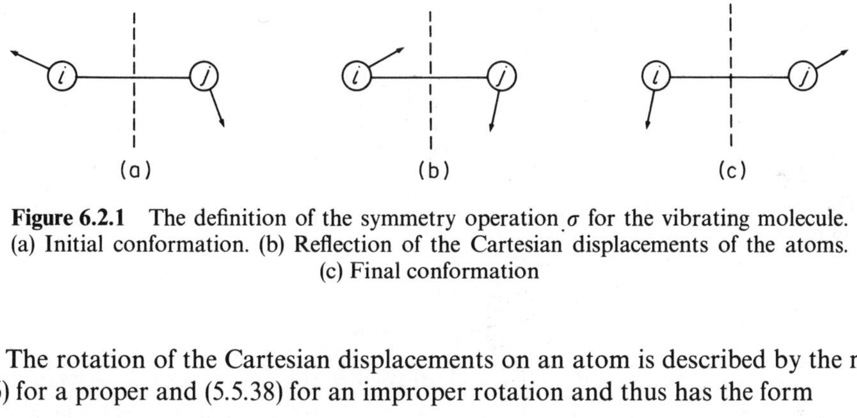

Figure 6.2.1 The definition of the symmetry operation σ for the vibrating molecule. (a) Initial conformation. (b) Reflection of the Cartesian displacements of the atoms. (c) Final conformation

plane. The rotation of the Cartesian displacements on an atom is described by the matrix (5.5.36) for a proper and (5.5.38) for an improper rotation and thus has the form

$$\mathbf{D}_c(\mathscr{R}) = \begin{pmatrix} \cos\theta & -\sin\theta & 0 \\ \sin\theta & \cos\theta & 0 \\ 0 & 0 & \pm 1 \end{pmatrix} \tag{6.2.14}$$

The shift from atom i to atom j gives the matrix $\mathbf{D}(\mathscr{R})$, for the operation $\mathscr{R}$ connecting the two atoms, the form

$$\mathbf{D}(\mathscr{R}) = \left(\begin{array}{c|c} \mathbf{0} & \mathbf{D}_c(\mathscr{R}) \\ \hline \mathbf{D}_c(\mathscr{R}) & \mathbf{0} \end{array}\right) \tag{6.2.15}$$

The matrix $\mathbf{D}(\mathscr{R})$ is therefore divided into 3×3 blocks. If an atom is sent into itself by a symmetry operation, the corresponding $\mathbf{D}_c(\mathscr{R})$ block will occur on the diagonal. If instead the atom is sent into an equivalent atom by the symmetry operation, the $\mathbf{D}_c(\mathscr{R})$ block will occur off the diagonal.

Example 6.2.1

A bent triatomic molecule like water (Figure 6.2.2) belongs to the C_{2v} point group. It possesses two mutually perpendicular planes of symmetry and a two-fold axis at their intersection. The C_{2v} group has four symmetry operations, the identity E, the rotation C_2 and the two reflections σ_1 and σ_2.

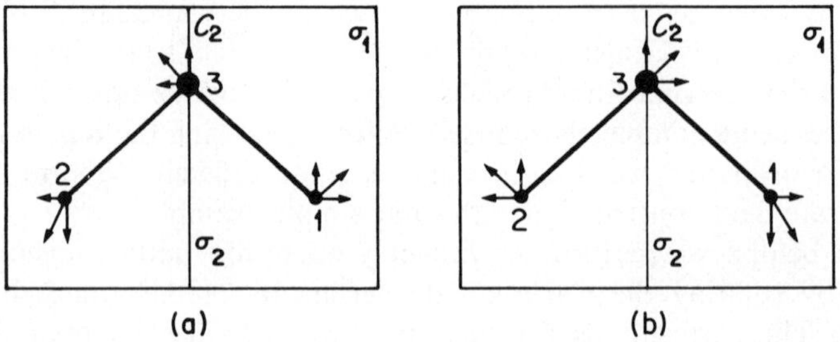

Figure 6.2.2 Effect of the symmetry operation C_2 on the Cartesian displacement coordinates of the water molecule

The Cartesian components of arbitrary displacements of the atoms are shown in Figure 6.2.2(a). The z-axis has been taken in the direction of the two-fold axis and the x-axis in the plane of the paper. If we apply to this distorted configuration the operation C_2, we obtain the configuration of Figure 6.2.2(b). It is easily seen from the figure that the matrix $\mathbf{D}(\mathscr{R})$ has the form

$$\mathbf{D}(C_2) = \begin{pmatrix} 0 & \mathbf{D}_c(C_2) & 0 \\ \mathbf{D}_c(C_2) & 0 & 0 \\ 0 & 0 & \mathbf{D}_c(C_2) \end{pmatrix}$$

where $\mathbf{D}_c(C_2)$ has the form (see equation 5.5.37)

$$\mathbf{D}_c(C_2) = \begin{pmatrix} -1 & 0 & 0 \\ 0 & -1 & 0 \\ 0 & 0 & 1 \end{pmatrix}$$

The matrices $\mathbf{D}(\mathscr{R})$ associated with the other three operations of the group are listed below

$$\mathbf{D}(E) = \begin{pmatrix} \mathbf{D}_c(E) & 0 & 0 \\ 0 & \mathbf{D}_c(E) & 0 \\ 0 & 0 & \mathbf{D}_c(E) \end{pmatrix} \quad \mathbf{D}(\sigma_1) = \begin{pmatrix} \mathbf{D}_c(\sigma_1) & 0 & 0 \\ 0 & \mathbf{D}_c(\sigma_1) & 0 \\ 0 & 0 & \mathbf{D}_c(\sigma_1) \end{pmatrix} \quad \mathbf{D}(\sigma_2) = \begin{pmatrix} 0 & \mathbf{D}_c(\sigma_2) & 0 \\ \mathbf{D}_c(\sigma_2) & 0 & 0 \\ 0 & 0 & \mathbf{D}_c(\sigma_2) \end{pmatrix}$$

where (see equations 5.5.37, 5.5.39 and the discussion below)

$$\mathbf{D}_c(E) = \begin{pmatrix} 1 & 0 & 0 \\ 0 & 1 & 0 \\ 0 & 0 & 1 \end{pmatrix} \quad \mathbf{D}_c(\sigma_1) = \begin{pmatrix} 1 & 0 & 0 \\ 0 & -1 & 0 \\ 0 & 0 & 1 \end{pmatrix} \quad \mathbf{D}_c(\sigma_2) = \begin{pmatrix} -1 & 0 & 0 \\ 0 & 1 & 0 \\ 0 & 0 & 1 \end{pmatrix}$$

For future reference we observe that only atoms which are unshifted in the symmetry operation $\mathscr{R}$ contribute to the character of the matrix $\mathbf{D}(\mathscr{R})$. The character of $\mathbf{D}(\mathscr{R})$ is thus simply obtained by multiplying the character of the submatrix $\mathbf{D}_c(\mathscr{R})$ by the number of atoms which are unshifted under the symmetry operation $\mathscr{R}$. For instance the character of $\mathbf{D}(C_2)$ is -1 since the character of $\mathbf{D}_c(C_2)$ is -1 and only one atom (atom 3) remains unshifted under the operation C_2. The character of the four matrices $\mathbf{D}(\mathscr{R})$ are listed below.

	E	C_2	σ_1	σ_2
Number of unshifted atoms	3	1	3	1
Character of $\mathbf{D}_c(\mathscr{R}) = \chi_c(\mathscr{R})$	3	-1	1	1
Character of $\mathbf{D}(\mathscr{R}) = \chi(\mathscr{R})$	9	-1	3	1

The reader can easily verify that the four matrices $\mathbf{D}(\mathscr{R})$ written above are orthogonal and obey the multiplication table of the C_{2v} point group.

Example 6.2.2

A planar tetratomic molecule like H_2CO, also belongs to the C_{2v} point group. Arbitrary displacements of the atoms are shown in Figure 6.2.3. From the figure it is seen that the

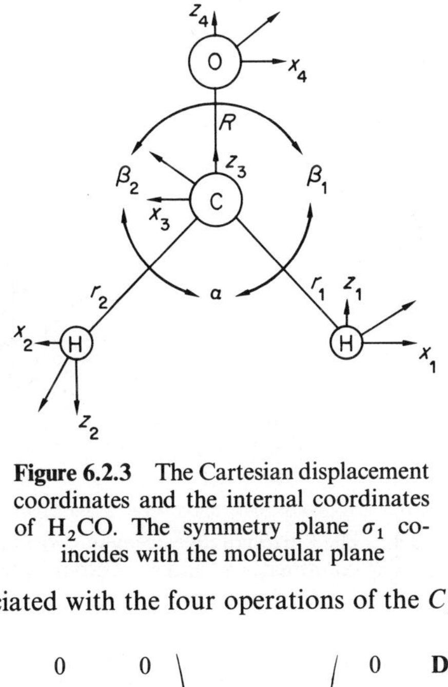

Figure 6.2.3 The Cartesian displacement coordinates and the internal coordinates of H_2CO. The symmetry plane σ_1 coincides with the molecular plane

four matrices $\mathbf{D}(\mathscr{R})$, associated with the four operations of the C_{2v} molecular group, have the form

$$\mathbf{D}(E) = \begin{pmatrix} \mathbf{D}_c(E) & 0 & 0 & 0 \\ 0 & \mathbf{D}_c(E) & 0 & 0 \\ 0 & 0 & \mathbf{D}_c(E) & 0 \\ 0 & 0 & 0 & \mathbf{D}_c(E) \end{pmatrix} \qquad \mathbf{D}(C_2) = \begin{pmatrix} 0 & \mathbf{D}_c(C_2) & 0 & 0 \\ \mathbf{D}_c(C_2) & 0 & 0 & 0 \\ 0 & 0 & \mathbf{D}_c(C_2) & 0 \\ 0 & 0 & 0 & \mathbf{D}_c(C_2) \end{pmatrix}$$

$$\mathbf{D}(\sigma_1) = \begin{pmatrix} \mathbf{D}_c(\sigma_1) & 0 & 0 & 0 \\ 0 & \mathbf{D}_c(\sigma_1) & 0 & 0 \\ 0 & 0 & \mathbf{D}_c(\sigma_1) & 0 \\ 0 & 0 & 0 & \mathbf{D}_c(\sigma_1) \end{pmatrix} \qquad \mathbf{D}(\sigma_2) = \begin{pmatrix} 0 & \mathbf{D}_c(\sigma_2) & 0 & 0 \\ \mathbf{D}_c(\sigma_2) & 0 & 0 & 0 \\ 0 & 0 & \mathbf{D}_c(\sigma_2) & 0 \\ 0 & 0 & 0 & \mathbf{D}_c(\sigma_2) \end{pmatrix}$$

The characters of the four matrices are given below.

	E	C_2	σ_1	σ_2
Number of unshifted atoms	4	2	4	2
Character of $\mathbf{D}_c(\mathscr{R})$	3	-1	1	1
Character of $\mathbf{D}(\mathscr{R})$	12	-2	4	2

It may be verified that the four $\mathbf{D}(\mathscr{R})$ matrices are orthogonal and are a representation of the C_{2v} point group.

Example 6.2.3

The ammonia molecule (Figure 6.2.4) belongs to the C_{3v} group. As seen in Example 5.2.3 there are six symmetry operations; $E, C_3, C_3^{-1}, \sigma_1, \sigma_2$ and σ_3. Atom 4 is sent into itself by all operations of the group. C_3 sends $1 \to 2, 2 \to 3, 3 \to 1$.

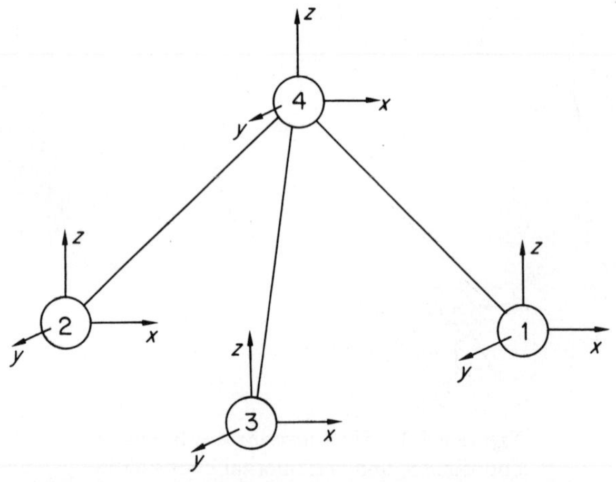

Figure 6.2.4. The Cartesian displacement coordinates of ammonia. The coordinates x_1 and z_1 of atom 1 are in the σ_1 plane

C_3^{-1} sends $1 \to 3, 3 \to 2, 2 \to 1$. σ_1 sends $2 \to 3; 3 \to 2$ and leaves atom 1 unshifted. σ_2 and σ_3 leave atom 2 and 3 unshifted and exchange the other two atoms.

The six $\mathbf{D}(\mathscr{R})$ matrices associated with the six symmetry operations of the group are

$$\mathbf{D}(E) = \begin{pmatrix} \mathbf{D}_c(E) & 0 & 0 & 0 \\ 0 & \mathbf{D}_c(E) & 0 & 0 \\ 0 & 0 & \mathbf{D}_c(E) & 0 \\ 0 & 0 & 0 & \mathbf{D}_c(E) \end{pmatrix} \qquad \mathbf{D}(C_3) = \begin{pmatrix} 0 & 0 & \mathbf{D}_c(C_3) & 0 \\ \mathbf{D}_c(C_3) & 0 & 0 & 0 \\ 0 & \mathbf{D}_c(C_3) & 0 & 0 \\ 0 & 0 & 0 & \mathbf{D}_c(C_3) \end{pmatrix}$$

$$\mathbf{D}(C_3^{-1}) = \begin{pmatrix} 0 & \mathbf{D}_c(C_3^{-1}) & 0 & 0 \\ 0 & 0 & \mathbf{D}_c(C_3^{-1}) & 0 \\ \mathbf{D}_c(C_3^{-1}) & 0 & 0 & 0 \\ 0 & 0 & 0 & \mathbf{D}_c(C_3^{-1}) \end{pmatrix} \quad \mathbf{D}(\sigma_1) = \begin{pmatrix} \mathbf{D}_c(\sigma_1) & 0 & 0 & 0 \\ 0 & 0 & \mathbf{D}_c(\sigma_1) & 0 \\ 0 & \mathbf{D}_c(\sigma_1) & 0 & 0 \\ 0 & 0 & 0 & \mathbf{D}_c(\sigma_1) \end{pmatrix}$$

$$\mathbf{D}(\sigma_2) = \begin{pmatrix} 0 & 0 & \mathbf{D}_c(\sigma_2) & 0 \\ 0 & \mathbf{D}_c(\sigma_2) & 0 & 0 \\ \mathbf{D}_c(\sigma_2) & 0 & 0 & 0 \\ 0 & 0 & 0 & \mathbf{D}_c(\sigma_2) \end{pmatrix} \quad \mathbf{D}(\sigma_3) = \begin{pmatrix} 0 & \mathbf{D}_c(\sigma_3) & 0 & 0 \\ \mathbf{D}_c(\sigma_3) & 0 & 0 & 0 \\ 0 & 0 & \mathbf{D}_c(\sigma_3) & 0 \\ 0 & 0 & 0 & \mathbf{D}_c(\sigma_3) \end{pmatrix}$$

where (see Example 5.5.2)

$$\mathbf{D}_c(E) = \begin{pmatrix} 1 & 0 & 0 \\ 0 & 1 & 0 \\ 0 & 0 & 1 \end{pmatrix} \quad \mathbf{D}_c(C_3) = \begin{pmatrix} -\frac{1}{2} & -\frac{\sqrt{3}}{2} & 0 \\ \frac{\sqrt{3}}{2} & -\frac{1}{2} & 0 \\ 0 & 0 & 1 \end{pmatrix} \quad \mathbf{D}_c(C_3^{-1}) = \begin{pmatrix} -\frac{1}{2} & \frac{\sqrt{3}}{2} & 0 \\ -\frac{\sqrt{3}}{2} & -\frac{1}{2} & 0 \\ 0 & 0 & 1 \end{pmatrix}$$

$$\mathbf{D}_c(\sigma_1) = \begin{pmatrix} 1 & 0 & 0 \\ 0 & -1 & 0 \\ 0 & 0 & 1 \end{pmatrix} \quad \mathbf{D}_c(\sigma_2) = \begin{pmatrix} -\frac{1}{2} & -\frac{\sqrt{3}}{2} & 0 \\ -\frac{\sqrt{3}}{2} & \frac{1}{2} & 0 \\ 0 & 0 & 1 \end{pmatrix} \quad \mathbf{D}_c(\sigma_3) = \begin{pmatrix} -\frac{1}{2} & \frac{\sqrt{3}}{2} & 0 \\ \frac{\sqrt{3}}{2} & \frac{1}{2} & 0 \\ 0 & 0 & 1 \end{pmatrix}$$

The characters of the six matrices, which are a representation of the C_{3v} point group, are given below.

	E	C_3	C_3^{-1}	σ_1	σ_2	σ_3
N = number of unshifted atoms	4	1	1	2	2	2
$\chi_c(\mathscr{R})$ character of $\mathbf{D}_c(\mathscr{R})$	3	0	0	1	1	1
$\chi(\mathscr{R})$ character of $\mathbf{D}(\mathscr{R})$	12	0	0	2	2	2

It is unnecessary to calculate the character of a representation for all matrices since, as seen in Section 5.7 (see Example 5.7.2), elements of the same class have the same characters. The previous table can be therefore written in the simpler form

	E	$2C_3$	3σ
N	4	1	2
$\chi_c(\mathscr{R})$	3	0	1
$\chi(\mathscr{R})$	12	0	2

The representation carried by the Cartesian displacement coordinates is not in the simplest possible form and is reducible by means of a suitable unitary transformation to block form. Since the Cartesian displacements vector $\mathbf{x}$ has $3N$ components (N atoms in the molecule) it furnishes a representation which describes the symmetry of all molecular motions, including the three translations and the three rotations of the molecule as a whole.

It is therefore convenient to use as a basis the internal coordinate basis which furnishes a description of the internal motions only. The representation carried by the internal coordinates (equivalent to that based on the Cartesian displacements) is still reducible, but represents a first step toward irreducibility. The internal coordinates do not mix with the six external coordinates, and they are themselves conveniently divided into sets. The members of each set are interchanged under a symmetry operation. Since an internal coordinate is sent by a symmetry operation into an equivalent one, with at most a reversal of sign, the structure of the matrices $\mathbf{T}(\mathcal{R})$ of equation (6.2.7) is very simple, the elements being either 0 or ± 1. Furthermore, because of the non-mixing of different sets, each set always carries a representation of the group (sometimes unfaithful) and can be analysed separately. This furnishes a description of the normal coordinates which is sometimes approximately determined by one set of internal coordinates (for example the CH stretching modes) and allows the determination of the number of such modes. Only the internal coordinates which remain unshifted for a given symmetry operation $\mathcal{R}$ contribute to the character of $\mathbf{T}(\mathcal{R})$ with the value ± 1 on the diagonal.

It is necessary to point out that it is not always possible to select $(3N - 6)$ independent coordinates and make full use of the molecular symmetry. In many cases, in order to use symmetrical sets of coordinates and physically reasonable potential functions, more internal coordinates are employed than the number of internal degrees of freedom. In this case the coordinate system is said to be redundant. The redundant coordinates, which can in some cases be rather complicated linear combinations of internal coordinates, are thus included in the description as 'zero frequency' coordinates. We shall analyse in detail the problem of redundant coordinates in Section 6.7.

Example 6.2.4

Let us consider the water molecule of Example 6.2.1 and use as internal coordinates Δs_1, Δs_2 and Δs_3 the stretches Δr_1 and Δr_2 of the bonds 3–1 and 3–2 and the deformation $\Delta\alpha$ of the angle α. $\Delta\alpha$ is sent into itself by all operations of the group whereas Δr_1 and Δr_2 are exchanged by C_2 and σ_2. The matrices $\mathbf{T}_s(\mathcal{R})$ are thus (the subscript s indicates that only the $(3N - 6)$ internal coordinates are used)

$$\mathbf{T}_s(E) = \left(\begin{array}{cc|c} 1 & 0 & 0 \\ 0 & 1 & 0 \\ \hline 0 & 0 & 1 \end{array}\right) \quad \mathbf{T}_s(C_2) = \left(\begin{array}{cc|c} 0 & 1 & 0 \\ 1 & 0 & 0 \\ \hline 0 & 0 & 1 \end{array}\right) \quad \mathbf{T}_s(\sigma_1) = \left(\begin{array}{cc|c} 1 & 0 & 0 \\ 0 & 1 & 0 \\ \hline 0 & 0 & 1 \end{array}\right) \quad \mathbf{T}_s(\sigma_2) = \left(\begin{array}{cc|c} 0 & 1 & 0 \\ 1 & 0 & 0 \\ \hline 0 & 0 & 1 \end{array}\right)$$

Since each set of coordinates forms a block which does not mix with other blocks, we can evaluate the character separately for each block. If we call $\chi_r(\mathcal{R})$ and $\chi_\alpha(\mathcal{R})$ respectively the characters of the blocks corresponding to the coordinates Δr_1, Δr_2 and $\Delta\alpha$ we have

	E	C_2	σ_1	σ_2
$\chi_r(\mathcal{R})$	2	0	2	0
$\chi_\alpha(\mathcal{R})$	1	1	1	1
$\chi_s(\mathcal{R})$	3	1	3	1

where $\chi_s(\mathscr{R})$ is the character of the $\mathbf{T}_s(\mathscr{R})$ matrices and is therefore the sum of $\chi_r(\mathscr{R})$ and $\chi_\alpha(\mathscr{R})$.

We know that if two representations are equivalent, corresponding matrices must have the same character. If we compare the characters $\chi_s(\mathscr{R})$ of the previous table with the characters $\chi(\mathscr{R})$ of Example 6.2.1 we notice that they differ. This difference is due to the fact that using the internal coordinate basis we have ignored the three translations and the three rotations of the molecule, which were automatically included in the Cartesian displacement basis.

In order to investigate the effect of a symmetry operation upon the translational and rotational modes we shall use the explicit form of the translational and rotational coordinates of equation (2.6.5). The three translational coordinates have the form

$$t_x = \frac{1}{M^{\frac{1}{2}}}\sum_\alpha m_\alpha \Delta x_\alpha \qquad t_y = \frac{1}{M^{\frac{1}{2}}}\sum_\alpha m_\alpha \Delta y_\alpha \qquad t_z = \frac{1}{M^{\frac{1}{2}}}\sum_\alpha m_\alpha \Delta z_\alpha$$

and thus they behave under a symmetry operation in the same way as the corresponding Cartesian displacements. Since these all have the same orientation, the symmetry behaviour of the translational coordinates is conveniently described by the Cartesian displacements of the centre of mass of the molecule. The effect of the symmetry operations

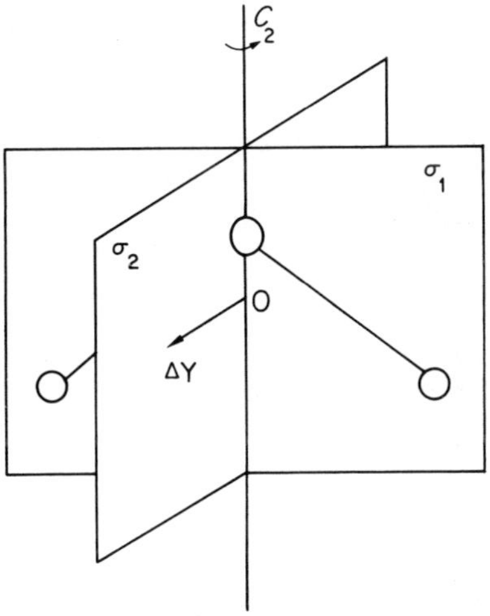

Figure 6.2.5 The Cartesian displacement ΔY of the mass-centre lies in the σ_2 symmetry plane and is left unchanged by a reflection in this plane. C_2 and σ_1 send ΔY into $-\Delta Y$

upon the t_y translation can easily be derived from Figure 6.2.5. The displacement ΔY of the centre of mass is left unchanged by E and by σ_2 whereas it is sent into $-\Delta Y$ by C_2 and by σ_1. In the same way one can see that ΔX is sent into $-\Delta X$ by C_2 and σ_2 and that ΔZ is left unchanged by all symmetry operations.

The four submatrices $\mathbf{T}_t(\mathscr{R})$, where the subscript t indicates the translational coordinates block, are therefore

$$\mathbf{T}_t(E) = \begin{pmatrix} 1 & 0 & 0 \\ 0 & 1 & 0 \\ 0 & 0 & 1 \end{pmatrix} \qquad \mathbf{T}_t(C_2) = \begin{pmatrix} -1 & 0 & 0 \\ 0 & -1 & 0 \\ 0 & 0 & 1 \end{pmatrix}$$

$$\mathbf{T}_t(\sigma_1) = \begin{pmatrix} 1 & 0 & 0 \\ 0 & -1 & 0 \\ 0 & 0 & 1 \end{pmatrix} \qquad \mathbf{T}_t(\sigma_2) = \begin{pmatrix} -1 & 0 & 0 \\ 0 & 1 & 0 \\ 0 & 0 & 1 \end{pmatrix}$$

and the corresponding characters are

	E	C_2	σ_1	σ_2
$\chi_t(\mathscr{R})$	3	−1	1	1

The three rotational coordinates have the form

$$R_x = \sum_\alpha \left(\frac{m_\alpha}{I_x^{\frac{1}{2}}}\right)(y_\alpha\,\Delta z_\alpha - z_\alpha\,\Delta y_\alpha) \qquad R_y = \sum_\alpha \left(\frac{m_\alpha}{I_y^{\frac{1}{2}}}\right)(z_\alpha\,\Delta x_\alpha - x_\alpha\,\Delta z_\alpha) \qquad R_z = \sum_\alpha \left(\frac{m_\alpha}{I_z^{\frac{1}{2}}}\right)(x_\alpha\,\Delta y_\alpha - y_\alpha\,\Delta x_\alpha)$$

since the Cartesian coordinates and the Cartesian displacements of the atoms are iso-oriented, they behave under a symmetry operation as the product of two coordinates $\rho\tau$ $(\rho, \tau = x, y, z)$.

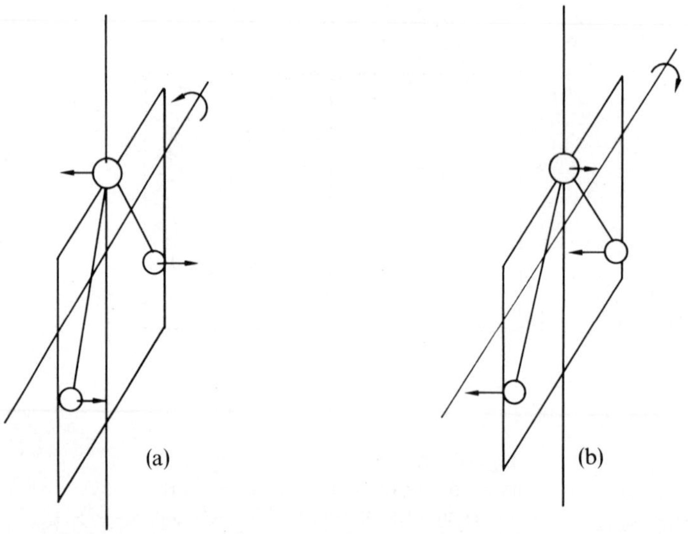

Figure 6.2.6 The effect of C_2 upon the rotation R_x for the water molecule. In (a) the oxygen atom moves toward the left and the hydrogen atom moves toward the right. After the symmetry operation C_2 (b) the oxygen atom moves toward the right and the hydrogen atoms toward the left. The sense of rotation is thus inverted

The effect of the operation C_2 upon the rotation R_x is shown in Figure 6.2.6. The behaviour of a rotational coordinate under a symmetry operation is correctly described by the behaviour of an axial vector, i.e. of a vector with an associated sense of rotation. In Figure 6.2.6 the axial vector is indicated schematically by means of a line segment pointing in the direction of a rotating corkscrew with a curved arrow which specifies the sense of rotation.

From equations (2.6.5), or from figures of the type of Figure 6.2.6, it is easily seen that R_z is left unchanged by E and C_2 and is sent into $-R_z$ by σ_1 and σ_2, R_x is left unchanged by E and σ_2 and is sent into $-R_x$ by C_2 and σ_1, whereas R_y is left unchanged by E and σ_1 and is sent into $-R_y$ by C_2 and σ_2. The four submatrices $\mathbf{T}_\mathrm{R}(\mathscr{R})$, where the subscript R indicates the rotational coordinates block, are thus

$$\mathbf{T}_\mathrm{R}(E) = \begin{pmatrix} 1 & 0 & 0 \\ 0 & 1 & 0 \\ 0 & 0 & 1 \end{pmatrix} \qquad \mathbf{T}_\mathrm{R}(C_2) = \begin{pmatrix} -1 & 0 & 0 \\ 0 & -1 & 0 \\ 0 & 0 & 1 \end{pmatrix}$$

$$\mathbf{T}_\mathrm{R}(\sigma_1) = \begin{pmatrix} -1 & 0 & 0 \\ 0 & 1 & 0 \\ 0 & 0 & -1 \end{pmatrix} \qquad \mathbf{T}_\mathrm{R}(\sigma_2) = \begin{pmatrix} 1 & 0 & 0 \\ 0 & -1 & 0 \\ 0 & 0 & -1 \end{pmatrix}$$

and the corresponding characters are

	E	C_2	σ_1	σ_2
$\chi_\mathrm{R}(\mathscr{R})$	3	−1	−1	−1

In Table 6.2.1 the characters $\chi_\mathrm{s}(\mathscr{R})$, $\chi_\mathrm{t}(\mathscr{R})$ and $\chi_\mathrm{R}(\mathscr{R})$, as well as their sum, are collected. If we now compare the character system of the **x**-representation with the character system of Table 6.2.1, we see that they are equal, as expected.

Table 6.2.1

	E	C_2	σ_1	σ_2
χ_s	3	1	3	1
χ_t	3	−1	1	1
χ_R	3	−1	−1	−1
$\chi = \chi_\mathrm{s} + \chi_\mathrm{t} + \chi_\mathrm{R}$	9	−1	3	1

Example 6.2.5

The H_2CO molecule of example 6.2.2 is planar. The $3N$ degrees of freedom can be divided into $2N$ degrees of freedom in the molecular plane and N degrees of freedom out of the plane. Since in the plane we have the two translations t_x and t_z and the rotation R_y and out of the plane the translation t_y and the rotations R_x and R_z, we need $2N - 3 = 5$ internal coordinates to describe the in-plane internal modes and $N - 3 = 1$ internal coordinates to describe the out-of-plane internal motion. As out-of-plane internal coordinate we can choose the angle θ (see Chapter 4, Section 4) formed by bond 1–4 with the plane formed by bonds 1–3 and 2–3. The coordinate $\Delta\theta = s_7$ is sent into itself by operations E and σ_2 whereas it is sent into $-\Delta\theta$ by C_2 and σ_1.

If we choose as in-plane internal coordinates the variation of the bond lengths and angles of Figure 6.2.3, i.e.

$$s_1 = \Delta r_1; \quad s_2 = \Delta r_2; \quad s_3 = \Delta R; \quad s_4 = \Delta\alpha; \quad s_5 = \Delta\beta_1; \quad s_6 = \Delta\beta_2$$

we see that we have one redundant coordinate. Keeping this in mind, let us calculate the elements of the matrices $\mathbf{T}_s(\mathscr{R})$. Since ΔR and $\Delta\alpha$ are left unchanged by all symmetry operations, whereas Δr_1 and $\Delta\beta_1$ are permuted with Δr_2 and $\Delta\beta_2$ respectively by C_2 and σ_2, we have

$$\mathbf{T}_s(E) = \begin{pmatrix} 1 & 0 & 0 & 0 & 0 & 0 & 0 \\ 0 & 1 & 0 & 0 & 0 & 0 & 0 \\ 0 & 0 & 1 & 0 & 0 & 0 & 0 \\ 0 & 0 & 0 & 1 & 0 & 0 & 0 \\ 0 & 0 & 0 & 0 & 1 & 0 & 0 \\ 0 & 0 & 0 & 0 & 0 & 1 & 0 \\ 0 & 0 & 0 & 0 & 0 & 0 & 1 \end{pmatrix} \qquad \mathbf{T}_s(C_2) = \begin{pmatrix} 0 & 1 & 0 & 0 & 0 & 0 & 0 \\ 1 & 0 & 0 & 0 & 0 & 0 & 0 \\ 0 & 0 & 1 & 0 & 0 & 0 & 0 \\ 0 & 0 & 0 & 1 & 0 & 0 & 0 \\ 0 & 0 & 0 & 0 & 0 & 1 & 0 \\ 0 & 0 & 0 & 0 & 1 & 0 & 0 \\ 0 & 0 & 0 & 0 & 0 & 0 & -1 \end{pmatrix}$$

$$\mathbf{T}_s(\sigma_1) = \begin{pmatrix} 1 & 0 & 0 & 0 & 0 & 0 & 0 \\ 0 & 1 & 0 & 0 & 0 & 0 & 0 \\ 0 & 0 & 1 & 0 & 0 & 0 & 0 \\ 0 & 0 & 0 & 1 & 0 & 0 & 0 \\ 0 & 0 & 0 & 0 & 1 & 0 & 0 \\ 0 & 0 & 0 & 0 & 0 & 1 & 0 \\ 0 & 0 & 0 & 0 & 0 & 0 & -1 \end{pmatrix} \qquad \mathbf{T}_s(\sigma_2) = \begin{pmatrix} 0 & 1 & 0 & 0 & 0 & 0 & 0 \\ 1 & 0 & 0 & 0 & 0 & 0 & 0 \\ 0 & 0 & 1 & 0 & 0 & 0 & 0 \\ 0 & 0 & 0 & 1 & 0 & 0 & 0 \\ 0 & 0 & 0 & 0 & 0 & 1 & 0 \\ 0 & 0 & 0 & 0 & 1 & 0 & 0 \\ 0 & 0 & 0 & 0 & 0 & 0 & 1 \end{pmatrix}$$

with the character system

	E	C_2	σ_1	σ_2
$\chi_s(\mathscr{R})$	7	1	5	3

The symmetry behaviour of the translational and rotational modes is independent of the shape of the molecule, being the same for all molecules which belong to the same point group. As we have seen in the previous example, the matrices associated with the translations and with the rotations can be derived by application of the symmetry operations of the point group to the three Cartesian displacements of the centre of mass and to the three axial vectors describing the rotations. The matrices $\mathbf{T}_t(\mathscr{R})$ and $\mathbf{T}_R(\mathscr{R})$ are thus identical with those of the previous example.

Table 6.2.2 summarizes the various character systems for the submatrices $\mathbf{T}_s(\mathscr{R})$, $\mathbf{T}_t(\mathscr{R})$ and $\mathbf{T}_R(\mathscr{R})$ as well as the character system of the Cartesian displacement basis discussed in Example 6.2.2. If we had no redundant coordinates, $\bar{\chi}(\mathscr{R})$ would be equal to $\chi(\mathscr{R})$ as in the previous example. Here there is one redundant coordinate, and the two character systems differ by just the character system of the redundant coordinate.

Table 6.2.2

	E	C_2	σ_1	σ_2
$\chi_s(\mathscr{R})$	7	1	5	3
$\chi_t(\mathscr{R})$	3	-1	1	1
$\chi_R(\mathscr{R})$	3	-1	-1	-1
$\bar{\chi}(\mathscr{R}) = \chi_s(\mathscr{R}) + \chi_t(\mathscr{R}) + \chi_R(\mathscr{R})$	13	-1	5	3
$\chi(\mathscr{R})$	12	-2	4	2
$\chi_\rho = \bar{\chi}(\mathscr{R}) - \chi(\mathscr{R})$	1	1	1	1

For a molecule like H_2CO the form of the redundant coordinate is easily found. Since (Figure 6.2.3)

$$\alpha + \beta_1 + \beta_2 = 360°$$

we have

$$\rho = \Delta\alpha + \Delta\beta_1 + \Delta\beta_2 = 0$$

which shows that $\Delta\alpha$, $\Delta\beta_1$ and $\Delta\beta_2$ are not independent. This relationship can be taken as the redundant coordinate since it is left unchanged by all symmetry operations of the group and has, therefore, the character system of the last row of the table.

Example 6.2.6

For the ammonia molecule of Example 6.2.3 we can choose as internal coordinates the variation of the three NH bond lengths Δr_1, Δr_2 and Δr_3 and of the three bond angles $\Delta\alpha_1$, $\Delta\alpha_2$ and $\Delta\alpha_3$. Since the characters are the same for the elements of a class, only one representative of each class will be considered.

The operation C_3 sends Δr_1 into Δr_2, Δr_2 into Δr_3 and Δr_3 into Δr_1, whereas σ_1 leaves Δr_1 unchanged and permutes Δr_2 and Δr_3. Since the coordinates $\Delta\alpha_i$ have been numbered so that $\Delta\alpha_i$ behaves under all symmetry operations like Δr_i, the matrices $T_s(\mathscr{R})$ are

$$\mathbf{T}_s(E) = \left(\begin{array}{ccc|ccc} 1 & 0 & 0 & 0 & 0 & 0 \\ 0 & 1 & 0 & 0 & 0 & 0 \\ 0 & 0 & 1 & 0 & 0 & 0 \\ \hline 0 & 0 & 0 & 1 & 0 & 0 \\ 0 & 0 & 0 & 0 & 1 & 0 \\ 0 & 0 & 0 & 0 & 0 & 1 \end{array}\right) \qquad \mathbf{T}_s(C_3) = \left(\begin{array}{ccc|ccc} 0 & 0 & 1 & 0 & 0 & 0 \\ 1 & 0 & 0 & 0 & 0 & 0 \\ 0 & 1 & 0 & 0 & 0 & 0 \\ \hline 0 & 0 & 0 & 0 & 0 & 1 \\ 0 & 0 & 0 & 1 & 0 & 0 \\ 0 & 0 & 0 & 0 & 1 & 0 \end{array}\right)$$

$$\mathbf{T}_s(\sigma_1) = \left(\begin{array}{ccc|ccc} 1 & 0 & 0 & 0 & 0 & 0 \\ 0 & 0 & 1 & 0 & 0 & 0 \\ 0 & 1 & 0 & 0 & 0 & 0 \\ \hline 0 & 0 & 0 & 1 & 0 & 0 \\ 0 & 0 & 0 & 0 & 0 & 1 \\ 0 & 0 & 0 & 0 & 1 & 0 \end{array}\right)$$

and the characters are

	E	$2C_3$	3σ
$\chi_r(\mathscr{R})$	3	0	1
$\chi_\alpha(\mathscr{R})$	3	0	1
$\chi_s(\mathscr{R})$	6	0	2

For the three translations and the three rotations we obtain from (2.6.5) and from Example 5.5.2

$$\mathbf{T}_t(E) = \left(\begin{array}{cc|c} 1 & 0 & 0 \\ 0 & 1 & 0 \\ \hline 0 & 0 & 1 \end{array}\right) \quad \mathbf{T}_t(C_3) = \left(\begin{array}{cc|c} -\frac{1}{2} & -\frac{\sqrt{3}}{2} & 0 \\ \frac{\sqrt{3}}{2} & -\frac{1}{2} & 0 \\ \hline 0 & 0 & 1 \end{array}\right) \quad \mathbf{T}_t(\sigma_1) = \left(\begin{array}{cc|c} 1 & 0 & 0 \\ 0 & -1 & 0 \\ \hline 0 & 0 & 1 \end{array}\right)$$

$$\mathbf{T}_R(E) = \left(\begin{array}{cc|c} 1 & 0 & 0 \\ 0 & 1 & 0 \\ \hline 0 & 0 & 1 \end{array}\right) \quad \mathbf{T}_R(C_3) = \left(\begin{array}{cc|c} -\frac{1}{2} & -\frac{\sqrt{3}}{2} & 0 \\ \frac{\sqrt{3}}{2} & -\frac{1}{2} & 0 \\ \hline 0 & 0 & 1 \end{array}\right) \quad \mathbf{T}_R(\sigma_1) = \left(\begin{array}{cc|c} -1 & 0 & 0 \\ 0 & 1 & 0 \\ \hline 0 & 0 & -1 \end{array}\right)$$

with the character systems

	E	$2C_3$	3σ
$\chi_t(\mathscr{R})$	3	0	1
$\chi_R(\mathscr{R})$	3	0	-1

The sum $\chi_t(\mathscr{R}) + \chi_R(\mathscr{R}) + \chi_s(\mathscr{R})$ is equal to $\chi(\mathscr{R})$ (see Example 6.2.3) as expected since a non-redundant set of internal coordinates has been used.

In principle it is easy to write down the matrices of the representations for any molecule and any possible point group. Since we need only the characters of the representations, it is unnecessary to go through all this work.

In Example 6.2.1 we have pointed out that only the atoms which are unshifted by the symmetry operation $\mathscr{R}$ contribute to the character of $\mathbf{D}(\mathscr{R})$. The character of $\mathbf{D}(\mathscr{R})$ is thus given by the product of the character $\mathbf{D}_c(\mathscr{R})$ multiplied by the number of unshifted atoms. We have thus for a proper or improper rotation respectively

$$\chi(\mathscr{R}) = N_R \chi_c(\mathscr{R}) = N_R(\pm 1 + 2\cos\theta) \tag{6.2.16}$$

where N_R is the number of atoms unshifted by operation $\mathscr{R}$ and θ is the angle of rotation (see 5.5.42).

The character of the representation carried by the three translational coordinates is simply

$$\chi_t(\mathscr{R}) = \pm 1 + 2\cos\theta \tag{6.2.17}$$

since, as discussed before, they behave as the Cartesian displacements of the centre of mass, and the corresponding matrix is just a rotation matrix.

The character of the representation carried by the three rotational coordinates can be easily obtained remembering that an axial vector behaves like an ordinary vector in a proper rotation ($\chi = 1 + 2\cos\theta$), whereas in an improper rotation the sense is reversed ($\chi = -(-1 + 2\cos\theta)$). The character is thus

$$\chi_R(\mathscr{R}) = 1 \pm 2\cos\theta \tag{6.2.18}$$

Finally the character of the representation carried by a symmetrically equivalent set of internal coordinates $\Delta\omega_1$, $\Delta\omega_2$, $\Delta\omega_n$ is given by the number of coordinates which remain unshifted in the symmetry operation N_ω, multiplied by $+1$ or -1 if the coordinate remains unchanged or changes sign respectively. Therefore

$$\chi(\mathscr{R}) = \pm N_\omega \tag{6.2.19}$$

6.3 The symmetry of normal coordinates

The whole application of group-theoretical methods to molecular vibrations is based upon the fact that the normal coordinates form the basis for a completely reduced representation of the molecular point group. In order to prove this result, let us consider the effect of a symmetry operation on the normal coordinate vector $\mathbf{Q}$. If we call $\overline{\mathbf{D}}(\mathscr{R})$ the matrix associated with the operation $\mathscr{R}$, which transforms the vector $\mathbf{Q}$ into the rotated vector $\mathbf{Q}'$, we have

$$\mathbf{Q}' = \overline{\mathbf{D}}(\mathscr{R})\mathbf{Q} \tag{6.3.1}$$

analogous to equation (6.2.1).

By substitution of (6.3.1) into the equations (2.4.6) and (2.4.7) which give the kinetic and the potential energies in terms of normal coordinates, we have

$$2T = \tilde{\dot{\mathbf{Q}}}'\dot{\mathbf{Q}}' = \tilde{\dot{\mathbf{Q}}}\tilde{\overline{\mathbf{D}}}(\mathscr{R})\overline{\mathbf{D}}(\mathscr{R})\dot{\mathbf{Q}} \tag{6.3.2}$$

and

$$2V = \tilde{\mathbf{Q}}'\mathbf{\Lambda}\mathbf{Q}' = \tilde{\mathbf{Q}}\tilde{\overline{\mathbf{D}}}(\mathscr{R})\mathbf{\Lambda}\overline{\mathbf{D}}(\mathscr{R})\mathbf{Q} \tag{6.3.3}$$

and thus

$$\tilde{\overline{\mathbf{D}}}(\mathscr{R})\overline{\mathbf{D}}(\mathscr{R}) = \mathbf{E} \tag{6.3.4}$$

$$\tilde{\overline{\mathbf{D}}}(\mathscr{R})\mathbf{\Lambda}\overline{\mathbf{D}}(\mathscr{R}) = \mathbf{\Lambda} \tag{6.3.5}$$

These last two equations, analogous to equations (6.2.6) describe the invariance of V and T under the symmetry operations of the molecular point groups. Owing to the special form of the potential and kinetic energies in normal coordinates, equations (6.3.4) and (6.3.5) put some special requirements on the form of the $\overline{\mathbf{D}}(\mathscr{R})$ matrices.

The generic ith diagonal elements of the products on the left of equations (6.3.4) and (6.3.5) have the form

$$\sum_k \bar{D}_{ki}(\mathscr{R})^2 = 1 \tag{6.3.6}$$

$$\sum_k \lambda_k \bar{D}_{ki}(\mathscr{R})^2 = \lambda_i \tag{6.3.7}$$

respectively. These equations are compatible only if

$$\bar{D}_{ki}(\mathscr{R}) = 0 \quad \text{when } \lambda_i \neq \lambda_k \tag{6.3.8}$$

i.e. if the matrices $\overline{\mathbf{D}}(\mathscr{R})$ are in block form and are completely reduced. For a non-degenerate normal coordinate Q_i, we have in fact, from (6.3.6) and (6.3.7),

$$\bar{D}_{ii}(\mathscr{R}) = \pm 1 \qquad \bar{D}_{ki}(\mathscr{R}) = 0 \qquad (k \neq i) \tag{6.3.9}$$

A non-degenerate normal coordinate therefore spans a one-dimensional subspace and carries a one-dimensional and hence irreducible representation. A symmetry operation cannot mix Q_i with any other normal coordinate Q_k and can only send Q_i into $\pm Q_i$.

For two-fold degenerate normal coordinates Q_i and Q_j, with $\lambda_i = \lambda_j$, we have

$$\begin{aligned} \bar{D}_{ii}(\mathscr{R})^2 + \bar{D}_{ji}(\mathscr{R})^2 &= 1 \qquad \bar{D}_{ki}(\mathscr{R}) = 0 \quad (k \neq i, j) \\ \bar{D}_{ij}(\mathscr{R})^2 + \bar{D}_{jj}(\mathscr{R})^2 &= 1 \qquad \bar{D}_{kj}(\mathscr{R}) = 0 \quad (k \neq j, i) \end{aligned} \tag{6.3.10}$$

Two-fold degenerate normal coordinates therefore span a two-dimensional subspace and carry a two-dimensional irreducible representation. A symmetry operation mixes the coordinates Q_i and Q_j among themselves but cannot mix them with any other normal coordinate Q_k. In general a set of d-fold degenerate normal coordinates carries a d-dimensional irreducible representation of the molecular group ($d = 1, 2, 3$).

It is important to notice that the degeneracy considered here is due to molecular symmetry and to the presence in the molecule of a three-fold or higher axis of symmetry. Accidental degeneracy, due to the fact that two or more normal frequencies of the molecule are approximately equal, because of special numerical values of the masses and force constants, can also occur. In this case, however, the normal coordinates corresponding to the accidentally degenerate frequencies do not mix under the symmetry operations of the group and span separate subspaces.

6.4 The classification of normal vibrations

As we have seen in the previous section, each non-degenerate normal coordinate carries a one-dimensional representation of the group, each pair of two-fold degenerate normal coordinates carries a bidimensional representation and each triplet of triply-degenerate normal coordinates, carries a tridimensional representation of the molecular group. The representations carried by two different sets of d-fold degenerate normal coordinates can, however, be the same, i.e. the two sets can behave in exactly the same way under the symmetry operations of the group. In this case the corresponding blocks of the $\overline{\mathbf{D}}(\mathscr{R})$ matrices, and the relative characters, are the same. Since the infrared and Raman activity of a normal mode is determined by its behaviour under the symmetry operations of the molecular group, it is important to know how many normal coordinates belong to the same irreducible representation. This is simply accomplished using equation (5.7.26)

$$n_\gamma = \frac{1}{g}\sum_r (\chi_r^\gamma)^* \chi_r$$

which tells us the number of times the γth irreducible representation, with character χ_r^γ for operation $\mathscr{R}$ occurs in the reducible representation with character χ_r.

Since the characters of elements of the same class are the same, equation (5.7.26) can be rewritten in the simpler form

$$n_\gamma = \frac{1}{g}\sum_h g_h(\chi_h^\gamma)^* \chi_h \tag{6.4.1}$$

where χ_h^γ is the character of the irreducible representation for the hth class, χ_h is the corresponding character of the reducible representation, g is the order of the group, g_h the number of elements in the hth class and the sum runs over all the classes of the group.

The characters of the irreducible representations are given in the tables of Appendix I. From the same tables we can obtain the order g of the group as well as the number of classes and the number of elements in each class.

The character system of the reducible representation carried by the Cartesian displacement coordinates can be obtained from equation (6.2.16) (see also Examples 6.2.1, 6.2.2 and 6.2.3). For the convenience of the reader the characters $\chi_c(\mathscr{R})$ are summarized in Table 6.4.1 for the most common symmetry operations.

Table 6.4.1 Characters of the submatrices $\mathbf{D}_c(\mathscr{R})$

	E	C_2	C_3	C_4	C_6	i	σ	S_3	S_4	S_6
$\chi_c(\mathscr{R})$	3	−1	0	1	2	−3	1	−2	−1	0

Some examples will illustrate to the reader the correct use of equation (6.4.1).

Example 6.4.1

The table of the characters of the irreducible representations of the C_{2v} point group (see Appendix I) is given in Table 6.4.2.

Table 6.4.2

	E	C_2	$\sigma_1(zx)$	$\sigma_2(yz)$	
A_1	1	1	1	1	T_z
A_2	1	1	−1	−1	R_z
B_1	1	−1	1	−1	T_x, R_y
B_2	1	−1	−1	1	T_y, R_x

There are four different classes (each operation is in this case a class by itself) and hence four different irreducible representations labelled by the symbols A_1, A_2, B_1 and B_2. The nomenclature of the irreducible representations, often called symmetry species, is given in Appendix I (see also Section 6.1).

For the water molecule of Example 6.2.1 we obtain from equation (6.4.1)

$$n_{A_1} = \tfrac{1}{4}(1.1\cdot 9 + 1.1.(-1) + 1.1.3 + 1.1.1) \qquad = 3$$

$$n_{A_2} = \tfrac{1}{4}(1.1.9 + 1.1.(-1) + 1.(-1).3 + 1.(-1).1) = 1$$

$$n_{B_1} = \tfrac{1}{4}(1.1.9 + 1.(-1).(-1) + 1.1.3 + 1.(-1).1 \;= 3$$

$$n_{B_2} = \tfrac{1}{4}(1.1.9 + 1.(-1).(-1) + 1.(-1).3 + 1.1.1 \;= 2$$

$$\text{Total} \quad 9$$

There are thus three modes of A_1 species, one mode of A_2 species, 3 modes of B_1 species and 2 modes of B_2 species. We shall say that the structure of the representation carried by the 9 Cartesian displacement coordinates of water is

$$3A_1 + A_2 + 3B_1 + 2B_2$$

This structure includes the three translations and the three rotations. In order to obtain the number of true vibrational modes within each symmetry species it is necessary to find the number of external coordinates falling in each species and subtract this number from the corresponding value of n_γ.

The symmetry species of the translations and of the rotations can be obtained easily from the matrices $\mathbf{T}_t(\mathscr{R})$ and $\mathbf{T}_R(\mathscr{R})$. Since these matrices were obtained from equations (2.6.5) which define true translational and rotational normal coordinates, they are already in reduced form. As discussed before, it is not necessary to carry out this calculation for each molecule since the symmetry of the translational and rotational modes is a characteristic of the point group and is independent of the molecular structure. Once obtained for one molecule, the symmetry of the external motions can be used for all molecules belonging to the same point group. For this reason the symmetry of the three translations and of the three rotations is listed in the character tables of Appendix I.

From the matrices $\mathbf{T}_t(\mathscr{R})$ of Example 6.2.4 we see, considering the four diagonal elements of the first row, that the translation T_x carries the representation

$$1 \quad -1 \quad 1 \quad -1$$

and belongs to the species B_1, whereas the translations T_y and T_z carry the representations $(1 \quad -1 \quad -1 \quad 1)$ and $(1 \quad 1 \quad 1 \quad 1)$ and belong to the B_2 and A_1 species respectively. From the matrices $\mathbf{T}_R(\mathscr{R})$ of the same example we obtain that R_x belongs to the B_2 species, R_y to the B_1 species and R_z to the A_2 species.

The number of vibrational modes falling in each symmetry species is thus

$$n_{A_1} = 3 - 1 = 2$$

$$n_{A_2} = 1 - 1 = 0$$

$$n_{B_1} = 3 - 2 = 1$$

$$n_{B_2} = 2 - 2 = 0$$

The structure of the representations carried by the vibrational normal coordinates of water is, therefore,

$$2A_1 + B_1$$

It is very convenient to know the type of internal coordinates involved in the motions of each symmetry species. This is conveniently done with equation (6.4.1), using as characters of the reducible representations the characters of the matrices $\mathbf{T}_s(\mathscr{R})$ relative to the internal coordinate basis.

From the first table of Example 6.2.4 we have

$$n_{A_1}(r) = \tfrac{1}{4}(1\,.\,1\,.\,2 + 1\,.\,1\,.\,0 + 1\,.\,1\,.\,2 + 1\,.\,1\,.\,0) = 1$$

$$n_{A_2}(r) = \tfrac{1}{4}(1\,.\,1\,.\,2 + 1\,.\,1\,.\,0 + 1\,.\,(-1)\,.\,2 + 1\,.\,(-1)\,.\,0) = 0$$

$$n_{B_1}(r) = \tfrac{1}{4}(1\,.\,1\,.\,2 + 1\,.\,(-1)\,.\,0 + 1\,.\,1\,.\,2 + 1\,.\,(-1)\,.\,0) = 1$$

$$n_{B_2}(r) = \tfrac{1}{4}(1\,.\,1\,.\,2 + 1\,.\,(-1)\,.\,0 + 1\,.\,(-1)\,.\,2 + 1\,.\,1\,.\,0) = 0$$

and

$$n_{A_1}(\alpha) = \tfrac{1}{4}(1\,.\,1\,.\,1 + 1\,.\,1\,.\,1 + 1\,.\,1\,.\,1 + 1\,.\,1\,.\,1) = 1$$

$$n_{A_2}(\alpha) = \tfrac{1}{4}(1\,.\,1\,.\,1 + 1\,.\,1\,.\,1 + 1\,.\,(-1)\,.\,1 + 1\,.\,(-1)\,.\,1) = 0$$

$$n_{B_1}(\alpha) = \tfrac{1}{4}(1\,.\,1\,.\,1 + 1\,.\,(-1)\,.\,1 + 1\,.\,1\,.\,1 + 1\,.\,(-1)\,.\,1) = 0$$

$$n_{B_2}(\alpha) = \tfrac{1}{4}(1\,.\,1\,.\,1 + 1\,.\,(-1)\,.\,1 + 1\,.\,(-1)\,.\,1 + 1\,.\,1\,.\,1) = 0.$$

where $n(r)$ and $n(\alpha)$ are the number of stretches (coordinates Δr_1 and Δr_2) and bends (coordinate $\Delta\alpha$) respectively in each symmetry species. We conclude, therefore, that the vibration of species B_1 is a stretching motion of the OH bonds whereas the two vibrations of species A_1 are a stretching and bending of the OH bonds respectively.

Example 6.4.2

The structure of the representation carried by the Cartesian coordinates of H_2CO is shown in Table 6.4.3 under the label N_c. By subtraction of the translations and rotations from the proper symmetry species, we obtain the column labelled N_v which lists the number of vibrational modes belonging to each symmetry species. The next column, labelled N_s, gives the structure of the representation carried by the internal coordinates (Example 6.2.5). The following columns give the contribution to each symmetry species of the different symmetrically equivalent sets of internal coordinates. The last column gives the symmetry species of the redundant coordinate.

Table 6.4.3

	N_c	N_v	N_s	N_r	N_R	N_α	N_β	N_θ	Redund.
A_1	4	3	4	1	1	1	1	0	1
A_2	1	0	0	0	0	0	0	0	0
B_1	4	2	2	1	0	0	1	0	0
B_2	3	1	1	0	0	0	0	1	0

The reader should have no difficulty in performing the calculations using equation (6.4.1) and following the schema of the previous example. We report in full the calculation for the contribution of $\Delta\theta$ to the B_2 species.

$$n_{B_2}(\theta) = \tfrac{1}{4}(1\,.\,1\,.\,1 + 1\,.\,(-1)\,.\,(-1) + 1\,.\,(-1)\,.\,(-1) + 1\,.\,1\,.\,1) = 1$$

Example 6.4.3

Ammonia belongs to the C_{3v} point group. The table of characters of the irreducible representations is given in Table 6.4.4.

Table 6.4.4

	E	$2C_3$	3σ	
A_1	1	1	1	T_z
A_2	1	1	-1	R_z
E	2	-1	0	$(T_x, T_y); (R_x, R_y)$

From the matrices $\mathbf{T}_t(\mathscr{R})$ of Example 6.2.6 we see that the T_z translation and the R_z rotation are non-degenerate and belong to the A_1 and A_2 species respectively, whereas T_x and T_y, as well as R_x and R_y, are doubly degenerate and belong to the E species.

From equation (6.4.1) we obtain, using Table 6.4.4 and the characters of the reducible representation carried by the Cartesian displacement coordinates (equation 6.2.16),

$$n_{A_1} = \tfrac{1}{6}(1\,.\,1\,.\,12 + 2\,.\,1\,.\,0 + 3\,.\,1\,.\,2) = 3 - 1 = 2$$

$$n_{A_2} = \tfrac{1}{6}(1\,.\,1\,.\,12 + 2\,.\,1\,.\,0 + 3\,.\,(-1)\,.\,2) = 1 - 1 = 0$$

$$n_E = \tfrac{1}{6}(1\,.\,2\,.\,12 + 2\,.\,(-1)\,.\,0 + 3\,.\,0\,.\,2) = 4 - 2 = 2$$

Using the internal coordinates we have

$$n_{A_1}(r) = \tfrac{1}{6}(1\,.\,1\,.\,3 + 2\,.\,1\,.\,0 + 3\,.\,1\,.\,1) = 1$$

$$n_{A_2}(r) = \tfrac{1}{6}(1\,.\,1\,.\,3 + 2\,.\,1\,.\,0 + 3\,.\,(-1)\,.\,1) = 0$$

$$n_E(r) = \tfrac{1}{6}(1\,.\,2\,.\,3 + 2\,.\,(-1)\,.\,0 + 3\,.\,0\,.\,1) = 1$$

$$n_{A_1}(\alpha) = \tfrac{1}{6}(1\,.\,1\,.\,3 + 2\,.\,1\,.\,0 + 3\,.\,1\,.\,1) = 1$$

$$n_{A_2}(\alpha) = \tfrac{1}{6}(1\,.\,1\,.\,3 + 2\,.\,1\,.\,0 + 3\,.\,(-1)\,.\,1) = 0$$

$$n_E(\alpha) = \tfrac{1}{6}(1\,.\,2\,.\,3 + 2\,.\,(-1)\,.\,0 + 3\,.\,0\,.\,1) = 1$$

We conclude, therefore, that two normal modes are non-degenerate and belong to the A_1 species whereas two pairs of degenerate modes occur in the E species.

6.5 Symmetry coordinates

The solution of the secular equation (4.2.14) can involve, even for medium-size molecules, the diagonalization of very large matrices. It is therefore convenient to devise a linear transformation of coordinates such that in the new coordinate basis the potential and the kinetic energy matrices are factorized to the maximum extent permitted by the molecular symmetry. We shall prove that any symmetry adapted basis, i.e. any coordinate basis which carries a completely reduced unitary representation of the molecular group, is able to factorize the kinetic and the potential energy matrices in block form.

Let us call $\mathbb{S}$ the vector formed by these symmetry coordinates, related to the vector $\mathbf{s}$ of the internal coordinates by the unitary transformation

$$\mathbb{S} = \mathbb{U}\mathbf{s} \tag{6.5.1}$$

The kinetic and potential energies in terms of the symmetry coordinates are easily obtained by substitution of (6.5.1) in the expressions (4.2.1) and (4.2.2) which give the potential and kinetic energies in terms of the internal coordinates. We have

$$2T = \tilde{\dot{\mathbf{s}}}\mathbf{G}^{-1}\dot{\mathbf{s}} = \tilde{\dot{\mathbb{S}}}\mathbb{U}\mathbf{G}^{-1}\tilde{\mathbb{U}}\dot{\mathbb{S}} = \tilde{\dot{\mathbb{S}}}\mathbb{G}^{-1}\dot{\mathbb{S}} \tag{6.5.2}$$

$$2V = \tilde{\mathbf{s}}\mathbf{F}\mathbf{s} = \tilde{\mathbb{S}}\mathbb{U}\mathbf{F}\tilde{\mathbb{U}}\mathbb{S} = \tilde{\mathbb{S}}\mathbb{F}\mathbb{S} \tag{6.5.3}$$

where

$$\mathbb{G}^{-1} = \mathbb{U}\mathbf{G}^{-1}\tilde{\mathbb{U}} \tag{6.5.4}$$

and

$$\mathbb{F} = \mathbb{U}\mathbf{F}\tilde{\mathbb{U}} \tag{6.5.5}$$

are the kinetic and potential energy matrices in terms of the symmetry coordinates.

Let us now consider the effect of a symmetry operation $\mathscr{R}$ on the vector $\mathbb{S}$. If we call $\mathbb{D}(\mathscr{R})$ the matrix associated with the operation $\mathscr{R}$ which transforms the vector $\mathbb{S}$ into the rotated vector $\mathbb{S}'$, we have

$$\mathbb{S}' = \mathbb{D}(\mathscr{R})\mathbb{S} \tag{6.5.6}$$

and, by substitution in (6.5.2) and (6.5.3)

$$\tilde{\mathbb{D}}(\mathscr{R})\mathbb{G}^{-1}\mathbb{D}(\mathscr{R}) = \mathbb{G}^{-1} \tag{6.5.7}$$

$$\tilde{\mathbb{D}}(\mathscr{R})\mathbb{F}\mathbb{D}(\mathscr{R}) = \mathbb{F} \tag{6.5.8}$$

which show the usual invariance property of the potential and kinetic energy under the symmetry operations of the molecular group. From (6.5.8) we obtain

$$\mathbb{F}\mathbb{D}(\mathscr{R}) = \mathbb{D}(\mathscr{R})\mathbb{F} \tag{6.5.9}$$

which shows that $\mathbb{F}$ commutes with all matrices $\mathbb{D}(\mathscr{R})$. The same is true for the $\mathbb{G}$-matrix.

Since $\mathbb{D}(\mathscr{R})$ is completely reduced by definition of symmetry coordinates, we can use Schur's lemma (Section 5.7) and its corollary to investigate the structure of $\mathbb{F}$. A given irreducible representation can appear several times, as we have seen in the example of Section 6.4, in the structure of $\mathbb{D}(\mathscr{R})$, i.e. it can happen that several different symmetry coordinates carry the same irreducible representation. We shall indicate by $\mathbb{D}_\alpha^\mu$ the μth block of $\mathbb{D}(\mathscr{R})$ carrying the αth irreducible representation. Correspondingly we shall consider the $\mathbb{F}$-matrix partitioned into similar blocks and indicate the generic block with the symbol $\mathbb{F}_{\alpha\beta}^{\mu\nu}$.

A generic term of the products of equation (6.5.9) is thus given by

$$\mathbb{F}_{\alpha\beta}^{\mu\nu}\mathbb{D}_\alpha^\mu = \mathbb{D}_\beta^\nu\mathbb{F}_{\alpha\beta}^{\mu\nu} \tag{6.5.10}$$

and from Schur's lemma and its corollary we have

$$\mathbb{F}_{\alpha\beta}^{\mu\nu} = \mathbf{0} \qquad \text{for } \Gamma_\alpha \text{ inequivalent to } \Gamma_\beta$$

$$\mathbb{F}_{\alpha\beta}^{\mu\nu} = F_{\mu\nu}^\alpha\mathbf{E}_{g_\alpha} \qquad \text{for } \Gamma_\alpha \text{ identical to } \Gamma_\beta$$

where $\mathbf{E}_{g_\alpha}$ is the unit matrix of dimension g_α.

We conclude, therefore, that the $\mathbb{F}$-matrix has off-diagonal elements only between corresponding basis vectors of identical representations. Suppose, for example, that $\mathbb{D}(\mathscr{R})$ contains the monodimensional irreducible representation A twice and the bidimensional representation E of the group C_3 twice (Appendix I). Equation (6.5.9) will have the form

$$\begin{pmatrix} \mathbb{D}_A^1 & 0 & 0 & 0 \\ 0 & \mathbb{D}_A^2 & 0 & 0 \\ 0 & 0 & \mathbb{D}_E^1 & 0 \\ 0 & 0 & 0 & \mathbb{D}_E^2 \end{pmatrix} \begin{pmatrix} \mathbb{F}_{AA}^{11} & \mathbb{F}_{AA}^{12} & \mathbb{F}_{AE}^{11} & \mathbb{F}_{AE}^{12} \\ \mathbb{F}_{AA}^{21} & \mathbb{F}_{AA}^{22} & \mathbb{F}_{AE}^{21} & \mathbb{F}_{AE}^{22} \\ \mathbb{F}_{EA}^{11} & \mathbb{F}_{EA}^{12} & \mathbb{F}_{EE}^{11} & \mathbb{F}_{EE}^{12} \\ \mathbb{F}_{EA}^{21} & \mathbb{F}_{EA}^{22} & \mathbb{F}_{EE}^{21} & \mathbb{F}_{EE}^{22} \end{pmatrix} = \begin{pmatrix} \mathbb{F}_{AA}^{11} & \mathbb{F}_{AA}^{12} & \mathbb{F}_{AE}^{11} & \mathbb{F}_{AE}^{12} \\ \mathbb{F}_{AA}^{21} & \mathbb{F}_{AA}^{22} & \mathbb{F}_{AE}^{21} & \mathbb{F}_{AE}^{22} \\ \mathbb{F}_{EA}^{11} & \mathbb{F}_{EA}^{12} & \mathbb{F}_{EE}^{11} & \mathbb{F}_{EE}^{12} \\ \mathbb{F}_{EA}^{21} & \mathbb{F}_{EA}^{22} & \mathbb{F}_{EE}^{21} & \mathbb{F}_{EE}^{22} \end{pmatrix} \begin{pmatrix} \mathbb{D}_A^1 & 0 & 0 & 0 \\ 0 & \mathbb{D}_A^2 & 0 & 0 \\ 0 & 0 & \mathbb{D}_E^1 & 0 \\ 0 & 0 & 0 & \mathbb{D}_E^2 \end{pmatrix}$$

If we write down the elements of the product matrix for the first row we obtain the following matrix equations

$$\mathbb{D}_A^1\mathbb{F}_{AA}^{11} = \mathbb{F}_{AA}^{11}\mathbb{D}_A^1 \qquad \mathbb{D}_A^1\mathbb{F}_{AA}^{12} = \mathbb{F}_{AA}^{12}\mathbb{D}_A^2 \qquad \mathbb{D}_A^1\mathbb{F}_{AE}^{11} = \mathbb{F}_{AE}^{11}\mathbb{D}_E^1 \qquad \mathbb{D}_A^1\mathbb{F}_{AE}^{12} = \mathbb{F}_{AE}^{12}\mathbb{D}_E^2$$

Since A and E are inequivalent representations we obtain from Schur's lemma, in the last two equations,

$$\mathbb{F}_{AE}^{11} = \mathbb{F}_{AE}^{12} = 0$$

whereas from the corollary we obtain from the first two equations

$$\mathbb{F}_{AA}^{11} = F_{11} \qquad \mathbb{F}_{AA}^{12} = F_{12}$$

In the same way we obtain

$$\mathbb{F}_{AE}^{21} = \mathbb{F}_{AE}^{22} = \mathbb{F}_{EA}^{11} = \mathbb{F}_{EA}^{12} = \mathbb{F}_{EA}^{21} = \mathbb{F}_{EA}^{22} = 0$$

and

$$\mathbb{F}_{AA}^{21} = F_{21}, \qquad \mathbb{F}_{AA}^{22} = F_{22}, \qquad \mathbb{F}_{EE}^{11} = \begin{pmatrix} F_{33} & 0 \\ 0 & F_{33} \end{pmatrix}, \qquad \mathbb{F}_{EE}^{12} = \begin{pmatrix} F_{34} & 0 \\ 0 & F_{34} \end{pmatrix}, \qquad \mathbb{F}_{EE}^{21} = \begin{pmatrix} F_{43} & 0 \\ 0 & F_{43} \end{pmatrix},$$

$$\mathbb{F}_{EE}^{22} = \begin{pmatrix} F_{44} & 0 \\ 0 & F_{44} \end{pmatrix}$$

The matrix $\mathbb{F}$ therefore has the form

$$\mathbb{F} = \left(\begin{array}{cc|cc|cc} F_{11} & F_{12} & 0 & 0 & 0 & 0 \\ F_{21} & F_{22} & 0 & 0 & 0 & 0 \\ \hline 0 & 0 & F_{33} & 0 & F_{34} & 0 \\ 0 & 0 & 0 & F_{33} & 0 & F_{34} \\ \hline 0 & 0 & F_{43} & 0 & F_{44} & 0 \\ 0 & 0 & 0 & F_{43} & 0 & F_{44} \end{array}\right)$$

and permuting the rows and columns

$$\mathbb{F} = \left(\begin{array}{cc|cc|cc} F_{11} & F_{12} & 0 & 0 & 0 & 0 \\ F_{21} & F_{22} & 0 & 0 & 0 & 0 \\ \hline 0 & 0 & F_{33} & F_{34} & 0 & 0 \\ 0 & 0 & F_{43} & F_{44} & 0 & 0 \\ \hline 0 & 0 & 0 & 0 & F_{33} & F_{34} \\ 0 & 0 & 0 & 0 & F_{43} & F_{44} \end{array}\right)$$

The same result is found for the matrix $\mathbb{G}$. We conclude therefore that, when a symmetry adapted basis is used, the potential and kinetic energy matrices are in block form. Furthermore, each degenerate block of dimension g_α appears g_α times on the diagonal. In the previous example, for instance, the non-degenerate block of species A appears once and the doubly degenerate block of species E appears twice.

We wish to point out that using symmetry coordinates, the $\mathbb{F}$ and $\mathbb{G}$ matrices and hence the secular equation are not reduced to diagonal form but only to block form, off-diagonal elements occurring between symmetry coordinates of the same species. A further remark is necessary. In the above proof of the block form of $\mathbb{F}$ in symmetry coordinates we have assumed that sets of symmetry coordinates of the same species α, carry identical representations. This is always true for non-degenerate irreducible representations. For degenerate representations this is also true, providing the symmetry coordinates have been properly chosen. We shall discuss this point further in the next section.

6.6 The construction of symmetry coordinates

In this section we shall discuss how to construct a symmetry-adapted coordinate basis, from an initial set of coordinates. The discussion in this section will assume an initial set of internal coordinates, but it applies equally well with an initial set of Cartesian displacement coordinates. The problem we wish to solve can be conveniently stated as follows. Given a basis of coordinates $\mathbf{S}$ which carries a reducible representation of the molecular group with matrices $\mathbf{D}(\mathscr{R})$, how can we find a new basis $\mathbb{S} = \mathbb{U}\mathbf{S}$, which carries a completely reduced representation with matrices $\mathbb{D}(\mathscr{R})$?

Let the component of the vector $\mathbb{S}$ belonging to the kth row of the αth irreducible representation be denoted by $\mathbb{S}_k^\alpha$. By definition, if we operate on $\mathbb{S}_k^\alpha$ with a symmetry operator $\mathscr{R}_r$ we obtain

$$\mathscr{R}_r(\mathbb{S}_k^\alpha) = \sum_{i=1}^{g_\alpha} \mathbb{S}_i^\alpha(\mathbb{D}_r^\alpha)_{ik} \tag{6.6.1}$$

where g_α is the dimension of the αth representation.

If we multiply equation (6.6.1) by $(\mathbb{D}_r^\beta)_{jl}^*$, sum over r (i.e. over all operations of the group) and use the orthogonality relation (5.7.17) we have

$$\sum_r (\mathbb{D}_r^\beta)_{jl}^*\mathscr{R}_r(\mathbb{S}_k^\alpha) = \sum_r \sum_{i=1}^{g_\alpha} \mathbb{S}_i^\alpha(\mathbb{D}_r^\beta)_{jl}^*(\mathbb{D}_r^\alpha)_{ik} = g/g_\alpha\delta_{\alpha\beta}\delta_{lk}\mathbb{S}_j^\alpha \tag{6.6.2}$$

and thus

$$(g_\alpha/g)\sum_r (\mathbb{D}_r^\beta)_{jl}^*\mathscr{R}_r(\mathbb{S}_k^\alpha) = \mathscr{P}_{jl}^\beta\mathbb{S}_k^\alpha = \delta_{\alpha\beta}\delta_{lk}\mathbb{S}_j^\alpha \tag{6.6.3}$$

where

$$\mathscr{P}_{jl}^\beta = (g_\alpha/g)\sum_r (\mathbb{D}_r^\beta)_{jl}^*\mathscr{R}_r \tag{6.6.4}$$

is an operator having the property, when applied to a component $\mathbb{S}_k^\alpha$ of the basis $\mathbb{S}$, of yielding zero unless the component belongs to the lth row of the βth representation Γ^β. Moreover, if this condition is satisfied, we obtain

$$\mathscr{P}_{jk}^\alpha\mathbb{S}_k^\alpha = \mathbb{S}_j^\alpha \tag{6.6.5}$$

and, if we put $j = k$,

$$\mathscr{P}_{kk}^\alpha\mathbb{S}_k^\alpha = \mathbb{S}_k^\alpha \tag{6.6.6}$$

i.e. $\mathbb{S}_k^\alpha$ is an eigenvector of $\mathscr{P}_{kk}^\alpha$ with eigenvalue unity. We observe also, from equation (6.6.6), that the operator $\mathscr{P}_{kk}^\alpha$ is an idempotent operator since

$$\mathscr{P}_{kk}^\alpha \mathscr{P}_{kk}^\alpha \mathbb{S}_k^\alpha = \mathscr{P}_{kk}^\alpha \mathbb{S}_k^\alpha \tag{6.6.7}$$

Let $\mathbf{V}$ be an arbitrary vector and let us express $\mathbf{V}$ in terms of the symmetry adapted basis $\mathbb{S}$

$$\mathbf{V} = \sum_\beta \sum_j C_j^\beta \mathbb{S}_j^\beta \tag{6.6.8}$$

If we apply the operator $\mathscr{P}_{kk}^\alpha$ to the vector $\mathbf{V}$, we obtain

$$\mathscr{P}_{kk}^\alpha \mathbf{V} = \sum_\beta \sum_j C_j^\beta \mathscr{P}_{kk}^\alpha \mathbb{S}_j^\beta = C_k^\alpha \mathbb{S}_k^\alpha \tag{6.6.9}$$

i.e. the operator $\mathscr{P}_{kk}^\alpha$ is able to extract from the vector $\mathbf{V}$ the component belonging to the kth row of the αth representation. This corresponds geometrically to a projection of the vector $\mathbf{V}$ in the subspace spanned by the vector $\mathbb{S}_k^\alpha$. For this reason the operator $\mathscr{P}_{kk}^\alpha$ is called a projection operator.

Let us now go back to the problem of the construction of symmetry coordinates from the set of internal coordinates. As discussed in Section 6.2 the vector $\mathbf{S}$ of the internal coordinates is divided into symmetrically equivalent subsets

$$\underbrace{S_{11} \quad S_{12} \ldots S_{1m}} \qquad \underbrace{S_{21} \quad S_{22} \ldots S_{2n}} \qquad \underbrace{S_{31} \quad S_{32} \ldots S_{3p}} \qquad \underbrace{S_{h1} \quad S_{h2} \ldots S_{hq}}$$

the internal coordinates within each set being permuted (sometimes with reversal of sign) by the symmetry operations of the group. For simplicity we shall use only one subscript in the following treatment for internal coordinates. Therefore S_i indicates the ith coordinate of a given set and it is understood that all formulas that follow apply separately to each set of symmetrically equivalent internal coordinates.

Let us now express one of the internal coordinates in terms of the symmetry adapted basis $\mathbb{S}$. From equation (6.5.1) we have

$$S_i = \sum_\beta \sum_j (\mathbb{U}_{ij}^\beta)^{-1} \mathbb{S}_j^\beta = \sum_\beta \sum_j \mathbb{U}_{ji}^{\beta *} \mathbb{S}_j^\beta \tag{6.6.10}$$

$\mathbb{U}$ being unitary. Here the label β runs over the representations and the label j counts the coordinates within each irreducible representation. If we apply to the coordinate S_i the operator $\mathscr{P}_{kk}^\alpha$, we have

$$\mathscr{P}_{kk}^\alpha S_i = \sum_\beta \sum_j \mathbb{U}_{ji}^{\beta *} \mathscr{P}_{kk}^\alpha \mathbb{S}_j^\beta = \mathbb{U}_{ki}^{\alpha *} \mathbb{S}_k^\alpha \tag{6.6.11}$$

Thus, $\mathbb{U}_{ki}^{\alpha *}$ being arbitrary except for normalization, we have

$$\mathbb{S}_k^\alpha = \mathscr{N} \sum_r (\mathbb{D}_r^\alpha)_{kk}^* \mathscr{R}_r S_i \tag{6.6.12}$$

which furnishes a simple and compact formula for the construction of symmetry coordinates.

In a non-degenerate representation the matrix elements $(\mathbb{D}_r^\alpha)_{kk}$ are simply the characters χ_α of the representation since the matrices $\mathbb{D}_r^\alpha$ are monodimensional. In a d-fold degenerate representation we have d diagonal elements in each matrix and by suitable choice of S_i we can obtain the d different symmetry coordinates from the same set of internal coordinates. Alternatively, in a degenerate representation we can use, instead of the projection operator $\mathscr{P}_{kk}^\alpha$ of equation (6.6.6), the projection operator $\mathscr{P}_{jk}^\alpha$ of equation (6.6.5). By application of this operator to the same initial coordinate S_i we obtain from (6.6.10)

$$\mathbb{S}_j^\alpha = \mathscr{N} \sum_r (\mathbb{D}_r^\alpha)_{jk}^* \mathscr{R}_r S_i \tag{6.6.13}$$

In both equations the notation $\mathscr{R}_r S_i$ means the coordinate in which S_i is permuted by the symmetry operator $\mathscr{R}_r$. Obviously equation (6.6.12) is a special case of (6.6.13) for $j = k$. Usually some choices of j will yield the same symmetry coordinate while others will yield zero. This is always true if the number of symmetry coordinates from the same set in each of the degenerate species is equal to one. The same does not occur for a regular representation.

The use of equations (6.6.12) and (6.6.13) requires in practice knowledge of the matrix elements $(\mathbb{D}_r^\alpha)_{jk}$, or at least of the diagonal elements. Since, however, the matrix elements are not always available,† it is convenient to rewrite equation (6.6.12) in terms of the characters of the representations, which are tabulated in many books. To do this we sum over k in equation (6.6.6) thus defining a new projection operator

$$\mathscr{P}^\alpha = \sum_k \mathscr{P}_{kk}^\alpha = (g_\alpha/g) \sum_r \chi_r^{\alpha *} \mathscr{R}_r \tag{6.6.14}$$

such that

$$\mathscr{P}^\alpha \mathbb{S}_k^\alpha = S_k^\alpha \tag{6.6.15}$$

Following the same arguments as before, we obtain

$$\mathbb{S}_k^\alpha = \mathscr{N} \sum_r \chi_r^{\alpha *} \mathscr{R}_r S_i \tag{6.6.16}$$

For non-degenerate species equation (6.6.16) is identical to equation (6.6.12). For d-fold degenerate species equation (6.6.16) furnishes only one of the d degenerate symmetry coordinates from the same set of internal coordinates.

Since the partners of degenerate symmetry coordinates give rise, as seen in Section 6.5, to identical factors in the secular equation, it is normally unnecessary to consider more than one representative of each degenerate set. For this reason equation (6.6.16) is generally used instead of equations (6.6.12) or (6.6.13).

If only one set of internal coordinates contributes to a degenerate species, the choice of the starting coordinate S_i on which the operator $\mathscr{P}^\alpha$ operates is completely arbitrary. If more than one set of internal coordinates contribute to the same degenerate species, it is necessary, once a given coordinate S_i^1 has been chosen for the first set, to orient the starting coordinates S_j^h for the other sets properly. This means that the coordinate S_j^h chosen for each additional set must behave exactly as S_i^1 under all symmetry operations of the group. This 'iso-orientation' is required only when the projection operator $\mathscr{P}^\alpha$, defined in terms of the characters, is utilized. If instead one uses the more general operators $\mathscr{P}_{jk}^\alpha$, defined in terms of the elements of the matrices of the irreducible representations, the iso-orientation is unnecessary and the choice of S_j^h is again arbitrary.

We shall now illustrate the use of the above defined projection operators with some selected examples.

Example 6.6.1

The structure of the representation carried by the internal coordinates of ammonia has been found in Example 6.4.3 to be

$$2A_1 + 2E$$

Using the table of characters of the C_{3v} point group given in Example 6.4.3 let us apply

† The matrix element of all point groups are tabulated in the book *Symmetry* by R. McWeeny, Pergamon Press, 1963.

equation (6.6.16) to the construction of the symmetry coordinates. Choosing Δr_1 as starting coordinate we have

$$\mathbb{S}_r^{A_1} = \mathcal{N}[\chi^{A_1}(E)\mathcal{R}_E\,\Delta r_1 + \chi^{A_1}(C_3)\mathcal{R}_{C_3}\,\Delta r_1 + \chi^{A_1}(C_3^{-1})\mathcal{R}_{C_3^{-1}}\,\Delta r_1 + \chi^{A_1}(\sigma_1)\mathcal{R}_{\sigma_1}\,\Delta r_1 + \chi^{A_1}(\sigma_2)\mathcal{R}_{\sigma_2}\,\Delta r_1 + \chi^{A_1}(\sigma_3)\mathcal{R}_{\sigma_3}\,\Delta r_1]$$

$$\mathbb{S}_r^{A_1} = \mathcal{N}(1\,.\,\Delta r_1 + 1\,.\,\Delta r_2 + 1\,.\,\Delta r_3 + 1\,.\,\Delta r_1 + 1\,.\,\Delta r_2 + 1\,.\,\Delta r_3)$$

and thus, after normalization,

$$\mathbb{S}_r^{A_1} = \tfrac{1}{\sqrt{3}}(\Delta r_1 + \Delta r_2 + \Delta r_3)$$

Choosing $\Delta\alpha_1$ we obtain in the same way

$$\mathbb{S}_\alpha^{A_1} = \tfrac{1}{\sqrt{3}}(\Delta\alpha_1 + \Delta\alpha_2 + \Delta\alpha_3)$$

For the E-block we have for the r-set

$$\mathbb{S}_r^{E} = \mathcal{N}[\chi^{E}(E)\mathcal{R}_E\,\Delta r_1 + \chi^{E}(C_3)\mathcal{R}_{C_3}\,\Delta r_1 + \chi^{E}(C_3^{-1})\mathcal{R}_{C_3^{-1}}\,\Delta r_1 + \chi^{E}(\sigma_1)\mathcal{R}_{\sigma_1}\,\Delta r_1 + \chi^{E}(\sigma_2)\mathcal{R}_{\sigma_2}\,\Delta r_1 + \chi^{E}(\sigma_3)\mathcal{R}_{\sigma_3}\,\Delta r_1]$$

$$\mathbb{S}_r^{E} = \mathcal{N}(2\,.\,\Delta r_1 - 1\,.\,\Delta r_2 - 1\,.\,\Delta r_3 + 0\,.\,\Delta r_1 + 0\,.\,\Delta r_2 + 0\,.\,\Delta r_3)$$

and after normalization

$$\mathbb{S}_r^{E} = \tfrac{1}{\sqrt{6}}(2\Delta r_1 - \Delta r_2 - \Delta r_3)$$

In this species, before forming the coordinate from the set α, we have to make sure that the starting coordinate adopted, has exactly the same symmetry properties as Δr_1. It is easily seen that $\Delta\alpha_1$ satisfies these conditions. It is sent into itself by E and σ_1 and is shifted to another coordinate by C_3, C_3^{-1}, σ_2 and σ_3, exactly as Δr_1. We can therefore safely use $\Delta\alpha_1$ for the construction of the symmetry coordinate and we obtain, in the same way as before,

$$\mathbb{S}_\alpha^{E} = \tfrac{1}{\sqrt{6}}(2\Delta\alpha_1 - \Delta\alpha_2 - \Delta\alpha_3)$$

As discussed before, use of equation (6.6.16) yields only one component for each degenerate normal coordinate. If one wishes to obtain the other component, equation (6.6.13) can be used. In this case we need the complete matrices of the irreducible representation. For the C_{3v} group these matrices are given in Table 5.4.1.

From the operator $\mathcal{P}_{11}^E$ we obtain, using as starting coordinate Δr_1,

$$\mathbb{S}_r^{E} = \mathcal{N}(\Delta r_1 - \tfrac{1}{2}\Delta r_2 - \tfrac{1}{2}\Delta r_3 + \Delta r_1 - \tfrac{1}{2}\Delta r_3 - \tfrac{1}{2}\Delta r_2) = \tfrac{1}{\sqrt{6}}(2\Delta r_1 - \Delta r_2 - \Delta r_3)$$

from the operator $\mathcal{P}_{12}^E$

$$\mathbb{S}_r^{E} = \mathcal{N}(0\,.\,\Delta r_1 - \tfrac{\sqrt{3}}{2}\Delta r_2 + \tfrac{\sqrt{3}}{2}\Delta r_3 + 0\,.\,\Delta r_1 - \tfrac{\sqrt{3}}{2}\Delta r_3 + \tfrac{\sqrt{3}}{2}\Delta r_2) = 0$$

(first E block)

from the operator $\mathcal{P}_{21}^E$

$$\mathbb{S}_r^{E} = \mathcal{N}(\Delta r_1\,.\,0 + \tfrac{\sqrt{3}}{2}\Delta r_2 - \tfrac{\sqrt{3}}{2}\Delta r_3 + 0\,.\,\Delta r_1 - \tfrac{\sqrt{3}}{2}\Delta r_3 + \tfrac{\sqrt{3}}{2}\Delta r_2) = \tfrac{1}{\sqrt{2}}(\Delta r_2 - \Delta r_3)$$

and from the operator $\mathcal{P}_{22}^E$

$$\mathbb{S}_r^{E} = \mathcal{N}(\Delta r_1 - \tfrac{1}{2}\Delta r_2 - \tfrac{1}{2}\Delta r_3 - \Delta r_1 + \tfrac{1}{2}\Delta r_3 + \tfrac{1}{2}\Delta r_2) = 0$$

(second E block)

In the same way for the set α we obtain

$$\mathbb{S}_\alpha^{E} = \tfrac{1}{\sqrt{6}}(2\Delta\alpha_1 - \Delta\alpha_2 - \Delta\alpha_3) \qquad \mathbb{S}_\alpha^{E} = 0 \qquad \text{first } E \text{ block}$$

$$\mathbb{S}_\alpha^{E} = \tfrac{1}{\sqrt{2}}(\Delta\alpha_2 - \Delta\alpha_3) \qquad \mathbb{S}_\alpha^{E} = 0 \qquad \text{second } E \text{ block}$$

The matrix equation $\mathbb{S} = \mathbb{U}\mathbf{S}$ for ammonia has the form

	Δr_1	Δr_2	Δr_3	$\Delta\alpha_1$	$\Delta\alpha_2$	$\Delta\alpha_3$
$\mathbb{S}_1^{A_1}$	$\frac{1}{\sqrt{3}}$	$\frac{1}{\sqrt{3}}$	$\frac{1}{\sqrt{3}}$	0	0	0
$\mathbb{S}_2^{A_1}$	0	0	0	$\frac{1}{\sqrt{3}}$	$\frac{1}{\sqrt{3}}$	$\frac{1}{\sqrt{3}}$
$\mathbb{S}_{11}^{E}$	$\frac{2}{\sqrt{6}}$	$-\frac{1}{\sqrt{6}}$	$\frac{1}{\sqrt{6}}$	0	0	0
$\mathbb{S}_{12}^{E}$	0	0	0	$\frac{2}{\sqrt{6}}$	$-\frac{1}{\sqrt{6}}$	$-\frac{1}{\sqrt{6}}$
$\mathbb{S}_{21}^{E}$	0	$\frac{1}{\sqrt{2}}$	$-\frac{1}{\sqrt{2}}$	0	0	0
$\mathbb{S}_{22}^{E}$	0	0	0	0	$\frac{1}{\sqrt{2}}$	$-\frac{1}{\sqrt{2}}$

Using this $\mathbb{U}$-matrix we can now factorize the **F**- and **G**-matrices. For the matrix **F** in internal coordinates we shall adopt the following nomenclature: diagonal elements are indicated by the symbol F_i where i labels the type of coordinate (r or α in this case). Off-diagonal elements are indicated by F_{ij}, where i and j label the two coordinates involved. If more than one type of force constant of the type F_{ij} occurs, they are indicated by F'_{ij}, F''_{ij} and so on. For ammonia we have

$$\mathbb{F} = \mathbb{U}\begin{pmatrix} F_r & F_{rr} & F_{rr} & F_{r\alpha} & F'_{r\alpha} & F'_{r\alpha} \\ F_{rr} & F_r & F_{rr} & F'_{r\alpha} & F_{r\alpha} & F'_{r\alpha} \\ F_{rr} & F_{rr} & F_r & F'_{r\alpha} & F'_{r\alpha} & F_{r\alpha} \\ F_{r\alpha} & F'_{r\alpha} & F'_{r\alpha} & F_\alpha & F_{\alpha\alpha} & F_{\alpha\alpha} \\ F'_{r\alpha} & F_{r\alpha} & F'_{r\alpha} & F_{\alpha\alpha} & F_\alpha & F_{\alpha\alpha} \\ F'_{r\alpha} & F'_{r\alpha} & F_{r\alpha} & F_{\alpha\alpha} & F_{\alpha\alpha} & F_\alpha \end{pmatrix}\tilde{\mathbb{U}}$$

and, performing the matrix multiplication,

$$\mathbb{F} = \begin{pmatrix} F_r + 2F_{rr} & F_{ra} + 2F'_{ra} & 0 & 0 & 0 & 0 \\ F_{r\alpha} + 2F'_{r\alpha} & F_\alpha + 2F_{\alpha\alpha} & 0 & 0 & 0 & 0 \\ 0 & 0 & F_r - F_{rr} & -F'_{r\alpha} & 0 & 0 \\ 0 & 0 & -F'_{r\alpha} & F_\alpha - F_{\alpha\alpha} & 0 & 0 \\ 0 & 0 & 0 & 0 & F_r - F_{rr} & -F'_{r\alpha} \\ 0 & 0 & 0 & 0 & -F'_{r\alpha} & F_\alpha - F_{\alpha\alpha} \end{pmatrix}$$

The first 2×2 block is of species A_1 and the other two identical blocks are the two degenerate blocks of species E.

The **G**-matrix obviously assumes the same block form.

The secular equation in symmetry coordinates

$$\mathbb{G}\mathbb{F}\,\mathbb{L} = \mathbb{L}\mathbf{\Lambda}$$

thus factors into three independent equations, one for the A_1 species of dimension 2×2 and two identical equations also of order 2×2 for the E species. Solution of the first equation yields the normal frequencies of A_1 species and solution of one of the other two equations yields the degenerate normal frequencies of species E.

Example 6.6.2

As a more complex example let us consider cyclopropane (Figure 6.6.1). The molecule has symmetry D_{3h}. The internal coordinates are labelled according to the numbering of the atoms shown in Figure 6.6.1. Representatives of each set of internal coordinates are shown in the figure.

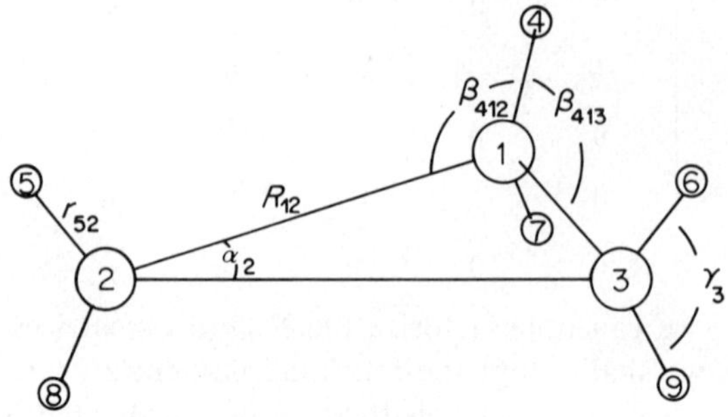

Figure 6.6.1 Internal coordinates and numbering of atoms in cyclopropane

The characters of the irreducible representations of the D_{3h} group are reported in Appendix I. The structure of the representation carried by the internal coordinates, found by the standard method of Section 6.4, is given in Table 6.6.1.

Table 6.6.1.

	Δr_{ij}	ΔR_{ij}	$\Delta\alpha_i$	$\Delta\gamma_i$	$\Delta\beta_{ijk}$	N_s	N_v	Red.
A_1'	1	1	1	1	1	5	3	2
A_2'	0	0	0	0	1	1	1	0
E'	1	1	1	1	2	6	4	2
A_1''	0	0	0	0	1	1	1	0
A_2''	1	0	0	0	1	2	2	0
E''	1	0	0	0	2	3	3	0

We observe that: (a) there are two redundant coordinates in the A_1' and two doubly degenerate redundant coordinates in the E' species; (b) the twelve HCC angle bending coordinates carry a regular representation, i.e. the number of symmetry coordinates originated from this set is equal to the character of the identity (1 for non-degenerate and 2 for degenerate species).

The symmetry coordinates for the non-degenerate blocks are easily written down using equation (6.6.16). They are given below without further comments.

A_1' species

$$\mathbb{S}_1 = \tfrac{1}{\sqrt{6}}(\Delta r_{41} + \Delta r_{71} + \Delta r_{52} + \Delta r_{82} + \Delta r_{63} + \Delta r_{93})$$

$$\mathbb{S}_2 = \tfrac{1}{\sqrt{3}}(\Delta R_{12} + \Delta R_{23} + \Delta R_{31})$$

$$\mathbb{S}_3 = \tfrac{1}{\sqrt{3}}(\Delta\alpha_1 + \Delta\alpha_2 + \Delta\alpha_3)$$

$$\mathbb{S}_4 = \tfrac{1}{\sqrt{3}}(\Delta\gamma_1 + \Delta\gamma_2 + \Delta\gamma_3)$$

$$\mathbb{S}_5 = \tfrac{1}{2\sqrt{3}}(\Delta\beta_{412} + \Delta\beta_{413} + \Delta\beta_{712} + \Delta\beta_{713} + \Delta\beta_{523} + \Delta\beta_{521} + \Delta\beta_{823} + \Delta\beta_{821} + \Delta\beta_{631} + \Delta\beta_{632} + \beta_{931} + \Delta\beta_{932})$$

A_2' species

$$\mathbb{S}_6 = \frac{1}{2\sqrt{3}}(\Delta\beta_{412} - \Delta\beta_{413} + \Delta\beta_{712} - \Delta\beta_{713} + \Delta\beta_{523} - \Delta\beta_{521} + \Delta\beta_{823} - \Delta\beta_{821} + \Delta\beta_{631} - \Delta\beta_{632} + \Delta\beta_{931} - \Delta\beta_{932})$$

A_1'' species

$$\mathbb{S}_{13} = \frac{1}{2\sqrt{3}}(\Delta\beta_{412} - \Delta\beta_{413} - \Delta\beta_{712} + \Delta\beta_{713} + \Delta\beta_{523} - \Delta\beta_{521} - \Delta\beta_{823} + \Delta\beta_{821} + \Delta\beta_{631} - \Delta\beta_{632} - \Delta\beta_{931} + \Delta\beta_{932})$$

A_2'' species

$$\mathbb{S}_{14} = \frac{1}{\sqrt{6}}(\Delta r_{41} - \Delta r_{71} + \Delta r_{52} - \Delta r_{82} + \Delta r_{63} - \Delta r_{93})$$

$$\mathbb{S}_{15} = \frac{1}{2\sqrt{3}}(\Delta\beta_{412} + \Delta\beta_{413} - \Delta\beta_{712} - \Delta\beta_{713} + \Delta\beta_{523} + \Delta\beta_{521} - \Delta\beta_{823} - \Delta\beta_{821} + \Delta\beta_{631} + \Delta\beta_{632} - \Delta\beta_{931} - \Delta\beta_{932})$$

In order to obtain the symmetry coordinates for the degenerate species using equation (6.6.13), we need the matrices of the irreducible representations. These matrices are listed below. The upper sign refers to the E' species and the lower sign refers to the E'' species.

$$\begin{array}{cccccc}
E & C_3 & C_3^{-1} & C_2^{(1)} & C_2^{(2)} & C_2^{(3)} \\
\begin{pmatrix} 1 & 0 \\ 0 & 1 \end{pmatrix} &
\begin{pmatrix} -\frac{1}{2} & -\frac{\sqrt{3}}{2} \\ \frac{\sqrt{3}}{2} & -\frac{1}{2} \end{pmatrix} &
\begin{pmatrix} -\frac{1}{2} & \frac{\sqrt{3}}{2} \\ -\frac{\sqrt{3}}{2} & -\frac{1}{2} \end{pmatrix} &
\begin{pmatrix} 1 & 0 \\ 0 & -1 \end{pmatrix} &
\begin{pmatrix} -\frac{1}{2} & -\frac{\sqrt{3}}{2} \\ -\frac{\sqrt{3}}{2} & \frac{1}{2} \end{pmatrix} &
\begin{pmatrix} -\frac{1}{2} & \frac{\sqrt{3}}{2} \\ \frac{\sqrt{3}}{2} & \frac{1}{2} \end{pmatrix}
\end{array}$$

$$\begin{array}{cccccc}
\sigma_h & S_3 & S_3^{-1} & \sigma_1 & \sigma_2 & \sigma_3 \\
\begin{pmatrix} \pm 1 & 0 \\ 0 & \pm 1 \end{pmatrix} &
\begin{pmatrix} \mp\frac{1}{2} & \mp\frac{\sqrt{3}}{2} \\ \pm\frac{\sqrt{3}}{2} & \mp\frac{1}{2} \end{pmatrix} &
\begin{pmatrix} \mp\frac{1}{2} & \pm\frac{\sqrt{3}}{2} \\ \mp\frac{\sqrt{3}}{2} & \mp\frac{1}{2} \end{pmatrix} &
\begin{pmatrix} \pm 1 & 0 \\ 0 & \mp 1 \end{pmatrix} &
\begin{pmatrix} \mp\frac{1}{2} & \mp\frac{\sqrt{3}}{2} \\ \mp\frac{\sqrt{3}}{2} & \pm\frac{1}{2} \end{pmatrix} &
\begin{pmatrix} \mp\frac{1}{2} & \pm\frac{\sqrt{3}}{2} \\ \pm\frac{\sqrt{3}}{2} & \pm\frac{1}{2} \end{pmatrix}
\end{array}$$

E' species

Starting with Δr_{41} we have

$$\left.\begin{aligned} \mathbb{S}_{7a_1} &= \frac{1}{2\sqrt{3}}(2\Delta r_{41} + 2\Delta r_{71} - \Delta r_{52} - \Delta r_{82} - \Delta r_{63} - \Delta r_{93}) \\ \mathbb{S}_{7a_2} &= 0 \end{aligned}\right\} \text{block } E_a'$$

$$\left.\begin{aligned} \mathbb{S}_{7b_1} &= \frac{1}{2}(\Delta r_{52} + \Delta r_{82} - \Delta r_{63} - \Delta r_{93}) \\ \mathbb{S}_{7b_2} &= 0 \end{aligned}\right\} \text{block } E_b'$$

Starting with ΔR_{23}

$$\left.\begin{aligned} \mathbb{S}_{8a_1} &= \frac{1}{\sqrt{6}}(2\Delta R_{23} - \Delta R_{12} - \Delta R_{31}) \\ \mathbb{S}_{8a_2} &= 0 \end{aligned}\right\} \text{block } E_a'$$

$$\left.\begin{aligned} \mathbb{S}_{8b_1} &= \frac{1}{\sqrt{2}}(\Delta R_{31} - \Delta R_{12}) \\ \mathbb{S}_{8b_2} &= 0 \end{aligned}\right\} \text{block } E_b'$$

Starting with $\Delta\alpha_1$

$$\left.\begin{aligned} \mathbb{S}_{9a_1} &= \frac{1}{\sqrt{6}}(2\Delta\alpha_1 - \Delta\alpha_2 - \Delta\alpha_3) \\ \mathbb{S}_{9a_2} &= 0 \end{aligned}\right\} \text{block } E_a'$$

$$\mathbb{S}_{9b_1} = \frac{1}{\sqrt{2}}(\Delta\alpha_2 - \Delta\alpha_3) \qquad \mathbb{S}_{9b_2} = 0 \quad \Big\}\ \text{block } E'_b$$

Starting with $\Delta\gamma_1$

$$\mathbb{S}_{10a_1} = \frac{1}{\sqrt{6}}(2\Delta\gamma_1 - \Delta\gamma_2 - \Delta\gamma_3) \qquad \mathbb{S}_{10a_2} = 0 \quad \Big\}\ \text{block } E'_a$$

$$\mathbb{S}_{10b_1} = \frac{1}{\sqrt{2}}(\Delta\gamma_2 - \Delta\gamma_3) \qquad \mathbb{S}_{10b_2} = 0 \quad \Big\}\ \text{block } E'_b$$

Starting with $\Delta\beta_{412}$

$$\left.\begin{aligned} \mathbb{S}_{11a} &= \frac{1}{2\sqrt{6}}(2\Delta\beta_{412} + 2\Delta\beta_{413} + 2\Delta\beta_{712} + 2\Delta\beta_{713} - \Delta\beta_{523} - \Delta\beta_{521} \\ &\quad - \Delta\beta_{823} - \Delta\beta_{821} - \Delta\beta_{631} - \Delta\beta_{632} - \Delta\beta_{931} - \Delta\beta_{932}) \\ \mathbb{S}_{12a} &= \frac{1}{2\sqrt{2}}(-\Delta\beta_{523} + \Delta\beta_{521} - \Delta\beta_{823} + \Delta\beta_{821} + \Delta\beta_{631} - \Delta\beta_{632} \\ &\quad + \Delta\beta_{931} - \Delta\beta_{932}) \end{aligned}\right\}\ \text{block } E'_a$$

$$\left.\begin{aligned} \mathbb{S}_{11b} &= \frac{1}{2\sqrt{2}}(\Delta\beta_{523} + \Delta\beta_{521} + \Delta\beta_{823} + \Delta\beta_{821} - \Delta\beta_{631} - \Delta\beta_{632} - \Delta\beta_{931} - \Delta\beta_{932}) \\ \mathbb{S}_{12b} &= \frac{1}{2\sqrt{6}}(2\Delta\beta_{412} - 2\Delta\beta_{413} + 2\Delta\beta_{712} - 2\Delta\beta_{713} - \Delta\beta_{523} + \Delta\beta_{521} - \Delta\beta_{823} \\ &\quad + \Delta\beta_{821} - \Delta\beta_{631} + \Delta\beta_{632} - \Delta\beta_{931} + \Delta\beta_{932}) \end{aligned}\right\}\ \text{block } E'_b$$

E″ species

Starting with Δr_{41}

$$\mathbb{S}_{16a_1} = 0 \qquad \mathbb{S}_{16a_2} = \frac{1}{2}(-\Delta r_{52} + \Delta r_{82} + \Delta r_{63} - \Delta r_{93}) \quad \Big\}\ \text{block } E''_a$$

$$\mathbb{S}_{16b_1} = 0 \qquad \mathbb{S}_{16b_2} = \frac{1}{2\sqrt{3}}(2\Delta r_{41} - 2\Delta r_{71} - \Delta r_{52} + \Delta r_{82} - \Delta r_{63} + \Delta r_{93}) \quad \Big\}\ \text{block } E''_b$$

Starting with $\Delta\beta_{412}$

$$\left.\begin{aligned} \mathbb{S}_{17a} &= \frac{1}{2\sqrt{6}}(2\Delta\beta_{412} - 2\Delta\beta_{413} - 2\Delta\beta_{712} + 2\Delta\beta_{713} - \Delta\beta_{523} + \Delta\beta_{521} + \Delta\beta_{823} \\ &\quad - \Delta\beta_{821} - \Delta\beta_{631} + \Delta\beta_{632} + \Delta\beta_{931} - \Delta\beta_{932}) \\ \mathbb{S}_{18a} &= \frac{1}{2\sqrt{2}}(-\Delta\beta_{523} - \Delta\beta_{521} + \Delta\beta_{823} + \Delta\beta_{821} + \Delta\beta_{631} + \Delta\beta_{632} - \Delta\beta_{931} \\ &\quad - \Delta\beta_{932}) \end{aligned}\right\}\ \text{block } E''_a$$

$$\left.\begin{aligned} \mathbb{S}_{17b} &= \frac{1}{2\sqrt{2}}(\Delta\beta_{523} - \Delta\beta_{521} - \Delta\beta_{823} + \Delta\beta_{821} - \Delta\beta_{631} + \Delta\beta_{632} + \Delta\beta_{931} \\ &\quad - \Delta\beta_{932}) \\ \mathbb{S}_{18b} &= \frac{1}{2\sqrt{6}}(2\Delta\beta_{412} + 2\Delta\beta_{413} - 2\Delta\beta_{712} - 2\Delta\beta_{713} - \Delta\beta_{523} - \Delta\beta_{521} + \Delta\beta_{823} \\ &\quad + \Delta\beta_{821} - \Delta\beta_{631} - \Delta\beta_{632} + \Delta\beta_{931} + \Delta\beta_{932}) \end{aligned}\right\}\ \text{block } E''_b$$

In this example we have always used equation (6.6.13) and thus we have not paid attention to the iso-orientation of the starting coordinates. We wish to point out that an alternative method, based on the use of equation (6.6.16) and on the correlation between the irreducible representations of a group and its subgroups can be used. In this case it is, however, necessary to take care of the iso-orientation of the starting coordinates. The method is discussed in Section 6.8.

We have pointed out before that there are two redundancies in the E' and two redundancies in the A_1' blocks. For the calculation of the vibrational frequencies, it is normally unnecessary to eliminate them since all that happens is that in both blocks there will be two additional zero roots. If one wishes to eliminate the redundancies, the method of the next section can be used.

6.7 Redundant coordinates

In Section 6.2 we have seen that, in order to use physically meaningful force fields and to take advantage of molecular symmetry, it is often necessary to use a number of internal coordinates greater than the number of internal degrees of freedom. In such a case the internal coordinates used are not linearly independent and relationships of the type

$$\sum_t a_t s_t = 0 \tag{6.7.1}$$

exist among them. The set of internal coordinates is then said to be redundant and the relationships (6.7.1) are called redundancy conditions.

If a redundancy condition exists among internal coordinates, the **G**-matrix becomes singular since then the rows and columns of **G**, corresponding to the internal coordinates involved in the redundancy, are linearly related. This can be easily seen as follows.[3] The internal coordinates are related to the Cartesian displacement coordinates by equation (4.1.2). An internal coordinate can thus be written as

$$s_t = \sum_{j=1}^{3N} B_{tj} x_j \tag{6.7.2}$$

and by substitution of this expression in (6.7.1) we have

$$\sum_{i=1}^{3N} \left(\sum_t a_t B_{ti} \right) x_i = 0$$

from which it follows, the x_i being independent, that

$$\sum_t a_t B_{ti} = 0 \qquad i = 1, 2, \ldots, 3N \tag{6.7.3}$$

Use of the definition of **G** (equation 4.5.1)

$$G_{tr} = \sum_i \mu_i B_{ti} B_{ri}$$

yields

$$\sum_t a_t G_{tr} = \sum_t a_t G_{rt} = \sum_i \mu_i B_{ri} \left(\sum_t a_t B_{ti} \right) = 0 \tag{6.7.4}$$

This proves that if a linear combination of internal coordinates vanishes, the corresponding linear combination of the elements of **G** also vanishes. The occurrence of a redundancy condition is thus a sufficient condition for the **G**-matrix to be singular. It can be proved that this is also a necessary condition,[5] i.e. that if the **G**-matrix is singular then a redundancy exists among the internal coordinates. If the **G**-matrix is singular it can in principle be rearranged by taking the appropriate combination of rows and columns so that, if m

redundancies exist among the coordinates, it is reduced to a matrix with m rows and columns containing zeros. If the same combinations of rows and columns are also made in the **F**-matrix, the rows and columns containing zeros in the **G**-matrix as well as the corresponding rows and columns of **F** are omitted. The secular equation is solved and has no zero roots.

The singularity of the **G**-matrix suggests a different approach[4] to the problem of finding the analytical form of the redundancy conditions. The method consists in considering the **G**-matrix as the matrix of coefficients of a set of homogeneous linear equations of the type

$$\sum_j G_{ij} A_{ij} = 0 \tag{6.7.5}$$

A_{ij} are the unknown coefficients of the transformation from a set of redundant symmetry coordinates to a new set of coordinates in which the first coordinate is just the redundancy condition (in symmetry coordinates)

$$\mathbb{S}_1 = \sum_k A_{ik} \mathbb{S}_k = 0 \tag{6.7.6}$$

The method obviously requires a spanning of all submatrices of **G**, by successive elimination of rows and columns, until an $m \times m$ submatrix $\mathbf{G}_m$ with $|\mathbf{G}_m| = 0$ is found, such that any possible submatrix $\mathbf{G}_{m-1}$ of $\mathbf{G}_m$ is non-singular. A similar, but more efficient method, well adapted for computer calculation, has been proposed by M. Gussoni and G. Zerbi.[5] Let **G** be a singular matrix because of one redundancy condition. If we diagonalize **G** we have

$$\mathbf{GD} = \mathbf{D\Gamma} \tag{6.7.7}$$

where **D** is the eigenvector matrix and **Γ** is the matrix of the eigenvalues of **G**. The eigenvector $\mathbf{D}_k$, belonging to the zero eigenvalue of **G** is given by

$$\mathbf{GD}_k = 0 \tag{6.7.8}$$

and may be written in the form

$$\mathbf{D}_k = \{G_{k1}, G_{k2}, \ldots, G_{km}\} \tag{6.7.9}$$

where the G_{ki} are the algebraic complements of the elements of the kth row or column of **G**.

The elements of the eigenvector belonging to the zero eigenvalue are then proportional to the coefficients of the redundancy and equation (6.7.1) can be rewritten in the form

$$\sum_t G_{kt} s_t = 0 \tag{6.7.10}$$

A deeper insight into the nature of the redundancy conditions may be gained starting from the so-called *primitive redundancies*, i.e. from the simple linear relationship existing between a set of vectors oriented along the bonds. These vectorial relationships are easily transformed into scalar relationships between bond lengths and angles by forming scalar products with appropriate vectors. The redundancies among the internal coordinates are then obtained by taking the derivatives of these primitive redundancies with respect to internal coordinates.

We shall distinguish between two types of primitive vectorial redundancies; *branching* and *cyclic* redundancies.

Branching redundancies[6,7,8] occur whenever a central atom is bound to more than three other atoms in the space (or to more than two in a plane) and arise from the fact that in an n-dimensional space only n independent vectors can be defined. Consider n bonds originating in a single central atom and define n unit vectors $\mathbf{e}_k$ ($k = 1, 2, \ldots, n$) along the n bonds.

Since the vectors are not independent, there exists among any four vectors a relation of the type

$$C_i\mathbf{e}_i + C_j\mathbf{e}_j + C_k\mathbf{e}_k + C_l\mathbf{e}_l = \mathbf{0} \tag{6.7.11}$$

where the coefficients C_n are determined by the molecular geometry. These coefficients can be found by multiplying (6.7.11) by four different unit vectors ($\mathbf{e}_p$, $\mathbf{e}_q$, $\mathbf{e}_s$, $\mathbf{e}_t$), to yield four linear equations in the unknown C_n. For a system of linear and homogeneous equations, nontrivial solutions exist if the determinant of the coefficient is zero. Hence we have

$$D^{ijkl}_{pqst} = \begin{vmatrix} \mathbf{e}_i \cdot \mathbf{e}_p & \mathbf{e}_j \cdot \mathbf{e}_p & \mathbf{e}_k \cdot \mathbf{e}_p & \mathbf{e}_l \cdot \mathbf{e}_p \\ \mathbf{e}_i \cdot \mathbf{e}_q & \mathbf{e}_j \cdot \mathbf{e}_q & \mathbf{e}_k \cdot \mathbf{e}_q & \mathbf{e}_l \cdot \mathbf{e}_q \\ \mathbf{e}_i \cdot \mathbf{e}_s & \mathbf{e}_j \cdot \mathbf{e}_s & \mathbf{e}_k \cdot \mathbf{e}_s & \mathbf{e}_l \cdot \mathbf{e}_s \\ \mathbf{e}_i \cdot \mathbf{e}_t & \mathbf{e}_j \cdot \mathbf{e}_t & \mathbf{e}_k \cdot \mathbf{e}_t & \mathbf{e}_l \cdot \mathbf{e}_t \end{vmatrix} = 0 \tag{6.7.12}$$

for any four bonds in the molecule.

It is convenient to select three bond vectors, say $\mathbf{e}_i$, $\mathbf{e}_j$ and $\mathbf{e}_k$, as a basis and to express all other bonds as functions of these three through relations of the type (6.7.11). For n non-coplanar bonds originating from a central atom we thus have $(n-3)$ equations of the type $D^{ijkl}_{ijkl} = 0$ plus $(n-3)(n-4)/2$ equations of the type $D^{ijkl}_{ijkp} = 0$, i.e. a total of $(n-2)(n-3)/2$ equations for the complete determination of all branching redundancies.

For a tetrahedral molecule like CH_4 or CH_3Cl we have, for instance, only one determinant, whereas for an octahedral molecule like SF_6 we have six determinants.

Example 6.7.1

Let us consider in more detail the case of a tetrahedral molecule. If we call r_i the bond from the central atom to atom i (see Figure 6.7.1), $\mathbf{e}_i$ the associated unit vector and α_{ij}

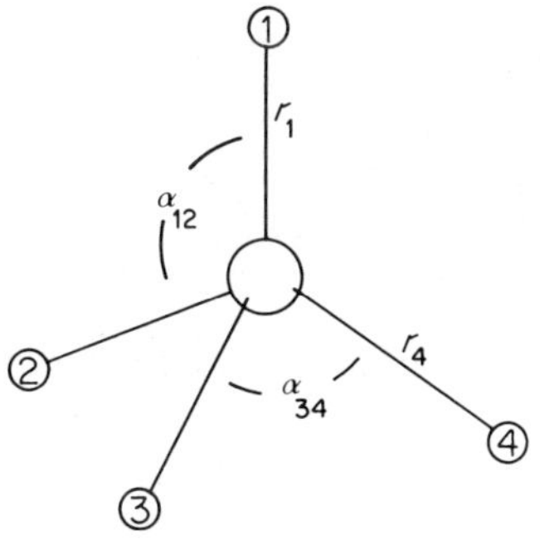

Figure 6.7.1 The internal coordinates of methane

the angle between bond r_i and bond r_j, equation (6.7.12) takes the form

$$\begin{vmatrix} 1 & \cos\alpha_{12} & \cos\alpha_{13} & \cos\alpha_{14} \\ \cos\alpha_{12} & 1 & \cos\alpha_{23} & \cos\alpha_{24} \\ \cos\alpha_{13} & \cos\alpha_{23} & 1 & \cos\alpha_{34} \\ \cos\alpha_{14} & \cos\alpha_{24} & \cos\alpha_{34} & 1 \end{vmatrix} = 0 \tag{6.7.13}$$

For small displacements of the atoms we can develop the trigonometric functions in (6.7.13) in series of the internal coordinates and write

$$\cos\alpha_{ij} = \cos(\alpha_{ij}^0 + \Delta\alpha_{ij}) = \cos\alpha_{ij}^0 - \sin\alpha_{ij}^0\Delta\alpha_{ij} - \tfrac{1}{2}\cos\alpha_{ij}^0\Delta\alpha_{ij}^2 \tag{6.7.14}$$

where the α_{ij}^0s are the equilibrium angles.

If we introduce (6.7.14) into (6.7.13) and retain only terms up to the second order in the internal coordinates, we get

$$\begin{aligned}\sum_{i<j} \theta_{ij}^0 \sin\alpha_{ij}^0\,\Delta\alpha_{ij} + \frac{1}{2}\Bigg\{ &\sum_{i<j} [\theta_{ij}^0 \cos\alpha_{ij} - \sin\alpha_{ij}^{0\,2} \sin\alpha_{kl}^{0\,2}]\,\Delta\alpha_{ij}^2 \\ &+ 2\sum_{\substack{i,j\\ j<k}} \sin\alpha_{ij}^0 \sin\alpha_{ik}^0[\cos\alpha_{jk}^0 - \cos\alpha_{jl}^0\cos\alpha_{kl}^0]\,\Delta\alpha_{ij}\,\Delta\alpha_{ik} \\ &+ 2\sum_{\substack{i<j\\ k<l}} \sin\alpha_{ij}^0 \sin\alpha_{kl}^0[2\cos\alpha_{ij}^0\cos\alpha_{kl}^0 - \cos\alpha_{il}^0\cos\alpha_{jk}^0 - \cos\alpha_{ik}^0\cos\alpha_{jl}^0]\,\Delta\alpha_{ij}\,\Delta\alpha_{kl}\Bigg\}\end{aligned} \tag{6.7.15}$$

where

$$\theta_{ij}^0 = \cos\alpha_{ij}^0 \sin\alpha_{kl}^{0\,2} - (\cos\alpha_{ik}^0\cos\alpha_{jk}^0 + \cos\alpha_{il}^0\cos\alpha_{jl}^0) + \cos\alpha_{kl}^0(\cos\alpha_{ik}^0\cos\alpha_{jl}^0 + \cos\alpha_{il}^0\cos\alpha_{jk}^0) \tag{6.7.16}$$

is the minor of the element $\cos\alpha_{ij}$.

For methane all equilibrium angles are tetrahedral (109° 28′), and $\cos\alpha = -\frac{1}{3}$, $\sin\alpha = \frac{2\sqrt{2}}{3}$. Introducing these values in (6.7.15) we obtain, after division by $\frac{2^6\sqrt{2}}{3}$

$$\sum_{i<j} \Delta\alpha_{ij} + \frac{\sqrt{2}}{8}\Bigg[3\sum_{i<j}\Delta\alpha_{ij}^2 + 4\sum_{\substack{i\neq j\\ j<k}}\Delta\alpha_{ij}\,\Delta\alpha_{ik}\Bigg] = 0 \tag{6.7.17}$$

This relation, first found by Shimanouchi,[9] is correct to the second order. For the moment we are interested only in the first-order redundancy which, from equation (6.7.17) is found to be

$$\sum_{i<j} \Delta\alpha_{ij} = 0 \tag{6.7.18}$$

Cyclic redundancies[10] occur in all cyclic molecules and arise from the fact that for a closed system of n vectors $\vec{\mathbf{l}}_i$

$$\mathbb{P} = \sum_i \vec{\mathbf{l}}_i = \mathbf{0} \tag{6.7.19}$$

$$\mathbb{Q} = \sum_{i<j} \vec{\mathbf{l}}_i \times \vec{\mathbf{l}}_j = \mathbf{0} \tag{6.7.20}$$

In a molecule with n atoms in a ring the most complete set of internal coordinates includes n bond stretching, n bond–bond angle bending and n bond torsion coordinates. Since the normal modes of the ring are $3n - 6$, six redundancies occur among these coordinates.

The analytical form of these redundancies, with the proper symmetry, can be found[11] by scalar multiplication of the two primitive vectorial redundancies by six linear combinations of the vectors $\vec{\mathbf{l}}_i$ or of the vectors $\vec{\mathbf{l}}_i \times \vec{\mathbf{l}}_j$ chosen so that the products belong to the irreducible representations to which the redundancies belong. The total differential of these six scalar primitive redundancies then yields the six redundancies among the internal coordinates to any desired order.

If the ring is planar a great simplification occurs since the redundancies involving the torsion coordinates are completely separated from the redundancies involving in-plane internal coordinates.

As a simple example we shall use equation (6.7.19) for the determination of the cyclic redundancies in cyclopropane.

Example 6.7.2

For the complete description of the normal modes of the cyclopropane ring (see Example 6.6.2) only three angle-bending and three bond-stretching coordinates are needed. This simplifies the problem very much, only equation (6.7.19) being required to find the redundancies. In Example 6.6.2 we have seen that the redundancies belong to the species A_1' and E'. Since the vectorial redundancy (6.7.19)

$$\vec{\mathbf{R}}_{12} + \vec{\mathbf{R}}_{23} + \vec{\mathbf{R}}_{31} = 0$$

belongs to the A_2' species and, since (see Table 7.5.3) $A_2' \otimes A_2' = A_1'$ and $A_2' \otimes E' = E'$, the products

$$\begin{aligned}
&(1)\ (\vec{\mathbf{R}}_{23} + \vec{\mathbf{R}}_{31} + \vec{\mathbf{R}}_{12}) \cdot (\vec{\mathbf{R}}_{23} + \vec{\mathbf{R}}_{31} + \vec{\mathbf{R}}_{12}) = 0 \quad A_1' \text{ species}\\
&(2)\ (\vec{\mathbf{R}}_{23} + \vec{\mathbf{R}}_{31} + \vec{\mathbf{R}}_{12}) \cdot (2\vec{\mathbf{R}}_{23} - \vec{\mathbf{R}}_{31} - \vec{\mathbf{R}}_{12}) = 0 \quad E_a' \\
&(3)\ (\vec{\mathbf{R}}_{23} + \vec{\mathbf{R}}_{31} + \vec{\mathbf{R}}_{12}) \cdot (\vec{\mathbf{R}}_{31} - \vec{\mathbf{R}}_{12}) = 0 \quad E_b'
\end{aligned} \Big\} \text{species}$$

will yield the three primitive scalar redundancies. Performing the scalar products, we obtain

$$\begin{aligned}
&(1)\ R_{23}^2 + R_{31}^2 + R_{12}^2 - 2R_{23}R_{31}\cos\alpha_3 - 2R_{23}R_{12}\cos\alpha_2 - 2R_{31}R_{12}\cos\alpha_1 = 0\\
&(2)\ 2R_{23}^2 - R_{31}^2 - R_{12}^2 - R_{23}R_{31}\cos\alpha_3 - R_{23}R_{12}\cos\alpha_2 + 2R_{31}R_{12}\cos\alpha_1 = 0\\
&(3)\ R_{31}^2 - R_{12}^2 - R_{23}R_{31}\cos\alpha_3 + R_{23}R_{12}\cos\alpha_2 = 0
\end{aligned}$$

and taking the total differentials

$$\begin{aligned}
&(1)\ (\Delta\alpha_1 + \Delta\alpha_2 + \Delta\alpha_3) = 0 && \text{i.e. } \mathbb{S}_3^{A'} = 0\\
&(2)\ \sqrt{3}(2\,\Delta R_{23} - \Delta R_{31} - \Delta R_{12}) - R(2\,\Delta\alpha_1 - \Delta\alpha_2 - \Delta\alpha_3) = 0 && \text{i.e. } \sqrt{3}\mathbb{S}_{8a}^{E'} - R\mathbb{S}_{9a}^{E'} = 0\\
&(3)\ \sqrt{3}(\Delta R_{31} - \Delta R_{12}) - R(\Delta\alpha_2 - \Delta\alpha_3) = 0 && \text{i.e. } \sqrt{3}\mathbb{S}_{8b}^{E'} - R\mathbb{S}_{9b}^{E'} = 0
\end{aligned}$$

where R is the equilibrium C–C bond length.

6.8 The correlation tables

In many problems of molecular dynamics, including isotopic substitution, construction of degenerate symmetry coordinates and interpretation of the infrared and Raman spectra of crystals, it is important to know the relationships between the irreducible representations of a given point group and those of one of its subgroups.

Consider for instance the ammonia molecule of Example 6.6.1. The structure of the representation carried by the normal coordinates was found to be

$$2A_1 + 2E$$

and the symmetry coordinates

$$\begin{aligned}
&\mathbb{S}_1^{A'} = \tfrac{1}{\sqrt{3}}(\Delta r_1 + \Delta r_2 + \Delta r_3) && \mathbb{S}_{3a}^{E} = \tfrac{1}{\sqrt{6}}(2\,\Delta r_1 - \Delta r_2 - \Delta r_3) && \mathbb{S}_{3b}^{E} = \tfrac{1}{\sqrt{2}}(\Delta r_2 - \Delta r_3)\\
&\mathbb{S}_2^{A'} = \tfrac{1}{\sqrt{3}}(\Delta\alpha_1 + \Delta\alpha_2 + \Delta\alpha_3) && \mathbb{S}_{4a}^{E} = \tfrac{1}{\sqrt{6}}(2\,\Delta\alpha_1 - \Delta\alpha_2 - \Delta\alpha_3) && \mathbb{S}_{4b}^{E} = \tfrac{1}{\sqrt{2}}(\Delta\alpha_2 - \Delta\alpha_3)
\end{aligned}$$

If the hydrogen atom 1 is substituted by a deuterium atom, some symmetry elements such as C_3, σ_2 and σ_3 are lost and the only symmetry element still present is the symmetry plane σ_1 (see Figure 6.2.4). The molecule H_2DN belongs then to the C_s point group and the structure of the representation carried by the normal coordinates becomes

$$4A' + 2A''$$

Consider now the character tables for the two groups C_{3v} and C_s, given below and, ignoring all symmetry elements of C_{3v} but E and σ_1, compare them.

C_{3v}	E	C_3	C_3^{-1}	σ_1	σ_2	σ_3
A_1	1	1	1	1	1	1
A_2	1	1	1	-1	-1	-1
E	2	-1	-1	0	0	0

C_s	E	σ
A'	1	1
A''	1	-1

From the character tables we realize at once that, as far as these two operations are concerned, the symmetry species A_1 and A_2 of C_{3v} are identical to the symmetry species A' and A'' of C_s respectively, whereas the symmetry species E is simply $A' + A''$. We can summarize this correlation between the point group C_{3v} and its subgroup C_s by means of the following correlation table:

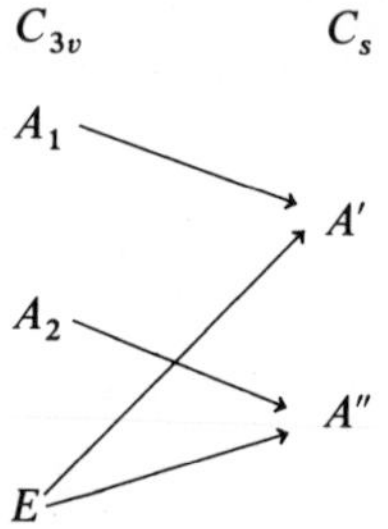

where the arrows mean 'goes into' when the symmetry is lowered from C_{3v} to C_s.

The six symmetry coordinates found for ammonia have the important property that they are also symmetry coordinates for a lower molecular symmetry. They were chosen to satisfy all the symmetry requirements of the C_{3v} group and therefore will satisfy part of them, i.e. those of the C_s subgroup.

The behaviour of the six symmetry coordinates under the symmetry operation σ shows, in agreement with the correlation table, that the coordinates $\mathbb{S}_1$, $\mathbb{S}_2$, $\mathbb{S}_{3a}$ and $\mathbb{S}_{4a}$ belong to the A' and the coordinates $\mathbb{S}_{3b}$ and $\mathbb{S}_{4b}$ belong to the A'' species of the C_s subgroup. It must be noticed that these are not exactly the symmetry coordinates that one would obtain from the character table of the C_s group,

$$\mathbb{S}_1^{A'} = \Delta r_1 \qquad \mathbb{S}_4^{A'} = \tfrac{1}{\sqrt{2}}(\Delta\alpha_2 + \Delta\alpha_3)$$

$$\mathbb{S}_2^{A'} = \tfrac{1}{\sqrt{2}}(\Delta r_2 + \Delta r_3) \qquad \mathbb{S}_5^{A''} = \tfrac{1}{\sqrt{2}}(\Delta r_2 - \Delta r_3)$$

$$\mathbb{S}_3^{A'} = \Delta\alpha_1 \qquad \mathbb{S}_6^{A''} = \tfrac{1}{\sqrt{2}}(\Delta\alpha_2 - \Delta\alpha_3)$$

but a linear combination of them.

This result suggests a method,[3] alternative to that of Section 6.6, for the construction of degenerate pairs of symmetry coordinates using only the characters of the irreducible representations. The method is based upon the construction of 'local symmetry coordinates', where by local symmetry we mean the subgroup of the molecular group which has the property of permuting among themselves the members of a given subset of a set of equivalent internal coordinates. In the case of ammonia, for instance, the subgroup $C_s\{E, \sigma_1\}$ has the property of permuting among themselves the coordinates Δr_2 and Δr_3. Using these coordinates and the character table of the C_s group we can construct the two mutually orthogonal local symmetry coordinates $\mathbb{S}_2^{A'}$ and $\mathbb{S}_5^{A''}$ given above. Now we use these coordinates as 'starting' coordinates for the construction of the degenerate pairs of symmetry coordinates of ammonia. Starting with $\Delta r_2 + \Delta r_3$ we have, using Table 6.4.4,

$$2(\Delta r_2 + \Delta r_3) - (\Delta r_3 + \Delta r_1) - (\Delta r_1 + \Delta r_2) = (-2\,\Delta r_1 + \Delta r_2 + \Delta r_3)$$

and, after normalization,

$$\mathbb{S}^E_{3a} = \frac{1}{\sqrt{6}}(2\,\Delta r_1 - \Delta r_2 - \Delta r_3)$$

Starting with $(\Delta r_2 - \Delta r_3)$ we have

$$2(\Delta r_2 - \Delta r_3) - (\Delta r_3 - \Delta r_1) - (\Delta r_1 - \Delta r_2) = (3\,\Delta r_2 - 3\,\Delta r_3)$$

and, after normalization,

$$\mathbb{S}^E_{3b} = \frac{1}{\sqrt{2}}(\Delta r_2 - \Delta r_3)$$

Exactly the same method can be applied for the construction of symmetry coordinates in the case of the regular representation.

Consider, for example, the β_{ijk} coordinates of cyclopropane (see Example 6.6.2). The four coordinates $\Delta\beta_{412}$, $\Delta\beta_{413}$, $\Delta\beta_{712}$ and $\Delta\beta_{713}$ are permuted among themselves by all symmetry operations of the subgroup $C_{2v}\{E, C_2^{(1)}, \sigma_1, \sigma_h\}$. Using the character table of the C_{2v} point group, we can construct the four local symmetry coordinates

$$\mathbb{S}_1^{A_1} = \tfrac{1}{2}(\Delta\beta_{412} + \Delta\beta_{413} + \Delta\beta_{712} + \Delta\beta_{713})$$

$$\mathbb{S}_1^{A_2} = \tfrac{1}{2}(\Delta\beta_{412} - \Delta\beta_{413} - \Delta\beta_{712} + \Delta\beta_{713})$$

$$\mathbb{S}_1^{B_1} = \tfrac{1}{2}(\Delta\beta_{412} + \Delta\beta_{413} - \Delta\beta_{712} - \Delta\beta_{713})$$

$$\mathbb{S}_1^{B_2} = \tfrac{1}{2}(\Delta\beta_{412} - \Delta\beta_{413} + \Delta\beta_{712} - \Delta\beta_{713})$$

where the label 1 on the symmetry coordinates indicates that the four coordinates have the atom 1 in common. Using these local symmetry coordinates we can construct, with the help of the character table of the D_{3h} point group, the symmetry coordinates of cyclopropane.

Using $\mathbb{S}_1^{A_1}$ as starting coordinates we obtain

$$\mathbb{S}_5^{A_1'} = \frac{1}{\sqrt{3}}(\mathbb{S}_1^{A_1} + \mathbb{S}_2^{A_1} + \mathbb{S}_3^{A_1})$$

$$\mathbb{S}_{11a}^{E'} = \frac{1}{\sqrt{6}}(2\mathbb{S}_1^{A_1} - \mathbb{S}_2^{A_1} - \mathbb{S}_3^{A_1})$$

and zero for all other species.

Using $\mathbb{S}_1^{A_2}$ we obtain

$$\mathbb{S}_{13}^{A_1''} = \frac{1}{\sqrt{3}}(\mathbb{S}_1^{A_2} + \mathbb{S}_2^{A_2} + \mathbb{S}_3^{A_2})$$

$$\mathbb{S}_{17a}^{E''} = \frac{1}{\sqrt{6}}(2\mathbb{S}_1^{A_2} - \mathbb{S}_2^{A_2} - \mathbb{S}_3^{A_2})$$

and zero for all other species.

Using $\mathbb{S}_1^{B_1}$ we obtain

$$\mathbb{S}_{15}^{A_2''} = \frac{1}{\sqrt{3}}(\mathbb{S}_1^{B_1} + \mathbb{S}_2^{B_1} + \mathbb{S}_3^{B_1})$$

$$\mathbb{S}_{18b}^{E''} = \frac{1}{\sqrt{6}}(2\mathbb{S}_1^{B_1} - \mathbb{S}_2^{B_1} - \mathbb{S}_3^{B_1})$$

and zero for all other species.

Using $\mathbb{S}_1^{B_2}$ we obtain

$$\mathbb{S}_6^{A_2'} = \frac{1}{\sqrt{3}}(\mathbb{S}_1^{B_2} + \mathbb{S}_2^{B_2} + \mathbb{S}_3^{B_2})$$

$$\mathbb{S}_{12b}^{E'} = \frac{1}{\sqrt{6}}(2\mathbb{S}_1^{B_2} - \mathbb{S}_2^{B_2} - \mathbb{S}_3^{B_2})$$

and zero for all other species.

In order to obtain the other two coordinates of species E', $\mathbb{S}_{11b}$ and $\mathbb{S}_{12a}$, and the other two coordinates of species E'', $\mathbb{S}_{17b}$ and $\mathbb{S}_{18a}$, we have to take care of orientation.

In the E' block, since the coordinates $\mathbb{S}_1^{B_2}$ and $\mathbb{S}_1^{A_1}$ are not iso-oriented, we can use as starting coordinates $(\mathbb{S}_2^{B_2} - \mathbb{S}_3^{B_2})$ which is iso-oriented with $\mathbb{S}_1^{A_1}$ and $(\mathbb{S}_2^{A_1} - \mathbb{S}_3^{A_1})$ which is iso-oriented with $\mathbb{S}_1^{B_2}$. We get

$$\mathbb{S}_{12a}^{E'} = \mathcal{N}[4(\mathbb{S}_2^{B_2} - \mathbb{S}_3^{B_2}) - 2(\mathbb{S}_3^{B_2} - \mathbb{S}_1^{B_2}) - 2(\mathbb{S}_1^{B_2} - \mathbb{S}_2^{B_2}) = \tfrac{1}{\sqrt{2}}(\mathbb{S}_2^{B_2} - \mathbb{S}_3^{B_2})$$

$$\mathbb{S}_{11b}^{E'} = \mathcal{N}[4(\mathbb{S}_2^{A_1} - \mathbb{S}_3^{A_1}) - 2(\mathbb{S}_3^{A_1} - \mathbb{S}_1^{A_1}) - 2(\mathbb{S}_1^{A_1} - \mathbb{S}_2^{A_1}) = \tfrac{1}{\sqrt{2}}(\mathbb{S}_2^{A_1} - \mathbb{S}_3^{A_1})$$

In the E'' block, since the coordinates $\mathbb{S}_1^{A_2}$ and $\mathbb{S}_1^{B_1}$ are not iso-oriented we can use $(\mathbb{S}_2^{B_1} - \mathbb{S}_3^{B_1})$ iso-oriented with $\mathbb{S}_1^{A_2}$ and $(\mathbb{S}_2^{A_2} - \mathbb{S}_3^{A_2})$ iso-oriented with $\mathbb{S}_1^{B_1}$. We get

$$\mathbb{S}_{17b}^{E''} = \mathcal{N}[4(\mathbb{S}_2^{A_2} - \mathbb{S}_3^{A_2}) - 2(\mathbb{S}_3^{A_2} - \mathbb{S}_1^{A_2}) - 2(\mathbb{S}_1^{A_2} - \mathbb{S}_2^{A_2})] = \tfrac{1}{\sqrt{2}}(\mathbb{S}_2^{A_2} - \mathbb{S}_3^{A_2})$$

$$\mathbb{S}_{18a}^{E''} = \mathcal{N}[4(\mathbb{S}_2^{B_1} - \mathbb{S}_3^{B_1}) - 2(\mathbb{S}_3^{B_1} - \mathbb{S}_1^{B_1}) - 2(\mathbb{S}_1^{B_1} - \mathbb{S}_2^{B_1})] = \tfrac{1}{\sqrt{2}}(\mathbb{S}_2^{B_1} - \mathbb{S}_3^{B_1})$$

The symmetry coordinates thus obtained are identical to those of Example 6.6.2. The correlation diagram between the group D_{3h} and the subgroup C_{2v}, taken from Appendix II, where the correlation tables are collected, furnishes a complete check of the correctness of the above given symmetry coordinates.

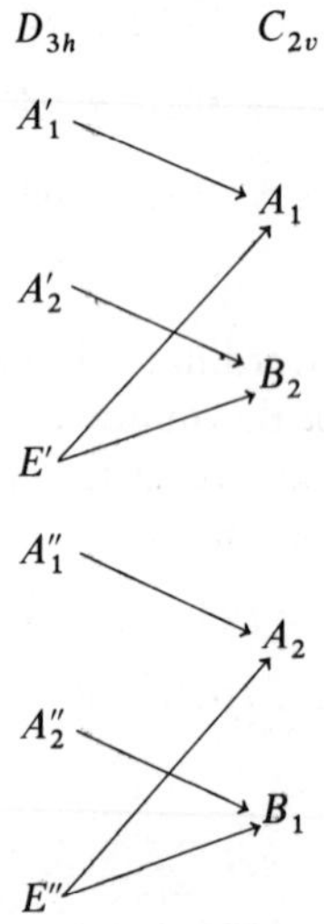

References

1. J. S. Lomont, *Applications of Finite Groups*, Academic Press, 1959.
2. R. McWeeny, *Quantum Mechanics*, Pergamon Press, 1973, Vol. II, *Methods and Basic Applications*.
3. E. B. Wilson jr., J. C. Decius and P. C. Cross, *Molecular Vibrations*, McGraw-Hill, 1955.
4. C. E. Sun, R. G. Parr and B. Crawford, *J. Chem. Phys.*, **17**, 840 (1949).
5. M. Gussoni and G. Zerbi, *Atti Acc. Naz. Lincei*, **40**, 1032 (1966).
6. I. N. Godnev and I. V. Orlova, *Optics and Spectroscopy*, **6**, 360 (1959).
7. Z. Chila and J. Pliva, *Coll. Czechoslov. Chem. Comm.*, **26**, 1903 (1961).
8. B. Crawford and J. Overend, *J. Mol. Spectrosc.*, **12**, 307 (1964).
9. T. Shimanouchi, *J. Chem. Phys.*, **17**, 245 (1949).
10. S. Califano and B. Crawford, *Z. Elektrochem.*, **64**, 571 (1960).
11. S. Califano, unpublished work.

Chapter 7
Symmetry Selection Rules

7.1 Symmetry invariance of μ and α matrix elements

In Chapter 3 we reached the conclusion that a transition from a vibrational level n to another vibrational level m is active in the infrared spectrum, i.e. an absorption band is found at frequency ν_{mn} only if one of the matrix elements in (7.1.1) is different from zero

$$\langle\psi_m|\mu_x|\psi_n\rangle \qquad \langle\psi_m|\mu_y|\psi_n\rangle \qquad \langle\psi_m|\mu_z|\psi_n\rangle \tag{7.1.1}$$

where μ_x, μ_y and μ_z are the components of the electric moment in a rotating molecular system. Similarly the transition is active in the Raman spectrum, i.e. a displaced Raman line is found at frequency $\nu_0 \pm \nu_{mn}$, only if one of the matrix elements in (7.1.2) is different from zero

$$\langle\psi_m|\alpha_{\rho\sigma}|\psi_n\rangle \qquad \rho, \sigma = x, y, z \tag{7.1.2}$$

where the $\alpha_{\rho\sigma}$ are the components of a tensor in a space-fixed molecular system.

In this chapter we shall discuss the behaviour of these matrix elements under symmetry operations and show how the knowledge of their transformation properties is sufficient to decide whether or not a matrix element can be different from zero. The validity of these selection rules is very general. We shall see in what follows that they are not restricted to the harmonic approximation nor to a specific mechanism of interaction between the molecule and the radiation. It must be understood, however, that the selection rules represent limiting conditions and give no information on the magnitude of the matrix elements when they are different from zero. This has two implications which often complicate the interpretation of vibrational spectra. The first is that a transition forbidden by symmetry can be actually observed in the spectrum as a weak band, when the molecular symmetry is perturbed by external factors (intermolecular forces, collisions, etc.) as often occurs in liquids and in gases at high pressure. The second is that, when a transition is allowed by symmetry, it is not necessarily observed in the spectrum. In many cases, in fact, the matrix elements (7.1.1) and (7.1.2) have vanishingly small values and therefore give rise to very weak bands. In polyatomic molecules this is probably the major source of confusion in the assignment of fundamental vibrations.

Despite these difficulties the symmetry selection rules have played a fundamental role in the development of vibrational spectroscopy and the interpretation of infrared and Raman spectra of symmetrical molecules rests largely on their use.

The symmetry selection rules are based on the property that the matrix elements (7.1.1) and (7.1.2) are invariant under a coordinate transformation and therefore under a symmetry operation.

This invariance can be rigorously stated in group-theoretical language as follows. Let Γ^{μ} be the representation carried by the wavefunction ψ_m, Γ^{ν} the representation carried by

ψ_n and Γ^γ the representation carried by one of the components of μ or of α. The representation carried by the matrix elements (7.1.1) and (7.1.2) is then, according to their definition

$$\Gamma^\varepsilon = \Gamma^\mu \otimes \Gamma^\gamma \otimes \Gamma^\nu \tag{7.1.3}$$

The following possibilities can occur

(a) Γ^ε is an irreducible representation of the molecular point group. In this case the matrix element has a non-vanishing value only if Γ^ε coincides with Γ^0 where Γ^0 is the totally symmetric representation of the group.

(b) Γ^ε is reducible. In this case the matrix element is different from zero only if the structure of Γ_ε is such that it contains Γ^0.

We shall see in the following sections that the first possibility occurs when Γ^μ, Γ^γ and Γ^ν are monodimensional representations while the second possibility occurs when two of them are degenerate.

In order to obtain the structure of the representation Γ^ε it is clearly necessary to know the structures of the representations Γ^μ, Γ^γ and Γ^ν. In Sections 7.2 and 7.3 we shall therefore discuss the representations carried by the components of $\boldsymbol{\mu}$ and $\boldsymbol{\alpha}$ respectively and in Sections 7.4 to 7.7 the representations carried by the vibrational wavefunctions. In Sections 7.8 and 7.9 we shall then discuss the infrared and Raman selection rules, on the basis of equation (7.1.3).

7.2 The symmetry of the electric dipole moment

The dipole moment of a molecule, referred to the rotating axes system, is a vector with components

$$\mu_x = \sum_\alpha e_\alpha x_\alpha \qquad \mu_y = \sum_\alpha e_\alpha y_\alpha \qquad \mu_z = \sum_\alpha e_\alpha z_\alpha \tag{7.2.1}$$

where e_α is the effective charge and x_α, y_α, z_α are the instantaneous Cartesian coordinates of atom α.

Let us now consider the effect of a symmetry operation $\mathscr{R}$ on μ_x. Since symmetry operations permute only equivalent atoms and since equivalent atoms have the same effective charge, $\mathscr{R}$ has no effect on the charge distribution. If the atoms of the molecule are divided into m equivalent sets, we can formally write μ_x in the form

$$\mu_x = e_1 \sum_{\alpha_1} x_{\alpha_1} + e_2 \sum_{\alpha_2} x_{\alpha_2} + \ldots + e_i \sum_{\alpha_i} x_{\alpha_i} + \ldots + e_m \sum_{\alpha_m} x_{\alpha m} \tag{7.2.2}$$

where the index α_i counts the atoms in the ith equivalent set and e_i is the effective charge of the atoms of the ith set. The effect of $\mathscr{R}$ on μ_x is then given by

$$\mathscr{R}\mu_x = e_1 \sum_{\alpha_1} \mathscr{R}x_{\alpha_1} + e_2 \sum_{\alpha_2} \mathscr{R}x_{\alpha_2} + \ldots + e_i \sum_{\alpha_i} \mathscr{R}x_{\alpha_i} + e_m \sum_{\alpha_m} \mathscr{R}x_{\alpha_m} \tag{7.2.3}$$

From Chapters 5 and 6 we know that the effect of a symmetry operation on the x Cartesian coordinate of an atom belonging to the ith set of equivalent atoms, can be written in the form

$$\mathscr{R}x_k = D_c(\mathscr{R})_{xx} x_l + D_c(\mathscr{R})_{xy} y_l + D_c(\mathscr{R})_{xz} z_l \tag{7.2.4}$$

where k and l are possible values of the index α_i and $D_c(\mathscr{R})_{\rho\sigma}$ are the elements of the first row of the $\mathbf{D}_c(\mathscr{R})$ matrix of equation (6.2.14). From (7.2.4) we obtain the relationship

$$\sum_{\alpha_i} \mathscr{R}x_{\alpha_i} = D_c(\mathscr{R})_{xx} \sum_{\alpha_i} x_{\alpha_i} + D_c(\mathscr{R})_{xy} \sum_{\alpha_i} y_{\alpha_i} + D_c(\mathscr{R})_{xz} \sum_{\alpha_i} z_{\alpha_i} \tag{7.2.5}$$

which, inserted into (7.2.3) yields, using (7.2.1),

$$\mathscr{R}\mu_x = D_c(\mathscr{R})_{xx}\mu_x + D_c(\mathscr{R})_{xy}\mu_y + D_c(\mathscr{R})_{xz}\mu_z \tag{7.2.6}$$

In the same way we obtain

$$\begin{aligned} \mathscr{R}\mu_y &= D_c(\mathscr{R})_{yx}\mu_x + D_c(\mathscr{R})_{yy}\mu_y + D_c(\mathscr{R})_{yz}\mu_z \\ \mathscr{R}\mu_z &= D_c(\mathscr{R})_{zx}\mu_x + D_c(\mathscr{R})_{zy}\mu_y + D_c(\mathscr{R})_{zz}\mu_z \end{aligned} \tag{7.2.7}$$

This shows that the components of the dipole moment vector behave under symmetry operations like the x, y- and z-axes. Since the three translations t_x, t_y and t_z also transform like the axes x, y and z, we conclude that the components of the dipole moment vector carry the same representations carried by the three translational coordinates. This result can be obtained simply by considering that the components of the dipole moment are vectors oriented along the x-, y-, and z-axes respectively and must therefore transform under symmetry operations like the axes themselves.

In Section 3.11 we have shown that it is not the dipole moment but only its variation in a molecular vibration that matters for the vibrational selection rules. In the above discussion we have actually used the components of the total dipole moment instead of the components of its vibrational part. It is obvious that the vibrational dipole moment with components

$$\mu_x^1 = \sum_\alpha e_\alpha \Delta x_\alpha \qquad \mu_y^1 = \sum_\alpha e_\alpha \Delta y_\alpha \qquad \mu_z^1 = \sum_\alpha e_\alpha \Delta z_\alpha \tag{7.2.8}$$

behaves exactly like the total dipole moment since the instantaneous Cartesian coordinates and the displacement coordinates are iso-oriented, being related by the relationships

$$x_\alpha = x_\alpha^0 + \Delta x_\alpha \qquad y_\alpha = y_\alpha^0 + \Delta y_\alpha \qquad z_\alpha = z_\alpha^0 + \Delta z_\alpha$$

x_α^0, y_α^0 and z_α^0 being the Cartesian coordinates of the equilibrium positions.

7.3 The symmetry of the scattering tensor

In the discussion of Raman scattering (Section 3.14) we have defined the scattering tensor

$$\boldsymbol{\alpha}_{mn} = \begin{pmatrix} (\alpha_{xx})_{mn} & (\alpha_{xy})_{mn} & (\alpha_{xz})_{mn} \\ (\alpha_{xy})_{mn} & (\alpha_{yy})_{mn} & (\alpha_{yz})_{mn} \\ (\alpha_{xz})_{mn} & (\alpha_{yz})_{mn} & (\alpha_{zz})_{mn} \end{pmatrix} \tag{7.3.1}$$

as a symmetric tensor (of rank two) whose components are the matrix elements

$$(\alpha_{\rho\sigma})_{mn} = \langle \psi_m | \alpha_{\rho\sigma} | \psi_n \rangle \qquad \rho, \sigma = x, y, z \tag{7.3.2}$$

where

$$\alpha_{\rho\sigma} = \sum_{e,i} \frac{2k_{ee_0} - k_{mn}}{(k_{mn}^2 - k_0^2) - k_{mn}(k_{ee_0} - k_0)} [(P_\rho^0)_{e_0e}(P_\sigma^i)_{ee_0} + (P_\rho^i)_{e_0e}(P_\sigma^0)_{ee_0}]Q_i \tag{7.3.3}$$

In order to obtain the Raman selection rules, we need to investigate the transformation properties of the components $\alpha_{\rho\sigma}$ of the tensor $\boldsymbol{\alpha}$, defined by the previous equation. For this, let us consider a generic tensor $\mathbf{T}$ which relates two vectors $\mathbf{r}$ and $\mathbf{v}$ in a Cartesian basis $\mathbf{e}$, through the equation

$$\mathbf{r} = \mathbf{T}\mathbf{v} \tag{7.3.4}$$

If we perform a symmetry operation $\mathscr{R}$ on the vectors **r** and **v**, the 'rotated' vectors **r**′ and **v**′ are given,† according to Section 6.2, by

$$\mathbf{r}' = \mathscr{R}\mathbf{r} = \mathbf{D}_c(\mathscr{R})\mathbf{r} \qquad \mathbf{v}' = \mathscr{R}\mathbf{v} = \mathbf{D}_c(\mathscr{R})\mathbf{v} \tag{7.3.5}$$

The rotated vectors **r**′ and **v**′ are now related by the equation

$$\mathbf{r}' = \mathbf{T}'\mathbf{v}' \tag{7.3.6}$$

and the relationship between the tensor **T** and the 'rotated' tensor **T**′ is easily obtained by substitution of (7.3.5) into (7.3.6), which yields

$$\mathbf{T}' = \mathbf{D}_c(\mathscr{R})\mathbf{T}\mathbf{D}_c(\mathscr{R})^{-1} = \mathbf{D}_c(\mathscr{R})\mathbf{T}\tilde{\mathbf{D}}_c(\mathscr{R}) \tag{7.3.7}$$

By performing this matrix multiplication we obtain that the generic element $\mathbf{T}'_{\rho\sigma}$ is given by

$$T'_{\rho\sigma} = \sum_{k,l} D_c(\mathscr{R})_{\rho k} D_c(\mathscr{R})_{\sigma l} T_{kl} = \sum_{k,l} S_{\rho\sigma,kl}(\mathscr{R}) T_{k,l} \tag{7.3.8}$$

We can formally consider the elements of the tensor **T** as the nine components in a nine-dimensional space of a vector **t** which is transformed by the symmetry operator $\mathscr{R}$ into a rotated vector **t**′ according to the equation

$$\mathbf{t}' = \mathscr{R}\mathbf{t} = \mathbf{S}(\mathscr{R})\mathbf{t} = (\mathbf{D}_c(\mathscr{R}) \otimes \mathbf{D}_c(\mathscr{R}))\mathbf{t} \tag{7.3.9}$$

where the matrix $\mathbf{S}(\mathscr{R})$ is the outer product of the matrix $\mathbf{D}_c(\mathscr{R})$ with itself. Since the matrices $\mathbf{D}_c(\mathscr{R})$ form, as discussed in Section 6.2, a representation of the group carried by the x-, y- and z-axes and also by the translations t_x, t_y and t_z, we conclude that the representation carried by the element $T_{\rho\sigma}$ of **T** is the direct product of the representation carried by t_ρ and of the representation carried by t_σ. Symbolically we write

$$\Gamma^{\rho\sigma} = \Gamma^\rho \otimes \Gamma^\sigma \tag{7.3.10}$$

The character of the representation $\Gamma^{\rho\sigma}$ is thus simply the product of the characters χ^{Γ_ρ} and χ^{Γ_σ} of the representations Γ^ρ and Γ^σ.

In this discussion we have assumed that the tensor **T** is asymmetric, i.e. that all nine components are different and that no relations exist between them. The tensors encountered in physical problems are, however, always symmetrical, i.e. the element $T_{\rho\sigma}$ is identical with the element $T_{\sigma\rho}$. In this case we have only six independent elements of **T** and the structure of the representation $\Gamma^{\rho\sigma}$ given by equation (7.3.10) is incorrect.

In the case of a symmetric tensor the generic element $T'_{\rho\sigma}$ is given by

$$T'_{\rho\sigma} = \sum_k D_c(\mathscr{R})_{\rho k} D_c(\mathscr{R})_{\sigma k} T_{kk} + \sum_{k<l} (D_c(\mathscr{R})_{\rho k} D_c(\mathscr{R})_{\sigma l} + D_c(\mathscr{R})_{\rho l} D_c(\mathscr{R})_{\sigma k}) T_{kl} \tag{7.3.11}$$

In this case equation (7.3.9) becomes

$$\mathbf{t}' = \mathscr{R}\mathbf{t} = [\mathbf{S}(\mathscr{R})]\mathbf{t} = [\mathbf{D}_c(\mathscr{R}) \otimes \mathbf{D}_c(\mathscr{R})]\mathbf{t} \tag{7.3.12}$$

where the square brackets indicate that the outer product of the matrix $\mathbf{D}_c(\mathscr{R})$ by itself must be made using (7.3.11).

We shall call the matrix $[\mathbf{D}_c(\mathscr{R}) \otimes \mathbf{D}_c(\mathscr{R})]$ the *symmetrized Kronecker* square of $\mathbf{D}_c(\mathscr{R})$. The elements of $\mathbf{S}(\mathscr{R})$ are, according to (7.3.11) and using double subscripts for rows and columns,

$$\begin{aligned} [S(\mathscr{R})]_{\rho\sigma,kk} &= D_c(\mathscr{R})_{\rho k} D_c(\mathscr{R})_{\sigma k} \\ [S(\mathscr{R})]_{\rho\sigma,kl} &= D_c(\mathscr{R})_{\rho k} D_c(\mathscr{R})_{\sigma l} + D_c(\mathscr{R})_{\rho l} D_c(\mathscr{R})_{\sigma k} \qquad (k \neq l) \end{aligned} \tag{7.3.13}$$

† We recall from Chapter 5 that this corresponds to an inverse rotation of the basis **e**, such that $\mathbf{e}' = \mathbf{e}\mathbf{D}(\mathscr{R})$.

The character of the representation carried by the elements $T_{\rho\sigma}$ of a symmetric tensor is no longer the product of the characters of Γ^ρ and Γ^σ. It can be obtained by summing all the diagonal elements of the $[\mathbf{S}(\mathscr{R})]$ matrix, i.e. by putting $\rho = k$ and $\sigma = l$ and summing. In this way we obtain

$$\mathrm{tr}\,[\mathbf{S}(\mathscr{R})] = \sum_{\rho<\sigma} (D_c(\mathscr{R})_{\rho\rho} D_c(\mathscr{R})_{\sigma\sigma} + D_c(\mathscr{R})_{\rho\sigma} D_c(\mathscr{R})_{\sigma\rho}) + \sum_{\rho} D_c^2(\mathscr{R})_{\rho\rho}$$

$$\mathrm{tr}\,[\mathbf{S}(\mathscr{R})] = \tfrac{1}{2}\sum_{\rho,\sigma} (D_c(\mathscr{R})_{\rho\rho} D_c(\mathscr{R})_{\sigma\sigma} + D_c(\mathscr{R})_{\rho\sigma} D_c(\mathscr{R})_{\sigma\rho}) \tag{7.3.14}$$

$$\mathrm{tr}\,[\mathbf{S}(\mathscr{R})] = \tfrac{1}{2}[(\mathrm{tr}\,D_c(\mathscr{R}))^2 + \mathrm{tr}\,D_c(\mathscr{R})^2]$$

The representation $\Gamma^{\rho\sigma}$ carried by the component $T_{\rho\sigma}$ of $\mathbf{T}$ is then said to be the *symmetrized direct* product of Γ^ρ and Γ^σ and is indicated by the symbol of the direct product in square brackets

$$\Gamma^{\rho\sigma} = [\Gamma^\rho \otimes \Gamma^\sigma] \tag{7.3.15}$$

The representation carried by the components of the tensor $\boldsymbol{\alpha}$ are thus simply obtained from equation (7.3.15) by forming the symmetrized direct product of the representations carried by the translations.

We shall now show that the result of equation (7.3.15) is also obtained if the symmetry of the components $\alpha_{\rho\sigma}$ of the tensor $\boldsymbol{\alpha}$ is calculated directly by means of equation (7.3.3) by forming the symmetrized direct product of the various terms. Since $(P_\rho^0)_{pq}$ is an electronic transition moment, while $(P_\sigma^i)_{pq}$ is a derivative with respect to the Q_i normal coordinate, we obtain from (7.3.3)

$$\begin{aligned}\Gamma^{\rho\sigma} &= \Gamma^\rho \otimes \Gamma^{eo} \otimes \Gamma^e \otimes \Gamma^\sigma \otimes \Gamma^{Q_i} \otimes \Gamma^e \otimes \Gamma^{eo} \otimes \Gamma^{Q_i}\\ &= [\Gamma^\rho \otimes \Gamma^\sigma] \otimes [\Gamma^{eo} \otimes \Gamma^{eo}] \otimes [\Gamma^e \otimes \Gamma^e] \otimes [\Gamma^{Q_i} \otimes \Gamma^{Q_i}] = [\Gamma^\rho \otimes \Gamma^\sigma]\end{aligned} \tag{7.3.16}$$

For non-degenerate electronic levels and normal coordinates, the direct products $[\Gamma^e \otimes \Gamma^e]$ and $[\Gamma^{Q_i} \otimes \Gamma^{Q_i}]$ are identical to Γ^0, where Γ^0 is the totally symmetric representation of the group. In this case the normal and the symmetrized direct products coincide and the representation (7.3.15) is easily obtained.

If the electronic levels or the normal coordinates are degenerate, equation (7.3.15) is still valid, although the products (7.3.16) become much more complicated. It will be seen in Section 7.6 that the direct product of a degenerate representation by itself, yields a reducible representation which always includes Γ^0. Incidentally, it must be noticed that, since (7.3.3) involves a sum over all electronic levels connected by non-zero matrix elements to the ground level, the occurrence of some degenerate levels plays a limited role in (7.3.3). We conclude therefore that the representation carried by the components of $\boldsymbol{\alpha}$ is correctly obtained by means of equation (7.3.15). In the case that t_ρ and t_σ carry a degenerate representation, the characters of the representation $\Gamma^{\rho\sigma}$ can be obtained from equation (7.3.14) and then the structure of the representation can be derived in the usual way. Alternatively one can use Table 7.6.2 directly, which lists the structures of the representations carried by the vibrational wavefunctions of degenerate overtone levels, since they are also obtained from symmetrized direct products.

Example 7.3.1

As a simple example consider a point group with non-degenerate representations such as the D_{2h} point group. The character table is given in Appendix I. The three translations t_x, t_y and t_z belong to the irreducible representations B_{3u}, B_{2u} and B_{1u} respectively. Using

(7.3.15) and Table 7.5.3 we obtain

$$\Gamma^{xx} = \Gamma^x \otimes \Gamma^x = B_{3u} \otimes B_{3u} = A_{1g}$$

$$\Gamma^{yy} = \Gamma^y \otimes \Gamma^y = B_{2u} \otimes B_{2u} = A_{1g}$$

$$\Gamma^{zz} = \Gamma^z \otimes \Gamma^z = B_{1u} \otimes B_{1u} = A_{1g}$$

$$\Gamma^{xy} = \Gamma^x \otimes \Gamma^y = B_{3u} \otimes B_{2u} = B_{1g}$$

$$\Gamma^{xz} = \Gamma^x \otimes \Gamma^z = B_{3u} \otimes B_{1u} = B_{2g}$$

$$\Gamma^{yz} = \Gamma^y \otimes \Gamma^z = B_{2u} \otimes B_{1u} = B_{3g}$$

Since these results are independent of the molecule considered, being characteristic of the point group, the elements of $\boldsymbol{\alpha}$ are listed in the tables of Appendix I for the various representations.

Example 7.3.2

As a more complex example let us consider the point group D_{6h}. The translation t_z belongs to the irreducible representation A_{2u} while the two translations t_x and t_y belong to the degenerate representation E_{1u} (Appendix I). From equation (7.3.15) and using Table 7.5.3 for non-degenerate species and Table 7.6.3 for degenerate species, we have

$$\alpha_{xx}, \alpha_{yy}, \alpha_{xy} \qquad \Gamma^{xx} = [\Gamma^x \otimes \Gamma^x] = [E_{1u} \otimes E_{1u}] = A_{1g} + E_{2g}$$

$$\alpha_{zz} \qquad \Gamma^{zz} = \Gamma^z \otimes \Gamma^z = A_{2u} \otimes A_{2u} = A_{1g}$$

$$\alpha_{xz}, \alpha_{yz} \qquad \Gamma^{xz} = \Gamma^x \otimes \Gamma^z = E_{1u} \otimes A_{2u} = E_{1g}$$

From the first relationship we see that α_{xx}, α_{yy} and α_{xy} are divided among the two species A_{1g} and E_{2g}. In order to discover which linear combination of them belongs to the A_{1g} and which to the E_{2g} species we can write down, using equations (7.3.13), the elements of the $\mathbf{S}(\mathscr{R})$ matrix. Since the transformation of the translations t_x and t_y has the form

$$\begin{pmatrix} t'_x \\ t'_y \end{pmatrix} = \mathbf{D}_c(\mathscr{R}) \begin{pmatrix} t_x \\ t_y \end{pmatrix} = \begin{pmatrix} \cos\beta & \sin\beta \\ -\sin\beta & \cos\beta \end{pmatrix} \begin{pmatrix} t_x \\ t_y \end{pmatrix}$$

for each operation $\mathscr{R}$ of the group, we obtain from (7.3.13)

$$\begin{pmatrix} \alpha'_{xx} \\ \alpha'_{xy} \\ \alpha'_{yy} \end{pmatrix} = \begin{pmatrix} \cos^2\beta & 2\sin\beta\cos\beta & \sin^2\beta \\ -\sin\beta\cos\beta & \cos^2\beta - \sin^2\beta & \sin\beta\cos\beta \\ \sin^2\beta & -2\sin\beta\cos\beta & \cos^2\beta \end{pmatrix} \begin{pmatrix} \alpha_{xx} \\ \alpha_{xy} \\ \alpha_{yy} \end{pmatrix}$$

It is easily seen that if we replace the three components of $\boldsymbol{\alpha}$ by their linear combinations $(\alpha_{xx} + \alpha_{yy})$, α_{xy}, $(\alpha_{xx} - \alpha_{yy})$, the matrix $\mathbf{S}(\mathscr{R})$ is partitioned in blocks, the first forming an A_{1g} and the second an E_{2g} representation:

$$\begin{pmatrix} (\alpha_{xx} + \alpha_{yy})' \\ \alpha'_{xy} \\ (\alpha_{xx} - \alpha_{yy})' \end{pmatrix} = \left(\begin{array}{c|cc} 1 & 0 & 0 \\ \hline 0 & \cos^2\beta - \sin^2\beta & -\sin\beta\cos\beta \\ 0 & 4\sin\beta\cos\beta & \cos^2\beta - \sin^2\beta \end{array}\right) \begin{pmatrix} \alpha_{xx} + \alpha_{yy} \\ \alpha_{xy} \\ \alpha_{xx} - \alpha_{yy} \end{pmatrix}$$

We conclude, therefore, that $\alpha_{xx} + \alpha_{yy}$ belongs to the A_{1g} while α_{xy} and $\alpha_{xx} - \alpha_{yy}$ belong to the E_{2g} species.

7.4 The symmetry of fundamental levels

In the present section we shall investigate the symmetry behaviour of the vibrational wavefunctions and show how they can be classified according to the irreducible representations of the molecular point groups. The symmetry of the vibrational wavefunctions will be discussed using harmonic oscillator functions. It is important to notice that, even if the harmonic approximation is dropped, the results are equally valid for anharmonic wavefunctions and are thus of great generality. The reason for this is that the harmonic wavefunctions form a complete set of functions and the wavefunctions describing anharmonic motions can always be written as linear combinations of harmonic functions

$$\psi_{\text{anharm}} = \sum_i a_i \psi_i \tag{7.4.1}$$

where all the ψ_i which enter a given linear combination must have the same symmetry behaviour under all symmetry operations of the point group. According to Section (3.2), the vibrational wavefunction of a molecule has the general form

$$\psi_v = \mathcal{N}\left[\prod_{i=1}^{3N-6} H_{v_i}(\gamma_i^{\frac{1}{2}} Q_i)\right] \exp\left[-\tfrac{1}{2}\sum_i \gamma_i Q_i^2\right] \tag{7.4.2}$$

where $\gamma_i = \lambda_i^{\frac{1}{2}}/\hbar$ and $H_{v_i}(\gamma_i^{\frac{1}{2}} Q_i)$ are the Hermite polynomials of degree v_i. The ground state wavefunction is the product of $(3N - 6)$ wavefunctions of the type (3.2.3), where all quantum numbers v_i are equal to zero. Since $H_0(\gamma_i^{\frac{1}{2}} Q_i) = 1$, the ground state wavefunction has the form

$$\psi_0 = \mathcal{N} \exp\left(-\frac{\gamma_1 Q_1^2}{2} - \frac{\gamma_2 Q_2^2}{2} - \ldots - \frac{\gamma_{3N-6} Q_{3N-6}^2}{2}\right) \tag{7.4.3}$$

It is easily seen that this function remains unchanged under all symmetry operations of the group and therefore belongs to the totally symmetric irreducible representation. If the Q_i are non-degenerate, a symmetry operation $\mathscr{R}$ transforms Q_i into $\pm Q_i$ (equation 6.3.9) and thus leaves Q_i^2 invariant. If two normal coordinates Q_{ia} and Q_{ib} are degenerate ($\lambda_{ia} = \lambda_{ib}$), we need to investigate the symmetry behaviour of the term $\gamma_i(Q_{ia}^2 + Q_{ib}^2)/2$ which appears in the exponential factor of (7.4.3). We recall from Section 6.3 that the 2×2 block of the unitary matrix $\mathbf{D}(\mathscr{R})$ associated with the operation $\mathscr{R}$, which transforms the coordinates Q_{ia} and Q_{ib} into the rotated coordinates Q'_{ia} and Q'_{ib} has the form

$$\begin{pmatrix} D(\mathscr{R})_{aa} & D(\mathscr{R})_{ab} \\ D(\mathscr{R})_{ba} & D(\mathscr{R})_{bb} \end{pmatrix}$$

and its elements satisfy the following relations

$$\begin{aligned} D^2(\mathscr{R})_{aa} + D^2(\mathscr{R})_{ba} &= 1 \\ D^2(\mathscr{R})_{ab} + D^2(\mathscr{R})_{bb} &= 1 \\ D(\mathscr{R})_{aa}D(\mathscr{R})_{ab} + D(\mathscr{R})_{ba}D(\mathscr{R})_{bb} &= 0 \end{aligned} \tag{7.4.4}$$

Now, since

$$\begin{aligned} \mathscr{R}Q_{ia} &= D(\mathscr{R})_{aa}Q_{ia} + D(\mathscr{R})_{ab}Q_{ib} \\ \mathscr{R}Q_{ib} &= D(\mathscr{R})_{ba}Q_{ia} + D(\mathscr{R})_{bb}Q_{ib} \end{aligned} \tag{7.4.5}$$

we obtain, using (7.4.4) and (7.4.5)

$$\begin{aligned} \mathscr{R}(Q_{ia}^2 + Q_{ib}^2) &= (\mathscr{R}Q_{ia})^2 + (\mathscr{R}Q_{ib})^2 = [D(\mathscr{R})_{aa}^2 + D(\mathscr{R})_{ba}^2]Q_{ia}^2 \\ &\quad + [D(\mathscr{R})_{ab}^2 + D(\mathscr{R})_{bb}^2]Q_{ib}^2 + 2[D(\mathscr{R})_{aa}D(\mathscr{R})_{ab} + D(\mathscr{R})_{ba}D(\mathscr{R})_{bb}]Q_{ia}Q_{ib} = Q_{ia}^2 + Q_{ib}^2 \end{aligned} \tag{7.4.6}$$

thus proving that a degenerate pair of coordinates Q_{ia} and Q_{ib} always transforms so that the sum $Q_{ia}^2 + Q_{ib}^2$ is invariant under all symmetry operations of the group.

Using the same arguments it is simple to prove that a triplet of triply degenerate normal coordinates Q_{ia}, Q_{ib} and Q_{ic} transforms so that $Q_{ia}^2 + Q_{ib}^2 + Q_{ic}^2$ is invariant. We conclude, therefore, that ψ_0 is invariant under all symmetry operations of the point group, i.e.

$$\mathscr{R}\psi_0 = \psi_0 \tag{7.4.7}$$

and thus belongs to the totally symmetric irreducible representation.

The wavefunction ψ_k corresponding to a fundamental level $E_{(k)}$ for which the quantum number $v_k = 1$ and all others are zero, has the form

$$\psi_k = \mathscr{N} H_{v_k=1}(\gamma_k^{\frac{1}{2}} Q_k) \exp\left[-\tfrac{1}{2}\sum_i \gamma_i Q_i^2\right] = \mathscr{N} 2\gamma_k^{\frac{1}{2}} Q_k \exp\left[-\tfrac{1}{2}\sum_i \gamma_i Q_i^2\right] \tag{7.4.8}$$

Since we know from the preceding discussion that the exponential factor is invariant under the symmetry operations of the group, it follows that ψ_k transforms under a symmetry operation in exactly the same way as Q_k, i.e. if Q_k is non-degenerate

$$\mathscr{R}\psi_k = \chi_i(\mathscr{R})\psi_k \tag{7.4.9}$$

where $\chi_i(\mathscr{R})$ is the character of the irreducible representation to which Q_k belongs.

If Q_{ia} and Q_{ib} are two degenerate coordinates we have, using (7.4.5) and putting

$$A = \exp\left(-\tfrac{1}{2}\sum_i \gamma_i Q_i^2\right) \tag{7.4.10}$$

$$\begin{aligned} \mathscr{R}\psi_{ia} &= \mathscr{N} 2\gamma_i^{\frac{1}{2}}(\mathscr{R}Q_{ia})A = \mathscr{N} 2\gamma_i^{\frac{1}{2}}(D(\mathscr{R})_{aa}Q_{ia} + D(\mathscr{R})_{ab}Q_{ib})A = D(\mathscr{R})_{aa}\psi_{ia} + D(\mathscr{R})_{ab}\psi_{ib} \\ \mathscr{R}\psi_{ib} &= \mathscr{N} 2\gamma_i^{\frac{1}{2}}(\mathscr{R}Q_{ib})A = \mathscr{N} 2\gamma_i^{\frac{1}{2}}(D(\mathscr{R})_{ba}Q_{ia} + D(\mathscr{R})_{bb}Q_{ib})A = D(\mathscr{R})_{ba}\psi_{ia} + D(\mathscr{R})_{bb}\psi_{ib} \end{aligned} \tag{7.4.11}$$

We therefore conclude that the wavefunctions corresponding to a two-fold degenerate fundamental level form a basis of a two-dimensional irreducible representation of the group exactly like the corresponding normal coordinates. With the same arguments it is easy to show that the wavefunctions belonging to a *d*-fold degenerate level form a basis for a *d*-dimensional representation of the group.

7.5 The symmetry of combination levels

The rule for determining the representation carried by the wavefunction $\psi_{(i)(j)\dots(l)}$, corresponding to the energy level $E_{(i)(j)\dots(l)}$ for which m quantum numbers $v_i, v_j, \dots, v_l$ are equal to unity, all others being zero, is rather simple. If the normal coordinates Q_i, $Q_j, \dots, Q_l$ carry the irreducible representations $\Gamma^\alpha, \Gamma^\beta, \dots, \Gamma^\lambda$ respectively, then the wavefunction $\psi_{(i)(j)\dots(l)}$, which according to (7.4.2) has the form

$$\psi_{(i)(j)\dots(l)} = \mathscr{N}' 2^m(\gamma_i^{\frac{1}{2}} \cdot \gamma_j^{\frac{1}{2}} \dots \gamma_l^{\frac{1}{2}})(Q_i \cdot Q_j \dots Q_l)A = \mathscr{N}(Q_i \cdot Q_j \dots Q_l)A \tag{7.5.1}$$

carries the representation

$$\Gamma = \Gamma^\alpha \otimes \Gamma^\beta \otimes \dots \otimes \Gamma^\lambda$$

i.e. a representation which is the direct product of the representations associated with the normal coordinates $Q_i, Q_j, \dots, Q_l$ and thus, according to the previous section, also by the fundamental level wavefunctions $\psi_{(i)}, \psi_{(j)}, \dots, \psi_{(l)}$.

To prove this we shall limit ourselves to binary combinations, i.e. to the case $v_i = 1$, $v_j = 1$, all other $v_k = 0$. The extension to the general case is obvious. For a binary combination, (7.4.2) has the form

$$\psi_{(i)(j)} = \mathcal{N}' 4\gamma_i^{\frac{1}{2}}\gamma_j^{\frac{1}{2}} Q_i Q_j A = \mathcal{N} Q_i Q_j A \tag{7.5.2}$$

and three cases are possible.

(a) Q_i and Q_j are both non-degenerate. In this case $\psi_{(i)(j)}$ and $E_{(i)(j)}$ are also non-degenerate and $\psi_{(i)(j)}$ carries a non-degenerate irreducible representation Γ^γ. The effect of a symmetry operation $\mathcal{R}$ on $\psi_{(i)(j)}$ is easily obtained from (6.3.9). Since

$$\begin{aligned} \mathcal{R}Q_i &= \chi_r^\alpha Q_i \\ \mathcal{R}Q_j &= \chi_r^\beta Q_j \end{aligned} \tag{7.5.3}$$

we have

$$\mathcal{R}\psi_{(i)(j)} = \chi_r^\alpha \chi_r^\beta \psi_{(i)(j)} \tag{7.5.4}$$

The representation Γ^γ carried by $\psi_{(i)(j)}$ is thus the direct product

$$\Gamma^\gamma = \Gamma^\alpha \otimes \Gamma^\beta$$

of the irreducible representations carried by Q_i and Q_j.

(b) Q_i is non-degenerate while Q_j is two-fold degenerate. In this case $\psi_{(i)(j)}$ carries a two-fold degenerate irreducible representation also. If Q_j is three-fold degenerate, $\psi_{(i)(j)}$ carries a three-fold degenerate irreducible representation.

If Q_{ja} and Q_{jb} are the two degenerate normal coordinates (for simplicity we assume two-fold degeneracy, but the same arguments apply for three-fold degeneracy) we have two degenerate wavefunctions

$$\begin{aligned} \psi_{(i)(ja)} &= \mathcal{N} Q_i Q_{ja} A \\ \psi_{(i)(jb)} &= \mathcal{N} Q_i Q_{jb} A \end{aligned} \tag{7.5.5}$$

associated† with the level $E_{(i)(j)}$. According to (7.4.5) we have‡

$$\mathcal{R}\psi_{(i)(ja)} = \mathcal{N}\chi_r^\alpha Q_i (D_{aa}^\beta Q_{ja} + D_{ab}^\beta Q_{jb})A = \chi_r^\alpha D_{aa}^\beta \psi_{(i)(ja)} + \chi_r^\alpha D_{ab}^\beta \psi_{(i)(jb)}$$

$$\mathcal{R}\psi_{(i)(jb)} = \mathcal{N}\chi_r^\alpha Q_i (D_{ba}^\beta Q_{ja} + D_{bb}^\beta Q_{jb})A = \chi_r^\alpha D_{ba}^\beta \psi_{(i)(ja)} + \chi_r^\alpha D_{bb}^\beta \psi_{(i)(jb)}$$

The degenerate wavefunctions $\psi_{(i)(ja)}$ and $\psi_{(i)(jb)}$ therefore carry a degenerate representation with matrices of the type

$$\begin{pmatrix} \chi_r^\alpha D_{aa}^\beta & \chi_r^\alpha D_{ab}^\beta \\ \chi_r^\alpha D_{ba}^\beta & \chi_r^\alpha D_{bb}^\beta \end{pmatrix} \tag{7.5.6}$$

for each operation $\mathcal{R}$. The matrix (7.5.6) has the character

$$\chi_r^\alpha [D_{aa}^\beta(\mathcal{R}) + D_{bb}^\beta(\mathcal{R})] \tag{7.5.7}$$

and is the outer product of the 1×1 matrix of the representation Γ^α and of the 2×2 matrix of the irreducible representation Γ^β, i.e.

$$\mathbf{D}^\gamma(\mathcal{R}) = \mathbf{D}^\alpha(\mathcal{R}) \otimes \mathbf{D}^\beta(\mathcal{R}) \tag{7.5.8}$$

The representation carried by $\psi_{(i)(ja)}$ and $\psi_{(i)(jb)}$ is thus again the direct product of Γ^α and Γ^β.

† One could also say, more rigorously, that the two levels $E_{(i)(ja)}$ and $E_{(i)(jb)}$ are degenerate.
‡ For simplicity $D_{aa}^\beta(\mathcal{R})$ etc. will be written D_{aa}^β etc., unless it is necessary to specify the dependence on $\mathcal{R}$.

(c) Q_i and Q_j are both degenerate. If we call Q_{ia} and Q_{ib}, Q_{ja} and Q_{jb} the two pairs of two-fold degenerate normal coordinates (again we shall limit ourselves to two-fold degeneracy) we have four degenerate wavefunctions $\psi_{(ia)(ja)}$, $\psi_{(ia)(jb)}$, $\psi_{(ib)(ja)}$ and $\psi_{(ib)(jb)}$ associated with the level $E_{(i)(j)}$. The representation carried by these four functions is not irreducible, but its structure can be determined using equation (5.7.26)

$$n^\gamma = \frac{1}{g}\sum_r (\chi_r^\gamma)^* \chi_r \tag{7.5.9}$$

once the character of the representation has been found.

Using (7.4.5) and (7.5.2) we find for $\psi_{(ia)(ja)}$

$$\begin{aligned}\mathscr{R}\psi_{(ia)(ja)} &= \mathscr{N}(\mathscr{R}Q_{ia})(\mathscr{R}Q_{ja}) = \mathscr{N}(D^\alpha_{aa}Q_{ia} + D^\alpha_{ab}Q_{ib})(D^\beta_{aa}Q_{ja} + D^\beta_{ab}Q_{jb})A \\ &= D^\alpha_{aa}D^\beta_{aa}\psi_{(ia)(ja)} + D^\alpha_{aa}D^\beta_{ab}\psi_{(ia)(jb)} + D^\alpha_{ab}D^\beta_{aa}\psi_{(ib)(ja)} + D^\alpha_{ab}D^\beta_{ab}\psi_{(ib)(jb)}\end{aligned}$$

with similar relations for the other three functions.

We find therefore that the four functions carry a representation with matrices (for each operation $\mathscr{R}$)

$$\begin{pmatrix} D^\alpha_{aa}D^\beta_{aa} & D^\alpha_{aa}D^\beta_{ab} & D^\alpha_{ab}D^\beta_{aa} & D^\alpha_{ab}D^\beta_{ab} \\ D^\alpha_{aa}D^\beta_{ba} & D^\alpha_{aa}D^\beta_{bb} & D^\alpha_{ab}D^\beta_{ba} & D^\alpha_{ab}D^\beta_{bb} \\ D^\alpha_{ba}D^\beta_{aa} & D^\alpha_{ba}D^\beta_{ab} & D^\alpha_{bb}D^\beta_{aa} & D^\alpha_{bb}D^\beta_{ab} \\ D^\alpha_{ba}D^\beta_{ba} & D^\alpha_{ba}D^\beta_{bb} & D^\alpha_{bb}D^\beta_{ba} & D^\alpha_{bb}D^\beta_{bb} \end{pmatrix} \tag{7.5.10}$$

which are the outer product of the matrices $\mathbf{D}^\alpha(\mathscr{R})$ and $\mathbf{D}^\beta(\mathscr{R})$ of the representations Γ^α and Γ^β. Again, therefore, the representation carried by the four degenerate wavefunctions is the direct product of Γ^α and Γ^β. The character of the matrices of the type (7.5.10) is

$$\chi_r = (D^\alpha_{aa}(\mathscr{R}) + D^\alpha_{bb}(\mathscr{R}))(D^\beta_{aa}(\mathscr{R}) + D^\beta_{bb}(\mathscr{R}))$$

i.e. is the product of the characters of the matrices of the Γ^α and Γ^β representations. It can be used, in equation (7.5.9), for the determination of the structure of the representation.

Some examples will now be discussed, to clarify the content of this section.

Example 7.5.1

Let us consider the energy levels of SO_2, shown in Figure 3.3.1. The character table for the C_{2v} point group is reproduced in Table 7.5.1. The normal coordinates Q_1 and Q_2 belong to the A_1 and the normal coordinate Q_3 belongs to the B_1 irreducible representation of C_{2v}. The wavefunctions $\psi_{(1)}$ and $\psi_{(2)}$ thus belong to the A_1 species and the wavefunction $\psi_{(3)}$ belongs to the B_1 species.

The symmetry of the wavefunction $\psi_{(1)(2)}$ is then

$$A_1 \otimes A_1 = A_1$$

the symmetry of the wavefunction $\psi_{(1)(3)}$ is

$$A_1 \otimes B_1 = B_1$$

the symmetry of the wavefunction $\psi_{(2)(3)}$ is

$$A_1 \otimes B_1 = B_1$$

and the symmetry of the wavefunction $\psi_{(1)(2)(3)}$ is

$$A_1 \otimes A_1 \otimes B_1 = B_1$$

as can be easily obtained from Table 7.5.1 by direct multiplication of the characters of the representations.

Table 7.5.1

C_{2v}	E	C_2	σ_1	σ_2
A_1	1	1	1	1
A_2	1	1	−1	−1
B_1	1	−1	1	−1
B_2	1	−1	−1	1
$A_1 \otimes A_1, A_2 \otimes A_2, B_1 \otimes B_1, B_2 \otimes B_2$	1	1	1	1
$A_1 \otimes A_2, B_1 \otimes B_2$	1	1	−1	−1
$A_1 \otimes B_1, A_2 \otimes B_2$	1	−1	1	−1
$A_1 \otimes B_2, A_2 \otimes B_1$	1	−1	−1	1

Example 7.5.2

As a more complex example let us consider the combination levels of methane. Methane belongs to the T_d point group (Table 7.5.2). The symmetry classification of the normal coordinates is as follows. Q_1 belongs to the species A_1, Q_2 to the species E, Q_3 and Q_4 to the species F_2.

Table 7.5.2
Character table of the T_d group

	E	$8C_3$	$6\sigma_v$	$6S_4$	$3C_2$
A_1	1	1	1	1	1
A_2	1	1	−1	−1	1
E	2	−1	0	0	2
F_1	3	0	−1	1	−1
F_2	3	0	1	−1	−1

We therefore obtain the following level classification

	A_1	A_2	E	F_1	F_2
Fundamental Levels	$E_{(1)}$		$E_{(2)}$		$E_{(3)}, E_{(4)}$
Binary Combination Levels			$E_{(1)(2)}$		$E_{(1)(3)}, E_{(1)(4)}$
				$E_{(2)(3)}$	
				$E_{(2)(4)}$	
	$E_{(3)(4)}$			$E_{(3)(4)}$	
Ternary Combination Levels				$E_{(1)(2)(3)}$	
				$E_{(1)(2)(4)}$	
	$E_{(1)(3)(4)}$			$E_{(1)(3)(4)}$	
	$E_{(2)(3)(4)}$			$2E_{(2)(3)(4)}$	

For instance, let us investigate the structure of the representation carried by the nine degenerate wavefunctions

$$\psi_{(3a)(4a)} \quad \psi_{(3a)(4b)} \quad \psi_{(3a)(4c)} \quad \psi_{(3b)(4a)} \quad \psi_{(3b)(4b)} \quad \psi_{(3b)(4c)} \quad \psi_{(3c)(4a)} \quad \psi_{(3c)(4b)} \quad \psi_{(3c)(4c)}$$

associated with the combination level $E_{(3)(4)}$. The representation Γ carried by these nine functions is given, according to the general rule, by the direct product

$$\Gamma = \Gamma^{F_2} \otimes \Gamma^{F_2}$$

The characters of Γ are the products of the characters of F_2 by themselves. We thus have

	E	$8C_3$	$6\sigma_v$	$6S_4$	$3C_2$
$\chi(F_2)$	3	0	1	-1	-1
$\chi(\Gamma)$	9	0	1	1	1

Using (7.5.9) we find

$$n^{A_1} = \tfrac{1}{24}(1.1.9 + 8.1.0 + 6.1.1 + 6.1.1 + 3.1.1) = 1$$

$$n^{A_2} = \tfrac{1}{24}(1.1.9 + 8.1.0 - 6.1.1 - 6.1.1 + 3.1.1) = 0$$

$$n^{E} = \tfrac{1}{24}(2.1.9 - 8.1.0 + 6.0.1 + 6.0.1 + 3.2.1) = 1$$

$$n^{F_1} = \tfrac{1}{24}(3.1.9 + 8.0.0 - 6.1.1 + 6.1.1 - 3.1.1) = 1$$

$$n^{F_2} = \tfrac{1}{24}(3.1.9 + 8.0.0 + 6.1.1 - 6.1.1 - 3.1.1) = 1$$

The structure of Γ is therefore $A_1 + E + F_1 + F_2$.

In performing the direct product of two representations, the rules given in Table 7.5.3 can be used to simplify the calculations.

Example 7.5.3

For O_h we have, for instance, from Table 7.5.3

$$F_{1g} \otimes F_{2u} = A_{2u} + E_u + F_{1u} + F_{2u}$$

$$F_{2u} \otimes F_{2u} = A_{1g} + E_g + F_{1g} + F_{2g}$$

And for D_{6h}

$$E_{1g} \otimes E_{2u} = B_{1u} + B_{2u} + E_{1u}$$

$$E_{2u} \otimes E_{2u} = A_{1g} + A_{2g} + E_{2g}$$

7.6 The symmetry of overtone levels

The structure of the representation carried by wavefunctions of the type $\psi_{n(i)}$, corresponding to overtone levels $E_{n(i)}$ with $v_i > 1$, is simple for non-degenerate wavefunctions but can be very complicated for degenerate wavefunctions and high quantum numbers. If the wavefunction $\psi_{(i)}$ is non-degenerate, the wavefunctions $\psi_{n(i)}$ are also non-degenerate.

Table 7.5.3

$A \otimes A = A$	$A \otimes E = E$	$A \otimes F = F$	$g \otimes g = g$	$' \otimes ' = '$	$A \otimes E_1 = E_1$	$1 \otimes 1 = 1$
$A \otimes B = B$	$B \otimes E = E$	$B \otimes F = F$	$u \otimes u = g$	$'' \otimes '' = '$	$B \otimes E_1 = E_2$	$1 \otimes 2 = 2$
$B \otimes B = A$[a]			$g \otimes u = u$	$' \otimes '' = ''$	$A \otimes E_2 = E_2$	$2 \otimes 2 = 1$
					$B \otimes E_2 = E_1$	

Doubly degenerate representations[b]

$C_3, C_{3h}, C_{3v}, D_3, D_{3h}, D_{3d}, C_6, C_{6h}, C_{6v}, D_6, D_{6h}, S_6, O, O_h, T, T_d, T_h$	$E_1 \otimes E_1 = E_2 \otimes E_2 = A_1 + A_2 + E_2$ $E_1 \otimes E_2 = B_1 + B_2 + E_1$
$C_4, C_{4v}, C_{4h}, D_{2d}, D_4, D_{4h}, S_4$	$E \otimes E = A_1 + A_2 + B_1 + B_2$

Triply degenerate representations[c]

T_d, O, O_h, T, T_h	$E \otimes F_1 = E \otimes F_2 = F_1 + F_2$	
	$F_1 \otimes F_1 = F_2 \otimes F_2 = A_1 + E + F_1 + F_2$	
	$F_1 \otimes F_2 = A_2 + E + F_1 + F_2$	
$D_{\infty h}, C_{\infty v}$	$\Sigma^+ \otimes \Sigma^+ = \Sigma^- \otimes \Sigma^- = \Sigma^+$	$\Sigma^+ \otimes \Sigma^- = \Sigma^-$
	$\Sigma^+ \otimes \Pi = \Sigma^- \otimes \Pi = \Pi$	$\Sigma^+ \otimes \Delta = \Delta$
	$\Pi \otimes \Pi = \Sigma^+ + \Sigma^- + \Delta$	$\Sigma^- \otimes \Delta = \Delta$ etc.
	$\Delta \otimes \Delta = \Sigma^+ + \Sigma^- + \Gamma$	
	...	
	$\Pi \otimes \Delta = \Pi + \Phi$	

[a] For D_2 and D_{2h} groups, $B_i \otimes B_j = B_k$ unless $i = j$ $(i, j, k = 1, 2, 3)$.
[b] For the groups $C_3, C_4, C_{3h}, C_{4h}, S_4, S_6, T, T_h$ omit subscripts 1 and 2.
For the groups C_6, C_{6h}, omit subscripts 1 and 2 of A and B.
For the groups $D_{2d}, D_{3d}, D_{3h}, T_d, O, O_h, D_3, D_4, C_{3v}, C_{4v}, D_{4h}$ omit subscripts 1 and 2 of E.
For the groups C_{3h}, D_{3h} add ′ and ″ superscripts.
For the groups $C_{4h}, D_{4h}, D_{6h}, S_6, D_{3d}, O_h, T_h$, add g and u subscripts.
[c] For T and T_h drop subscripts 1 and 2. For T_h and O_h add subscripts u and g.

From equations (7.4.2) and (3.2.10) an overtone wavefunction has the form

$$\psi_{n(i)} = \mathcal{N} H_{v_i}(\gamma_i^{\frac{1}{2}} Q_i) A \tag{7.6.1}$$

where $H_{v_i}(\gamma_i^{\frac{1}{2}} Q_i)$ is the Hermite polynomial of degree v_i. From (3.2.12) and (3.2.13) it is easily seen that the Hermite polynomials are even functions of Q_i if v_i is even and odd functions if v_i is odd as shown in Table 7.6.1.

Now, if Q_i is a non-degenerate normal coordinate belonging to the Γ^α irreducible representation, such that

$$\mathscr{R} Q_i = \chi_r^\alpha Q_i \tag{7.6.2}$$

Table 7.6.1

v_i	$H_{v_i}(\gamma_i^{\frac{1}{2}}Q_i)$	v_i	$H_{v_i}(\gamma_i^{\frac{1}{2}}Q_i)$
1	$2\gamma_i^{\frac{1}{2}}Q_i$	2	$4\gamma_i Q_i^2 - 2$
3	$8\gamma_1^{\frac{3}{2}}Q_i^3 - 12\gamma_i^{\frac{1}{2}}Q_i$	4	$16\gamma_i^2 Q_i^4 - 48\gamma_i Q_i^2 + 12$
5	$32\gamma_i^{\frac{5}{2}}Q_i^5 - 160\gamma_i^{\frac{3}{2}}Q_i^3 + 120\gamma_i^{\frac{1}{2}}Q_i$	6	$64\gamma_i^3 Q_i^6 - 480\gamma_i^2 Q_i^4 + 720\gamma_i Q_i^2 - 120$

we have, for n even

$$\mathscr{R}\psi_{n(i)} = \psi_{n(i)} \tag{7.6.3}$$

and for n odd

$$\mathscr{R}\psi_{n(i)} = \chi_r^\alpha \psi_{n(i)} \tag{7.6.4}$$

Even overtones of non-degenerate fundamentals belong, therefore, to the totally symmetric irreducible representation Γ^0 of the molecular point group, whereas odd overtones belong to the same representation Γ^α to which the fundamental belongs.

The overtones of degenerate fundamentals are more tedious to handle. The structure of the representation is different from that obtained from combination wavefunctions, discussed in the previous section, for the following reason. Compare (assuming two-fold degeneracy) the number of independent wavefunctions associated with the combination level $E_{(i)(j)}$ ($v_i = 1$, $v_j = 1$) and the overtone level $E_{2(i)}$ ($v_i = 2$). If we call Q_{ia} and Q_{ib} the degenerate normal coordinates, we have in the former case four wavefunctions

$$\psi_{(ia)(ja)} \quad \psi_{(ia)(jb)} \quad \psi_{(ib)(ja)} \quad \psi_{(ib)(jb)}$$

but only three in the second

$$\psi_{2(ia)} \quad \psi_{(ia)(ib)} \quad \psi_{2(ib)}$$

The representation carried by the set of wavefunctions associated with the combination level ($v_i = 1$, $v_j = 1$) is thus formed by 4×4 matrices and can be obtained, as shown in the previous section, by forming the direct product of the representations carried separately by ψ_{ia}, ψ_{ib} and by ψ_{ja}, ψ_{jb}. In the case of the overtones the representation carried by the three wavefunctions is formed by 3×3 matrices and thus has a different structure.

For three-fold degeneracy the number of independent wavefunctions would obviously be 9 and 6 respectively for the two cases. The reader will realize that this is the same problem encountered in the discussion of the representations carried by the components of a symmetric tensor. The correct representation carried by overtone wavefunctions in the case of degeneracy is thus obtained by forming the symmetrized direct product of the representation carried by the fundamental wavefunctions by itself so many times as indicated by the value of the quantum number. We shall now discuss the problem for some specific examples.

Consider first the level $v_i = 2$ of a two-fold degenerate fundamental and for simplicity call Q_a and Q_b the degenerate pair of normal coordinates associated with this level. As discussed before we have three degenerate wavefunctions

$$\begin{aligned} \psi_{20} &= \mathscr{N}_{20}(4\gamma_i Q_a^2 - 2)A \\ \psi_{11} &= \mathscr{N}_{11}(4\gamma_i Q_a Q_b)A \\ \psi_{02} &= \mathscr{N}_{02}(4\gamma_i Q_b^2 - 2)A \end{aligned} \tag{7.6.5}$$

with the same energy. The effect of a symmetry operation $\mathscr{R}$ on these functions is simply obtained using (7.4.5). We have

$$\mathscr{R}\psi_{20} = \mathscr{N}_{20}[4\gamma_i(D_{aa}^2Q_a^2 + 2D_{aa}D_{ab}Q_aQ_b + D_{ab}^2Q_b^2) - 2]A$$

$$\mathscr{R}\psi_{11} = \mathscr{N}_{11}4\gamma_i[D_{aa}D_{ba}Q_a^2 + (D_{aa}D_{bb} + D_{ab}D_{ba})Q_aQ_b + D_{ab}D_{bb}Q_b^2]A \tag{7.6.6}$$

$$\mathscr{R}\psi_{02} = \mathscr{N}_{02}[4\gamma_i(D_{ba}^2Q_a^2 + 2D_{ba}D_{bb}Q_aQ_b + D_{bb}^2Q_b^2) - 2]A$$

and thus, since $\mathbf{D}(\mathscr{R})$ is unitary and since from (3.2.11) $\sqrt{2}\mathscr{N}_{20} = \sqrt{2}\mathscr{N}_{02} = \mathscr{N}_{11}$, we obtain

$$\mathscr{R}\psi_{20} = D_{aa}^2\psi_{20} + \sqrt{2}\,D_{aa}D_{ab}\psi_{11} + D_{ab}^2\psi_{02}$$

$$\mathscr{R}\psi_{11} = \sqrt{2}\,D_{aa}D_{ba}\psi_{20} + (D_{aa}D_{bb} + D_{ab}D_{ba})\psi_{11} + \sqrt{2}\,D_{ab}D_{bb}\psi_{02} \tag{7.6.7}$$

$$\mathscr{R}\psi_{02} = D_{ba}^2\psi_{20} + \sqrt{2}\,D_{ba}D_{bb}\psi_{11} + D_{bb}^2\psi_{02}$$

If we call $\chi_{v_i}(\mathscr{R})$ the character of the representation carried by the set of wavefunctions associated with the level $E_{n(v_i)}$, we have from the previous equation

$$\chi_2(\mathscr{R}) = D_{aa}^2 + D_{aa}D_{bb} + D_{ab}D_{ba} + D_{bb}^2 \tag{7.6.8}$$

It is worthwhile observing that the same result can be obtained using equation (7.3.14) which yields the character of a representation obtained by forming the symmetrized direct product of two representations. In fact the method outlined above can be used to generalize the results of equation (7.3.14). Consider, for instance, the level $v_i = 3$ of the same two-fold degenerate fundamental. In this case we have four degenerate wavefunctions

$$\psi_{30} = \mathscr{N}_{30}(8\gamma_i^{\frac{3}{2}}Q_a^3 - 12\gamma_i^{\frac{1}{2}}Q_a)A = A_{30}Q_a^3 - B_{30}Q_a$$

$$\psi_{21} = \mathscr{N}_{21}[(4\gamma_iQ_a^2 - 2)2\gamma_i^{\frac{1}{2}}Q_b]A = A_{21}Q_a^2Q_b - B_{21}Q_b$$

$$\psi_{12} = \mathscr{N}_{12}[2\gamma_i^{\frac{1}{2}}Q_a(4\gamma_iQ_b^2 - 2)]A = A_{12}Q_aQ_b^2 - B_{12}Q_a \tag{7.6.9}$$

$$\psi_{03} = \mathscr{N}_{03}(8\gamma_i^{\frac{3}{2}}Q_b^3 - 12\gamma_i^{\frac{1}{2}}Q_b)A = A_{03}Q_b^3 - B_{03}Q_b$$

associated with this level. The effect of a symmetry operation $\mathscr{R}$ on these functions can be obtained using (7.4.5). For instance, we have for ψ_{30}

$$\mathscr{R}\psi_{30} = D_{aa}^3A_{30}Q_a^3 + 3D_{aa}^2D_{ab}A_{30}Q_a^2Q_b + 3D_{aa}D_{ab}^2A_{30}Q_aQ_b^2 + D_{ab}^3A_{30}Q_b^3$$
$$-D_{aa}B_{30}Q_a - D_{ab}B_{30}Q_b \tag{7.6.10}$$

where

$$A_{30} = \mathscr{N}_{30}8\gamma_i^{\frac{3}{2}}A, \qquad B_{30} = \mathscr{N}_{30}12\gamma_i^{\frac{1}{2}}A, \text{ etc.} \tag{7.6.11}$$

Since $\mathbf{D}(\mathscr{R})$ is unitary, we have

$$D_{aa}^2 + D_{ab}^2 = 1$$

and, multiplying by D_{aa} or by D_{ab}, we obtain respectively

$$D_{aa}^3 + D_{aa}D_{ab}^2 = D_{aa}$$
$$D_{ab}D_{aa}^2 + D_{ab}^3 = D_{ab} \tag{7.6.12}$$

By substitution of (7.6.12) in the last two terms of (7.6.10) and making use of the following relations

$$\sqrt{3}\,\mathscr{N}_{30} = \sqrt{3}\,\mathscr{N}_{03} = \mathscr{N}_{21} = \mathscr{N}_{12}$$

$$\sqrt{3}\,A_{30} = A_{21} = A_{12} = \sqrt{3}\,A_{03} \qquad B_{30} = \sqrt{3}\,B_{21} = \sqrt{3}\,B_{12} = B_{03}$$

easily obtained from (3.2.11) and (7.6.9), we obtain

$$\mathscr{R}\psi_{30} = D^3_{aa}\psi_{30} + \sqrt{3}\, D^2_{aa}D_{ab}\psi_{21} + \sqrt{3}\, D_{aa}D^2_{ab}\psi_{12} + D^3_{ab}\psi_{03} \qquad (7.6.13)$$

In the same way we can obtain the effect of $\mathscr{R}$ on ψ_{21}, ψ_{12} and ψ_{30}. The complete transformation has the form

$$\begin{pmatrix}\psi_{30}\\ \psi_{21}\\ \psi_{12}\\ \psi_{03}\end{pmatrix} = \begin{pmatrix} D^3_{aa} & \sqrt{3}\, D^2_{aa}D_{ab} & \sqrt{3}\, D_{aa}D^2_{ab} & D^3_{ab} \\ \sqrt{3}\, D^2_{aa}D_{ba} & D_{aa}(2D_{ba}D_{ab} + D_{bb}D_{aa}) & D_{ab}(2D_{aa}D_{bb} + D_{ab}D_{ba}) & \sqrt{3}\, D^2_{ab}D_{bb} \\ \sqrt{3}\, D_{aa}D^2_{ba} & D_{ba}(2D_{aa}D_{bb} + D_{ab}D_{ba}) & D_{bb}(2D_{ab}D_{ba} + D_{aa}D_{bb}) & \sqrt{3}\, D_{ab}D^2_{bb} \\ D^3_{ba} & \sqrt{3}\, D^2_{ba}D_{bb} & \sqrt{3}\, D_{ba}D^2_{bb} & D^3_{bb} \end{pmatrix} \begin{pmatrix}\psi_{30}\\ \psi_{21}\\ \psi_{12}\\ \psi_{03}\end{pmatrix}$$

with character

$$\chi_3(\mathscr{R}) = D^3_{aa} + (2D_{ab}D_{ba} + D_{aa}D_{bb})(D_{aa} + D_{bb}) + D^3_{bb} \qquad (7.6.14)$$

The character $\chi_3(\mathscr{R})$ of the overtone $v_i = 3$ and the character $\chi_2(\mathscr{R})$ of the overtone $v_i = 2$ are related by the equation

$$\chi_3(\mathscr{R}) = \tfrac{1}{2}\chi_2(\mathscr{R})\chi(\mathscr{R}) + \chi(\mathscr{R}^3) \qquad (7.6.15)$$

where $\chi(\mathscr{R})$ and $\chi(\mathscr{R}^3)$ are the characters, for operations $\mathscr{R}$ and $\mathscr{R}^3$ respectively, of the irreducible representation carried by the fundamental wavefunctions. Equation (7.6.15) can be generalized without difficulty. The character of the representation carried by the set of wavefunctions associated to the $v_i = n$ overtone level of a two-fold degenerate fundamental is given by

$$\chi_n(\mathscr{R}) = \tfrac{1}{2}[\chi_{n-1}(\mathscr{R})\chi(\mathscr{R}) + \chi(\mathscr{R}^n)] \qquad (7.6.16)$$

Once the character of the transformation is known, it can be used for the determination of the structure of the reducible representation using, as usual, equation (7.5.9).

Example 7.6.1

Ammonia possesses two non-degenerate modes of A_1 species and two two-fold degenerate modes of E-species (Example 6.6.1). The overtone levels of the A_1 fundamentals are obviously all of A_1 species. The symmetry of the overtone levels of the degenerate E-modes can be obtained with the method outlined before, using the character system of Example 6.4.3. The results are summarized in Table 7.6.2.

Table 7.6.2

	E	$2C_3$	3σ	
$\mathscr{R}$	E	C_3	σ	
$\mathscr{R}^2$	E	C_3	E	
$\mathscr{R}^3$	E	E	σ	
$\chi(\mathscr{R})$	2	−1	0	
$\chi(\mathscr{R}^2)$	2	−1	2	
$\chi(\mathscr{R}^3)$	2	2	0	
$\chi_2(\mathscr{R})$	3	0	1	$A_1 + E$
$\chi_3(\mathscr{R})$	4	1	0	$A_1 + A_2 + E$

The overtones of triply degenerate fundamentals can be handled in much the same way. A recursion formula for the characters analogous to (7.6.16), can be obtained with the same procedure used for two-fold degenerate wavefunctions. In this case we have

$$\chi_n(\mathscr{R}) = \tfrac{1}{3}[2\chi(\mathscr{R})\chi_{n-1}(\mathscr{R}) + \tfrac{1}{2}(\chi(\mathscr{R}^2) - \chi(\mathscr{R})^2)\chi_{n-2}(\mathscr{R}) + \chi(\mathscr{R}^n)] \qquad (7.6.17)$$

In principle (7.6.16) and (7.6.17) can be used any time the structure of the representation carried by degenerate overtone wavefunctions is required. In practice these results can be put in the form of simple recipes such as those summarized in Table 7.6.3, for the most important groups.

7.7 The symmetry of a general level

The structure of the representation carried by the wavefunctions associated with the most general level $E_{a(i)b(j)c(k)\ldots}$ in which m quantum numbers are excited assuming the values $v_i = a$, $v_j = b$, $v_k = c$, etc. respectively, can be obtained from the methods of the two previous sections. Symbolically we can write

$$\Gamma = (\Gamma_1)^a \otimes (\Gamma_2)^b \otimes (\Gamma_3)^c \ldots \otimes (\Gamma_m)^n \qquad (7.7.1)$$

where $(\Gamma_i)^l$ means that a mode belonging to the Γ_i irreducible representation has been excited with quantum number $v = l$. The procedure to follow is to compute first the structure of the multiply excited levels according to the method of Section 7.6 and then to perform all the direct products in equation (7.7.1) according to the method of Section 7.5.

Example 7.7.1

Consider a molecule like sulphur hexafluoride which belongs to the O_h point group. The structure of the representation carried by the normal coordinates is

$$A_{1g} + E_g + 2F_{1u} + F_{2g} + F_{2u}$$

Consider now an excited level in which the A_{1g} mode is singly excited, the E_g mode is doubly excited, the F_{2g} mode is doubly excited and the F_{2u} mode is singly excited. In this case equation (7.7.1) reads

$$\Gamma = (A_{1g}) \otimes (E_g)^2 \otimes (F_{2g})^2 \otimes (F_{2u})$$

From Table 7.6.3 we obtain

$$\Gamma = (A_{1g}) \otimes (A_{1g} + E_g) \otimes (A_{1g} + E_g + F_{2g}) \otimes (F_{2u})$$

Since the totally symmetric species A_{1g} behaves like unity in the direct product, we have from Table 7.5.3

$$\Gamma = (A_{1g} + E_g + F_{2g}) \otimes (F_{2u}) + (E_g) \otimes (A_{1g} + E_g + F_{2g}) \otimes (F_{2u})$$

$$\Gamma = F_{2u} + F_{1u} + F_{2u} + A_{1u} + E_u + F_{1u} + F_{2u} + (E_g + A_{1g} + A_{2g} + E_g + F_{1g} + F_{2g}) \otimes F_{2u}$$

$$\Gamma = A_{1u} + E_u + 2F_{1u} + 3F_{2u} + F_{1u} + F_{2u} + F_{2u} + F_{1u} + F_{1u} + F_{2u} + A_{2u} + E_u + F_{1u} + F_{2u}$$
$$+ A_{1u} + E_u + F_{1u} + F_{2u}$$

$$\Gamma = 2A_{1u} + A_{2u} + 3E_u + 7F_{1u} + 8F_{2u}$$

Table 7.6.3 Overtones of degenerate fundamentals

Level	Resulting states	Level	Resulting states
$(\Gamma_3)^2$	$\Gamma_1 + \Gamma_3$	$(\Gamma_6)^2$	$\Gamma_1 + \Gamma_3$
$(\Gamma_3)^3$	$\Gamma_1 + \Gamma_2 + \Gamma_3$	$(\Gamma_6)^3$	$\Gamma_4 + \Gamma_5 + \Gamma_6$
$(\Gamma_3)^4$	$\Gamma_1 + 2\Gamma_3$	$(\Gamma_6)^4$	$\Gamma_1 + 2\Gamma_3$
$(\Gamma_3)^5$	$\Gamma_1 + \Gamma_2 + 2\Gamma_3$	$(\Gamma_6)^5$	$\Gamma_4 + \Gamma_5 + 2\Gamma_6$
$(\Gamma_3)^6$	$2\Gamma_1 + \Gamma_2 + 2\Gamma_3$	$(\Gamma_6)^6$	$2\Gamma_1 + \Gamma_2 + 2\Gamma_3$

	T, C_3	D_3, C_{3v} O, T_d	C_{3h}	D_{3h}	T_h, S_6	D_{3d}, O_h	D_6, C_{6v}	D_{6h} [a]	C_{6h} [a]	C_6
Γ_1	A	A_1	A'	A_1'	A_g	A_{1g}	A_1	A_{1g}	A_g	A
Γ_2	A	A_2	A'	A_2'	A_g	A_{2g}	A_2	A_{2g}	A_g	A
Γ_3	E	E	E	E	E_g	E_g	E_2	E_{2g}	E_{2g}	E_2
Γ_4	A	A_1	A'	A_1''	A_u	A_{1u}	B_1	B_{1g}	B_g	B
Γ_5	A	A_2	A'	A_2''	A_u	A_{2u}	B_2	B_{2g}	B_g	B
Γ_6	E	E	E'	E''	E_u	E_u	E_1	E_{1g}	E_{1g}	E_1

Level	Resulting states	Level	Resulting states
$(\Gamma_3)^2$	$\Gamma_1 + \Gamma_4 + \Gamma_5$	$(\Gamma_6)^2$	$\Gamma_1 + \Gamma_4 + \Gamma_5$
$(\Gamma_3)^3$	$2\Gamma_3$	$(\Gamma_6)^3$	$2\Gamma_6$
$(\Gamma_3)^4$	$2\Gamma_1 + \Gamma_2 + \Gamma_4 + \Gamma_5$	$(\Gamma_6)^4$	$2\Gamma_1 + \Gamma_2 + \Gamma_4 + \Gamma_5$
$(\Gamma_3)^5$	$3\Gamma_3$	$(\Gamma_6)^5$	$3\Gamma_6$
$(\Gamma_3)^6$	$2\Gamma_1 + \Gamma_2 + 2\Gamma_4 + 2\Gamma_5$	$(\Gamma_6)^6$	$2\Gamma_1 + \Gamma_2 + 2\Gamma_4 + 2\Gamma_5$

	D_{4h}	D_4, C_{4v}	D_{2d}	C_{4h}	S_4, C_4
Γ_1	A_{1g}	A_1	A_{1g}	A_g	A
Γ_2	A_{2g}	A_2	A_{2g}	B_g	B
Γ_3	E_g	E	E_g	E_g	E
Γ_4	B_{1g}	B_1	A_{1u}	A_u	A
Γ_5	B_{2g}	B_2	A_{2u}	B_u	B
Γ_6	E_u	E	E_u	E_u	E

Level	Resulting states	Level	Resulting states	Level	Resulting states	Level	Resulting states
$(\Gamma_3)^2$	$\Gamma_1 + \Gamma_4$	$(\Gamma_4)^2$	$\Gamma_1 + \Gamma_4$	$(\Gamma_7)^2$	$\Gamma_1 + \Gamma_4$	$(\Gamma_8)^2$	$\Gamma_1 + \Gamma_4$
$(\Gamma_3)^3$	$\Gamma_3 + \Gamma_4$	$(\Gamma_4)^3$	$\Gamma_3 + \Gamma_4$	$(\Gamma_7)^3$	$\Gamma_7 + \Gamma_8$	$(\Gamma_8)^3$	$\Gamma_7 + \Gamma_8$
$(\Gamma_3)^4$	$\Gamma_1 + \Gamma_3 + \Gamma_4$	$(\Gamma_4)^4$	$\Gamma_1 + \Gamma_3 + \Gamma_4$	$(\Gamma_7)^4$	$\Gamma_1 + \Gamma_3 + \Gamma_4$	$(\Gamma_8)^4$	$\Gamma_1 + \Gamma_3 + \Gamma_4$
$(\Gamma_3)^5$	$\Gamma_1 + \Gamma_2 + \Gamma_3 + \Gamma_4$	$(\Gamma_4)^5$	$\Gamma_1 + \Gamma_2 + \Gamma_3 + \Gamma_4$	$(\Gamma_7)^5$	$\Gamma_5 + \Gamma_6 + \Gamma_7 + \Gamma_8$	$(\Gamma_8)^5$	$\Gamma_5 + \Gamma_6 + \Gamma_7 + \Gamma_8$
$(\Gamma_3)^6$	$\Gamma_1 + 2\Gamma_3 + \Gamma_4$	$(\Gamma_4)^6$	$\Gamma_1 + 2\Gamma_4 + \Gamma_3$	$(\Gamma_7)^6$	$\Gamma_1 + 2\Gamma_3 + \Gamma_4$	$(\Gamma_8)^6$	$\Gamma_1 + \Gamma_3 + 2\Gamma_4$

	D_{5h}	D_5, C_{5v}	C_{5h}	C_5
Γ_1	A_1'	A_1	A'	A
Γ_2	A_2'	A_2	A'	A
Γ_3	E_1'	E_1	E_1'	E_1
Γ_4	E_2'	E_2	E_2'	E_2
Γ_5	A_1''	A_1	A''	A
Γ_6	A_2''	A_2	A''	A
Γ_7	E_1''	E_1	E_1''	E_1
Γ_8	E_2''	E_2	E_2''	E_2

Level	Resulting states
$(\Gamma)^2$	$\Gamma_1 + \Gamma_3$
$(\Gamma)^3$	$\Gamma_2 + \Gamma_4$
$(\Gamma)^4$	$\Gamma_1 + \Gamma_3 + \Gamma_5$
$(\Gamma)^5$	$\Gamma_2 + \Gamma_4 + \Gamma_6$
$(\Gamma)^6$	$\Gamma_1 + \Gamma_3 + \Gamma_5 + \Gamma_7$

	$D_{\infty h}$[b]	$C_{\infty v}$
Γ_1	Σ_g^+	Σ^+
Γ_2	Π_g	Π
Γ_3	Δ_g	Δ
Γ_4	Φ_g	Φ
Γ_5	Γ_g	Γ
Γ_6	H_g	H
Γ_7	I_g	I

Level	Resulting states	Level	Resulting states
$(\Gamma_4)^2$	$\Gamma_1 + \Gamma_3 + \Gamma_5$	$(\Gamma_5)^2$	$\Gamma_1 + \Gamma_3 + \Gamma_5$
$(\Gamma_4)^3$	$\Gamma_2 + 2\Gamma_4 + \Gamma_5$	$(\Gamma_5)^3$	$\Gamma_1 + \Gamma_4 + 2\Gamma_5$
$(\Gamma_4)^4$	$2\Gamma_1 + 2\Gamma_3 + \Gamma_4 + 2\Gamma_5$	$(\Gamma_5)^4$	$2\Gamma_1 + 2\Gamma_3 + \Gamma_4 + 2\Gamma_5$
$(\Gamma_4)^5$	$\Gamma_2 + \Gamma_3 + 4\Gamma_4 + 2\Gamma_5$	$(\Gamma_5)^5$	$\Gamma_1 + \Gamma_3 + 2\Gamma_4 + 4\Gamma_5$
$(\Gamma_4)^6$	$3\Gamma_1 + \Gamma_2 + 3\Gamma_3 + 2\Gamma_4 + 4\Gamma_5$	$(\Gamma_5)^6$	$3\Gamma_1 + \Gamma_2 + 3\Gamma_3 + 2\Gamma_4 + 4\Gamma_5$

	O_h[a]	O, T_d	T_h[a]	T
Γ_1	A_{1g}	A_1	A_g	A
Γ_2	A_{2g}	A_2	A_g	A
Γ_3	E_g	E	E_g	E
Γ_4	F_{1g}	F_1	F_g	F
Γ_5	F_{2g}	F_2	F_g	F

[a] For the $(\Gamma)^v$ overtone levels of u states for O_h, T_h, D_{6h} and C_{6h} groups, change the subscript g into u for odd values of v only.

[b] For the overtone levels $(\Pi_u)^v$ of $D_{\infty h}$, change the subscript g into u for odd values of v.

7.8 Infrared selection rules

The symmetry selection rules for the infrared spectrum are based, as discussed in Section 7.1, on the property that the matrix elements

$$(\mu_\rho)_{mn} = \langle \psi_m | \mu_\rho | \psi_n \rangle \qquad \rho = x, y, z \tag{7.8.1}$$

are invariant under symmetry operations and in general under any coordinate transformation. This simply reflects the fact that the numerical value of a physical observable is independent of the coordinate system used.

The invariance of (7.8.1) under all symmetry operations, requires that for any operation $\mathscr{R}$ of the group

$$\mathscr{R}(\mu_\rho)_{mn} = (\mu_\rho)_{mn} \tag{7.8.2}$$

In order to fulfil this condition, the matrix elements (7.8.1) can have a non-vanishing value only if they carry the totally symmetric irreducible representation Γ^0 of the group or a reducible representation which contains Γ^0 in its structure.

Since we know, from the previous sections, how to determine the representations Γ^{ψ_m}, Γ^{μ_ρ} and Γ^{ψ_n} carried separately by ψ_m, μ_ρ and by ψ_n, we have no difficulty in determining the representation Γ carried by $(\mu_\rho)_{mn}$, by forming the direct product

$$\Gamma = \Gamma^{\psi_m} \otimes \Gamma^{\mu_\rho} \otimes \Gamma^{\psi_n} \tag{7.8.3}$$

If Γ is not identical with Γ^0 (or does not contain Γ^0, if reducible) there must be in the group at least one operation for which

$$\mathscr{R}(\mu_\rho)_{mn} = \chi(\mathscr{R})(\mu_\rho)_{mn} \tag{7.8.4}$$

with $\chi(\mathscr{R}) \neq 1$, this being obviously possible only if $(\mu_\rho)_{mn} = 0$. Transitions for which $(\mu_\rho)_{mn} = 0$ by symmetry are called infrared inactive or forbidden. If instead the symmetry puts no restrictions on the value of $(\mu_\rho)_{mn}$ the transitions are called infrared active or allowed.

From (7.8.3) we obtain that a transition is allowed only if the direct product

$$\Gamma^{\psi_m} \otimes \Gamma^{\psi_n}$$

is equal to Γ^{μ_ρ} or contains (if reducible) Γ^{μ_ρ} in its structure. Since the ground vibrational wavefunction always belongs to the totally symmetric representation of the group, transitions involving the ground state (fundamentals, combinations and overtones) are then active in the infrared spectrum only if the excited level carries the same representation as one of the components of μ, i.e. the same representation as one of the translations.

Example 7.8.1

Ethylene belongs to the D_{2h} point group and its 12 fundamental levels are distributed among the irreducible representations of D_{2h} as

$$3A_g + 2B_{1g} + B_{2g} + A_u + B_{1u} + 2B_{2u} + 2B_{3u}$$

From the character table of the D_{2h} group (Appendix I) we see that the translations t_x, t_y and t_z belong to the representations B_{3u}, B_{2u} and B_{1u} respectively. We conclude therefore that only five fundamentals of ethylene (one B_{1u}, two B_{2u} and two B_{3u}) are active in the infrared. The reader can verify that use of the general formula (7.8.3) leads to the same result. For simplicity we list the direct products (7.8.3) for the μ_x component.

$$A_g \quad A_g \otimes B_{3u} \otimes A_g = B_{3u} \quad \text{inactive}$$

$$B_{1g} \quad B_{1g} \otimes B_{3u} \otimes A_g = B_{2u} \quad \text{inactive}$$

B_{2g}	$B_{2g} \otimes B_{3u} \otimes A_g = B_{1u}$	inactive
A_u	$A_u \otimes B_{3u} \otimes A_g = B_{3g}$	inactive
B_{1u}	$B_{1u} \otimes B_{3u} \otimes A_g = B_{2g}$	inactive
B_{2u}	$B_{2u} \otimes B_{3u} \otimes A_g = B_{1g}$	inactive
B_{3u}	$B_{3u} \otimes B_{3u} \otimes A_g = A_g$	active

Example 7.8.2

Methane has four fundamental levels, one of symmetry A_1, one of symmetry E and two of symmetry F_2. Since the three translations belong to the F_2 species, only transitions from the ground state to the two F_2 levels will be infrared active. The two fundamentals of species A_1 and E are inactive.

Consider now the combination levels given in Example 7.5.2. The level $E_{(1)(2)}$ is of species E and therefore the combination $\nu_1 + \nu_2$ is inactive in the infrared. All other combinations are infrared active since they belong either to $F_2(E_{(1)(2)}, E_{(1)(4)})$ or to a reducible representation which contains F_2.

As far as overtone levels are concerned we can use Table 7.6.3 to obtain the representations carried by the overtone wavefunctions. These are

ν_1	A_1	ν_2	E	$\nu_i = \nu_3, \nu_4$	F_2
$2\nu_1$	A_1	$2\nu_2$	$A_1 + E$	$2\nu_i$	$A_1 + E + F_2$
$3\nu_1$	A_1	$3\nu_2$	$A_1 + A_2 + E$	$3\nu_i$	$A_1 + F_1 + 2F_2$
$4\nu_1$	A_1	$4\nu_2$	$A_1 + 2E$	$4\nu_i$	$2A_1 + 2E + F_1 + 2F_2$
$5\nu_1$	A_1	$5\nu_2$	$A_1 + A_2 + 2E$	$5\nu_i$	$A_1 + E + 2F_1 + 4F_2$

Thus all overtones of ν_1 and ν_2 are infrared inactive and all overtones of ν_3 and ν_4 are infrared active.

7.9 The Raman selection rules

The symmetry selection rules for the Raman spectrum are based on the invariance of the matrix elements

$$\langle \psi_m | \alpha_{\rho\sigma} | \psi_n \rangle \qquad \rho, \sigma = x, y, z \tag{7.9.1}$$

under symmetry operations. These matrix elements can be discussed in exactly the same way as the matrix elements which control the infrared transitions. The invariance of (7.9.1) can be stated in terms of group theory by saying that a matrix element can have a non-vanishing value only if it carries a representation which contains in its structure the totally symmetric representation of the group. In order to fulfil this condition, expressed as usual in the form

$$\Gamma^{\psi_m} \otimes \Gamma^{\alpha_{\rho\sigma}} \otimes \Gamma^{\psi_n} \tag{7.9.2}$$

the direct product $\Gamma^{\psi_m} \otimes \Gamma^{\psi_n}$ must be equal to $\Gamma^{\alpha_{\rho\sigma}}$ or contain it in its structure.

Again then, transitions involving the ground state will be active in Raman only if the excited level belongs to a representation carried by one of the components of $\boldsymbol{\alpha}$.

Example 7.9.1

For the D_{2h} group, the components of $\boldsymbol{\alpha}$ (see Appendix I) belong to the species A_{1g}, B_{1g}, B_{2g} and B_{3g}. Therefore ethylene has only six Raman active fundamentals ($3A_g + 2B_{1g} + B_{2g}$).

Example 7.9.2

For the T_d group, the components of $\boldsymbol{\alpha}$ belong to the species A_1, E and F_2. The fundamentals of methane are thus Raman active; all combinations are active in the Raman spectrum. All overtones are also Raman active.

Chapter 8
Potential Functions

8.1 Types of potential function

In this chapter we shall discuss in some detail the validity of the harmonic model discussed in the previous chapters and present a critical survey of the large body of force-constant calculations accumulated until now in this approximation. Furthermore, we shall illustrate the various problems arising in the handling of the experimental data and in the calculation of significant force fields. Finally, we shall present some recent calculations of force constants from the electronic structure of the molecule.

The potential energy of a molecule is usually written as a Taylor expansion in terms of a set of displacement coordinates (see Chapter 2, Section 3)

$$V = V_0 + \sum_i f_i s_i + \tfrac{1}{2}\sum_{i,j} f_{ij} s_i s_j + \tfrac{1}{6}\sum_{i,j,k} f_{ijk} s_i s_j s_k + \tfrac{1}{24}\sum_{ijkl} f_{ijkl} s_i s_j s_k s_l + \ldots \tag{8.1.1}$$

where the constants

$$f_i = \left(\frac{\partial V}{\partial s_i}\right)_0 \qquad f_{ij} = \left(\frac{\partial^2 V}{\partial s_i\,\partial s_j}\right)_0 \qquad f_{ijk} = \left(\frac{\partial^3 V}{\partial s_i\,\partial s_j\,\partial s_k}\right)_0 \qquad f_{ijkl} = \left(\frac{\partial^4 V}{\partial s_i\,\partial s_j\,\partial s_k\,\partial s_l}\right)_0$$

are called linear, quadratic, cubic, quartic, etc., force constants and the s_i designate any suitable set of displacement coordinates (Cartesian, internal, symmetry, etc.). As discussed in Chapter 2, the zero of the energy scale can always be chosen so that V_0 is zero. Furthermore, since the molecule in the equilibrium configuration must be at a minimum of energy, we have

$$\sum_i f_i s_i = 0 \tag{8.1.2}$$

It is worthwhile to point out that, if the coordinates s_i are independent, i.e. if none of them can be expressed as a function of some of the others, equation (8.1.2) implies that each f_i is equal to zero.

The number of cubic and quartic force constants is, even for a small molecule, very large and only in very few cases is it possible to collect enough spectroscopic data for their calculation. Fortunately, their contribution to the total energy (8.1.1) is small for small displacements, if compared to the quadratic terms. For these reasons they are generally neglected in the treatment of polyatomic molecules and the potential energy is assumed to be a pure quadratic form in the displacement coordinates, of the type

$$2V = \sum_{i,j} f_{ij} s_i s_j \tag{8.1.3}$$

In this case the potential function is called harmonic, since insertion of such a potential in the equations of motion leads to solutions which describe harmonic vibrations of the nuclei around their equilibrium positions (see Chapter 2).

If terms higher than the quadratic ones are retained in (8.1.1), the potential function is called anharmonic, since the equations of motion have in this case different solutions which describe anharmonic motions of the nuclei. We shall discuss the anharmonic types of potential in Chapter 9.

The harmonic force constants of a molecule can be calculated by the method described in Chapter 4. The greatest difficulty encountered in such calculations is that the number of force constants is in general much larger than the number of experimental data so that, depending upon the size of the molecule and the kind of data accessible by experiment, assumptions must be made on the form of the force field (8.1.3). The experimental data that can be used for the evaluation of force constants are

(1) Vibrational frequencies.
(2) Coriolis coupling constants.
(3) Centrifugal distortion constants.
(4) Mean-square amplitudes of vibration.

These different experimental data furnish different kinds of information not being equally sensitive to the different types of force constants. A critical evaluation of the experimental data is therefore of paramount importance in the determination of force fields. We shall analyse the experimental information in the next sections. It can be anticipated however that Coriolis constants, centrifugal distortion constants and mean-square amplitudes can be determined only for relatively small molecules. The calculation of the force constants of large molecules thus relies entirely on the vibrational frequencies, which are by far the most important experimental source of force fields.

We recall from Chapter 2 that the number of independent force constants for a molecule without symmetry with n degrees of internal freedom is equal to $n(n+1)/2$. The occurrence of symmetry can considerably reduce this number since many elements of the **F**-matrix become identical in this case. For small molecules with high symmetry the number of force constants becomes comparable to the number of vibrational frequencies. In this case with the aid of the Coriolis constants and the centrifugal distortion constants it is often possible to find a unique solution for the force field and to determine unequivocally all the force constants of (8.1.3). For large molecules this is never possible. The number of force constants is always much larger than the number of vibrational frequencies and the only way to reach some conclusions on the force field is by introducing more or less drastic simplifications in the potential function, in order to keep the number of force constants within an acceptable limit. Simplified potential functions, containing a limited number of force constants have therefore been developed by several authors, often with considerable success, essentially using two different approaches.

The first approach avoids any type of assumption on the nature of the force field and uses only mathematical criteria in order to find a unique solution for the vibrational problem. Along this line, the simplest method of contraction of the number of force constants is probably that of leaving the responsibility of the elimination of terms in the potential function to the Jacobian matrix of Section 4.7. Force constants, for which a small influence on the vibrational frequencies is found, through the numerical values of the corresponding Jacobian elements, are disregarded[1] in the calculations. This method implicitly assumes that force constants that have a small effect on the vibrational frequencies also have small numerical values, a fact that is not necessarily true in theory. A disadvantage of this procedure is that it becomes very time consuming for large molecules since in principle it involves the analysis of all elements of the complete Jacobian matrix, which can have very large dimensions. In addition it is often not sufficiently restrictive for practical use, since too many force constants may pass this 'test'.

Another method is that of introducing constraints among the force constants[2–5] on the basis of parametric representations of the **F**-matrix and of a successive choice of appropriate parameters which lead to mathematically acceptable solutions. Most of these constraints methods can be reduced to the following schema. According to equation 4.6.2 the **G**-matrix can be transformed by means of a similarity transformation to diagonal form

$$\tilde{\mathbf{A}}\mathbf{G}\mathbf{A} = \mathbf{T} \qquad (8.1.4)$$

from which it follows that

$$\mathbf{G} = \mathbf{L}\tilde{\mathbf{L}} = (\mathbf{A}\mathbf{T}^{\frac{1}{2}})(\widetilde{\mathbf{A}\mathbf{T}^{\frac{1}{2}}}) \qquad (8.1.5)$$

and therefore

$$\mathbf{L} = (\mathbf{A}\mathbf{T}^{\frac{1}{2}})\mathbf{C} \qquad (8.1.6)$$

where **C** is any orthogonal matrix. The matrix **C** can always be written as the product of $N(N-1)/2$ matrices $\mathbf{C}_{ij}$, each depending upon a single parameter ϕ_{ij},

$$\mathbf{C} = \prod \mathbf{C}_{ij} \qquad i,j = 1,2,\ldots,N \qquad i<j \qquad (8.1.7)$$

of the type

$$\begin{vmatrix} 1 & & & & & \\ & 1 & & 0 & & \\ & & \ddots & & & \\ & & & \mathbf{C}_{ij} & & \\ & 0 & & & 1 & \\ & & & & & \ddots \; 1 \end{vmatrix}$$

with

$$\mathbf{C}_{ij} = \begin{vmatrix} \cos\phi_{ij} & -\sin\phi_{ij} \\ \sin\phi_{ij} & \cos\phi_{ij} \end{vmatrix} \qquad (8.1.8)$$

A given choice of the $N(N-1)/2$ ϕ_{ij} parameters corresponds to a given possible **F**-matrix, whose elements can be determined by means of equations (4.2.9) and (8.1.6)

$$\mathbf{F} = (\widetilde{\mathbf{A}\mathbf{T}^{\frac{1}{2}}})^{-1}\tilde{\mathbf{C}}^{-1}\mathbf{\Lambda}\mathbf{C}^{-1}(\mathbf{A}\mathbf{T}^{\frac{1}{2}})^{-1} \qquad (8.1.9)$$

The constraints among the $N(N+1)/2$ elements of the **F**-matrix can be now imposed simply by making a reasonable choice of the $N(N-1)/2$ parameters ϕ_{ij}. Several types of constraints have been proposed. For instance Strey[6] has selected for triatomic molecules a value of ϕ which maximizes the stretching force constant or minimizes the bending force constant. Several authors[7,8,9] have proposed that all the $\phi_{i,j}$ parameters should be set equal to zero, a solution that corresponds to the assumption, according to equation (8.1.6), that the normal coordinates are defined in purely kinematic terms, i.e. $\mathbf{C} \equiv \mathbf{E}$, the identity matrix. Another convenient type of constraint is that chosen by Herranz and Castano[10] and discussed in detail by Freeman[11] which maximizes the trace of the **L**-matrix by choosing $\mathbf{C} = \mathbf{A}$, so that the matrix **L** assumes the symmetric form

$$\mathbf{L} = \mathbf{A}\mathbf{T}^{\frac{1}{2}}\tilde{\mathbf{A}} \qquad (8.1.10)$$

An equivalent treatment has also been proposed by Pulay and Török.[12]

Recently Redington and Aljibury[13] have adopted a series of constraints obtained by maximizing the restoring forces on the atoms and assuming that these forces are all directed along the internal coordinates. A detailed discussion of these methods can be found in References 11 and 13.

The constraints methods have the advantage of always yielding a unique solution for the $N(N+1)/2$ force constants of the potential function (8.1.3). They have the disadvantage of not being easily applicable to large molecules and of giving different solutions for different isotopic states.

It must be pointed out, however, that such methods have been noticeably successful for small molecules[2–13] and that in principle they can be very efficient in yielding approximate starting values of force constants to be used in the iterative refinement procedure of Section 4.7.

The second type of approach, which is much more popular than the first, is that of building a model force field either by assuming a specific system of coupled springs or by correlation with a qualitative picture of the electronic distribution in the molecule. The simplest model one can think of is that of a system of springs attached to the internal coordinates which do not interact with one another. Such a model is directly obtained from equation (8.1.3) by putting all off-diagonal terms, i.e. all terms of the type f_{ij} $(i \neq j)$, equal to zero. The only force constants taken into account are those corresponding to stretching (Δr) bending $(\Delta\alpha)$ and torsion $(\Delta\tau)$ of chemical bonds, so that equation (8.1.3) takes the form

$$2V = \sum_i K_i \Delta r_i^2 + \sum_j H_j \Delta\alpha_j^2 + \sum_l Y_l \Delta\tau_l^2 \qquad (8.1.11)$$

This type of force field, called VFF (valence force field), which includes very few force constants, assumes implicitly that the variation in one internal coordinate causes no redistribution of the electrons in the molecule, a fact that is in evident contrast to the well known mobility of valence electrons. Owing to the small number of adjustable parameters (the non-zero force constants), in general it gives a poor fit of the observed vibrational spectrum and is therefore very little used.

The next step in looking for a more flexible potential function, is to assume that the interaction between two valence coordinates is important only if they lie close together in the molecule. This model, that we shall call SGVFF (simplified general valence force field), assumes therefore that only next-neighbouring coordinates can interact and includes in the force field, in addition to diagonal force constants, several other interaction constants F_{ij} $(i \neq j)$, for example those of the types

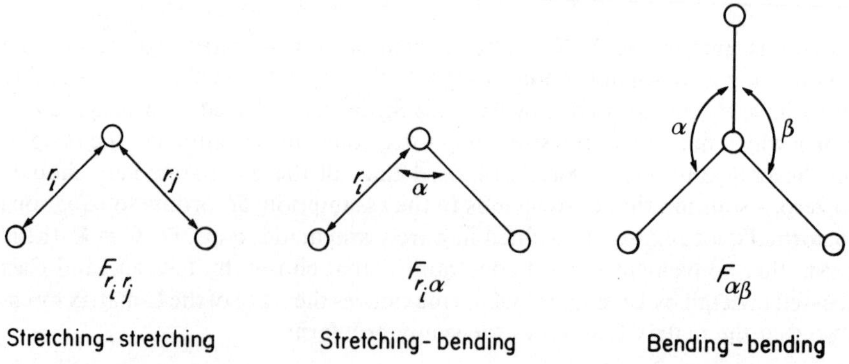

This type of force field is much more realistic than the previous one and for large molecules can include an appreciable number of interaction terms. For this reason it possesses a considerable degree of flexibility and can yield a very good fit of vibrational frequencies for molecules with relatively localized bonds. For molecules with delocalized

π-electrons, interactions between non-adjacent coordinates may be relevant and therefore need to be included for a good frequency fit. This leads us to the next approximation to the SGVFF, which is simply obtained by adding to the previous force constants the minimum number of interaction terms between non-adjacent coordinates which is required to reproduce correctly the observed vibrational spectrum. The reader will observe that this form of the SGVFF does not differ significantly from the 'Jacobian' procedure described before. As a matter of fact the two approaches almost completely overlap for situations in which the structure of the vibrating system is too complex to be accounted for in terms of a simple model, and thus the criterion of best frequency becomes the only significant one.

For a limited number of molecules it is possible to collect enough experimental data (vibrational frequencies for different isotopic species, Coriolis and distortion constants, etc.) to determine all the force constants of the potential function (8.1.3), which is called GVFF (general valence force field) in the internal coordinate basis.

A basically different model is the so-called 'central force field (CFF)' in which only stretching types of coordinates are used, so that it takes the form

$$2V = \sum_i K_i \Delta s_i^2 + \sum_{i<j} F_{ij} \Delta q_{ij}^2 \tag{8.1.12}$$

The coordinates Δs_i are in this case the usual stretching coordinates describing a variation in the length of a chemical bond, whereas the coordinates Δq_{ij} are variations of non-bonded interatomic distances. This force field has only historical significance since it is too simple to yield acceptable results. The idea of using non-bonded atom interactions has been reconsidered by Urey and Bradley,[14] who proposed a new type of force field that can be considered as the sum of the simple valence force field and the central force field. In the Urey–Bradley force field (UBFF) the diagonal force constants are the usual VFF diagonal force constants (stretching, bending, torsion, etc.), whereas the interaction force constants are expressed in terms of non-bonded atom interactions. The form of the UBFF is then

$$2V = \sum_i K_i \Delta r_i^2 + \sum_j H_j \Delta\alpha_j^2 + \sum_l Y_l \Delta\tau_l^2 + \sum_{i<j} F_{ij} \Delta q_{ij}^2 \tag{8.1.13}$$

In the above expressions we have used the symbols K, H and Y for stretching, bending and torsion force constants respectively and the symbol F for the interaction force constants, following a well established nomenclature.

Although in principle non-bonded interactions should be taken into account for any pair of non-bonded atoms, only those of the type

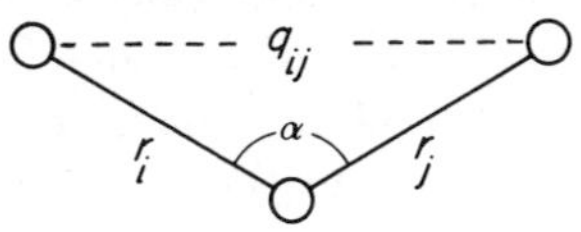

between two atoms both connected to the same atom, are normally considered, all others being expected from known semiempirical interaction potentials to be very small.

Since q_{ij}, r_i, r_j and α are connected by a simple geometrical relationship, the coordinate Δq_{ij} is redundant and thus the potential function (8.1.13) can be expressed in terms of valence coordinates only. Once this is done, one obtains a particular type of simplified valence force field in which interaction constants of the type F_{rr} and $F_{r\alpha}$ are both functions of the same non-bonded interaction.

The UBFF has been widely used and it is known to yield results comparable to those obtained with the SGVFF. The vibrational spectra of several hundreds of molecules have been calculated by means of this potential, essentially by Shimanouchi and his group.[15,16,17]

As with the SGVFF, the UBFF fails when interactions between non-adjacent coordinates are important as in π-bonded systems. In this case it is necessary to add to the standard UB force constants further interactions taken from the GVFF. This version of the UB field is often referred to as MUBFF (Modified Urey–Bradley force field).

Another type of force field which uses non-bonded interactions, is the so-called 'consistent force field', developed by Lifson and Warshel.[18,19,20] In this approach the potential function is chosen to fit not only vibrational frequencies, but also conformational structures and energies, the word 'consistent' just meaning that the potential function is consistent with both sets of data.

Useful information on some terms of the force field can also be obtained from a qualitative description of how the electronic distribution changes during a molecular vibration. A great number of authors have attempted to interpret force constants in terms of electronic structure. In some cases these interpretations have been framed into model force fields in the sense that they can help in the choice of important interaction terms, in the determination of the sign of such terms and in finding relationships among them. Examples are the so-called 'orbital following force field' (OFFF) proposed by Heath and Linnett,[21] the 'hybrid bond orbital force field' (HOFF) of Mills[22] and the 'Kekule' potential for aromatic hydrocarbons developed by Scherer and Overend.[23]

8.2 Analysis of the experimental data

The experimental data of interest for the determination of molecular force fields are of various types. In this section we discuss their utilization in the calculation of force constants and analyse the kind of information they can furnish.

Since vibrational frequencies represent the largest body of information that can be collected for the determination of a force field, most of this section will be devoted to them. Coriolis and distortion constants as well as the mean-square amplitudes of vibrations provide additional information, in the sense that they can be of great help in determining interaction force constants but are not, except in a very few particular cases, sufficient to determine a force field.

(A) *Vibrational frequencies*

The vibrational frequencies of a molecule are determined, experimentally, from the analysis of infrared and Raman spectra. With the aid of modern spectrographs vibrational frequencies are determined with uncertainties of about 0·5 cm^{-1}. For polyatomic molecules with a large number of vibrational modes, some bands overlap more or less strongly. Depending upon the relative intensity and the peak separation, the uncertainty in the frequency values can then be much greater.

For the treatment of molecular vibrations described in this book one should utilize, in principle, only vibrational frequencies of isolated molecules in the gaseous state since no interaction terms between different molecules are included in the potential function. Unfortunately, it is not always possible to measure the infrared and Raman spectra of a compound in the gaseous state and thus many actual calculations are based on solid or liquid spectra. Bands observed in liquid and crystal spectra are shifted with respect to those observed in the gaseous state by an amount which depends upon the type of vibration as well as upon the strength of the intermolecular forces. Van der Waals forces in molecular crystals produce small frequency shifts which are normally within 10–20 cm^{-1}. Strong directional forces, and in particular hydrogen bonds, have, however, a much stronger influence on the vibrational spectrum and can lead to shifts in some vibrational modes of more than a hundred cm^{-1}. In general, stretching vibrations are shifted toward lower frequencies, while angle deformation vibrations are shifted toward higher frequencies.

In order to investigate the information that the vibrational frequencies give on force constants, we shall consider the specific case of a 3×3 block of the secular equation of a given molecule. Let Q_a, Q_b and Q_c be the three normal coordinates associated with the three roots λ_a, λ_b and λ_c and let S_p, S_q and S_r be the three symmetry coordinates in this block. Let us further assume that the coordinate S_p is almost completely uncoupled from the other two coordinates, so that the normal coordinate Q_a is almost purely an S_p type of coordinate. The **L**-matrix then has the form

$$\begin{pmatrix} L_{pa} & 0 & 0 \\ 0 & L_{qb} & L_{qc} \\ 0 & L_{rb} & L_{rc} \end{pmatrix}$$

where the elements L_{pb}, L_{pc}, L_{qa}, L_{ra}, being small, have been neglected. In Section 4.7 we have derived the form of the Jacobian elements of $\mathbf{\Lambda}$ with respect to **F**, which are given by

$$\frac{\partial\lambda_i}{\partial F_{kk}} = (L_0)_{ki}^2$$

$$\frac{\partial\lambda_i}{\partial F_{kl}} = 2(L_0)_{ki}(L_0)_{li}$$

In the case considered we have then

$$\frac{\partial\lambda_a}{\partial F_{pp}} = L_{pa}^2 \quad \frac{\partial\lambda_a}{\partial F_{pq}} = 0 \quad \frac{\partial\lambda_a}{\partial F_{pr}} = 0 \quad \frac{\partial\lambda_a}{\partial F_{qq}} = 0 \quad \frac{\partial\lambda_a}{\partial F_{qr}} = 0 \quad \frac{\partial\lambda_a}{\partial F_{rr}} = 0$$

$$\frac{\partial\lambda_b}{\partial F_{pp}} = 0 \quad \frac{\partial\lambda_b}{\partial F_{pq}} = 0 \quad \frac{\partial\lambda_b}{\partial F_{pr}} = 0 \quad \frac{\partial\lambda_b}{\partial F_{qq}} = L_{qb}^2 \quad \frac{\partial\lambda_b}{\partial F_{qr}} = 2L_{qb}L_{rb} \quad \frac{\partial\lambda_b}{\partial F_{rr}} = L_{rb}^2$$

$$\frac{\partial\lambda_c}{\partial F_{pp}} = 0 \quad \frac{\partial\lambda_c}{\partial F_{pq}} = 0 \quad \frac{\partial\lambda_c}{\partial F_{pr}} = 0 \quad \frac{\partial\lambda_c}{\partial F_{qq}} = L_{qc}^2 \quad \frac{\partial\lambda_c}{\partial F_{qr}} = 2L_{qc}L_{rc} \quad \frac{\partial\lambda_c}{\partial F_{rr}} = L_{rc}^2$$

We conclude therefore that the vibrational frequencies give no information in such a case on the force constants F_{pq} and F_{pr}.

In practice, situations of the type considered above for S_p, i.e. no coupling with other coordinates, occur very frequently. As a consequence the vibrational frequencies always give information on the diagonal force constants but give no information on the off-diagonal force constants, unless the corresponding coordinates are strongly coupled.

(B) *Anharmonicity correction*

In the use of experimental frequencies for the calculation of force constants, a complication arises from the fact that molecular vibrations are anharmonic whereas the theoretical model used is purely harmonic. The experimental frequencies therefore need to be corrected for the anharmonic contribution. This correction can be quite important for vibrations involving the displacement of light atoms, such as hydrogen stretching motions which are of rather large amplitude, but is normally small for other kinds of vibrations.

The correction for anharmonicity is treated in great detail in Herzberg's book.[24] In what follows we shall give only a brief summary of the treatment, which is intended to furnish the reader with the necessary basis for the correct handling of experimental data. A more detailed treatment of anharmonicity will be given in Chapter 9.

We recall from Section 3.3 that in the harmonic approximation, i.e. using a potential function of the type (8.1.3), the energy levels of a polyatomic molecule are given by

$$E(v_1, \ldots, v_{3N-6}) = \sum_{i=1}^{3N-6} hc\nu_i(v_i + \tfrac{1}{2}) \tag{8.2.1}$$

A direct consequence of this expression is that the levels associated with a given quantum number have a constant energy spacing and therefore overtones have a frequency which is exactly twice, three times, etc., the frequency of the fundamental vibration. In the same way combination levels corresponding to the nth excitation of the quantum number v_i and to the mth excitation of the quantum number v_j have energies which are exactly n times the energy of the fundamental level E_i plus m times the energy of the fundamental level E_j.

Experimentally it is found that overtones and combination bands do not obey equation (8.2.1) exactly, but occur in the great generality of the cases at frequencies which are lower than those calculated from this equation. These deviations from the harmonic model can be easily accounted for if a potential function of the type (8.1.1), including cubic and possibly higher terms, is used in place of the simple harmonic potential. The energy levels of the anharmonic oscillator are then found to be given by the equation

$$E(v_1, \ldots, v_{3N-6}, l_1, \ldots, l_{3N-6}) = \sum_i hc\omega_i(v_i + \tfrac{1}{2}) + \sum_i \sum_{j \geq i} hcx_{ij}\left(v_i + \frac{d_i}{2}\right)\left(v_j + \frac{d_j}{2}\right) + \sum_{tt'} hcx_{l_t l_{t'}} l_t l_{t'} + \ldots \tag{8.2.2}$$

where ω_i are the (theoretical) harmonic frequencies that the system would have for infinitesimal displacements of the nuclei, x_{ij} the first-order correction factors due to the anharmonicity and d_k the degeneracy of the vibrations. The term $hcx_{l_t l_{t'}} l_t l_{t'}$ accounts for a break-down of the degeneracy due to the interaction of the degenerate modes with other vibrations, l_t and $l_{t'}$ being quantum numbers which take the values

$$l_k = v_k, v_k - 2, v_k - 4, \ldots, 1 \text{ or } 0 \tag{8.2.3}$$

and t, t' those values of i for which the vibrations are degenerate. For the moment we shall not discuss the derivation of (8.2.2) and the correlation of the anharmonic factors x_{ij} with the cubic and higher terms of the potential function. This problem will be discussed in Chapter 9. Here instead we shall concentrate our attention on the determination of the anharmonicity factors x_{ij} from the experimental data and on the calculation of the harmonic frequencies ω_i.

Example 8.2.1

Consider first the simple example of a diatomic molecule. In this case there is only one vibration and no degeneracy. We have therefore $d_i = d_j = 1$ and $x_{l_t l_{t'}} = 0$ and thus equation (8.2.2) reduces to

$$E(v) = hc\omega(v + \tfrac{1}{2}) + hcx(v + \tfrac{1}{2})^2 + \ldots \tag{8.2.4}$$

The frequencies of the fundamental and of the overtones are then, according to Bohr's rule $\nu = [E(v + 1) - E(v)]/hc$

$$\begin{aligned} \nu_{0-1} &= \omega + 2x \\ \nu_{0-2} &= 2\omega + 6x \\ \nu_{0-3} &= 3\omega + 12x \\ \nu_{0-4} &= 4\omega + 20x \\ &\ldots \end{aligned} \tag{8.2.5}$$

If we now form the difference $\Delta\nu$ between two successive frequencies, then the difference $\Delta^2\nu$ between two successive differences and so forth, we obtain

$$\begin{aligned}
\Delta\nu_1 &= \nu_{0-2} - \nu_{0-1} = \omega + 4x \qquad & \Delta^2\nu_1 &= 2x \\
\Delta\nu_2 &= \nu_{0-3} - \nu_{0-2} = \omega + 6x & \Delta^2\nu_2 &= 2x \\
\Delta\nu_3 &= \nu_{0-4} - \nu_{0-3} = \omega + 8x & \Delta^2\nu_3 &= 2x \\
\Delta\nu_4 &= \nu_{0-5} - \nu_{0-4} = \omega + 10x & & \\
&\ldots & &
\end{aligned}$$

We can then calculate the anharmonicity factor x from any of the $\Delta^2\nu_i$ differences. In particular if equation (8.2.4) were exact, all the $\Delta^2\nu_i$ differences would coincide, as shown by the previous equations. Experimentally it is actually found that they are not exactly equal and that a second-order anharmonicity correction is necessary[24] (see Herzberg, Vol. I, p. 92). This second-order anharmonicity factor is in general very small and is introduced only for diatomic molecules. If it is ignored, the values of x and of ω can be calculated through a best fit of equations (8.2.5).

Example 8.2.2

Consider now as a further example, a bent triatomic molecule. In this case ($x_{l_t l_{t'}} = 0$) equation (8.2.2) reduces to

$$E(v_1, v_2, v_3) = \sum_{i=1}^{3} hc\omega_i(v_i + \tfrac{1}{2}) + \sum_{i=1}^{3} hcx_{ii}(v_i + \tfrac{1}{2})^2 + \sum_{i=1}^{3}\sum_{j>i} hcx_{ij}(v_i + \tfrac{1}{2})(v_j + \tfrac{1}{2}) + \ldots \qquad (8.2.6)$$

From this equation we obtain, by means of Bohr's rule, the following expressions for the fundamentals, for the overtones and for the differences $\Delta\nu$ and $\Delta^2\nu$ ($i, k, l = 1, 2, 3$)

$$\begin{aligned}
\nu_i &= \omega_i + 2x_{ii} + \tfrac{1}{2}x_{ik} + \tfrac{1}{2}x_{il} \qquad & \Delta\nu_1 &= \omega_i + 4x_{ii} + \tfrac{1}{2}(x_{ik} + x_{il}) \qquad & \Delta^2\nu_i &= 2x_{ii} \\
2\nu_i &= 2\omega_i + 6x_{ii} + x_{ik} + x_{il} & \Delta\nu_2 &= \omega_i + 6x_{ii} + \tfrac{1}{2}(x_{ik} + x_{il}) & & \\
3\nu_i &= 3\omega_i + 12x_{ii} + \tfrac{3}{2}x_{ik} + \tfrac{3}{2}x_{il} & &\ldots & & \\
&\ldots & &\ldots & & \\
n\nu_i &= n\omega_i + n(n+1)x_{ii} + \frac{n}{2}x_{ik} + \frac{n}{2}x_{il} & & & &
\end{aligned} \qquad (8.2.7)$$

and therefore we can obtain from a best fit of the $\Delta^2\nu$ differences the values of x_{11}, x_{22} and x_{33}.

For the calculation of the three anharmonicity factors x_{12}, x_{13} and x_{23} we can use the combination frequencies which, according to (8.2.6) are given by

$$\begin{aligned}
\nu_i + \nu_k &= \omega_i + \omega_k + 2x_{ii} + 2x_{kk} + 2x_{ik} + \tfrac{1}{2}x_{il} + \tfrac{1}{2}x_{kl} \\
2\nu_i + \nu_k &= 2\omega_i + \omega_k + 6x_{ii} + 2x_{kk} + \tfrac{7}{2}x_{ik} + x_{il} + \tfrac{1}{2}x_{kl} \\
&\ldots \\
n\nu_i + m\nu_k &= n\omega_i + m\omega_k + n(n+1)x_{ii} + m(m+1)x_{kk} + \left(mn + \frac{m+n}{2}\right)x_{ik} + \frac{n}{2}x_{il} + \frac{m}{2}x_{kl}
\end{aligned} \qquad (8.2.8)$$

By forming appropriate differences between these combination bands and the fundamentals or the overtones, for example

$$(n\nu_i + m\nu_k) - n\nu_i - m\nu_k = nmx_{ik} \qquad (8.2.9)$$

it is possible to evaluate these anharmonic factors. Again a best fit method can be used. Although it should be self-evident, we point out that in all previous expressions the symbol ν_i was used for the three fundamentals ν_{100}, ν_{010} and ν_{001}, the symbol $n\nu_i$ for the overtones ν_{n00}, ν_{0n0} and ν_{00n} and the symbol $(n\nu_i + m\nu_k)$ for the combinations ν_{nm0}, ν_{n0m} and ν_{0nm}.

Once the six anharmonicity factors have been determined, the three harmonic frequencies ω_i can easily be calculated by means of equation (8.2.7). It is obvious from the above examples that in order to be able to obtain the anharmonic factors x_{ij} for a polyatomic molecule it is necessary to identify in the spectrum a great number of overtones and combination bands. This is possible for small molecules, but becomes very difficult for large molecules with many normal modes of vibration. It must therefore be kept in mind, in comparing force constants, that only for small molecules are the data corrected for anharmonicity. Table 8.2.1 shows the influence of anharmonicity on the vibrational frequencies of some simple molecules.

Table 8.2.1

H_2O[25]	$\nu_1 = 3656{\cdot}65$	$\nu_2 = 1594{\cdot}59$	$\nu_3 = 3755{\cdot}79$				
	$\omega_1 = 3832{\cdot}17$	$\omega_2 = 1648{\cdot}47$	$\omega_3 = 3942{\cdot}53$				
	$x_{11} = -42{\cdot}576$	$x_{22} = -16{\cdot}813$	$x_{33} = -47{\cdot}566$	$x_{12} = -15{\cdot}933$	$x_{13} = -165{\cdot}824$	$x_{23} = -20{\cdot}332$	
D_2O[25]	$\nu_1 = 2671{\cdot}46$	$\nu_2 = 1178{\cdot}33$	$\nu_3 = 2788{\cdot}05$				
	$\omega_1 = 2763{\cdot}80$	$\omega_2 = 1206{\cdot}39$	$\omega_3 = 2888{\cdot}78$				
	$x_{11} = -22{\cdot}58$	$x_{22} = -9{\cdot}18$	$x_{33} = -26{\cdot}15$	$x_{12} = -7{\cdot}58$	$x_{13} = -87{\cdot}15$	$x_{23} = -10{\cdot}61$	
SO_2[26]	$\nu_1 = 1151{\cdot}38$	$\nu_2 = 517{\cdot}69$	$\nu_3 = 1361{\cdot}76$				
	$\omega_1 = 1167{\cdot}60$	$\omega_2 = 526{\cdot}27$	$\omega_3 = 1380{\cdot}91$				
	$x_{11} = -3{\cdot}99$	$x_{22} = -3{\cdot}00$	$x_{33} = -5{\cdot}17$	$x_{12} = -2{\cdot}05$	$x_{13} = -1{\cdot}71$	$x_{23} = -3{\cdot}90$	
CH_4[27]	$\nu_1 = 2916{\cdot}5$	$\nu_2 = 1534$	$\nu_3 = 3018{\cdot}7$	$\nu_4 = 1306$	$x_{12} = x_{14} = x_{22} = 0$		
	$\omega_1 = 3143$	$\omega_2 = 1573$	$\omega_3 = 3154$	$\omega_4 = 1357$			
	$x_{11} = -64{\cdot}6$	$x_{13} = -65{\cdot}0$	$x_{23} = -15{\cdot}0$	$x_{24} = -11{\cdot}2$	$x_{33} = -17{\cdot}5$	$x_{34} = -12{\cdot}0$	$x_{44} = -6{\cdot}0$
CD_4[27]	$\nu_1 = 2107{\cdot}8$	$\nu_2 = 1092{\cdot}2$	$\nu_3 = 2258{\cdot}6$	$\nu_4 = 996{\cdot}0$	$x_{12} = x_{14} = x_{22} = 0$		
	$\omega_1 = 2224{\cdot}0$	$\omega_2 = 1113$	$\omega_3 = 2333$	$\omega_4 = 1027$			
	$x_{11} = -32{\cdot}4$	$x_{13} = -34{\cdot}0$	$x_{23} = -7{\cdot}8$	$x_{24} = -6{\cdot}0$	$x_{33} = -9{\cdot}6$	$x_{34} = -6{\cdot}7$	$x_{44} = -3{\cdot}4$
CH_3F[28]	$\nu_1 = 2930$	$\nu_2 = 1464$	$\nu_3 = 1048{\cdot}6$	$\nu_4 = 3005{\cdot}8$	$\nu_5 = 1467$	$\nu_6 = 1182$	
	$\omega_1 = 3055$	$\omega_2 = 1500$	$\omega_3 = 1067$	$\omega_4 = 3165$	$\omega_5 = 1510$	$\omega_6 = 1212$	
CH_3Cl[29]	$\nu_i = 2937$	$\nu_2 = 1354{\cdot}9$	$\nu_3 = 732{\cdot}1$	$\nu_4 = 3041{\cdot}8$	$\nu_5 = 1454{\cdot}6$	$\nu_6 = 1015{\cdot}0$	
	$\omega_1 = 3062$	$\omega_2 = 1398$	$\omega_3 = 733$	$\omega_4 = 3209$	$\omega_5 = 1520$	$\omega_6 = 1049$	
CH_3Br[29]	$\nu_1 = 2935$	$\nu_2 = 1305$	$\nu_3 = 611$	$\nu_4 = 3056{\cdot}8$	$\nu_5 = 1444$	$\nu_6 = 954$	
	$\omega_1 = 3061$	$\omega_2 = 1332$	$\omega_3 = 614$	$\omega_4 = 3213$	$\omega_5 = 1506$	$\omega_6 = 975$	
CH_3I[29]	$\nu_1 = 2933$	$\nu_2 = 1251{\cdot}5$	$\nu_3 = 532{\cdot}8$	$\nu_4 = 3060{\cdot}3$	$\nu_5 = 1440{\cdot}3$	$\nu_6 = 880$	
	$\omega_1 = 3060$	$\omega_2 = 1288$	$\omega_3 = 533$	$\omega_4 = 3229$	$\omega_5 = 1504$	$\omega_6 = 906$	

It is evident from the data of the table that by far the largest anharmonicity correction is found for the vibrations which involve large displacements of H atoms and in particular stretching modes of the X–H bonds. For instance, in the case of bent triatomic molecules the difference between the harmonic frequency ω_1 and the experimental gas frequency ν_1 is 175, 113 and 17 cm^{-1} for H_2O, D_2O and SO_2 respectively.

For larger molecules for which it is practically impossible to determine the anharmonicity factors, Arnett and Crawford[30] have proposed that only the X–H stretching vibrations should be corrected for anharmonicity using the so-called 'product rule' (see later) to estimate the amount of the correction and treating this particular vibration as a vibration of a diatomic molecule.

(C) *Isotopic substitution*

As discussed in Section 8.1, the number of harmonic force constants, even for a small molecule, is much greater than the number of normal frequencies. It is thus necessary to find additional data in order to be able to refine a force field. Since the force constants depend upon the electronic structure and not upon the masses of the atoms (in the limits of the Born–Oppenheimer approximation) two different isotopic species of the same molecule will have different vibrational frequencies only because the masses of some atoms (and therefore their kinetic energies) are different. We can then use the frequencies of isotopically substituted molecules for the refinement of the same force field. In particular for m different isotopic species we have m different secular equations

$$\begin{aligned} \mathbf{G}_1\mathbf{F}\mathbf{L}_1 &= \mathbf{L}_1\mathbf{\Lambda}_1 \\ \mathbf{G}_2\mathbf{F}\mathbf{L}_2 &= \mathbf{L}_2\mathbf{\Lambda}_2 \\ &\dots \\ \mathbf{G}_m\mathbf{F}\mathbf{L}_m &= \mathbf{L}_m\mathbf{\Lambda}_m \end{aligned} \tag{8.2.10}$$

in which all the **G**, **L** and **Λ** matrices are different whereas the **F**-matrix is the same.

The m isotopic species would therefore furnish $m(3N - 6)$ experimental frequencies for the refinement of the force field, if these data were all independent. Since, however, the frequencies of isotopic species are related by some 'isotopic rules', the number of independent data is much smaller.

The best known isotopic rule is the so-called 'Teller[31] and Redlich[32] product rule' which relates the vibrational frequencies of two isotopically substituted molecules through the relation

$$\frac{\omega_1 . \omega_2 \dots \omega_{3N-6}}{\omega'_1 . \omega'_2 \dots \omega'_{3N-6}} = \left(\frac{m'_1}{m_1} . \frac{m'_2}{m_2} \dots \frac{m'_{3N}}{m_{3N}}\right)^{\frac{1}{2}} \left(\frac{\mathscr{M}}{\mathscr{M}'}\right)^{\frac{3}{2}} \left(\frac{I_x . I_y . I_z}{I'_x . I'_y . I'_z}\right)^{\frac{1}{2}} \tag{8.2.11}$$

where ω_i indicates the harmonic frequencies, m_i the masses of the atoms, $\mathscr{M}$ the mass of the molecule and I_x, I_y, I_z the moments of inertia. Quantities relative to one of the two isotopic species are labelled with a prime, whereas those relative to the other have no such label.

In order to prove the Teller–Redlich product rule consider the secular equation (4.2.16a)

$$|\mathbf{GF} - \lambda\mathbf{E}| = 0 \tag{8.2.12}$$

If we expand this determinant we obtain a polynomial of order n in λ of the type

$$\lambda^n + a_1\lambda^{n-1} + a_2\lambda^{n-2} + \dots + a_{n-1}\lambda + a_n = 0 \tag{8.2.13}$$

According to well-known theorems on matrices the constant term a_n is related to the roots of the polynomial by the equation

$$(-1)^n a_n = \lambda_1 . \lambda_2 \dots \lambda_n = (4\pi^2)^n \omega_1^2 \cdot \omega_2^2 \dots \omega_n^2 \tag{8.2.14}$$

and to the **GF** matrix by the equation

$$(-1)^n a_n = |\mathbf{GF}| = |\mathbf{G}| . |\mathbf{F}| \tag{8.2.15}$$

we have therefore

$$\frac{\omega_1 . \omega_2 \dots \omega_n}{\omega'_1 . \omega'_2 \dots \omega'_n} = \frac{|\mathbf{G}|^{\frac{1}{2}} . |\mathbf{F}|^{\frac{1}{2}}}{|\mathbf{G}'|^{\frac{1}{2}} . |\mathbf{F}|^{\frac{1}{2}}} = \left(\frac{|\mathbf{G}|}{|\mathbf{G}'|}\right)^{\frac{1}{2}} \tag{8.2.16}$$

This relationship is true for any coordinate basis in which the **G**- and **F**-matrices have been written. In particular, if the coordinates used are symmetry coordinates, equation (8.2.16) relates the frequencies of two isotopically substituted molecules in each symmetry block. The number of independent frequencies is thus equal to

$$2 \times (3N - 6) - h$$

where h is the number of symmetry blocks for the molecule considered.

Equation (8.2.16) is the general form of the product rule and is very suitable for computer calculations. In order to derive equation (8.2.11) from it, we recall that in a Cartesian coordinate basis, the **G**-matrix is equal to the inverse of the diagonal matrix **M** of the masses (see Section 4.2.1). We have then, from (8.2.16),

$$\frac{\omega_1 . \omega_2 \ldots \omega_{3N}}{\omega'_1 . \omega'_2 \ldots \omega'_{3N}} = \left(\frac{|\mathbf{M}^{-1}|}{|\mathbf{M}^{-1'}|}\right)^{\frac{1}{2}} = \left(\frac{m'_1 . m'_2 \ldots m'_{3N}}{m_1 . m_2 \ldots m_{3N}}\right)^{\frac{1}{2}} \tag{8.2.17}$$

Since in the Cartesian basis the **G**-matrix is of dimension $3N \times 3N$, the last six frequencies (from ω_{3N-7} to ω_{3N}), corresponding to the three translations and to the three rotations, are equal to zero and therefore the left-hand side of the equation is indeterminate. To determine the ratio ω_T/ω'_T between two translations and ω_R/ω'_R between two rotations we can use a simple mathematical trick. Assume that a system of weak external forces acts upon the molecule. Under these conditions the translations and the rotations become low-frequency oscillations. Let the external forces now go to zero. In the limit of vanishing forces these low-frequency vibrations become uncoupled from each other as well as from the internal vibrations and can be treated separately. We then have

$$\frac{\omega_T}{\omega'_T} = \left(\frac{\mathcal{M}'}{\mathcal{M}}\right)^{\frac{1}{2}} \qquad \frac{\omega_{R\sigma}}{\omega'_{R\sigma}} = \left(\frac{I'_\sigma}{I_\sigma}\right)^{\frac{1}{2}} \tag{8.2.18}$$

By substitution of (8.2.18) into (8.2.17) we obtain equation (8.2.11). In order to use the product rule for each symmetry block in the form (8.2.11) it is necessary to determine from the character tables the number of translations and rotations which belong to the given irreducible representation Γ^γ as well as the number of times each ratio $(m'_k/m_k)^{\frac{1}{2}}$ of individual atomic masses should appear in equation (8.2.11).

For each set of equivalent atoms, this number is obviously equal to the number of vibrations that the set s contributes to the symmetry block considered, which can be obtained by use of the formula

$$n_\gamma^s = \frac{1}{g}\sum_r \chi_r^{\gamma *}\chi_r \tag{8.2.19}$$

applied to the specific set. For a given symmetry block of dimension n, equation (8.2.11) then becomes

$$\frac{\omega_1 . \omega_2 \ldots \omega_n}{\omega'_1 . \omega'_2 \ldots \omega'_n} = \left[\left(\frac{m'_1}{m_1}\right)^{n_{s1}^\gamma} \cdot \left(\frac{m'_2}{m_2}\right)^{n_{s2}^\gamma} \cdots \left(\frac{m'_r}{m_r}\right)^{n_{sr}^\gamma} \left(\frac{\mathcal{M}}{\mathcal{M}'}\right)^t \left(\frac{I_x}{I'_x}\right)^{d_x} \left(\frac{I_y}{I'_y}\right)^{d_y} \left(\frac{I_z}{I'_z}\right)^{d_z}\right]^{\frac{1}{2}} \tag{8.2.20}$$

$n_{s1}^\gamma, n_{s2}^\gamma, \ldots, n_{sr}^\gamma$ are the number of vibrations that the r sets $s_1, s_2, \ldots, s_r$ contribute to the species Γ; $m_1, m_2, \ldots, m_r$ are the masses of the representative atoms of the sets; t is the number of translations which belong to the irreducible representation Γ; d_x, d_y, d_z are either 0 or 1 depending on whether the rotations R_x, R_y and R_z belong to the representation Γ or not. Degenerate vibrations are counted only once. In the same way if I_ρ and I_σ belong to a degenerate species, only one ratio (I_ρ/I'_ρ) is used in the right-hand side of equation (8.2.20). The same is true for the translations.

Example 8.2.3

Ethylene belongs to the D_{2h} point group and its normal modes of vibration may be classified as (see Section 6.4)

$$3Ag + 2B_{1g} + B_{2g} + A_u + B_{1u} + 2B_{2u} + 2B_{3u}$$

whereas the translations and the rotations have symmetries

$$R_z(B_{1g}) \qquad R_y(B_{2g}) \qquad R_x(B_{3g}) \qquad T_z(B_{1u}) \qquad T_y(B_{2u}) \qquad T_x(B_{3u})$$

The six atoms of ethylene are divided into two sets of equivalent atoms, one including the two carbon atoms (s_C) and one including the four hydrogen atoms (s_H).

Using (8.2.19) we obtain the number of vibrations that each set contributes to a given representation. For instance for the representation A_g, the number of vibrations due to the set s_H is equal to

$$n_{s_H}^{A_g} = \tfrac{1}{8}(3.4.1 + 0 + 0 + 0 + 0 + 1.4.1 + 0 + 0) = 2$$

and the number due to the set s_C is

$$n_{s_C}^{A_g} = \tfrac{1}{8}(3.2.1 + 0 + 0 - 1.2.1 + 0 + 1.2.1 + 1.2.1 + 0) = 1$$

In this way we obtain that the vibrations produced by the two sets belong to the following representations of D_{2h}

$$s_H = 2A_g + 2B_{1g} + B_{2g} + B_{3g} + A_u + B_{1u} + 2B_{2u} + 2B_{3u}$$
$$s_C = A_g + B_{1g} + B_{2g} + B_{1u} + B_{2u} + B_{3u}$$

If we apply equation (8.2.20) to the two isotopic molecules C_2H_4 and C_2D_4, we have the following equations

$$A_g \quad \frac{\omega_1 . \omega_2 . \omega_3}{\omega_1' . \omega_2' . \omega_3'} = \frac{m_D}{m_H}$$

$$B_{1g} \quad \frac{\omega_4 . \omega_5}{\omega_4' . \omega_5'} = \left(\frac{m_D}{m_H}\right) . \left(\frac{I_z^H}{I_z^D}\right)^{\frac{1}{2}}$$

$$B_{2g} \quad \frac{\omega_6}{\omega_6'} = \left(\frac{m_D}{m_H}\right)^{\frac{1}{2}} . \left(\frac{I_y^H}{I_y^D}\right)^{\frac{1}{2}}$$

$$A_u \quad \frac{\omega_7}{\omega_7'} = \left(\frac{m_D}{m_H}\right)^{\frac{1}{2}}$$

$$B_{1u} \quad \frac{\omega_8}{\omega_8'} = \left(\frac{m_D}{m_H}\right)^{\frac{1}{2}} . \left(\frac{\mathcal{M}^H}{\mathcal{M}^D}\right)^{\frac{1}{2}}$$

$$B_{2u} \quad \frac{\omega_9 \cdot \omega_{10}}{\omega_9' . \omega_{10}'} = \left(\frac{m_D}{m_H}\right) . \left(\frac{\mathcal{M}^H}{\mathcal{M}^D}\right)^{\frac{1}{2}}$$

$$B_{3u} \quad \frac{\omega_{11} . \omega_{12}}{\omega_{11}' . \omega_{12}'} = \left(\frac{m_D}{m_H}\right) . \left(\frac{\mathcal{M}^H}{\mathcal{M}^D}\right)^{\frac{1}{2}}$$

where the prime indicates C_2D_4 frequencies.

In some cases one of the isotopic species has lower symmetry than the other. In this case the isotopic rule equations must be applied using the lower symmetry. Degenerate modes, if the degeneracy is broken because of the symmetry lowering, may appear more than once.

Example 8.2.4

The molecule $C^{35}Cl_4$ has T_d symmetry whereas the isotopic species $C^{35}Cl_3{}^{37}Cl$ has C_{3v} symmetry and the isotopic species $C^{35}Cl_2{}^{37}Cl_2$ has C_{2v} symmetry. From the correlation tables of Appendix II we have

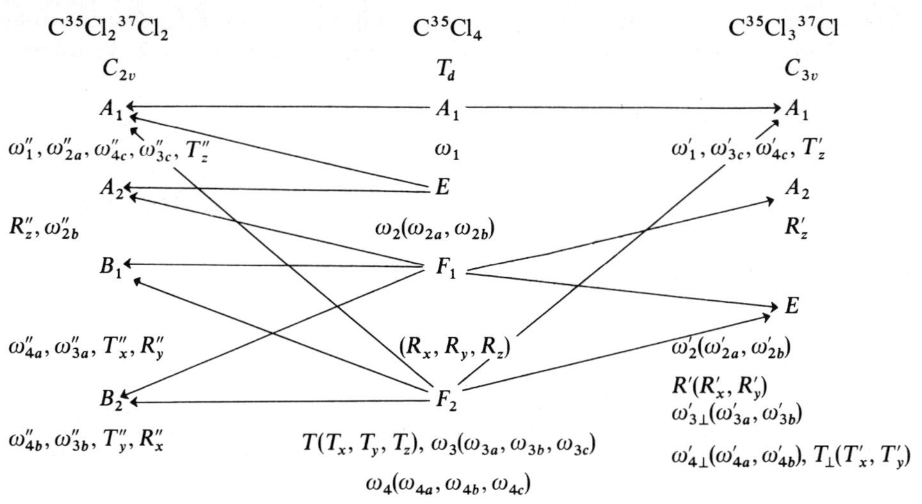

Equation (8.2.20) yields then for the two cases

C_{3v}

$$A_1 \quad \frac{\omega_1 \cdot \omega_4 \cdot \omega_3}{\omega_1' \cdot \omega_{4c}' \cdot \omega_{3c}'} = \left(\frac{37}{35}\right)^{\frac{1}{2}} \left(\frac{\mathcal{M}}{\mathcal{M}'}\right)^{\frac{1}{2}}$$

$$E \quad \frac{\omega_2 \cdot \omega_3 \cdot \omega_4}{\omega_{2a}' \cdot \omega_{3a}' \cdot \omega_{4a}'} = \left(\frac{37}{35}\right)^{\frac{1}{2}} \left(\frac{\mathcal{M}}{\mathcal{M}'}\right)^{\frac{1}{2}} \left(\frac{I}{I'}\right)$$

C_{2v}

$$A_1 \quad \frac{\omega_1 \cdot \omega_2 \cdot \omega_3 \cdot \omega_4}{\omega_1'' \cdot \omega_2'' \cdot \omega_3'' \cdot \omega_4''} = \left(\frac{37}{35}\right) \left(\frac{\mathcal{M}}{\mathcal{M}''}\right)^{\frac{1}{2}}$$

$$A_2 \quad \frac{\omega_2}{\omega_{2b}''} = \left(\frac{37}{35}\right)^{\frac{1}{2}} \cdot \left(\frac{I}{I_z''}\right)^{\frac{1}{2}}$$

$$B_1 \quad \frac{\omega_3 \cdot \omega_4}{\omega_{3b} \cdot \omega_{4b}} = \left(\frac{37}{35}\right)^{\frac{1}{2}} \left(\frac{\mathcal{M}}{\mathcal{M}''}\right)^{\frac{1}{2}} \left(\frac{I}{I_y}\right)^{\frac{1}{2}}$$

$$B_2 \quad \frac{\omega_3 \cdot \omega_4}{\omega_{3b} \cdot \omega_b} = \left(\frac{37}{35}\right)^{\frac{1}{2}} \left(\frac{\mathcal{M}}{\mathcal{M}''}\right)^{\frac{1}{2}} \left(\frac{I}{I_x}\right)^{\frac{1}{2}}$$

The Teller–Redlich product rule is not the only relation existing among the vibrational frequencies of isotopic molecules. Consider the polynomial of order n in λ obtained by direct expansion of the determinant (8.2.12).

$$\lambda^n + a_1\lambda^{n-1} + a_2\lambda^{n-2} + \ldots + a_{n-1}\lambda + a_n = 0$$

The coefficients of the polynomial are rèlated to the roots of the secular equation and to the elements of the matrix $\mathbf{H} = \mathbf{GF}$ by the well-known equations

$$\begin{aligned}
(-1)a_1 &= \sum_{i=1}^{n} \lambda_i = \sum_{i=1}^{n} H_{ii} \\
(-1)^2 a_2 &= \sum_{i=1}^{n} \sum_{j>i} \lambda_i\lambda_j = \sum_{i=1}^{n} \sum_{j>i} (H_{ii}H_{jj} - H_{ij}H_{ji}) \\
&\ldots \\
(-1)^{n-1}a_{n-1} &= \sum_{i=1}^{n} \left(\prod_{j=1}^{n} \lambda_j/\lambda_i \right) = \sum_{i=1}^{n} M_{ii} \\
(-1)^n a_n &= \prod_{i=1}^{n} \lambda_i = |\mathbf{H}|
\end{aligned} \tag{8.2.21}$$

where M_{ii} is the determinant obtained from $\mathbf{H}$ by omitting the ith row and the ith column. We have already seen in the previous discussion how to obtain the product rule from the last of these equations. From the first one of them we can derive a simple relation between the vibrational frequencies of three isotopic molecules.

Consider a molecule which contains among others, a set of r equivalent atoms (e.g. C_2H_4). Using Cartesian coordinates we can write the first equation (8.2.21) in the form

$$\sum_{i=1}^{3N-6} \lambda_i = \sum_{i=1}^{3r} \mu_i F_{ii} + \sum_{j=3r+1}^{3N} \mu_j F_{jj} \tag{8.2.22}$$

which can be rewritten, because of the symmetry which exchanges the r atoms of the set

$$\sum_{i=1}^{3N-6} \lambda_i = r\mu_1(F_{xx} + F_{yy} + F_{zz}) + \sum_{j=3r+1}^{3N} \mu_j F_{jj} \tag{8.2.23}$$

where $\mu_1 = 1/m_1$ and F_{xx}, F_{yy}, F_{zz} are the inverse-mass and the three diagonal $\mathbf{F}$-matrix elements of a representative atom of the set respectively.

Consider now an isotopic derivative of the molecule in which the r atoms of the set have been substituted with a different isotopic species (e.g. C_2D_4) of mass m_2, so that the original symmetry is retained. For this second molecule we have

$$\sum_{i=1}^{3N-6} \lambda_i' = r\mu_2(F_{xx} + F_{yy} + F_{zz}) + \sum_{j=3r+1}^{3N} \mu_j F_{jj} \tag{8.2.24}$$

and thus by subtracting (8.2.24) from (8.2.23)

$$\sum_{i=1}^{3N-6} (\lambda_i - \lambda_i') = r(\mu_1 - \mu_2)(F_{xx} + F_{yy} + F_{zz}) \tag{8.2.25}$$

Consider now a second isotopic derivative of the molecule in which only s of the r equivalent atoms ($s < r$) have been substituted with the isotope of mass m_2, so that the original symmetry is lost (e.g. $C_2H_2D_2$). In this case equation (8.2.22) reads

$$\sum_{i=1}^{3N-6} \lambda_i'' = [s\mu_2 + (r - s)\mu_1](F_{xx} + F_{yy} + F_{zz}) + \sum_{j=3r+1}^{3N} \mu_j F_{jj} \tag{8.2.26}$$

and by subtraction from (8.2.23)

$$\sum_{i=1}^{3N-6} (\lambda_i - \lambda_i'') = s(\mu_1 - \mu_2)(F_{xx} + F_{yy} + F_{zz}) \tag{8.2.27}$$

The ratio of (8.2.25) and (8.2.27) is then

$$\frac{\sum_i (\lambda_i - \lambda_i')}{\sum_i (\lambda_i - \lambda_i'')} = \frac{\sum_i (\omega_i^2 - \omega_i'^2)}{\sum_i (\omega_i^2 - \omega_i''^2)} = \frac{r}{s} \tag{8.2.28}$$

which can be easily rewritten in the form

$$\sum_{i=1}^{3N-6} [(r - s)\omega_i^2 + s\omega_i'^2 - r\omega_i''^2] = 0 \tag{8.2.29}$$

Equation (8.2.29), which relates the sum of the squares of the vibrational frequencies (with appropriate coefficients) for three isotopic species of the same molecule, is called the sum rule and has been found independently by Decius and Wilson[33] and by Sverdlov.[34] The treatment given here is slightly different from that used by these authors and is essentially identical to that developed by Heicklen.[35] It is obvious from the above given discussion that the sum rule will hold for more than three isotopic species as well and that it may involve, when several isotopic species are considered, substitution of different types of atoms. We shall not discuss these cases any further.

It must be noticed that if the three isotopic molecules have the same symmetry, the isotopic sum rule applies to every symmetry block separately. If however the three isotopic species have different symmetry, as normally happens in the great majority of cases for at least one of them, the factoring is limited to the number of blocks of the molecule with the lowest symmetry.

From the other equations (8.2.21) higher-order sum rules[35,36] can be obtained. These isotopic rules, rather cumbersome to discuss, are of very limited interest and will not be treated here. The reader interested in the subject is referred to the original papers. It is, however, necessary to point out that, by combining sum and product rules, Brodersen and Langseth[37,38] have established a so-called 'complete isotopic rule' which permits, in some cases and for definite classes of molecules, the calculation of the vibrational frequencies of an isotopic species from the knowledge of the frequencies of the other two isotopic species of the same molecule. Applications of the complete isotopic rule to deuterated acetylenes, ethylenes and benzenes can be found in References 37 and 38. Additional applications are given in References 39 and 40.

Among the various possible isotopic substitutions, by far the most important in vibrational spectroscopy is the substitution of hydrogen with deuterium. Since the mass of deuterium is twice that of hydrogen, large frequency shifts occur in the vibrational spectrum for those vibrational modes which mostly involve displacement of hydrogen atoms. As a consequence of the changes in the relative position in frequency of the normal modes of vibration of the molecule, the elements of the **L**-matrix also change. The elements of the Jacobian matrix (4.7.11) for the deuterated molecule will therefore be different in general from those of the original molecule and will then furnish significant equations for the determination of the force constants. If only the fully deuterated compound is considered, the number of new equations is, as discussed previously,

$$2(3N - 6) - h$$

where h is the number of symmetry blocks. If, in addition, we use other partially deuterated molecules, we have to subtract for each of them, from the additional $3N - 6$ frequencies, the number of product rule relations with the original molecule, plus the number of sum rules which exist among all the isotopic molecules considered.

The importance of isotopic substitution in determining force constants is best illustrated by a two-dimensional problem. Consider a 2×2 symmetry block of a molecule (with a suitable symmetry) and let s_1 and s_2 be the two symmetry coordinates in this block. The secular determinant

$$|\mathbf{GF} - \lambda\mathbf{E}| = 0 \tag{8.2.30}$$

then yields the two equations

$$\begin{aligned} G_{11}F_{11} + 2G_{12}F_{12} + G_{22}F_{22} &= \lambda_1 + \lambda_2 & \text{(a)} \\ (G_{11}G_{22} - G_{12}^2)(F_{11}F_{22} - F_{12}^2) &= \lambda_1 . \lambda_2 & \text{(b)} \end{aligned} \tag{8.2.31}$$

in the three unknowns F_{11}, F_{22} and F_{12}. In order to determine these three force constants we need another equation. For the isotopically substituted compound, in which the original symmetry is retained, we have two other equations

$$\begin{aligned} G'_{11}F_{11} + 2G'_{12}F_{12} + G'_{22}F_{22} &= \lambda'_1 + \lambda'_2 & \text{(a)} \\ (G'_{11}G'_{22} - G'^2_{12})(F_{11}F_{22} - F_{12}^2) &= \lambda'_1 . \lambda'_2 & \text{(b)} \end{aligned} \tag{8.2.32}$$

only one of which is independent from the first two because of the product rule relation

$$\frac{|\mathbf{G}|}{|\mathbf{G}'|} = \frac{G_{11}G_{22} - G_{12}^2}{G'_{11}G'_{22} - G'^2_{22}} = \frac{\lambda_1 . \lambda_2}{\lambda'_1 . \lambda'_2} \tag{8.2.33}$$

which is directly obtained by forming the ratio of equations (8.2.31b) and (8.2.32b). We then have three equations for the determination of the three unknown force constants. Since one of these equations is quadratic in F_{12}, there are always two sets of values of F_{11}, F_{22} and F_{12} which satisfy the equations equally well. This is conveniently seen in graphic form by means of the so-called 'display method' first outlined by Person and Crawford[41] in which the diagonal force constants F_{11} and F_{22} are plotted in a graph as function of F_{12}. If we eliminate, for instance, F_{22} from equations (8.2.31) for the normal molecule and from equations (8.2.32) for the isotopic molecule, we obtain the equations of two ellipses in the F_{11}, F_{12} space, which intersect in two points as shown schematically in Figure 8.2.1.

In many cases one of the two sets of force constants is found to be physically unrealistic in terms of known ranges of values of force constants. In some other cases, however, both sets seem equally reasonable and it is difficult to select the correct one by inspection of the numerical values of the force constant. Recently McKean and Duncan[42] have analysed in

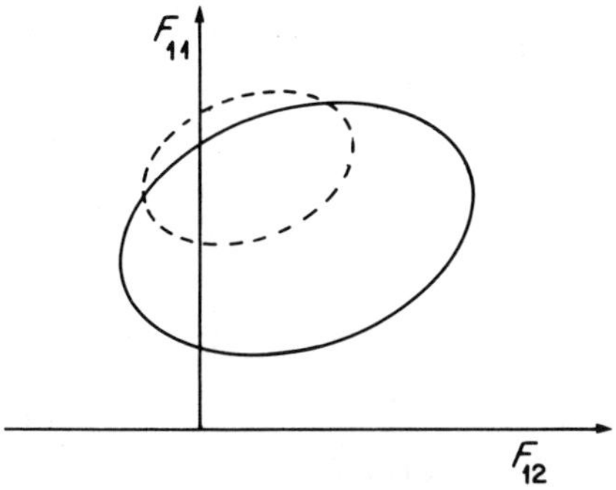

Figure 8.2.1 Schematic display of F_{11} as a function of F_{12} for a two-dimensional case. Solid and broken lines refer to the normal and deuterated molecule respectively

detail the problem of the two alternative sets of force constants for the two-dimensional case and have presented a useful table of selected examples which distinguishes between cases in which it is possible to choose a unique set of force constants, and those in which it is not.

An additional independent equation which would permit a choice between the two solutions can be obtained in principle by means of a different type of isotopic substitution, involving a different set of equivalent atoms. For example, in the case of a molecule like HCN, if the two isotopic species used before were $H^{12}C^{14}N$ and $D^{12}C^{14}N$, the third isotopic species could be $H^{13}C^{14}N$ or $H^{12}C^{15}N$. McKean and Duncan have, however, pointed out that, as consequence of the generalized sum rule,[37,38] this new isotopic substitution gives rise to an independent equation only if the molecule contains at least three sets of equivalent atoms. It follows then that for molecules like HCN (linear), SF_4Cl_2 (octahedral) H_2CO (planar), etc., the frequencies of the third isotopic species can be used to obtain a new independent equation, whereas for molecules like NH_3, CF_4, C_2H_2, C_2H_4, C_6H_6, etc., they give no new information.

In some special cases, however, asymmetric isotopic substitution, i.e. substitution of only part of the atoms of a set of equivalent ones, can supplement the additional equation required for a unique choice of the force field. Asymmetric substitution produces a lowering of the original symmetry and thus mixes together blocks of the secular equation which were separated in the original molecule. McKean and Duncan have shown that if the symmetry lowering mixes together two blocks, each with one degree of freedom for the atoms of the set, then the frequencies of the asymmetric isotopic species furnish independent information. If instead there is altogether only one degree of freedom for the atoms of the set in the two (or more) blocks, then the frequencies of the asymmetric species yield no new information.

These conclusions can easily be extrapolated to larger symmetry blocks. In general the greater the number of isotopic molecules available, the lower is the indeterminacy in the calculated force constants.

(D) *Small isotopic shifts*

Apart from the case of deuterium–hydrogen substitution discussed before, all other isotopic substitutions involve small differences in the atomic masses and thus small frequency shifts. A simple first-order perturbation treatment[43] can be very convenient for the calculations of these small isotopic shifts.

For the normal molecule we shall write the secular equation

$$\mathbf{G}_0\mathbf{F}\mathbf{L}_0 = \mathbf{L}_0\mathbf{\Lambda}_0 \tag{8.2.34}$$

in the equivalent form (4.2.15)

$$\mathbf{F}\mathbf{G}_0\tilde{\mathbf{L}}_0^{-1} = \tilde{\mathbf{L}}_0^{-1}\mathbf{\Lambda}_0 \tag{8.2.35}$$

which becomes by simple manipulation

$$\tilde{\mathbf{L}}_0\mathbf{F}\mathbf{G}_0\tilde{\mathbf{L}}_0^{-1} = \mathbf{\Lambda}_0 \tag{8.2.36}$$

For the isotopically substituted molecule, since the mass changes are small we can write $\mathbf{G}$ in the form

$$\mathbf{G} = \mathbf{G}_0 + \Delta\mathbf{G} \tag{8.2.37}$$

By substitution of (8.2.37) in the secular equation of the isotopic molecule, we have

$$\tilde{\mathbf{L}}\mathbf{F}\mathbf{G}\tilde{\mathbf{L}}^{-1} = \tilde{\mathbf{L}}\mathbf{F}(\mathbf{G}_0 + \Delta\mathbf{G})\tilde{\mathbf{L}}^{-1} = \tilde{\mathbf{L}}\mathbf{F}\mathbf{G}\tilde{\mathbf{L}}^{-1} + \tilde{\mathbf{L}}\mathbf{F}\,\Delta\mathbf{G}\tilde{\mathbf{L}}^{-1} = \mathbf{\Lambda} \tag{8.2.38}$$

Since the frequency shifts are small we can replace $\mathbf{L}$ by $\mathbf{L}_0$ in equation (8.2.38). The matrix $\mathbf{L}_0$ will not, however, diagonalize the $\mathbf{FG}$ matrix of the isotopic molecule completely, as $\mathbf{L}$ would. We can therefore write (8.2.38) in the form

$$\tilde{\mathbf{L}}_0\mathbf{F}\mathbf{G}\mathbf{L}_0^{-1} = \tilde{\mathbf{L}}_0\mathbf{F}\mathbf{G}_0\tilde{\mathbf{L}}_0^{-1} + \tilde{\mathbf{L}}_0\mathbf{F}\,\Delta\mathbf{G}\tilde{\mathbf{L}}_0^{-1} = \mathbf{\Lambda} + \mathbf{s} \tag{8.2.39}$$

where $\mathbf{s}$ is a non-diagonal matrix which accounts for the non-complete diagonalization achieved by substituting $\mathbf{L}$ by $\mathbf{L}_0$. By substitution of (8.2.36) into this equation, we obtain

$$\mathbf{\Lambda}_0 + \tilde{\mathbf{L}}_0\mathbf{F}\,\Delta\mathbf{G}\tilde{\mathbf{L}}_0^{-1} = \mathbf{\Lambda} + \mathbf{s} \tag{8.2.40}$$

Since (see equation 4.2.9)

$$\tilde{\mathbf{L}}_0\mathbf{F} = \mathbf{\Lambda}_0\mathbf{L}_0^{-1} \tag{8.2.41}$$

equation (8.2.40) can be rewritten in the form

$$\mathbf{\Lambda}_0 + \mathbf{\Lambda}_0(\mathbf{L}_0^{-1}\,\Delta\mathbf{G}\tilde{\mathbf{L}}_0^{-1}) = \mathbf{\Lambda} + \mathbf{s} \tag{8.2.42}$$

or

$$\mathbf{\Delta} = \mathbf{\Lambda}_0^{-1}\,\Delta\mathbf{\Lambda} = \mathbf{\Lambda}_0^{-1}(\mathbf{\Lambda} - \mathbf{\Lambda}_0) = \mathbf{L}_0^{-1}\,\Delta\mathbf{G}\tilde{\mathbf{L}}_0^{-1} - \mathbf{s}' \tag{8.2.43}$$

with elements

$$\begin{aligned} \Delta_{ii} &= \frac{\Delta\lambda_{ii}}{\lambda_i^0} = (L_0^{-1}\,\Delta G\tilde{L}_0^{-1})_{ii} - s'_{ii} = \sum_{k,l} (L_0^{-1})_{ik}(L_0^{-1})_{il}\,\Delta G_{kl} - s'_{ii} \\ \Delta_{ij} &= \frac{\Delta\lambda_{ij}}{\lambda_i^0} = (L_0^{-1}\,\Delta G\tilde{L}_0^{-1})_{ij} - s'_{ij} = \sum_{k,l} (L_0^{-1})_{ik}(L_0^{-1})_{jl}\,\Delta G_{kl} - s'_{ij} \end{aligned} \tag{8.2.44}$$

Since the elements of the $\mathbf{s}'$-matrix are very small, we can neglect their contribution to $\mathbf{\Delta}$ in (8.2.43). It is, however, necessary to remember that the $(\mathbf{L}_0^{-1}\,\Delta\mathbf{G}\tilde{\mathbf{L}}_0^{-1})$ matrix is non-diagonal and thus that the Δ_{kl} elements are different from zero.

The role of small isotopic shifts in determining force constants has been investigated by various authors and in particular by Duncan,[42] McKean[44,45] and Tsuboi.[46] It has been pointed out by these authors that the relative shifts Δ_{kk} are very significant for the determination of force constants, even more than the frequencies themselves, provided the frequencies are known with very high accuracy. In order to use this information for the refinement of force fields it is necessary to obtain the Jacobian elements $\partial\Delta_{kk}/\partial F_{ij}$. Equation (8.2.43) shows that a change in the force field affects the Δ_{kk} elements, to the first order, only through the changes in the elements of the **L**-matrix of the normal molecule.

According to Section 4.7, the changes in the **L**-matrix, due to a variation $\Delta\mathbf{F}$ in the force field, are given by the equation

$$\mathbf{L} = \mathbf{L}_0(\mathbf{E} + \mathbf{B}) \tag{8.2.45}$$

and thus, if $\boldsymbol{\Delta} + \delta\boldsymbol{\Delta}$ is the isotopic shift corresponding to the force field $\mathbf{F} + \Delta\mathbf{F}$, we obtain (to the first order) that the variation $\delta\boldsymbol{\Delta}$ of the shift is obtained, neglecting $\mathbf{s}'$ terms, through the equations

$$\boldsymbol{\Delta} + \delta\boldsymbol{\Delta} = (\mathbf{E} + \mathbf{B}^{-1})\mathbf{L}_0^{-1}\,\Delta\mathbf{G}\tilde{\mathbf{L}}_0^{-1}(\mathbf{E} + \tilde{\mathbf{B}}^{-1}) \tag{8.2.46}$$

i.e. using (8.2.43)

$$\delta\boldsymbol{\Delta} = \mathbf{B}^{-1}\boldsymbol{\Delta} + \boldsymbol{\Delta}\tilde{\mathbf{B}}^{-1} \tag{8.2.47}$$

Therefore (see equations (4.7.14) to (4.7.18)) the diagonal elements of $\delta\boldsymbol{\Delta}$ are

$$\delta\Delta_{ii} = 2\sum_{j\neq i} A_{ji}\Delta_{ij} = 2\sum_{j\neq i}\Delta_{ij}\sum_{k,l}\frac{(L_0)_{kj}(L_0)_{li}}{\lambda_i^0 - \lambda_j^0}\Delta F_{kl} \tag{8.2.48}$$

Letting $\delta\boldsymbol{\Delta} \to 0$ and $\Delta\mathbf{F} \to 0$ we obtain the Jacobian elements of $\boldsymbol{\Delta}$ with respect to $\mathbf{F}$ as the coefficients of ΔF_{kl}

$$\begin{aligned}\frac{\partial\Delta_{ii}}{\partial F_{kl}} &= 2\sum_{j\neq i}\Delta_{ij}\frac{(L_0)_{kj}(L_0)_{li} + (L_0)_{ki}(L_0)_{lj}}{\lambda_i^0 - \lambda_j^0}\\ \frac{\partial\Delta_{ii}}{\partial F_{kk}} &= 2\sum_{j\neq i}\Delta_{ij}\frac{(L_0)_{ki}(L_0)_{kj}}{\lambda_i^0 - \lambda_j^0}\end{aligned} \tag{8.2.49}$$

Consider again the 3×3 block of the secular equation, discussed in Section 8.2.A. In this case the Jacobian elements are

$$\frac{\partial\Delta_{aa}}{\partial F_{pp}} = 0 \qquad \frac{\partial\Delta_{bb}}{\partial F_{pp}} = 0 \qquad \frac{\partial\Delta_{cc}}{\partial F_{pp}} = 0$$

$$\frac{\partial\Delta_{aa}}{\partial F_{pq}} = \frac{2\Delta_{ab}}{\lambda_a - \lambda_b}L_{pa}L_{qb} + \frac{2\Delta_{ac}}{\lambda_a - \lambda_c}L_{pa}L_{qc} \qquad \frac{\partial\Delta_{bb}}{\partial F_{pq}} = \frac{2\Delta_{ab}}{\lambda_b - \lambda_a}L_{pa}L_{qb} \qquad \frac{\partial\Delta_{cc}}{\partial F_{pq}} = \frac{2\Delta_{ac}}{\lambda_c - \lambda_a}L_{pa}L_{qc}$$

$$\frac{\partial\Delta_{aa}}{\partial F_{pr}} = \frac{2\Delta_{ab}}{\lambda_a - \lambda_b}L_{pa}L_{rb} + \frac{2\Delta_{ac}}{\lambda_a - \lambda_c}L_{pa}L_{rc} \qquad \frac{\partial\Delta_{bb}}{\partial F_{pr}} = \frac{2\Delta_{ab}}{\lambda_b - \lambda_a}L_{pa}L_{rb} \qquad \frac{\partial\Delta_{cc}}{\partial F_{pr}} = \frac{2\Delta_{ac}}{\lambda_c - \lambda_a}L_{pa}L_{rc}$$

$$\frac{\partial\Delta_{aa}}{\partial F_{qq}} = 0 \qquad \frac{\partial\Delta_{bb}}{\partial F_{qq}} = \frac{2\Delta_{bc}}{\lambda_b - \lambda_c}L_{qb}L_{qc} \qquad \frac{\partial\Delta_{cc}}{\partial F_{qq}} = \frac{2\Delta_{bc}}{\lambda_c - \lambda_b}L_{qb}L_{qc}$$

$$\frac{\partial\Delta_{aa}}{\partial F_{qr}} = 0 \qquad \frac{\partial\Delta_{bb}}{\partial F_{qr}} = \frac{2\Delta_{bc}}{\lambda_b - \lambda_c}(L_{qb}L_{rc} + L_{qc}L_{rb}) \qquad \frac{\partial\Delta_{cc}}{\partial F_{qr}} = \frac{2\Delta_{bc}}{\lambda_c - \lambda_b}(L_{qb}L_{rc} + L_{qc}L_{rb})$$

$$\frac{\partial\Delta_{aa}}{\partial F_{rr}} = 0 \qquad \frac{\partial\Delta_{bb}}{\partial F_{rr}} = \frac{2\Delta_{bc}}{\lambda_b - \lambda_c}L_{rb}L_{rc} \qquad \frac{\partial\Delta_{cc}}{\partial F_{rr}} = \frac{2\Delta_{bc}}{\lambda_c - \lambda_b}L_{rb}L_{rc}$$

It is important to notice that the Jacobian elements $(\partial\Delta_{aa}/\partial F_{pq})$ and $(\partial\Delta_{aa}/\partial F_{pr})$ are different from zero even if the matrix elements L_{qa} and L_{ra} are zero. Isotopic shifts are therefore important to obtain information on the off-diagonal force constants for which the vibrational frequencies are ineffective.

8.3 Coriolis coupling constants

Coriolis coupling constants furnish another type of useful information for the refinement of intramolecular force fields. These constants are obtained experimentally from vibro-rotational or rotational spectra. For this reason they can be determined only for relatively small molecules. The basic theory of Coriolis constants has been discussed in Sections 2.7 and 4.8. Some important properties of the ζ-matrices were, however, neglected there, since their discussion requires a knowledge of group-theoretical methods. We shall discuss them presently. An important property of the ζ_{ij}^{σ} elements follows from the symmetry properties of equation (2.7.7)

$$\mathbf{\Omega}_{\sigma} = \tilde{\mathbf{Q}}\boldsymbol{\zeta}^{\sigma}\mathbf{Q} \tag{8.3.1}$$

Jahn[47,48] has shown that $\mathbf{\Omega}_{\sigma}$ transforms, under the symmetry operations of the molecular group, as the rotation R_{σ}. A proof of Jahn's theorem can also be found in Reference 60. From these symmetry properties follows the so-called 'Jahn rule' which states that an element ζ_{ij}^{σ} can be different from zero only if the direct product $\Gamma^{Q_i} \otimes \Gamma^{Q_j}$ of the irreducible representations carried by the normal coordinates Q_i and Q_j contains the representation $\Gamma^{R_{\sigma}}$ carried by the rotation R_{σ}.

The condition specified by Jahn's rule is necessary but not sufficient for ζ_{ij}^{σ} to be different from zero. According to equation (4.8.7) the elements of the ζ-matrix are given by

$$\zeta_{ij}^{\sigma} = \sum_{\alpha} (\mathscr{L}_{\alpha i}^{\tau}\mathscr{L}_{\alpha j}^{\rho} - \mathscr{L}_{\alpha i}^{\rho}\mathscr{L}_{\alpha j}^{\tau}) \qquad \sigma, \tau, \rho = x, y, z \tag{8.3.2}$$

where the sum extends over all atoms α of the molecule and where $\mathscr{L}_{\alpha i}^{\tau}$ is the element of the $\mathscr{L}$-matrix of equation (2.4.5) connecting the mass-weighted Cartesian displacement coordinate q_{α}^{τ} of atom α ($\tau = x, y, z$) with the normal coordinate Q_i. If we consider the three $\mathscr{L}$-matrix elements $\mathscr{L}_{\alpha i}^{\tau}$, $\mathscr{L}_{\alpha i}^{\rho}$ and $\mathscr{L}_{\alpha i}^{\sigma}$ relative to the atom α as the components of a vector $\mathbf{l}_{\alpha i}$, we can rewrite the preceding equation in the form

$$\zeta_{ij}^{\sigma} = \sum_{\alpha} (\mathbf{l}_{\alpha i} \times \mathbf{l}_{\alpha j}) \cdot \mathbf{e}_{\sigma} \tag{8.3.3}$$

where $\mathbf{e}_{\sigma}$ is a unit vector in the σ-direction. From this equation it follows that ζ_{ij}^{σ} is equal to zero either if $\mathbf{l}_{\alpha i}$ and $\mathbf{l}_{\alpha j}$ are parallel or if one of them is parallel to the σ-axis. Since these conditions occur very often, the number of non-zero ζ-elements is relatively small.

Another important property of the ζ-matrix follows directly from equation (2.7.8)

$$\boldsymbol{\zeta}^{\sigma} = \tilde{\mathscr{L}}\mathscr{M}^{\sigma}\mathscr{L}$$

which shows that the $\boldsymbol{\zeta}^{\sigma}$ and the $\mathscr{M}^{\sigma}$ matrices, being related by a similarity transformation, have the same properties. Since $\mathscr{M}^{\sigma}$ is skew-symmetric, $\boldsymbol{\zeta}^{\sigma}$ is also skew-symmetric and thus

$$\tilde{\boldsymbol{\zeta}}^{\sigma} = -\boldsymbol{\zeta}^{\sigma} \tag{8.3.4}$$

Further important properties shared by $\mathscr{M}^{\sigma}$ and $\boldsymbol{\zeta}^{\sigma}$ are the commutation relations

$$\boldsymbol{\zeta}^{\sigma}\tilde{\boldsymbol{\zeta}}^{\tau} - \boldsymbol{\zeta}^{\tau}\tilde{\boldsymbol{\zeta}}^{\sigma} = \boldsymbol{\zeta}^{\rho} \tag{8.3.5}$$

and the normalization relation

$$\boldsymbol{\zeta}^{x}\tilde{\boldsymbol{\zeta}}^{x} + \boldsymbol{\zeta}^{y}\tilde{\boldsymbol{\zeta}}^{y} + \boldsymbol{\zeta}^{z}\tilde{\boldsymbol{\zeta}}^{z} = 2\mathbf{E} \tag{8.3.6}$$

which can be easily verified by the reader.

Useful relations among the elements of the ζ-matrices can be obtained from some of the properties of the $\boldsymbol{\zeta}\tilde{\boldsymbol{\zeta}}$ and $\mathbf{E} - \boldsymbol{\zeta}\tilde{\boldsymbol{\zeta}}$ matrices, since these have simpler properties than the ζ-matrices themselves and their characteristic roots are real. We shall not discuss these properties since they are of little interest in vibrational spectroscopy. The reader interested in this argument is referred to the original paper by Meal and Polo[60] and also to References 49 and 56.

Using the properties of ζ described above it is easy to determine which Coriolis constant is different from zero for a given molecule.

Consider for instance a bent triatomic molecule of the XY_2 type (C_{2v} point group), with two A_1 and one B_1 normal modes, and assume that the z-axis coincides with the C_2 axis and that the y-axis is perpendicular to the molecular plane.

Since the rotation R_y is of species B_1, we obtain from Jahn's rule and from equation (8.3.4) that the only non-vanishing ζ-matrix is ζ^y, which has the form

$$\zeta^y = \begin{vmatrix} 0 & 0 & \zeta^y_{13} \\ 0 & 0 & \zeta^y_{23} \\ -\zeta^y_{13} & -\zeta^y_{23} & 0 \end{vmatrix}$$

In order to investigate the role that Coriolis coupling constants play in the refinement of force fields, we shall construct the Jacobian elements of ζ with respect to $\mathbf{F}$ (see Section 4.8)

$$\frac{\partial \zeta^\sigma_{ij}}{\partial F_{kl}} = \sum_{t \neq i} \zeta^\sigma_{jt} \frac{L_{ki}L_{lt} + L_{li}L_{kt}}{\lambda_t - \lambda_i} - \sum_{t \neq j} \zeta^\sigma_{it} \frac{L_{kj}L_{lt} + L_{lj}L_{kt}}{\lambda_t - \lambda_j}$$

$$\frac{\partial \zeta^\sigma_{ij}}{\partial F_{kk}} = \sum_{t \neq i} \zeta^\sigma_{jt} \frac{L_{ki}L_{kt}}{\lambda_t - \lambda_i} - \sum_{t \neq j} \zeta^\sigma_{it} \frac{L_{kj}L_{kt}}{\lambda_t - \lambda_j}$$

for the above discussed problem of the XY_2 molecule.

Since the $\mathbf{L}$- and $\mathbf{F}$-matrices have the form

$$\begin{vmatrix} L_{11} & L_{12} & 0 \\ L_{21} & L_{22} & 0 \\ 0 & 0 & L_{33} \end{vmatrix} \qquad \begin{vmatrix} F_{11} & F_{12} & 0 \\ F_{12} & F_{22} & 0 \\ 0 & 0 & F_{33} \end{vmatrix}$$

we obtain

$$\frac{\partial \zeta^y_{13}}{\partial F_{11}} = \zeta^y_{23} \frac{L_{11}L_{12}}{\lambda_1 - \lambda_2} \quad \frac{\partial \zeta^y_{13}}{\partial F_{12}} = \zeta^y_{23} \frac{(L_{11}L_{22} + L_{12}L_{21})}{\lambda_1 - \lambda_2} \quad \frac{\partial \zeta^y_{13}}{\partial F_{22}} = \zeta^y_{23} \frac{L_{21}L_{22}}{\lambda_1 - \lambda_2} \quad \frac{\partial \zeta^y_{13}}{\partial F_{33}} = 0$$

$$\frac{\partial \zeta^y_{23}}{\partial F_{11}} = \zeta^y_{13} \frac{L_{11}L_{12}}{\lambda_1 - \lambda_2} \quad \frac{\partial \zeta^y_{23}}{\partial F_{12}} = \zeta^y_{13} \frac{(L_{11}L_{22} + L_{12}L_{21})}{\lambda_1 - \lambda_2} \quad \frac{\partial \zeta^y_{23}}{\partial F_{22}} = \zeta^y_{13} \frac{L_{21}L_{22}}{\lambda_1 - \lambda_2} \quad \frac{\partial \zeta^y_{23}}{\partial F_{33}} = 0$$

It is important to notice from these equations that even in the absence of coupling between s_1 and s_2 in the A_1 block, the Jacobian elements $(\partial \zeta^y_{13}/\partial F_{12})$ and $(\partial \zeta^y_{23}/\partial F_{12})$ are different from zero. The Coriolis coupling constants therefore furnish information on the off-diagonal terms of the $\mathbf{F}$-matrix that the vibrational frequencies cannot give.

Coriolis coupling constants have been measured for a large variety of small molecules. A collection of calculated ζ-elements can be found in Cyvin's book.[49] Observed values of ζ-constants, measured from vibro-rotational infrared bands or from microwave spectra, are reported in References 78–104. The Coriolis constants obey sum rules which depend upon the molecular geometry and the atomic masses, but not upon the molecular force field. The most familiar sum rules are the 'linear sum rule' for symmetric[50–54] and spherical[54,55,56] top molecules and the 'quadratic sum rule' for any type of molecule[57,58,59,60] which are widely used in the determination of Coriolis constants.

8.4 Centrifugal distortion constants

The centrifugal stretching constants, when available, constitute another source of useful information for the refinement of the force field of simple polyatomic molecules. These constants are determined experimentally from the analysis of rotational or vibro-rotational transitions in microwave and infrared spectra and are thus limited to small molecules. The experimental data need some further manipulation before they can be expressed in terms of constants, called 'τ-constants' which are directly related to the elements of the inverse F-matrix.

The general theory of the centrifugal distortion constants has been developed by Kivelson and Wilson.[61] Recent developments are discussed in the paper by Kirchhoff[62] and also in References 63 and 64. A brief summary of the main results of the Kivelson–Wilson treatment is given below. For the derivation of the various equations the reader is referred to the original papers.

The energy levels of a non-rigid rotor are given, to the first order, by

$$E = E_0 + A_1E_0^2 + A_2E_0J(J+1) + A_3J^2(J+1)^2 + A_4J(J+1)\langle P_z^2\rangle + A_5\langle P_z^4\rangle + A_6E_0\langle P_z^2\rangle \quad (8.4.1)$$

where E_0 is the energy level of the corresponding rigid rotor, $\langle P_z^2\rangle$ and $\langle P_z^4\rangle$ are the average values of the operators P_z^2 and P_z^4 respectively, J is the rotational quantum number and $A_1, A_2, \ldots, A_6$ are constants independent of J.

Kivelson and Wilson have shown that these six constants are conveniently expressed in terms of some other constants τ defined in the relationship

$$\tau_{\alpha\beta\gamma\delta} = -\frac{1}{2I^0_{\alpha\alpha}I^0_{\beta\beta}I^0_{\gamma\gamma}I^0_{\delta\delta}}\sum_{k,l}\left(\frac{\partial I_{\alpha\beta}}{\partial s_k}\right)_0\left(\frac{\partial I_{\gamma\delta}}{\partial s_l}\right)_0(F^{-1})_{kl} \quad (8.4.2)$$

where s_k and s_l designate internal or symmetry coordinates, $\alpha, \beta, \gamma, \delta$ Cartesian coordinates and $I^0_{\alpha\alpha}$, $I_{\alpha\beta}$, etc., are the components of the inertia tensor

$$\begin{aligned} I_{\alpha\alpha} &= \sum_i m_i(\beta^2 + \gamma^2) \\ I_{\alpha\beta} &= -\sum_i m_i\alpha_i\beta_i \end{aligned} \quad (8.4.3)$$

The constants $A_1, A_2, \ldots, A_6$ and the constants $\tau_{\alpha\beta\gamma\delta}$ are related by the following equations

$$\begin{aligned} A_1 &= 16R_6/(B-C)^2 \\ A_2 &= -[16R_6(B+C)/(B-C)^2 + 4\delta_J/(B-C)] \\ A_3 &= -D_J + 2R_6 + 16R_6BC/(B-C)^2 + 2\delta_J(B+C)/(B-C) \\ A_4 &= -[D_{Jk} - 2\delta_J\sigma - 16R_6(A^2 - BC)/(B-C)^2 + 4R_6\sigma^2 + 4R_5(B+C)/(B-C)] \\ A_5 &= -[D_k + 4R_5\sigma + 2R_6 - 4R_6\sigma^2] \\ A_6 &= (8R_5 - 16R_6\sigma)/(B-C) \end{aligned} \quad (8.4.4)$$

where

$$\begin{aligned} A &= \hbar^2/2I^0_{zz} + \hbar^4(3\tau_{xyxy} - 2\tau_{zxzx} - 2\tau_{yzyz})/4 \\ B &= \hbar^2/2I^0_{xx} + \hbar^4(3\tau_{yzyz} - 2\tau_{xyxy} - 2\tau_{zxzx})/4 \\ C &= \hbar^2/2I^0_{yy} + \hbar^4(3\tau_{zxzx} - 2\tau_{xyxy} - 2\tau_{yzyz})/4 \\ R_5 &= -\hbar^4[\tau_{xxxx} - \tau_{yyyy} - 2(\tau_{xxzz} + 2\tau_{xzxz}) + 2(\tau_{yyzz} + 2\tau_{yzyz})]/32 \\ R_6 &= \hbar^4[\tau_{xxxx} + \tau_{yyyy} - 2(\tau_{xxyy} + 2\tau_{xyxy})]/64 \\ D_J &= -\hbar_4[3\tau_{xxxx} + 3\tau_{yyyy} + 2\tau_{xxyy} + 4\tau_{xyxy}]/32 \end{aligned} \quad (8.4.5)$$

$$D_K = D_J - \hbar^4[\tau_{zzzz} - \tau_{zzxx} - \tau_{yyzz} - 2\tau_{xzxz} - 2\tau_{yzyz}]/4$$
$$D_{JK} = -D_J - D_K - \hbar_4\tau_{zzzz}/4$$
$$\sigma = (2A - B - C)/(B - C)$$
$$\delta_J = -\hbar^4(\tau_{xxxx} - \tau_{yyyy})/16$$

The expectation values of $\langle P_z^4\rangle$ and $\langle P_z^2\rangle$, necessary for the use of equation (8.4.1), can be evaluated in a straightforward way, following the method described by Kivelson and Wilson.

For symmetric top molecules equation (8.4.1) can be considerably simplified. In this case it is easily seen from the above relations that $B = C$ and that $R_5 = R_6 = \delta_j = 0$. It can be shown[61] that the factors multiplying R_5, R_6 and δ_j in (8.4.4) are finite and since for the symmetric top $\langle P_z^4\rangle = \langle P_z^2\rangle^2 = K^4$, where K is the quantum number associated with the operator P_z, equation (8.4.1) reduces to

$$E = E_0 - D_J J^2(J + 1)^2 - D_{JK}J(J + 1)K^2 - D_K K^4 \tag{8.4.6}$$

The energy levels of linear and spherical tops are given by the much simpler expression

$$E = E_0 - DJ^2(J + 1)^2 \tag{8.4.7}$$

The non-vanishing $\tau_{\alpha\beta\gamma\delta}$ constants can be easily determined for each molecule from symmetry considerations.[61] From equations (8.4.3) it is clear that the components of the inertia tensor transform under symmetry operations as the components of the dyadic product formed by translation vectors. Thus, if $\mathscr{R}$ is a symmetry operator of the molecular group,

$$\begin{aligned} \mathscr{R}I_{\alpha\beta} &\equiv \mathscr{R}(\alpha\beta) \\ \mathscr{R}I_{\alpha\alpha} &\equiv \mathscr{R}(\beta^2 + \gamma^2) \end{aligned} \tag{8.4.8}$$

This behaviour is exactly the same as that discussed in Section 7.3 for the components of the polarizability tensor α and can thus be obtained directly from the tables of Appendix I.

Let us now expand a component of the inertia tensor in terms of a set of independent symmetry coordinates

$$I_{\alpha\beta} = I_{\alpha\beta}^0 + \sum_k \left(\frac{\partial I_{\alpha\beta}}{\partial s_k}\right) s_k + \ldots + \ldots = I_{\alpha\beta}^0 + \sum_k J_{\alpha\beta}^k s_k \tag{8.4.9}$$

If $I_{\alpha\beta}$ carries the irreducible representation $\Gamma^{\alpha\beta}$, only the symmetry coordinates which belong to the same irreducible representation can have non-vanishing coefficients in equation (8.4.9). We conclude then that the only non-zero $J_{\alpha\beta}^k$ elements are those for which the symmetry coordinates belong to the same irreducible representation to which $I_{\alpha\beta}$ belongs. Furthermore, since the $\mathbf{F}^{-1}$ matrix is also factorized in symmetry blocks, all elements F_{kl}^{-1} of equation (8.4.2) vanish unless the symmetry coordinates s_k and s_l belong to the same irreducible representation.

From these considerations it follows directly that the only non-vanishing $\tau_{\alpha\beta\gamma\delta}$ elements (or linear combinations of the τs for degenerate species) belong to the totally symmetric representation of the molecular group.

Consider for instance the bent triatomic molecule XY_2 discussed in the previous section. From Appendix I one obtains that the components of the inertia tensor carry the irreducible representations shown below

I_{xx}, I_{yy}, I_{zz}	A_1
I_{xy}	A_2
I_{xz}	B_1
I_{yz}	B_2

Since two symmetry coordinates belong to the A_1 species (S_1 and S_2) and one to the B_1 species (S_3), we obtain that the only non-vanishing elements are

$$\tau_{xxxx},\ \tau_{yyyy},\ \tau_{zzzz}$$

$$\tau_{xxyy} = \tau_{yyxx}$$

$$\tau_{xxzz} = \tau_{zzxx}$$

$$\tau_{yyzz} = \tau_{zzyy}$$

$$\tau_{xzxz} = \tau_{xzzx} = \tau_{zxzx} = \tau_{zxxz}$$

These seven non-zero τ elements are actually not all independent. Since the molecule is planar, one finds

$$I_{yy} = I_{xx} + I_{zz} \tag{8.4.10}$$

From this relation, by means of (8.4.2) it is easy to show that

$$\begin{aligned} I^4_{yy}\tau_{yyyy} &= I^4_{zz}\tau_{zzzz} + I^4_{xx}\tau_{xxxx} + 2I^2_{xx}I^2_{zz}\tau_{xxzz} \\ I^2_{yy}I^2_{xx}\tau_{yyxx} &= I^4_{xx}\tau_{xxxx} + I^2_{xx}I^2_{zz}\tau_{zzxx} \\ I^2_{yy}I^2_{zz}\tau_{yyzz} &= I^4_{zz}\tau_{zzzz} + I^2_{xx}I^2_{zz}\tau_{xxzz} \end{aligned}$$

There are therefore only four independent constants, i.e. τ_{xxxx}, τ_{zzzz}, τ_{xxzz} and τ_{xzxz}.

In order to investigate the role that centrifugal distortion constants play in the refinement of a force field, we recall the analytical form of the Jacobian elements of the τ constants with respect to **F**, given by equations (4.9.11)

$$\begin{aligned} \left(\frac{\partial\tau_{\alpha\beta\gamma\delta}}{\partial F_{rt}}\right) &= \frac{1}{2I_{\alpha\alpha}I_{\beta\beta}I_{\gamma\gamma}I_{\delta\delta}}\sum_{k,l} J^k_{\alpha\beta}J^l_{\gamma\delta}[(F^{-1})_{kr}(F^{-1})_{tl} + (F^{-1})_{kt}(F^{-1})_{rl}] \\ \left(\frac{\partial\tau_{\alpha\beta\gamma\delta}}{\partial F_{rr}}\right) &= \frac{1}{2I_{\alpha\alpha}I_{\beta\beta}I_{\gamma\gamma}I_{\delta\delta}}\sum_{k,l} J^k_{\alpha\beta}J^l_{\gamma\delta}(F^{-1})_{kr}(F^{-1})_{rl} \end{aligned} \tag{8.4.11}$$

where

$$J^k_{\alpha\beta} = \left(\frac{\partial I_{\alpha\beta}}{\partial s_k}\right)_0 \qquad J^l_{\gamma\delta} = \left(\frac{\partial I_{\gamma\delta}}{\partial s_l}\right)_0 \tag{8.4.12}$$

For the use of these equations it is unnecessary to evaluate the elements of the $\mathbf{F}^{-1}$ matrix by inversion of the **F**-matrix. From equation (4.2.9)

$$\tilde{\mathbf{L}}\mathbf{F}\mathbf{L} = \mathbf{\Lambda}$$

we obtain

$$\mathbf{F}^{-1} = \mathbf{L}\mathbf{\Lambda}^{-1}\tilde{\mathbf{L}} \tag{8.4.13}$$

and thus the element $(F^{-1})_{kr}$ is given by

$$(F^{-1})_{kr} = \sum_i \lambda_i^{-1}L_{ki}L_{ir} \tag{8.4.14}$$

The derivatives (8.4.11) will have an appreciable value only if the elements $J^k_{\alpha\beta} = (\partial I_{\alpha\beta}/\partial s_k)$ and $J^l_{\gamma\delta} = (\partial I_{\gamma\delta}/\partial s_l)$ are large. Kivelson and Wilson[61] have given the recipes for the construction of these derivatives for the various types of internal coordinates (stretching, bending, torsion, etc.). Since the treatment is complex we shall not enter into the details of the calculations. Recently, Pulay and Savodny[63] have proposed a simpler method whose essential features can be easily summarized.

From equations (8.4.12) and (8.4.3) we have

$$J^k_{\alpha\alpha} = \left(\frac{\partial I_{\alpha\alpha}}{\partial s_k}\right)_0 = 2\sum_i m_i\left(\beta_i\frac{\partial \beta_i}{\partial s_k} + \gamma_i\frac{\partial \gamma_i}{\partial s_k}\right) \qquad \alpha, \beta, \gamma = x, y, z \tag{8.4.15}$$
$$J^k_{\alpha\beta} = \left(\frac{\partial I_{\alpha\beta}}{\partial s_k}\right)_0 = -\sum_i m_i\left(\alpha_i\frac{\partial \beta_i}{\partial s_k} + \beta_i\frac{\partial \alpha_i}{\partial s_k}\right)$$

The only problem in these equations is to evaluate the derivatives $(\partial\alpha_i/\partial s_k)$, etc. The inverse derivatives $(\partial s_k/\partial\alpha_i)$ are, however, simply the elements of the **B**-matrix since (see equation 4.1.2)

$$\mathbf{s} = \mathbf{Bx} \tag{8.4.16}$$

The **B**-matrix is not square and thus cannot be inverted. We can, however, always use the six Eckart conditions as dummy coordinates to define a $\mathbf{B}^{-1}$ matrix which is then given, using (4.2.5) and (4.2.11), by

$$\mathbf{B}^{-1} = \mathbf{M}^{-1}\tilde{\mathbf{B}}\mathbf{G}^{-1} = \mathbf{M}^{-1}\tilde{\mathbf{B}}\tilde{\mathbf{L}}^{-1}\mathbf{L}^{-1} \tag{8.4.17}$$

For computer calculations the derivatives of the inertia tensor components can be more conveniently expressed by means of auxiliary matrices which reduce the two equations (8.4.15) to a single matrix expression. Pulay and Sawodny[63] and Sørensen, Hagen and Cyvin[64] have discussed the form of these auxiliary matrices.

8.5 Mean-square amplitudes of vibrations

Mean-square amplitudes are also quantities related to the force fields and thus can be used for the determination of force constants, if obtained by some independent method. Actually, mean-square amplitudes can be obtained from electron diffraction techniques for the gaseous state. This technique is limited for the moment to relatively small molecules. It must be pointed out, however, that X-ray diffraction spectra constitute in principle an immense source of information of this kind for crystalline materials.

Although the mean-square amplitudes are not directly measured in electron diffraction experiments, but rather calculated through complicated procedures, they are normally referred to as 'observed' in the broad sense that they can be considered as additional data for the calculation of force constants.

The theory of mean-square amplitudes, as well as its application to a great number of molecules, is described in detail in the book by S. J. Cyvin[49] entitled *Molecular Vibrations and Mean Square Amplitudes*. In this section we shall limit ourselves to a brief summary of the theoretical treatment, necessary to understand how these quantities can be used for the evaluation of a force field.

In the harmonic approximation, the mean-square amplitude of a normal coordinate Q_i is given by

$$\delta_i = \langle Q_i^2 \rangle = \frac{h}{8\pi^2 c\omega_i}\coth\left(\frac{hc\omega_i}{2KT}\right) \tag{8.5.1}$$

where ω_i is the (harmonic) vibrational frequency in cm^{-1}, h is Planck's constant, K Boltzmann's constant and T the absolute temperature. For the derivation of this relation, first obtained by Bloch[65] and by James,[66] see Cyvin's book.[49] Since we have also[49]

$$\langle Q_i Q_j^* \rangle = 0 \quad \text{for} \quad i \neq j \tag{8.5.2}$$

the mean-square amplitudes of the normal coordinates can be arranged in the form of a diagonal matrix

$$\boldsymbol{\delta} = \langle \mathbf{Q}\tilde{\mathbf{Q}} \rangle \tag{8.5.3}$$

whose non-zero elements are given by (8.5.1).

The mean-square amplitude matrices for other coordinates are simply obtained from (8.5.3) by using the linear transformations between the normal and the other coordinates. Thus, from the relations

$$\begin{aligned} \mathbf{s} &= \mathbf{L}\mathbf{Q} && \text{(a)} \\ \mathbf{q} &= \mathscr{L}\mathbf{Q} && \text{(b)} \quad (8.5.4) \\ \mathbf{x} &= \mathbf{M}^{-\frac{1}{2}}\mathscr{L}\mathbf{Q} && \text{(c)} \end{aligned}$$

we obtain

$$\begin{aligned} \boldsymbol{\Sigma}^{s} &= \langle \mathbf{s}\tilde{\mathbf{s}} \rangle = \mathbf{L}\langle \mathbf{Q}\tilde{\mathbf{Q}} \rangle \tilde{\mathbf{L}} && = \mathbf{L}\boldsymbol{\delta}\tilde{\mathbf{L}} && \text{(a)} \\ \boldsymbol{\Sigma}^{q} &= \langle \mathbf{q}\tilde{\mathbf{q}} \rangle = \mathscr{L}\langle \mathbf{Q}\tilde{\mathbf{Q}} \rangle \tilde{\mathscr{L}} && = \mathscr{L}\boldsymbol{\delta}\tilde{\mathscr{L}} && \text{(b)} \quad (8.5.5) \\ \boldsymbol{\Sigma}^{x} &= \langle \mathbf{x}\tilde{\mathbf{x}} \rangle = \mathbf{M}^{-\frac{1}{2}}\mathscr{L}\langle \mathbf{Q}\tilde{\mathbf{Q}} \rangle \tilde{\mathscr{L}}\mathbf{M}^{-\frac{1}{2}} && = \mathbf{M}^{-\frac{1}{2}}\mathscr{L}\boldsymbol{\delta}\tilde{\mathscr{L}}\mathbf{M}^{-\frac{1}{2}} && \text{(c)} \end{aligned}$$

From equation (8.5.1) we can now obtain an expression which relates the mean-square amplitudes and the **F**-matrix, in the following way. For the hyperbolic cotangent we can use the expansion[49]

$$x \coth x = 1 + \frac{x^2}{3} - \frac{x^4}{45} + \ldots \tag{8.5.6}$$

with $x = hc\omega_i/2KT$.

From (8.5.1) and (8.5.6) we easily obtain for the ith diagonal element of the $\boldsymbol{\delta}$-matrix

$$\langle Q_i^2 \rangle = KT\left[\frac{1}{\lambda_i} + \frac{1}{48}\left(\frac{h}{\pi KT}\right)^2 - \frac{1}{11520}\left(\frac{h}{\pi KT}\right)^4 \lambda_i + \ldots\right] \tag{8.5.7}$$

Equation (8.5.3) becomes then

$$\boldsymbol{\delta} = KT\left[\boldsymbol{\Lambda}^{-1} + \frac{1}{48}\left(\frac{h}{\pi KT}\right)^2 \mathbf{E} - \frac{1}{11520}\left(\frac{h}{\pi KT}\right)^4 \boldsymbol{\Lambda} + \ldots\right] \tag{8.5.8}$$

Correspondingly the matrix $\boldsymbol{\Sigma}^{s}$ in internal coordinates, given by (8.5.5a), becomes

$$\boldsymbol{\Sigma}^{s} = \mathbf{L}\boldsymbol{\delta}\tilde{\mathbf{L}} = KT\left[\mathbf{F}^{-1} + \frac{1}{48}\left(\frac{h}{\pi KT}\right)^2 \mathbf{G} - \frac{1}{11520}\left(\frac{h}{\pi KT}\right)^4 \mathbf{GFG} + \ldots\right] \tag{8.5.9}$$

The diagonal elements of (8.5.9) are the mean-square amplitudes of the internal coordinates used. Off-diagonal terms are called interaction mean-square amplitudes.

If symmetry coordinates are used equation (8.5.9) becomes

$$\boldsymbol{\Sigma} = \mathbb{L}\boldsymbol{\delta}\tilde{\mathbb{L}} = KT\left[\mathbb{F}^{-1} + \frac{1}{48}\left(\frac{h}{\pi KT}\right)^2 \mathbb{G} - \frac{1}{11520}\left(\frac{h}{\pi KT}\right)^4 \mathbb{GFG} + \ldots\right] \tag{8.5.10}$$

where $\boldsymbol{\Sigma}$ is now the matrix of the mean-square amplitudes in symmetry coordinates. In Cyvin's book the reader can find detailed calculations of the mean-square symmetry amplitudes for individual molecules.

The use of equations (8.5.8), (8.5.9) and (8.5.10) is rather complicated if all terms of the hyperbolic cotangent expansion have to be included. Many authors have tried to cut the expansion at the first two or three terms but the results turn out to be too inaccurate at temperatures below 1000 K except for molecules with very low vibrational frequencies. For this reason Morino *et al.*[67] have proposed the approximate expression

$$\coth x \approx 1/x + x^2/4 \tag{8.5.11}$$

for the hyperbolic cotangent.

Using this approximation equation (8.5.9) becomes

$$\mathbf{\Sigma}^s = KT\mathbf{F}^{-1} + \frac{h^2}{64\pi^2 KT}\mathbf{G} \tag{8.5.12}$$

The mean amplitude of the internal coordinate s_k describing the variation in the distance d_{ij} between two (bonded or non-bonded) atoms i and j is thus given by

$$\langle s_k^2 \rangle = \langle d_{ij}^2 \rangle = KTF_{kk}^{-1} + \frac{h^2}{64\pi^2 KT}(\mu_i + \mu_j) \tag{8.5.13}$$

where $(\mu_i + \mu_j)$ is the **G**-matrix element for the coordinate $s_k = d_{ij}$. In order to construct the Jacobian elements of the mean-square amplitudes with respect to the **F**-matrix elements we can use the treatment given in Section 4.9 for the centrifugal distortion constants which are also related to the $\mathbf{F}^{-1}$ elements.

Using equation (4.9.9) we see that a variation $\Delta\mathbf{F}$ in the force field produces a variation $\Delta\langle d_{ij}^2 \rangle$ in the mean-square amplitude, given by

$$\Delta\langle d_{ij}^2 \rangle = -KT \sum_{rt} F_{kr}^{-1} F_{tk}^{-1} \, \Delta F_{rt} \tag{8.5.14}$$

and therefore the Jacobian elements of $\langle d_{ij}^2 \rangle$ with respect to **F** are given by

$$\begin{aligned} \frac{\partial\langle d_{ij}^2 \rangle}{\partial F_{rr}} &= -KTF_{kr}^{-1}F_{rk}^{-1} \\ \frac{\partial\langle d_{ij}^2 \rangle}{\partial F_{rt}} &= -KT[F_{kr}^{-1}F_{tk}^{-1} + F_{kt}^{-1}F_{rk}^{-1}] \end{aligned} \tag{8.5.15}$$

8.6 The calculation of force constants

In Section 4.7 we have discussed the calculation of force constants in terms of internal coordinates and presented a first-order perturbation treatment for the refinement of an initial trial **F**-matrix (in internal coordinates) using as data the vibrational frequencies. In following sections we have extended the perturbation method to include in the refinement additional data such as Coriolis coupling constants (Sections 4.8 and 8.3), centrifugal distortion constants (Sections 4.9 and 8.4) and mean-square amplitudes of vibration (Section 8.5). In Section 6.5 we have then introduced symmetry coordinates and shown how a symmetry adapted basis factorizes the secular equation.

In this section we shall first give a unified treatment of the whole process and then discuss the problems that arise in practical cases. For the convenience of the reader we shall proceed by successive steps, recalling at each stage the necessary equations and the sections in which they have been discussed.

Step 1. *Choice of an internal coordinate basis*

This step does not require a detailed discussion. The internal coordinates are of the type described in Section 4.1 (stretching, bending, torsion, etc.). The number of internal coordinates to use is seldom equal to the number of internal degrees of freedom. In general a certain number of redundant coordinates (Section 6.7) are included. The number of redundancies increases rapidly with the molecular dimension.

Step 2. *Construction of the* **G**-*matrix*

If m internal coordinates have been chosen the **G**-matrix will be of dimension $m \times m$. The construction of the **G**-matrix is a standard procedure described in Section 4.5. If redundancies are included in the basic set of internal coordinates the **G**-matrix is singular, some rows (or columns) being linear combinations of others (Section 6.7). This produces as a consequence the occurrence of $m - (3N - 6)$ zero roots in the secular equation.

Step 3. *Construction of the* **F**-*matrix*

The **F**-matrix is also of dimension $m \times m$. If a given model force field is used (Section 8.1) not all elements of the **F**-matrix are different from zero. For some models, such as the Urey–Bradley force field (Section 8.1) the elements of the **F**-matrix in internal coordinates are linear combinations of some few force constants. In order to be able to use any kind of potential function with any kind of constraint among the force constants, it is convenient to use the following method.

Let $\boldsymbol{\phi}$ be a vector of the non-zero force constants one wishes to use either by choice or because the model potential adopted requires it. From this vector we can construct a new vector $\overline{\mathbf{F}}$ whose elements are simply the elements of the upper triangular part of the **F**-matrix, through the relation

$$\overline{\mathbf{F}} = \mathbf{Z}\boldsymbol{\phi} \tag{8.6.1}$$

where **Z** is a matrix with non-zero elements only in the rows corresponding to non-zero **F**-matrix elements.

The order of the **F**-matrix elements in the $\overline{\mathbf{F}}$-vector is normally chosen so that the first m elements of $\overline{\mathbf{F}}$ are the elements of the first row of **F** (in successive order), the next $m - 1$ elements are the elements of the second row except the first, the next $m - 2$ elements are the elements of the third row of **F** except the first two and so on up to the last element of **F**. The $\overline{\mathbf{F}}$-vector thus has $m(m + 1)/2$ elements and the **Z**-matrix has $m(m + 1)/2 \times r$ elements, if r is the number of elements of the $\boldsymbol{\phi}$-vector.

The construction of the **Z**-matrix is very simple and is normally accomplished by inspection of the terms of the Taylor expansion of the potential function. In the limiting case in which a force field expanded in internal coordinates is used, the elements of **Z** are all zero except one element in each row corresponding to a non-zero **F**-matrix element, which is equal to one. If instead the potential used requires

$$F_{ij} = \sum_{k=1}^{l} Z_{ij}^{k}\phi_k \tag{8.6.2}$$

then there are l non-zero elements in the row of **Z** corresponding to the element F_{ij}.

Step 4. *Symmetrization of* **F** *and* **G**

If the molecule possesses any symmetry, the computational problem is greatly simplified by use of symmetry coordinates (Section 6.5). The transformation from internal to symmetry coordinates is of the type

$$\mathbb{S} = \mathbb{U}\mathbf{s} \tag{8.6.3}$$

where $\mathbb{S}$ is the vector of the symmetry coordinates and $\mathbb{U}$ is a unitary matrix. The construction of the elements of the $\mathbb{U}$-matrix is described in Section 6.6.

The **F**- and **G**-matrices are then symmetrized by means of the similarity transformations

$$\mathbb{F} = \mathbb{U}\mathbf{F}\tilde{\mathbb{U}} \tag{8.6.4}$$

$$\mathbb{G} = \mathbb{U}\mathbf{G}\tilde{\mathbb{U}} \tag{8.6.5}$$

as discussed in Section 6.5.

The $\mathbb{F}$- and $\mathbb{G}$-matrices are in block form, the number and the dimension of the blocks depending upon the molecular dimension and symmetry. As a consequence the product $\mathbb{G}\mathbb{F}$ is also in block form. If we call $\mathbb{G}^\alpha$ the block of the $\mathbb{G}$-matrix belonging to the Γ^α irreducible representation and $\mathbb{F}^\alpha$ the corresponding block of $\mathbb{F}$, then the matrix $\mathbb{H}^\alpha = \mathbb{G}^\alpha\mathbb{F}^\alpha$ satisfies the secular equation

$$\mathbb{G}^\alpha\mathbb{F}^\alpha\mathbb{L}^\alpha = \mathbb{L}^\alpha\Lambda^\alpha \tag{8.6.6}$$

The secular equation in internal coordinates factorizes, therefore, into a number of secular equations of smaller dimension; this number is obviously equal to the number of blocks into which **G** and **F** are factorized.

Step 5. *Solution of the secular equation*

The secular equations (8.6.6) can be solved using any of the several methods of numerical analysis available for the calculation of the eigenvectors and eigenvalues of a matrix. Since methods for the diagonalization of symmetric matrices are fast, it is convenient to reduce each $\mathbb{G}^\alpha\mathbb{F}^\alpha$ matrix to symmetric form according to the methods of Section 4.7.

In symmetry coordinates the equations of Section 4.7 become, for the α-block,

$$\mathbb{G}^\alpha = \mathbb{W}^\alpha\tilde{\mathbb{W}}^\alpha \tag{8.6.7}$$

$$(\tilde{\mathbb{W}}^\alpha\mathbb{F}^\alpha\mathbb{W}^\alpha)\mathbb{C}^\alpha = \mathbb{C}^\alpha\mathbf{\Lambda}^\alpha \tag{8.6.8}$$

$$\mathbb{L}^\alpha = \mathbb{W}^\alpha\mathbb{C}^\alpha \tag{8.6.9}$$

Solution of the secular equation (8.6.8) therefore yields n_α eigenvalues (n_α = dimension of the α-block) λ_i^α as diagonal elements of the matrix $\mathbf{\Lambda}^\alpha$, as well as the $\mathbb{C}^\alpha$ matrix of dimension $n_\alpha \times n_\alpha$. From the $\mathbb{C}^\alpha$ matrix the $\mathbb{L}^\alpha$ matrix is easily obtained through equation (8.6.9).

Step 6. *Calculation of the potential energy distribution*

A convenient type of information on the contribution of each force constant F_{ij} to the normal frequencies of vibration is furnished by the so-called potential energy distribution (PED). We recall from Section 2.4 that in a normal coordinate basis the potential energy is given by

$$2V = \sum_i \lambda_i Q_i^2 \tag{8.6.10}$$

Each λ_i can thus be considered as a measure of the potential energy for a unit displacement of the corresponding normal coordinate Q_i, all other normal coordinates being at rest. According to equation (4.2.9) each normal frequency parameter λ_i is given by

$$\lambda_i = \sum_{jk} L_{ji}L_{ki}F_{jk} \tag{8.6.11}$$

from which it follows that

$$\sum_{jk} (L_{ji}L_{ki}F_{jk})/\lambda_i = 1 \tag{8.6.12}$$

We conclude therefore that the terms

$$2L_{ji}L_{ki}F_{jk}/\lambda_i \tag{8.6.13}$$

and

$$L_{ji}^2F_{jj}/\lambda_i$$

give the fractional contribution of the off-diagonal and of the diagonal elements respectively of the **F**-matrix to the potential energy. In matrix form equations (8.6.13) can be written

$$\text{PED} = \mathbf{\Lambda}^{-1}\mathbf{JF} \tag{8.6.14}$$

where **J** is the Jacobian matrix described in Section 4.7.

Step 7. *Cartesian displacement*

It is often desirable to obtain the Cartesian displacements of the atoms of a molecule in a normal vibration. From Section 4.2 we recall that

$$\mathbf{s} = \mathbf{LQ}$$

$$\mathbf{s} = \mathbf{Bx}$$

from which it follows that

$$\mathbf{x} = \mathbf{B}^{-1}\mathbf{LQ} = \mathbf{TQ} \tag{8.6.15}$$

In order to avoid inversion of the **B**-matrix we can rewrite (8.6.15) in the form

$$\mathbf{x} = \mathbf{B}^{-1}\mathbf{L}\tilde{\mathbf{L}}\tilde{\mathbf{L}}^{-1}\mathbf{Q} = \mathbf{B}^{-1}\mathbf{G}\tilde{\mathbf{L}}^{-1}\mathbf{Q}$$

and since (see Section 4.5)

$$\mathbf{G} = \mathbf{BM}^{-1}\tilde{\mathbf{B}}$$

we obtain

$$\mathbf{x} = \mathbf{M}^{-1}\tilde{\mathbf{B}}\tilde{\mathbf{L}}^{-1}\mathbf{Q} = \mathbf{TQ} \tag{8.6.16}$$

and thus

$$\mathbf{T} = \mathbf{M}^{-1}\tilde{\mathbf{B}}\tilde{\mathbf{L}}^{-1} \tag{8.6.17}$$

Equation (8.6.16) shows that the ith column of the **T**-matrix gives the atomic displacements for a unit displacement of the normal coordinate Q_i. If $\mathbf{x}_0$ is the vector of the Cartesian coordinates of the atoms in the equilibrium position, the atomic displacements for a unit variation of the normal coordinate Q_i are

$$\mathbf{x}^i = \mathbf{x}_0 + \mathbf{T}^i \tag{8.6.18}$$

where $\mathbf{T}^i$ is the ith column of **T**.

For the calculation of the **T**-matrix of equations (8.6.17) it is unnecessary to construct the $\tilde{\mathbf{L}}^{-1}$ matrix. From equation (8.4.13) we obtain in fact

$$\tilde{\mathbf{L}}^{-1} = \mathbf{FL\Lambda}^{-1} \tag{8.6.19}$$

and thus equation (8.6.17) becomes

$$\mathbf{T} = \mathbf{M}^{-1}\tilde{\mathbf{B}}\mathbf{FL\Lambda}^{-1} \tag{8.6.20}$$

Once the vector $\mathbf{x}^i$ has been calculated, a picture of the normal vibration Q_i is easily obtained. The elements of the $\mathbf{T}^i$ vector, taken in groups of three, are the components of vectors which give the displacements of the atoms. The lengths of the displacement vectors indicate the relative amplitude for each atom.

Step 8. *Calculation of the Coriolis constants*

According to Section 4.8 the $\boldsymbol{\zeta}^\sigma$ matrices are given by

$$\boldsymbol{\zeta}^\sigma = \tilde{\mathscr{L}}\mathscr{M}^\sigma\mathscr{L} \tag{8.6.21}$$

where $\mathscr{L}$ is the transformation matrix from mass-weighted Cartesian to normal coordinates and $\mathscr{M}^\sigma$ are the auxiliary matrices of equation (4.8.4) introduced by Meal and Polo.

From the relations

$$\mathbf{q} = \mathscr{L}\mathbf{Q}$$
$$\mathbf{q} = \mathbf{M}^{\frac{1}{2}}\mathbf{x}$$
$$\mathbf{x} = \mathbf{T}\mathbf{Q}$$

it follows that

$$\mathscr{L} = \mathbf{M}^{\frac{1}{2}}\mathbf{T} \tag{8.6.22}$$

and thus

$$\boldsymbol{\zeta}^\sigma = \tilde{\mathbf{T}}\mathbf{M}^{\frac{1}{2}}\mathscr{M}^\sigma\mathbf{M}^{\frac{1}{2}}\mathbf{T} \tag{8.6.23}$$

Step 9. *Calculation of the mean-square amplitudes*

Mean-square amplitudes can be calculated either using equation (8.5.5a)

$$\boldsymbol{\Sigma}^s = \mathbf{L}\,\boldsymbol{\delta}\tilde{\mathbf{L}} \tag{8.6.24}$$

where $\boldsymbol{\delta}$ is a diagonal matrix with elements

$$\delta_i = \frac{h}{8\pi^2 c\omega_i}\coth\left(\frac{hc\omega_i}{2KT}\right) \tag{8.6.25}$$

or using the approximate equation (8.5.12)

$$\boldsymbol{\Sigma}^s = KT\mathbf{F}^{-1} + \frac{h^2}{64\pi^2 KT}\mathbf{G} \tag{8.6.26}$$

Step 10. *Calculation of the centrifugal distortion constants*

Centrifugal distortion constants are calculated using equation (4.9.5)

$$\tau_{\alpha\beta\gamma\delta} = -\frac{1}{2I^0_{\alpha\alpha}I^0_{\beta\beta}I^0_{\gamma\gamma}I^0_{\delta\delta}}\sum_{k,l} J^k_{\alpha\beta}(F^{-1})_{kl}J^l_{\gamma\delta} \tag{8.6.27}$$

The $J^k_{\alpha\beta}$ and $J^l_{\gamma\delta}$ terms are evaluated using equation (8.4.15).

Step 11. *Refinement of the force constants*

Let R^{obs} be a vector whose elements are the experimental quantities one wishes to reproduce using an appropriate force field, and let $\mathbf{R}^{\text{calc}}$ be a vector of the corresponding quantities, in the same order, calculated using a trial **F**-matrix. In particular $\mathbf{R}^{obs}$ and $\mathbf{R}^{calc}$ are of the type

$$\left|\begin{matrix}\mathbf{R}_\lambda \\ \mathbf{R}_\zeta \\ \mathbf{R}_\tau \\ \mathbf{R}_{\delta^2}\end{matrix}\right|$$

where $\mathbf{R}_\lambda$ is the vector of the vibrational frequency parameters, $\mathbf{R}_\zeta$ the vector of the Coriolis coupling constants, $\mathbf{R}_\tau$ the vector of the centrifugal distortion constants and $\mathbf{R}_{\delta^2}$ the vector of the mean-square amplitudes.

In the least-squares adjustment[1,28,68,69,70] of a set of parameters one tries to minimize the sum of the weighted squares of the errors

$$S = \sum_i P_i(R_i^{obs} - R_i^{calc})^2 \tag{8.6.28}$$

which, in matrix notation, can be written

$$S = \widetilde{\Delta\mathbf{R}}\mathbf{P}\,\Delta\mathbf{R} \tag{8.6.29}$$

where $\Delta\mathbf{R}$ is the vector whose elements are the differences between observed and calculated quantities and **P** is a diagonal matrix whose elements are the weights assigned to each datum. If σ_i is the probable error of the ith experimental datum R_i^{obs}, the corresponding weight P_i is normally chosen in order to obey the equation

$$\sigma_1^2 P_1 = \sigma_2^2 P_2 = \ldots = \sigma_i^2 P_i = \ldots = \sigma^2(R) \tag{8.6.30}$$

where $\sigma^2(R)$ is the variance of a datum of unit weight, given in the statistical theory of errors by

$$\sigma^2(R) = \frac{S}{(N-M)} = \frac{1}{(N-M)}\widetilde{\Delta\mathbf{R}}\mathbf{P}\,\Delta\mathbf{R} \tag{8.6.31}$$

$(N - M)$ being the difference between the number of independent data and the number of unknowns (in our case the number of independent force constants).

Since each calculated quantity R_i is some function of the force constants, we can develop S in a Taylor series of the variations in the force constants

$$S = S_0 + \sum_k \left(\frac{\partial S}{\partial \phi_k}\right)\Delta\phi_k + \frac{1}{2}\sum_{k,l}\left(\frac{\partial^2 S}{\partial\phi_k\,\partial\phi_l}\right)\Delta\phi_k\,\Delta\phi_l + \ldots \tag{8.6.32}$$

Using equation (8.6.28) we obtain

$$\left(\frac{\partial S}{\partial \phi_k}\right) = -2\sum_i \Delta R_i P_i \left(\frac{\partial R_i}{\partial \phi_k}\right) \tag{8.6.33}$$

and

$$\left(\frac{\partial^2 S}{\partial\phi_k\,\partial\phi_l}\right) = 2\sum_i P_i\left[\left(\frac{\partial R_i}{\partial\phi_k}\right)\left(\frac{\partial R_i}{\partial\phi_l}\right) - \Delta R_i\left(\frac{\partial^2 R_i}{\partial\phi_k\,\partial\phi_l}\right)\right] \tag{8.6.34}$$

from which it follows that

$$\Delta S = S - S_0 = -2\sum_k\sum_i \Delta R_i P_i J_k^i\,\Delta\phi_k + \sum_{k,l}\sum_i P_i[J_k^i J_l^i - \Delta R_i A_{kl}^i]\,\Delta\phi_k\,\Delta\phi_l \tag{8.6.35}$$

where

$$J^i_k = \left(\frac{\partial R_i}{\partial \phi_k}\right) \qquad A^i_{k,l} = \left(\frac{\partial^2 R_i}{\partial \phi_k \, \partial \phi_l}\right) \tag{8.6.36}$$

If we assume that each R can be written as a linear function of the $\Delta\phi_k$ elements when the sum of the weighted square errors is close to the minimum, we can neglect all A^i_{kl} terms. In this 'linear approximation' equation (8.6.35) becomes, using matrix notation

$$\Delta S = -(2\widetilde{\Delta \mathbf{R}}\mathbf{P}\mathbf{J}_\phi)\,\Delta\boldsymbol{\phi} + \widetilde{\Delta\boldsymbol{\phi}}\tilde{\mathbf{J}}_\phi\mathbf{P}\mathbf{J}_\phi\,\Delta\boldsymbol{\phi} \tag{8.6.37}$$

The condition for the minimum value of S is that

$$\left(\frac{\partial S}{\partial \phi}\right) = 0 \tag{8.6.38}$$

and thus we obtain from (8.6.37)

$$\widetilde{\Delta \mathbf{R}}\mathbf{P}\mathbf{J}_\phi = \widetilde{\Delta\boldsymbol{\phi}}\tilde{\mathbf{J}}_\phi\mathbf{P}\mathbf{J}_\phi \tag{8.6.39}$$

or, by taking the transpose

$$\tilde{\mathbf{J}}_\phi\mathbf{P}\,\Delta\mathbf{R} = \tilde{\mathbf{J}}_\phi\mathbf{P}\mathbf{J}_\phi\,\Delta\boldsymbol{\phi} \tag{8.6.40}$$

the vector $\Delta\boldsymbol{\phi}$ of the corrections to the trial $\boldsymbol{\phi}$-vector is thus given by

$$\Delta\boldsymbol{\phi} = (\tilde{\mathbf{J}}_\phi\mathbf{P}\mathbf{J}_\phi)^{-1}\tilde{\mathbf{J}}_\phi\mathbf{P}\,\Delta\mathbf{R} \tag{8.6.41}$$

Equation (8.6.41) is a generalization of equation (4.7.12) which includes in the $\Delta\mathbf{R}$ vector all observed quantities. The matrix $\mathbf{J}_\phi$ which occurs in equation (8.6.41) is a Jacobian matrix whose elements are the derivatives of the elements of the vector $\mathbf{R}$ with respect to the elements of the vector $\boldsymbol{\phi}$.

It is however more convenient to use a Jacobian $\mathbf{J}$, whose elements are the derivatives of the elements of $\mathbf{R}$ with respect to the elements of the vector $\overline{\mathbf{F}}$ of equation (8.6.1), since, as discussed in Section 4.7, the Jacobian elements with respect to internal force constants are easily constructed from the eigenvectors of the secular equation. Since

$$\left(\frac{\partial R_i}{\partial \phi_k}\right) = \sum_{r \leqslant t}\left(\frac{\partial R_i}{\partial F_{rt}}\right)\left(\frac{\partial F_{rt}}{\partial \phi_k}\right) = \sum_{r \leqslant t}\left(\frac{\partial R_i}{\partial F_{rt}}\right)Z^k_{rt} \tag{8.6.42}$$

we have

$$\mathbf{J}_\phi = \mathbf{J}\mathbf{Z} \tag{8.6.43}$$

and thus

$$\Delta\boldsymbol{\phi} = [(\tilde{\mathbf{Z}}\tilde{\mathbf{J}})\mathbf{P}(\mathbf{J}\mathbf{Z})]^{-1}\tilde{\mathbf{Z}}\tilde{\mathbf{J}}\mathbf{P}\,\Delta\mathbf{R} \tag{8.6.44}$$

Equation (8.6.44) has been used with success in the refinement of a large number of force fields. Since it is based, according to the previous discussion, on the assumption of a linear relationship between the elements of the $\Delta\mathbf{R}$ vector and the force constants, it will furnish meaningful corrections only for small variations of the force constants.

It is therefore common practice to apply only a fraction of the calculated corrections to the vector $\boldsymbol{\phi}$ when the sum of the weighted square errors is large. The corrected $\boldsymbol{\phi}$-vector is then used to start a new cycle (from step 3) and the process is repeated until the error between observed and calculated quantities is minimized.

Problems of non-convergence of the refinement of the force field often arise when equation (8.6.44) is used for the calculation of the least-square corrections. One possibility is that the trial $\boldsymbol{\phi}$-vector used is not a good choice and therefore that the very large corrections required break down the linear approximation. In this case the use of fractional corrections can result in a very slow convergence or in no convergence at all. It is therefore advisable in such cases to use different trial $\boldsymbol{\phi}$-vectors before starting the refinement.

A second possibility is near-singularity of the $(\tilde{\mathbf{Z}}\tilde{\mathbf{J}})\mathbf{P}(\mathbf{JZ})$ matrix which implies an ill-conditioning of the problem, i.e. a number of data insufficient to determine all the elements of the $\boldsymbol{\phi}$-vector. Even if the number of data is larger than the number of force constants, it may happen that the data are nearly independent from some force constants or that two or more force constants are correlated each to the others.

In the first case we find that the elements of the columns of $(\mathbf{JZ})$, corresponding to the ill-determined force constants, are vanishingly small, causing the determinant of $(\tilde{\mathbf{Z}}\tilde{\mathbf{J}})\mathbf{P}(\mathbf{JZ})$ to be nearly zero. Under these conditions, large round-off errors will affect the elements of the inverse matrix $[(\tilde{\mathbf{Z}}\tilde{\mathbf{J}})\mathbf{P}(\mathbf{JZ})]^{-1}$, and thus also the calculated corrections, causing divergence of the refinement. In order to obtain a convergence refinement it is then necessary to eliminate the ill-defined force constants as well as the corresponding columns of the $(\mathbf{JZ})$ matrix from equation (8.6.44). This is conveniently done by using a reduced $\boldsymbol{\phi}$-vector and a reduced $\mathbf{Z}$-matrix for the refinement process. The $\boldsymbol{\phi}$-vector and the $\mathbf{Z}$-matrix must be then reintegrated for the construction of the $\overline{\mathbf{F}}$-vector of equation (8.6.1). The elements of $\boldsymbol{\phi}$ which are not refined, are held constant.

In the second case we find that the columns of $(\mathbf{JZ})$, corresponding to correlated force constants, are linearly related causing the determinant of $(\tilde{\mathbf{Z}}\tilde{\mathbf{J}})\mathbf{P}(\mathbf{JZ})$ to vanish. Again in this case it is necessary to eliminate $l - 1$ of the l correlated force constants from the refinement. These $l - 1$ force constants are then given by the $l - 1$ equations of constraint as linear functions of the one that has been refined.

In order to avoid unnecessary calculations, it is desirable to use some tests for singularity of the $(\tilde{\mathbf{Z}}\tilde{\mathbf{J}})\mathbf{P}(\mathbf{JZ})$ matrix. A convenient test can be made by comparing the product of the diagonal elements with the determinant of the matrix. Since $(\tilde{\mathbf{Z}}\tilde{\mathbf{J}})\mathbf{P}(\mathbf{JZ})$ is a symmetric positive definite matrix, if

$$\prod_u [(\tilde{\mathbf{Z}}\tilde{\mathbf{J}})\mathbf{P}(\mathbf{JZ})]_{uu} \gg |(\tilde{\mathbf{Z}}\tilde{\mathbf{J}})\mathbf{P}(\mathbf{JZ})|$$

the matrix is close to singularity. Several methods have been proposed for the identification of ill-defined force constants. Since they are more or less equivalent, we shall discuss only those most commonly used in standard computer programs.

From the standard deviation $\sigma^2(R)$ in the observed data, we can construct the matrix

$$\boldsymbol{\Sigma}(\phi) = \sigma^2(R)[(\tilde{\mathbf{Z}}\tilde{\mathbf{J}})\mathbf{P}(\mathbf{JZ})]^{-1} \tag{8.6.45}$$

The diagonal elements of $\boldsymbol{\Sigma}(\phi)$

$$\sigma^2(\phi_i) = \sigma^2(R)[(\tilde{Z}\tilde{J})P(JZ)]_{ii}^{-1} \tag{8.6.46}$$

give the standard deviations in the force constants whereas the off-diagonal elements give the correlation between the errors. We can use[113] then the standard deviations in the force constants to obtain a valid criterion for their inclusion in the refinement. If

$$\sigma(\phi_i) - \phi_i < \varepsilon$$

where ε is an arbitrary constant, properly chosen for each case, the force constant ϕ_i can be retained in the refinement. If instead $\sigma(\phi_i)$ is bigger than the force constant, then the frequencies are very little influenced by it and can be ignored. In general force constants with high standard deviations may give trouble in the refinement and it is safer to assign them a constant value.

Since equation (8.6.46) is based on a statistical analysis, it is really meaningful only if a large number of data are considered for a relatively small number of force constants. Furthermore, it assumes a random distribution of errors among the data, whereas it is well known that errors due to the neglect of anharmonicity are more important for some data than for others. Another method, which was shown later[71] to be equivalent to the previous one, has been proposed by Bruton and Woodward.[72] These authors suggest the use of the eigenvectors and the eigenvalues of the $(\tilde{\mathbf{Z}}\tilde{\mathbf{J}})\mathbf{P}(\mathbf{JZ})$ matrix to single out ill-defined force constants and have given a geometrical interpretation of the refinement process.

Since inversion of the $(\tilde{\mathbf{Z}}\tilde{\mathbf{J}})\mathbf{P}(\mathbf{JZ})$ matrix is a delicate step in the refinement procedure, the method of steepest descent, which avoids the matrix inversion, has been proposed by N. Neto[73] and Fletcher and Powell[74,75,76] for the refinement of a force field.

In the limits of the linear approximation

$$\left(\frac{\partial^2 R_i}{\partial \phi_k\, \partial \phi_l}\right) = 0 \tag{8.6.47}$$

on which equations (8.6.37) to (8.6.44) are based, the elements of the $\Delta\mathbf{R}$ vector can be written as a linear function of the $\Delta\phi_k$ elements.

$$\Delta R_i = \sum_k \left(\frac{\partial R_i}{\partial \phi_k}\right) \Delta\phi_k = \sum_k J_k^i\, \Delta\phi_k \tag{8.6.48}$$

or in matrix form (see equation 4.7.10)

$$\Delta\mathbf{R} = \mathbf{J}_\phi\, \Delta\boldsymbol{\phi} = \mathbf{JZ}\, \Delta\boldsymbol{\phi} \tag{8.6.49}$$

Because of the approximation involved in (8.6.49) the correction $\Delta\boldsymbol{\phi}$ calculated by use of this formula does not minimize the weighted square error. If S_i^2 and S_{i+1}^2 are the weighted square errors before and after one cycle of refinement, we have

$$S_i^2 = \widetilde{\Delta\mathbf{R}}\mathbf{P}\, \Delta\mathbf{R}$$
$$S_{i+1}^2 = (\widetilde{\Delta\mathbf{R}} - \widetilde{\Delta\boldsymbol{\phi}}\tilde{\mathbf{Z}}\tilde{\mathbf{J}})\mathbf{P}(\Delta\mathbf{R} - \mathbf{JZ}\, \Delta\boldsymbol{\phi})$$

and using (8.6.44)

$$\Delta S^2 = S_{i+1}^2 - S_i^2 = -\widetilde{\Delta\mathbf{R}}\mathbf{PJZ}\, \Delta\boldsymbol{\phi} \tag{8.6.50}$$

In a convergent refinement process ΔS^2 must always be negative (i.e. $S_{i+1}^2 < S_i^2$). According to the deepest-descent method, the path along which ΔS^2 decreases more rapidly, is that of the negative gradient. It is therefore sufficient to take for $\Delta\boldsymbol{\phi}$ the expression

$$\Delta\boldsymbol{\phi} = N\tilde{\mathbf{Z}}\tilde{\mathbf{J}}\mathbf{P}\, \Delta\mathbf{R} \tag{8.6.51}$$

to ensure convergence of the refinement. The factor N in (8.6.51) is a scaling term which can be used to reduce the amount of the correction in order to avoid oscillations around the minimum.

Since, if we adjust only the force constant ϕ_k, the reduction in the weighted square error sum is given by

$$\Delta S_k^2 = -N[\widetilde{\Delta\mathbf{R}}\mathbf{PJZ}]_k^2 \tag{8.6.52}$$

the elements of $[\widetilde{\Delta\mathbf{R}}\mathbf{PJZ}]^2$ can be used as a test of the importance of the force constants in the refinement.

The second-order refinement in which terms of the type

$$\left(\frac{\partial^2 R_i}{\partial \phi_k\, \partial \phi_l}\right) \neq 0$$

are included, has been discussed by Neto,[73] Gans[75,76] and by Fletcher and Powell.[74] In particular, these last authors have proposed an iterative procedure for the inclusion of the second-order terms in (8.6.35), which is claimed to furnish excellent results.

8.7 The general valence force field (GVFF)

As discussed in Section 8.1, the main problem in force constant calculations is that the number of force constants is normally larger than the number of vibrational frequencies. For several small molecules, however, the molecular geometry is well known, the vibrational assignment is well established, often for several isotopic species, and additional data such as Coriolis and centrifugal distortion constants are available. In such cases it is then possible to calculate the complete harmonic force field (GVFF) and to ensure that the solution found is unique.

In order to illustrate the problems involved in the determination of a complete force field and also to make the reader acquainted with numerical calculations, in this section we shall discuss a few selected examples taken from the recent literature.

Example 8.7.1. The GVFF of H_2O using isotopic frequencies

The GVFF of a triatomic molecule like H_2O can be calculated[77] using the nine vibrational frequencies of the three isotopic species H_2O, D_2O and HDO, listed in Table 8.7.1. H_2O and D_2O have a C_{2v} symmetry. Two normal modes (ω_1 and ω_2) belong to the A_1 species and one (ω_3) to the B_1 species. HDO has C_s symmetry and all three normal modes belong to the same species A'.

Table 8.7.1 Harmonic frequencies in cm^{-1} of H_2O, HDO and D_2O

Symmetry		ω_1	ω_2	ω_3
H_2O	C_{2v}	3832·2	1648·5	3942·5
D_2O		2763·8	1206·4	2888·8
HDO	C_s	3889·8	1441·4	2824·3

As internal coordinates we can use (see Figure 4.1.1) Δr_1, Δr_2 and $\Delta\alpha$. In this way we obtain the **G**-matrix given in Section 4.5. In order to have the same dimensions for all coordinates and thus for all force constants, it is sometimes convenient to scale the angle bending coordinates with the equilibrium bond length and to use as internal coordinates Δr_1, Δr_2 and $r\Delta\alpha$. In this example we shall adopt the latter choice of coordinates. In this way we obtain the **G**-matrix

	Δr_1	Δr_2	$r\,\Delta\alpha$		
Δr_1	$\mu_X + \mu_O$	$\mu_O \cos\alpha$	$-\mu_O \sin\alpha$	X = Y = H	for H_2O
Δr_2	$\mu_O \cos\alpha$	$\mu_Y + \mu_O$	$-\mu_O \sin\alpha$	X = Y = D	for D_2O
$r\,\Delta\alpha$	$-\mu_O \sin\alpha$	$-\mu_O \sin\alpha$	$\mu_X + \mu_Y + 2\mu_O(1 - \cos\alpha)$	X = H, Y = D	for HDO

and the **F**-matrix

	Δr_1	Δr_2	$r\Delta\alpha$
Δr_1	f_r	f_{rr}	$f_{r\alpha}$
Δr_2	f_{rr}	f_r	$f_{r\alpha}$
$r\,\Delta\alpha$	$f_{r\alpha}$	$f_{r\alpha}$	f_α

where

$$\mu_O = 1/m_O = 1/16{\cdot}000 \text{ (a.m.u.)}$$
$$\mu_H = 1/m_H = 1/1{\cdot}008 \text{ (a.m.u.)}$$
$$\mu_D = 1/m_D = 1/2{\cdot}016 \text{ (a.m.u.)}$$

and where all elements of the **F**-matrix are expressed in units of mdyne/Å. We can now use symmetry coordinates to factorize the **G**- and **F**-matrices in block form. From the method of Section 6.6, we get for H_2O and for D_2O

$$A_1 \text{ species} \quad \mathbb{S}_1 = \tfrac{1}{\sqrt{2}}(\Delta r_1 + \Delta r_2), \quad \mathbb{S}_2 = r\,\Delta\alpha$$

$$B_1 \text{ species} \quad \mathbb{S}_3 = \tfrac{1}{\sqrt{2}}(\Delta r_1 - \Delta r_2)$$

Using equations (8.6.4) and (8.6.5) we obtain the **G**- and **F**-matrices in symmetry coordinates with elements

$$\begin{aligned}
\mathbb{G}_{11} &= \mu_x + \mu_O(1 + \cos\alpha) & \mathbb{F}_{11} &= f_r + f_{rr} \\
\mathbb{G}_{12} &= -\sqrt{2}\,\mu_O \sin\alpha & \mathbb{F}_{12} &= \sqrt{2}\,f_{r\alpha} \\
\mathbb{G}_{22} &= 2\mu_x + 2\mu_O(1 - \cos\alpha) & \mathbb{F}_{22} &= f_\alpha \\
\mathbb{G}_{33} &= \mu_x + \mu_O(1 - \cos\alpha) & \mathbb{F}_{33} &= f_r - f_{rr} \\
\mathbb{G}_{13} &= \mathbb{G}_{23} = 0 & \mathbb{F}_{13} &= \mathbb{F}_{23} = 0
\end{aligned}$$

If we use the symmetry coordinates of the symmetric isotopic species to transform the **G**- and **F**-matrices of the unsymmetrical species HDO, we obtain

$$\begin{aligned}
\mathbb{G}_{11} &= \tfrac{1}{2}(\mu_H + \mu_D) + \mu_O(1 + \cos\alpha) & \mathbb{F}_{11} &= f_r + f_{rr} \\
\mathbb{G}_{12} &= -\sqrt{2}\,\mu_O \sin\alpha & \mathbb{F}_{12} &= \sqrt{2}\,f_{r\alpha} \\
\mathbb{G}_{13} &= \tfrac{1}{2}(\mu_H - \mu_D) & \mathbb{F}_{13} &= 0 \\
\mathbb{G}_{22} &= \mu_H + \mu_D + 2\mu_O(1 - \cos\alpha) & \mathbb{F}_{22} &= f_\alpha \\
\mathbb{G}_{23} &= 0 & \mathbb{F}_{23} &= 0 \\
\mathbb{G}_{33} &= \tfrac{1}{2}(\mu_H + \mu_D) + \mu_O(1 - \cos\alpha) & \mathbb{F}_{33} &= f_r - f_{rr}
\end{aligned}$$

The reader will notice that the **F**-matrix is again in block form whereas the **G**-matrix remains unfactorized. The reason for this different behaviour of the kinetic and potential energy matrices is that we have assumed that the same force field is valid for all three isotopic species. The potential energy thus has a C_{2v} symmetry even for the HDO molecule whereas the kinetic energy matrix has a lower symmetry.

As a consequence of the factorization the secular equation for H_2O and for D_2O splits into one equation of dimension 1×1 and one equation of dimension 2×2. The secular equation for HDO remains of dimension 3×3. We have then one 3×3, two 2×2 and two 1×1 secular equations for the determination of the four independent force constants $\mathbb{F}_{11}$, $\mathbb{F}_{12}$, $\mathbb{F}_{22}$ and $\mathbb{F}_{33}$. Because of the existence of product and sum rules (see Section 8.2) the nine vibrational frequencies are not all independent. In particular there is one product rule equation between the B_1 blocks and one between the A_1 blocks of H_2O and D_2O, two product rule equations between the 3×3 block of HDO and those of H_2O and D_2O respectively and one sum rule equation of the type

$$(\omega_1^2 + \omega_2^2 + \omega_3^2)_{H_2O} + (\omega_1^2 + \omega_2^2 + \omega_3^2)_{D_2O} - 2(\omega_1^2 + \omega_2^2 + \omega_3^2)_{HDO} = 0$$

We thus have at our disposal $9 - 5 = 4$ independent data for the determination of the four force constants and thus we shall be able in principle to determine them unambiguously. It is necessary to point out that the use of the vibrational frequencies of the unsymmetrical species HDO does not introduce any additional datum. If we had used only the frequencies of H_2O and D_2O, we would still have $6 - 2 = 4$ independent data for the determination of the four force constants. In this latter case however, as discussed in Section 8.2, two indistinguishable solutions for $\mathbb{F}_{11}$, $\mathbb{F}_{12}$ and $\mathbb{F}_{22}$ would be found. The importance of the use of the frequencies of HDO lies in the fact that they allow a distinction between the two sets.

Nibler and Pimentel[77] have calculated the force constants of H_2O using the least-square refinement method of the previous section and alternatively the 'display' method discussed in Section 8.2. They have also shown that the display method can be extended to blocks of dimension 3×3 in some particular cases such as that of HDO. The force constants obtained with the two methods are in excellent agreement.

The best set of force constants obtained by these authors using $1/\lambda_i$ weighting **P**-matrix elements is

$$\begin{aligned}
\mathbb{F}_{11} &= 8{\cdot}353 \pm 0{\cdot}005 \text{ mdyne/Å}\\
\mathbb{F}_{12} &= 0{\cdot}332 \pm 0{\cdot}048 \text{ mdyne/Å}\\
\mathbb{F}_{22} &= 0{\cdot}760 \pm 0{\cdot}004 \text{ mdyne/Å}\\
\mathbb{F}_{33} &= 8{\cdot}553 \pm 0{\cdot}004 \text{ mdyne/Å}
\end{aligned}$$

From these force constants we can calculate the internal force constants by means of the relations

$$\begin{aligned}
f_r &= \tfrac{1}{2}(\mathbb{F}_{11} + \mathbb{F}_{33}) = 8{\cdot}453 \text{ mdyne/Å}\\
f_{rr} &= \tfrac{1}{2}(\mathbb{F}_{11} - \mathbb{F}_{33}) = 0{\cdot}100 \text{ mdyne/Å}\\
f_\alpha &= \mathbb{F}_{22} \qquad\qquad\quad = 0{\cdot}760 \text{ mdyne/Å}\\
f_{r\alpha} &= \tfrac{1}{\sqrt{2}}\mathbb{F}_{12} \qquad\quad\;\; = 0{\cdot}235 \text{ mdyne/Å}
\end{aligned}$$

Example 8.7.2. The GVFF of NF_3 from vibrational frequencies, Coriolis and centrifugal distortion constants

The GVFF of NF_3 has been discussed by several authors.[78–83] The present example is essentially based on the treatment given in References 81 and 83. The reader is referred to the original papers for further details. NF_3 is a pyramidal molecule of C_{3v} symmetry. The six normal vibrations classify as $2A_1$ and $2E$ (doubly degenerate).

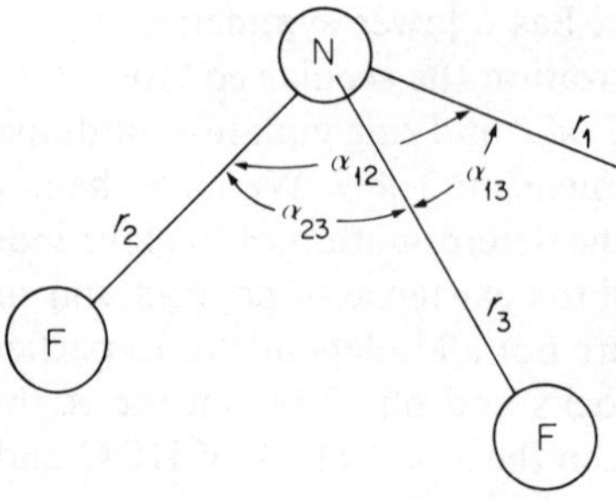

Figure 8.7.1 Internal coordinates for NF_3

The internal coordinates are shown in Figure 8.7.1. Symmetry coordinates are easily constructed from them and are of the type

$$\mathbb{S}_1 = \frac{1}{\sqrt{3}}(\Delta r_1 + \Delta r_2 + \Delta r_3)$$
$$\mathbb{S}_2 = \frac{1}{\sqrt{3}}(\Delta\alpha_{12} + \Delta\alpha_{13} + \Delta\alpha_{23})$$
$$\mathbb{S}_{3a} = \frac{1}{\sqrt{6}}(2\Delta r_1 - \Delta r_2 - \Delta r_3) \qquad \mathbb{S}_{3b} = \frac{1}{\sqrt{2}}(\Delta r_2 - \Delta r_3)$$
$$\mathbb{S}_{4a} = \frac{1}{\sqrt{6}}(2\Delta\alpha_{23} - \Delta\alpha_{13} - \Delta\alpha_{12}) \qquad \mathbb{S}_{4b} = \frac{1}{\sqrt{2}}(\Delta\alpha_{13} - \Delta\alpha_{12})$$

The secular equation then factorizes into one 2 × 2 block of species A_1 and two identical 2 × 2 blocks of species E.

For this molecule the data shown in Table 8.7.2 are available.

Table 8.7.2

$\nu_1 = 1031{\cdot}91\ \text{cm}^{-1}$	$\zeta_3 = 0{\cdot}81$	$D_J = 14{\cdot}534$ Kc/s
$\nu_2 = 647{\cdot}26\ \text{cm}^{-1}$	$\zeta_4 = -0{\cdot}90$	$D_{JK} = -22{\cdot}694$ Kc/s
$\nu_3 = 908{\cdot}40\ \text{cm}^{-1}$		
$\nu_4 = 492{\cdot}62\ \text{cm}^{-1}$		

The **F**-matrix in internal coordinates has the form

	Δr_1	Δr_2	Δr_3	$\Delta\alpha_{23}$	$\Delta\alpha_{13}$	$\Delta\alpha_{12}$
Δr_1	f_r	f_{rr}	f_{rr}	$f_{r\alpha}$	$f'_{r\alpha}$	$f'_{r\alpha}$
Δr_2		f_r	f_{rr}	$f'_{r\alpha}$	$f_{r\alpha}$	$f'_{r\alpha}$
Δr_3			f_r	$f'_{r\alpha}$	$f'_{r\alpha}$	$f_{r\alpha}$
$\Delta\alpha_{23}$	symm.			f_α	$f_{\alpha\alpha}$	$f_{\alpha\alpha}$
$\Delta\alpha_{13}$					f_α	$f_{\alpha\alpha}$
$\Delta\alpha_{12}$						f_α

and thus contains six independent force constants. In symmetry coordinates it becomes (only one of the two identical degenerate **E**-blocks is shown)

$\mathbb{F}_{11}$	$\mathbb{F}_{12}$	0	0
	$\mathbb{F}_{22}$	0	0
symm		$\mathbb{F}_{33}$	$\mathbb{F}_{34}$
			$\mathbb{F}_{44}$

The **G**- and $\mathbb{G}$-matrices have corresponding elements and are not reported for brevity.

From the two blocks of the secular equation we obtain

$$\mathbb{G}_{11}\mathbb{F}_{11} + 2\mathbb{G}_{12}\mathbb{F}_{12} + \mathbb{G}_{22}\mathbb{F}_{22} = \lambda_1 + \lambda_2$$
$$(\mathbb{G}_{11}\mathbb{G}_{22} - \mathbb{G}_{12}^2)(\mathbb{F}_{11}\mathbb{F}_{22} - \mathbb{F}_{12}^2) = \lambda_1 \cdot \lambda_2$$
$$\mathbb{G}_{33}\mathbb{F}_{33} + 2\mathbb{G}_{34}\mathbb{F}_{34} + \mathbb{G}_{44}\mathbb{F}_{44} = \lambda_3 + \lambda_4$$
$$(\mathbb{G}_{33}\mathbb{G}_{44} - \mathbb{G}_{34}^2)(\mathbb{F}_{33}\mathbb{F}_{44} - \mathbb{F}_{34}^2) = \lambda_3 \cdot \lambda_4$$

The general form of the distortion constants D_J and D_{JK} has been given in Section 8.4. Because NF_3 is a symmetric top molecule ($I^0_{xx} = I^0_{yy}$) these equations can be considerably simplified. From the symmetry properties of the inertia tensor components we see (see Section 8.4) that the non-zero $\tau_{\alpha\beta\gamma\delta}$ elements for a molecule with C_{3v} symmetry[61] are (the Z-axis is taken in the direction of the symmetry axis C_3)

$$\tau_{zzzz}$$
$$\tau_{xxxx} = \tau_{yyyy}$$
$$\tau_{xxyy} = \tau_{xxxx} - 2\tau_{xyxy}$$
$$\tau_{zzxx} = \tau_{zzyy}$$
$$-\tau_{xyzx} = \tau_{yyyz} = -\tau_{xxyz}$$
$$\tau_{zxzx} = \tau_{zyzy}$$

Using these relations we obtain from equations (8.4.5)

$$D_J = -\frac{\hbar^4}{4}\tau_{xxxx} = \hbar^4 \frac{1}{8(I^0_{xx})^4} \sum_{k,l} (\partial I_{xx}/\partial \mathbb{S}_k)(\partial I_{xx}/\partial \mathbb{S}_l)\mathbb{F}^{-1}_{kl}$$

$$D_{JK} = -2D_J - \frac{\hbar^4}{4}(2\tau_{zzxx} + 4\tau_{zxzx}) = -2D_J + \frac{\hbar^4}{4(I^0_{xx})^2(I^0_{zz})^2} \sum_{k,l} \left[\left(\frac{\partial I_{xx}}{\partial \mathbb{S}_k}\right)\left(\frac{\partial I_{zz}}{\partial \mathbb{S}_l}\right) + 2\left(\frac{\partial I_{xz}}{\partial \mathbb{S}_k}\right)\left(\frac{\partial I_{xz}}{\partial \mathbb{S}_l}\right)\right]\mathbb{F}^{-1}_{kl}$$

Since the $\mathbb{F}^{-1}$ matrix is also in block form, we can write D_J and D_{JK} as

$$D_J = D_J^{A_1} + D_J^E$$
$$D_{JK} = D_{JK}^{A_1} + D_{JK}^E$$

where $k, l = 1, 2$ for $D_J^{A_1}$ and $D_{JK}^{K_1}$ and $k, l = 3, 4$ for D_J^E and D_{JK}^E.

The $J^k_{\alpha\beta} = (\partial I_{\alpha\beta}/\partial \mathbb{S}_k)$ elements, necessary for the calculation of D_J and D_{JK} can be obtained, according to Section 8.4 using the method of Kivelson and Wilson[61] or alternatively the method of Pulay and Savodny.[63] Dowling, Gold and Meister[84] have used the first method to evaluate the $J^k_{\alpha\beta}$ elements for a ZY_3X molecule of C_{3v} symmetry. These expressions apply equally well to a ZY_3 molecule. The same calculations have been made by Pulay and Savodny[63] with the second method.

The Coriolis coupling constants ζ_3 and ζ_4 can be obtained using equation (4.8.9). The C^σ matrix elements for a ZY_3 molecule are given in the paper by Meal and Polo.[60] In particular these authors have also derived for the ZY_3 molecule simple formulae which permit the calculation of ζ_3 and ζ_4 without using the $\mathbf{L}^{-1}$ matrix

$$\zeta_3 + \zeta_4 = B/2A - 1 = I^0_{zz}/2I^0_{xx} - 1$$

$$\mathbb{F}_{33} - \frac{2N}{r_0}\mathbb{F}_{34} + \frac{3}{r_0^2}\mathbb{F}_{44} = \mu_Y[\lambda_3(1 - \zeta_3) + \lambda_4(1 - \zeta_4)]$$

where

$$N = [(1 - \cos\alpha)/(1 + \cos\alpha)]^{\frac{1}{2}}$$

and r_0 is the equilibrium bond length.

The reader will notice that we have four experimental data (ν_3, ν_4, ζ_3 and ζ_4) which depend on the three force constants $\mathbb{F}_3$, $\mathbb{F}_4$ and $\mathbb{F}_{34}$, two experimental data (ν_1 and ν_2) which depend on the three force constants $\mathbb{F}_{11}$, $\mathbb{F}_{12}$ and $\mathbb{F}_{22}$ and two (D_J and D_{JK}) which depend on both sets of force constants. The process to follow is thus to calculate first the set of force constants of E species from ν_3, ν_4, ζ_3 and ζ_4. With these force constants, the contributions D_J^E and D_{JK}^E to D_J and D_{JK} are then evaluated. Finally, the force constants of species A_1 are calculated from ν_1, ν_2, $D_J^{A_1}$ and $D_{JK}^{A_1}$.

The force constants of species E which give the best fit of the experimental data are shown in Table 8.7.3. From these force constants Savodny *et al.* have calculated values of 9·835 and −18·963 Kc/s for D_J^E and D_{JK}^E respectively, thus yielding values of 4·699 and −3·731 Kc/s for $D_J^{A_1}$ and $D_{JK}^{A_1}$. With the help of these data it is then possible to calculate the force constants of A_1 species shown in Table 8.7.3.

Table 8.7.3 Symmetry force constants of NF_3

	Savodny *et al.*	Otake *et al.*	Units
$\mathbb{F}_{11}$	6·131 ± 0·221	6·14 ± 0·15	mdyne/Å
$\mathbb{F}_{12}$	0·866 ± 0·058	0·84 ± 0·04	mdyne
$\mathbb{F}_{22}$	2·421 ± 0·077	2·41 ± 0·05	mdyne . Å
$\mathbb{F}_{33}$	3·404 ± 0·101	3·39 ± 0·05	mdyne/Å
$\mathbb{F}_{34}$	−0·451 ± 0·059	−0·45 ± 0·03	mdyne
$\mathbb{F}_{44}$	1·690 ± 0·042	1·67 ± 0·02	mdyne . Å

Otake, Hirota and Morino[83] have also determined the GVFF force constants of NF_3 using a different set of experimental data. In particular these authors have used the four vibrational frequencies of $^{14}NF_3$, two vibrational frequencies (ν_1 and ν_3) of $^{15}NF_3$ and the ζ_4 Coriolis constant of $^{14}NF_3$ and $^{15}NF_3$. The force field determined in this way is shown in the second column of Table 8.7.3 and coincides, within the limits of the error, with that obtained by Savodny *et al.*

Example 8.7.3. The GVFF of CF_3H from vibrational frequencies, isotopic shifts, Coriolis constants and centrifugal distortion constants

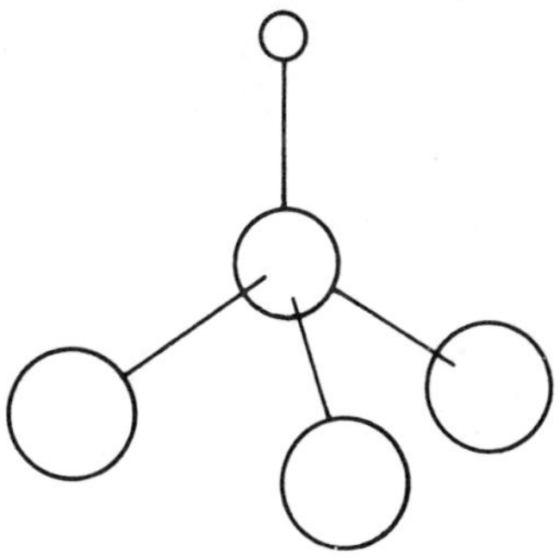

Figure 8.7.2 The molecule CF_3H

CF_3H also belongs to the C_{3v} point group and possesses 9 normal modes of vibrations, three of species A_1 and three doubly degenerate of species E. For this molecule vibrational frequencies, Coriolis constants and centrifugal distortion constants for both CF_3H and CF_3D are available. In a very recent paper D'Cunha[85] has refined a GVFF for this molecule using all the data at his disposal. The refinement has been made by the method discussed in Section 8.6. The results of the calculations are shown in Table 8.7.4. It must be noticed that for this molecule the refinement is based on the experimental frequencies and not on the harmonic frequencies as in previous examples.

The GVFF has been calculated for several other molecules. A list of references dealing with such calculations is given in Table 8.7.5.

Table 8.7.4 The general valence force constants of CF_3H

	CF_3H			CF_3D		
	ν_{obs} [b]	ν_{calc}	σ	obs	calc	σ
ν_1	3051	3034	18	2257	2255	8
ν_2	1140	1141	7	1109	1108	4
ν_3	700	701	4	693	692	2
ν_4	1372	1376	10	1207	1203	4
ν_5	1152	1155	8	974	972	3
ν_6	507	508	3	504	503	2
ζ_4	0·95	0·99	0·003	0·84	0·90	0·025
ζ_5	0·79	0·74	0·023	0·74	0·73	0·031
ζ_6	−0·81	−0·81	0·024	−0·76	−0·76	0·016
$D_J \times 10^6$	0·337	0·378	0·010	0·329	0·319	0·005
$D_{JK} \times 10^6$	−0·602	−0·604	0·012	—	−0·489	0·005

		GVFF[a]	σ(F)
A_1	F_{11}	5·0050	0·07
	F_{12}	0·5253	0·86
	F_{13}	0·1082	0·26
	F_{22}	8·1805	0·64
	F_{23}	0·3388	0·07
	F_{33}	0·9162	0·09
E	F_{44}	5·3681	0·11
	F_{45}	−0·4550	0·04
	F_{46}	0·4348	0·03
	F_{55}	0·8307	0·03
	F_{56}	−0·0847	0·03
	F_{66}	0·4526	0·01

[a] All force constants in mdyne/Å. Angle bending coordinates scaled with r_{CF}.
[b] All observed and calculated data are in cm^{-1} except the ζ-values which are dimensionless.

Table 8.7.5 References on the GVFF

Molecule	Reference	Molecule	Reference	Molecule	Reference
CO_2	104	NO_2	86, 45	PF_3	80, 96
CS_2	110	NOF	88, 89	AsF_3	80
COS	103	NOCl	87	H_2S_2	97
N_2O	102	NOBr	87	NO_2F	100
HCN	101, 108	NSF	62, 94	CH_3F	69
FCN	95	SO_2	62, 90	CH_3Cl	69
ClCN	95	SeO_2	91	CH_3Br	69
BrCN	95	SiF_2	62	CH_3I	69
ICN	95	O_3	92	CHF_3	85
HCP	108	F_2O	62, 93	CF_4	44
H_2O	77	BF_3	45, 106	SiF_4	45
H_2S	77	BCl_3	106	$SiCl_4$	98
H_2Se	77	BBr_3	106	GeF_4	99
		SO_3	109	$SnCl_4$	107
		NH_3	111	C_2H_4	42
		NF_3	80, 81, 82, 83		

8.8 The simplified general valence force field (SGVFF)

In the previous section we have shown how to evaluate the complete harmonic valence force field for small molecules, where by 'small molecules' we mean all the molecules for which it is possible to accumulate at least as many independent experimental data (vibrational frequencies, Coriolis and centrifugal distortion constants, etc.) as force constants in the complete potential function.

For large molecules this is not possible since, in addition to the fact that the number of force constants is much larger than the number of frequencies, the additional data available for small molecules are not accessible by experiment. For large molecules with complicated vibrational spectra, even the identification of the fundamental frequencies becomes a serious problem. Although the vibrational assignment of several molecules is well established today, for most of the large polyatomic molecules it is still incomplete in the sense that some fundamental frequencies are not identified in the spectrum whereas others are assigned to observed bands without definite evidence. Under these conditions the vibrational problem can be just the opposite of that encountered for small molecules. Instead of using the experimental data to obtain a force field, one needs to utilize a force field to predict the vibrational spectrum and to identify the fundamental modes on the basis of these indications. This can be done safely only if one has at one's disposal a simple but reliable force field which is known to be transferable from one molecule to another, even with minor adjustments and only within a class of similar molecules. Transferability thus becomes an important property of force constants and is often the only criterion to use in judging their value.

In Section 8.1 we have discussed the various forms of approximate force fields that have been used for polyatomic molecules. Among them, by far the most powerful is 'the simplified general valence force field' (SGVFF), i.e. a simplified form of the GVFF which contains the minimum possible number of interaction constants compatible with a good fit of the spectra. In such approximate force fields all interaction terms which have a small influence on the vibrational frequencies are always neglected. If this is not sufficient to reduce the number of independent force constants below an acceptable limit, further simplifications are introduced either by transferring some force constants from related force fields or by assuming the same numerical value for force constants which refer to similar types of interactions.

It is obvious that approximate force fields of this type have no claim to uniqueness. An infinite number of approximate solutions, all capable of reproducing the vibrational frequencies within a given range of error, exist for each molecule and none of them is in principle preferable to the others. Most of the physical meaning of force constants is thus lost and they assume essentially the role of spectroscopic parameters which translate in a mechanical language the vibrational information contained in the infrared and Raman spectra. The correlation with the exact force fields determined for small molecules is actually of great help in choosing solutions including numerical values of the force constants consistent with the general pattern of force constant–electronic structure relationships established for small molecules.

The amount of work accumulated on simplified force fields of this type is so large that it is practically impossible to discuss all papers dealing with this subject and to analyse the wide variety of different approximations used in actual cases. Since, however, some force fields with more or less general validity within classes of molecules have been proposed by several authors, we shall review them briefly in this section.

A force field for aliphatic hydrocarbons was obtained by Schachtschneider and Snyder[70] by fitting with about 30 force constants several hundred vibrational frequencies of *n*-alkanes. The same type of force field was then extended by the same authors[112] to cover branched and unstrained cyclic hydrocarbons.

The force field for aliphatic hydrocarbons is given in Table 8.8.1. The nomenclature of internal coordinates for the labelling of force constants is shown schematically in Figure 8.8.1. In this as well as in following tables, stretching force constants are indicated by the letter K, bending constants by the letter H and interaction constants by the letters F and f.

Table 8.8.1 Simplified force field for aliphatic hydrocarbons

Force constant	Numerical value	Force constant	Numerical value	Force constant	Numerical value	Force constant	Numerical value
K_r	4·699	H_ζ	0·657	$F'_{R\gamma}$	0·079	$f'^{g}_{\gamma\gamma}$	0·009
K_d	4·554	H_ω	1·130	$F_{R\omega}$	0·417	$f''^{t}_{\gamma\gamma}$	−0·014
K_s	4·588	H_ϕ	1·084	$F_{\beta\beta}$	−0·012	$f''^{g}_{\gamma\gamma}$	−0·025
K_R	4·387	H_Λ	1·086	$F_{\gamma\gamma}$	−0·021	$f^{t}_{\gamma\omega}$	0·049
$K_{R'}$	4·337	H_τ	0·024	$F'_{\gamma\gamma}$	0·012	$f^{g}_{\gamma\omega}$	−0·052
$K_{R''}$	4·534	H_Γ	0·024	$F_{\phi\phi}$	−0·041	$f^{t}_{\omega\omega}$	−0·011
H_α	0·540	F_{rr}	0·043	$F_{\gamma\omega}$	−0·031	$f^{g}_{\omega\omega}$	0·011
H_β	0·645	F_{dd}	0·006	$f^{t}_{\gamma\gamma}$	0·127		
H_δ	0·550	F_{RR}	0·101	$f^{g}_{\gamma\gamma}$	−0·005		
H_γ	0·656	$F_{R\gamma}$	0·328	$f'^{t}_{\gamma\gamma}$	0·002		

Stretching force constants in units of mdyne/Å.
Stretch–bend interactions in units of mdyne/rad.
Bending force constants in units of mdyne Å/rad.
Approximations used: $F'_{\zeta\zeta} = F'_{\gamma\gamma}$, $F_{\Lambda\Lambda} = F_{\phi\phi}$, $F''_{\Lambda\Lambda} = 0$.

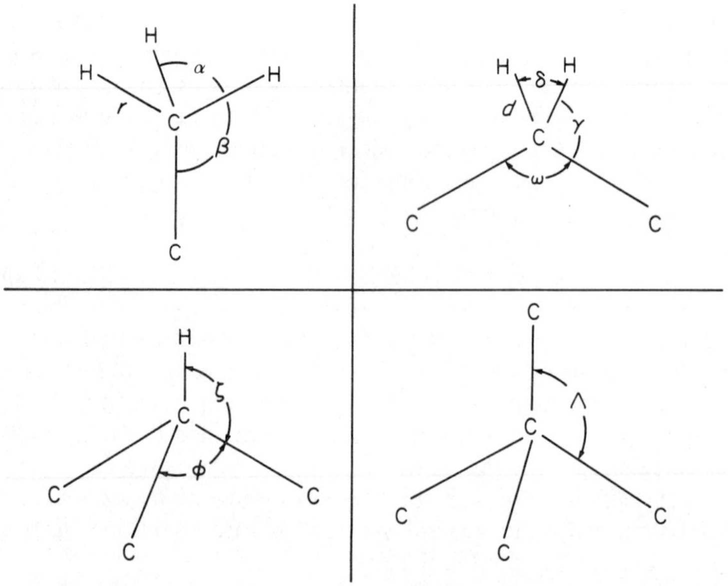

Figure 8.8.1 Labelling of internal coordinates in the alkanes for Table 8.8.1

Owing to the fact that the nomenclature of some interaction terms is often rather obscure, a schematic representation is given in Figure 8.8.2 for each type of interaction term. The approximations involved in the force field are also indicated in Table 8.8.2 by showing the interaction terms to which the same numerical value was assigned in the refinement. Additional approximations are indicated at the bottom of Table 8.8.1.

The force field of Snyder and Schachtschneider, as all simplified force fields of practical use, is refined by fitting observed vibrational frequencies rather than harmonic frequencies. A version of the same force field, adapted to the harmonic frequencies of propane and deuterated propanes, has been proposed by Gayles, King and Schachtschneider.[113] The excellent performances of the force field of Table 8.8.1 have encouraged many authors to

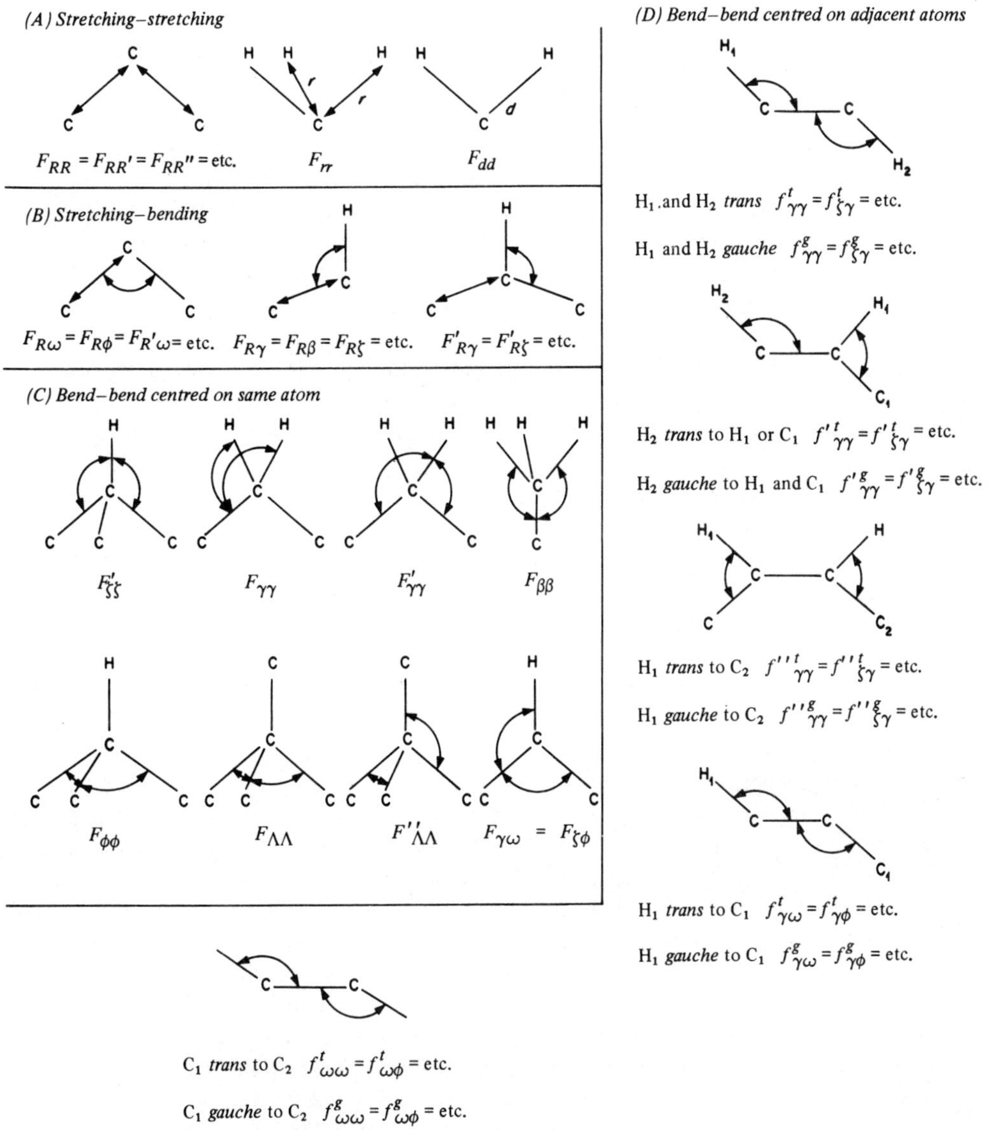

Figure 8.8.2 Definition of the interaction force constants for the force field of Table 8.8.1

the derivation of related force fields for molecules containing different substituents in the aliphatic chain, such as ethers, chlorides, nitriles, etc. Snyder and Zerbi[114] have used 204 observed frequencies of aliphatic ethers to refine a force field including 25 force constants transferred from the paraffins and 35 new force constants of various types, involving either two or one coordinate at least which is adjacent to the C–O bond. Schachtschneider[116] has derived a force field for primary aliphatic chlorides and Opaskar and Krimm[117] a related force field for secondary chlorides with 41 force constants, 23 of which were transferred from the paraffins, 13 from the primary chlorides and 5 were either new or modified from the other force fields. Yamadera and Krimm[118] have derived a similar force field for aliphatic nitriles with 40 force constants, 29 of which were transferred from the paraffins and 11 added or modified to obtain the best frequency fit.

Since these force fields contain a large number of force constants which involve very many types of coordinates, their tabulation would be very laborious and space consuming. For this reason we shall not give them here and the reader interested is referred to the original papers.

A simplified force field for ethylenic derivatives has been obtained by Schachtschneider.[119] A combination of this force field and of the aliphatic force field discussed before was used by Zerbi and Gussoni[120] to calculate the molecular vibrations of 1,2-polybutadiene and by Neto *et al.*[121] to predict the vibrational spectrum of cyclohexene. A cyclenic force field was then derived by these authors and used with minor changes to calculate the spectra of cycloheptene,[122] cyclooctene,[123] *trans-trans-trans* cyclododecatriene,[124] 1,3-cyclohexadiene[125] and 1,4-*trans* polybutadiene.[126]

Several simplified force fields have been proposed for aromatic hydrocarbons. Since the aromatic hydrocarbons are planar, the force field for the planar modes of vibration is separated from the force field for the non-planar vibrations, the first being symmetric and the second antisymmetric with respect to a reflection in the plane.

Apart from the classical works of Whiffen[127] and Albrecht,[128] who have calculated the complete force field of benzene and deutero benzenes in symmetry coordinates, simplified planar force fields have been proposed by Scherer[129] by refinement of a 13-term potential using the frequencies of several chlorobenzenes (including benzene itself) and of the corresponding deutero compounds, by Neto, Scrocco and Califano[130] by refinement of a 34-term potential by fitting the frequencies of benzene, naphthalene, anthracene and of the deutero analogues, by Duinker and Mills[131] by refinement of a 14-term potential on the frequencies of benzene and deuterobenzenes and by Kovner and coworkers[132,133,134,138] by fitting a 26-term potential to the frequencies of benzene and deuterobenzenes.

Simplified force fields for the non-planar vibrations of aromatic hydrocarbons have been proposed by several authors. Among them we shall quote those proposed by Scherer,[135] by Evans and Scully,[136] by Kakiuti and Shimanouchi[137] and by Kovner.[134]

A detailed bibliography of papers dealing with planar and non-planar force fields of substituted benzenes can be found in the book by Varsanyi entitled *Vibrational Spectra of Benzene Derivatives.*[138] The heuristic power of these force fields, i.e. their ability in predicting spectra of other molecules has been well tested[1] and it is worth pointing out that the general pattern of the vibrational spectra of large aromatic hydrocarbons such as phenanthrene,[139] pyrene[140] and triphenylene[141] has been predicted from them.

A large amount of work has also been accumulated on the simplified valence force fields of polymers and of inorganic substances. The reader is referred to a review by Zerbi[115] for the polymers and to the books by Nakamoto[105] and by Ross[173] for inorganic compounds.

8.9 The Urey–Bradley force field (UBFF)

The diffusion of the UBFF as a simplified potential function in normal coordinate analysis is comparable to that of the SGVFF, thanks to the monumental work of Shimanouchi and his school. Most of the work dealing with the UBFF has actually been discussed by Shimanouchi in reviews[16,17] which cover the state of the art up to 1970. In this section we shall thus only discuss the essential features of the UBFF and illustrate the merits and the limitations of this force field. The UBFF uses, in addition to the diagonal valence force constants of the stretching, bending and torsion type, non-bonded atom–atom repulsions. In order to introduce these force constants in the potential function, it is necessary to employ the variation in non-bonded atom distances as additional coordinates.

For instance, for the bent XYZ triatomic molecule of Figure 8.9.1 one utilizes as coordinates Δr_1, Δr_2, $\Delta\alpha$ and Δq, where q is the distance between the non-bonded atoms X

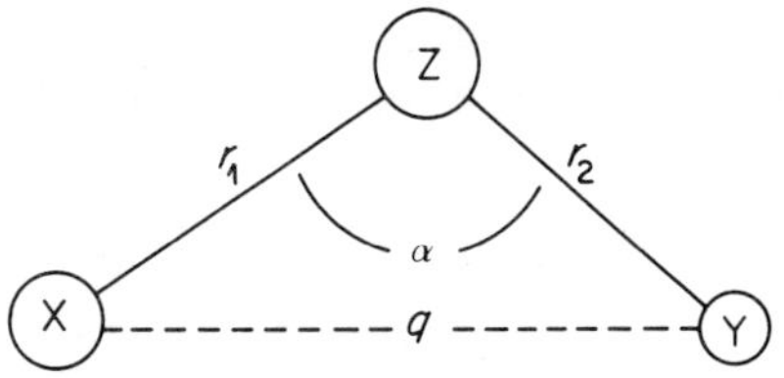

Figure 8.9.1 Non-bonded distance q in a triatomic molecule

and Y. Since a redundancy condition exists among these four coordinates, they do not constitute an independent set, in the sense that it is impossible to change one of the coordinates arbitrarily without changing at least one of the others. As discussed in Section 8.1 and more extensively in Section 8.10, the equilibrium conditions for the molecule do not necessarily imply that each $(\partial V/\partial s_i)_0$ is zero for a set of non-independent coordinates. For this reason the UBFF will include linear terms in its basic form. In the case of the XZY molecule the UBFF is therefore

$$2V = K_1\,\Delta r_1^2 + K_2\,\Delta r_2^2 + H(r_1\,\Delta\alpha)^2 + F\,\Delta q^2 + 2(K_1' r_1\,\Delta r_1 + K_2' r_2\,\Delta r_2 + H' r_1^2\,\Delta\alpha + F' q\,\Delta q) \quad (8.9.1)$$

where K_1, K_2, H and F are quadratic force constants, K_1', K_2', H' and F' are the corresponding linear terms and where all terms are expressed in the same units (mdyne/Å). Since the distance q between the non-bonded atoms X and Y is related to the other molecular parameters by the well-known relationship

$$q^2 = r_1^2 + r_2^2 - 2r_1 r_2 \cos\alpha \quad (8.9.2)$$

we can express the redundant coordinate Δq in terms of the valence coordinates, using a Taylor expansion

$$\Delta q = \sum_i \left(\frac{\partial q}{\partial s_i}\right) \Delta s_i + \frac{1}{2}\sum_{ij}\left(\frac{\partial^2 q}{\partial s_i\,\partial s_j}\right) \Delta s_i\,\Delta s_j \quad (8.9.3)$$

which yields

$$\Delta q = s_1(\Delta r_1) + s_2(\Delta r_2) + (t_1 t_2)^{\frac{1}{2}} R^{\frac{1}{2}}(r_1\,\Delta\alpha) + \frac{1}{2q}[t_1^2(\Delta r_1)^2 + t_2^2(\Delta r_2)^2 - s_1 s_2 R(r_1\,\Delta\alpha)^2$$
$$+ 2t_1 t_2(\Delta r_1\,\Delta r_2) + 2t_1 s_2 R(\Delta r_1 r_1\,\Delta\alpha) + 2t_2 s_1(\Delta r_2 r_1\,\Delta\alpha)] \quad (8.9.4)$$

where

$$s_1 = \frac{r_1 - r_2\cos\alpha}{q} \qquad t_1 = \frac{r_2\sin\alpha}{q}$$
$$R = \frac{r_2}{r_1} \quad (8.9.5)$$
$$s_2 = \frac{r_2 - r_1\cos\alpha}{q} \qquad t_2 = \frac{r_1\sin\alpha}{q}$$

By substitution of (8.9.4) in equation (8.9.1) and by neglecting terms higher than those of the second order we obtain the potential function expressed in terms of the valence coordinates Δr_1, Δr_2 and $(r_1\,\Delta\alpha)$

$$2V = 2[K_1' r_1 + q s_1 F'](\Delta r_1) + 2[K_2' r_2 + q s_2 F'](\Delta r_2) + 2[H' r_1 + q(R t_1 t_2)^{\frac{1}{2}} F'](r_1\,\Delta\alpha)$$
$$+ [K_1 + t_1^2 F' + s_1^2 F](\Delta r_1)^2 + [K_2 + t_2^2 F' + s_2^2 F](\Delta r_2)^2$$
$$+ [H - R s_1 s_2 F' + R t_1 t_2 F](r_1\,\Delta\alpha)^2 + 2[-t_1 t_2 F' + s_1 s_2 F](\Delta r_1\,\Delta r_2)$$
$$+ 2[R t_1 s_2 F' + s_1 (R t_1 t_2)^{\frac{1}{2}} F](\Delta r_1 r_1\,\Delta\alpha)$$
$$+ 2[t_2 s_1 F' + s_2 (R t_1 t_2)^{\frac{1}{2}} F](\Delta r_2 r_1\,\Delta\alpha) \quad (8.9.6)$$

The potential function (8.9.6) is now expressed in terms of a set of independent coordinates and therefore each linear term must be equal to zero in order to satisfy the equilibrium conditions. We have then

$$F' = -(1/q)(r_1/s_1)K_1' = -(1/q)(r_2/s_2)K_2' = -(1/q)[r_1/(Rt_1t_2)^{\frac{1}{2}}]H' \qquad (8.9.7)$$

and the potential function assumes the form

$$2V = K_{r_1}(\Delta r_1)^2 + K_{r_2}(\Delta r_2)^2 + H_\alpha(r_1\,\Delta\alpha)^2 + F_{rr}(\Delta r_1\,\Delta r_2) + F_{r_1\alpha}(\Delta r_1 r_1\,\Delta\alpha) + F_{r_2\alpha}(\Delta r_2 r_1\,\Delta\alpha) \qquad (8.9.8)$$

where

$$\begin{aligned} K_{r_1} &= K_1 + t_1^2F' + s_1^2F & F_{rr} &= s_1s_2F - t_1t_2F' \\ K_{r_2} &= K_2 + t_2^2F' + s_2^2F & F_{r_1\alpha} &= s_1(Rt_1t_2)^{\frac{1}{2}}F + Rt_1s_2F' \\ H_\alpha &= H + Rt_1t_2F - Rs_1s_2F' & F_{r_2\alpha} &= s_2(Rt_1t_2)^{\frac{1}{2}}F + t_2s_1F' \end{aligned} \qquad (8.9.9)$$

The elimination of the redundant coordinate Δq has the effect of introducing the linear term F' into the coefficients of the quadratic terms. Although the basic UBFF includes five force constants for the XZY molecule, a relationship between F and F' is easily established by assuming a specific analytical form of the interaction potential between non-bonded atoms. Thus, if we use a Lennard-Jones potential in the form

$$V = -A/q^6 + B/q^{12} \qquad (8.9.10)$$

we obtain

$$F' = -F/13[1 - (A/2B)q^6][1 - (7/13)(A/2B)q^6]^{-1} \qquad (8.9.11)$$

For the short range of distances occurring between two atoms bonded to a central one, the attractive part of the potential is negligible with respect to the repulsive part. Putting $A = 0$ we obtain

$$F' = -F/13 \qquad (8.9.12)$$

In the practice of UBFF calculations it is customary to assume a repulsive potential of the type

$$V = B/q^9 \qquad (8.9.13)$$

which gives rise to the relationship

$$F' = -0{\cdot}1F \qquad (8.9.14)$$

normally adopted in this kind of calculation.

In some cases, the interaction constant for the non-bonded distance between two atoms not bonded to the same atom has also been included in the UBFF. The coefficients for both the linear and the quadratic term for the non-bonded coordinate ΔQ of Figure 8.9.2 can

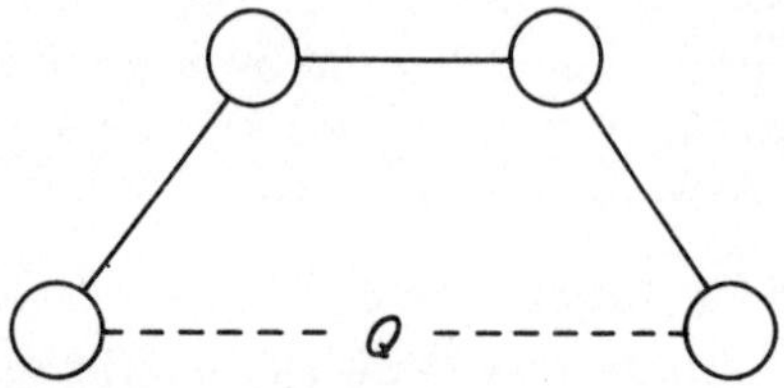

Figure 8.9.2 Non-bonded distance Q between two atoms separated by three bonds

be found in Reference 68. Owing to the fact that the interaction constant F_Q is very small, it is normal practice to neglect this interaction in actual calculation.

It is easily seen from equation (8.9.8) that the UBFF, when properly transformed in a valence coordinate basis, is equivalent to a special form of the SGUFF in which only interaction constants between adjacent coordinates occur and furthermore the interaction constants of the type F_{rr} and $F_{r\alpha}$ are not independent. It is therefore obvious that the UBFF will be a good approximation for all molecules for which other force constants can be neglected.

It turns out that the UBFF gives excellent results for molecules with heavy atoms and with non-delocalized π-electrons. It is, however, less adequate for vibrations of the hydrogen atoms and fails completely for molecules with delocalized electron systems. Some attempts have been made to modify the UBFF so that it gives a reasonable frequency fit in these cases also. This is accomplished by adding to the basic UBFF interaction constants some GVFF interaction terms which have a specific influence on the poorly fitted vibrational frequencies. Examples of the modified Urey–Bradley force field (MUBFF) are those used by Shimanouchi[17] for the CH_3 and CH_2 groups and by Overend and Scherer[68] for benzene.

The case of benzene is so characteristic of the situation that it is worth discussing it in some detail. The basic UBFF correctly reproduces all vibrational frequencies of benzene except the vibration ν_{15} of species B_{2u} which is observed at 1309 cm^{-1} and invariably calculated[142] around 1600 cm^{-1}. Overend and Scherer[68] suggested that the failure of the UBFF in this case is due to the absence of independent stretching–stretching force constants between the C–C bonds in *ortho*, *meta* and *para* position. By considering the stabilization of the Kekulé structures during a normal vibration, these authors reached the conclusion that the *ortho* and *para* force constants should be positive whereas the *meta* force constant should be negative. Introduction of these force constants in the form of a single 'Kekulé' parameter, is sufficient for a correct frequency fit. The same method has been later extended to naphthalene,[143] anthracene,[144] pyridine[145] and pyrazine.[146]

8.10 Redundant coordinates and harmonic potential functions

The discussion of the harmonic force field given in the previous sections was based on the assumption that a set of $3N - 6$ independent internal coordinates is used to set up the vibrational problem. In this case, since the molecule in the equilibrium configuration must be at a minimum of the energy, the derivatives of the potential with respect to each internal coordinate, taken at equilibrium, must vanish

$$f_i = \left(\frac{\partial V}{\partial s_i}\right)_0 \qquad i = 1, 2, \ldots, 3N - 6 \tag{8.10.1}$$

and the potential function, in the harmonic approximation, assumes a quadratic form of the type (8.1.3).

In many cases, however, it is convenient to use a number of internal coordinates greater than the number of internal degrees of freedom, either to make full use of symmetry or to obtain transferable and physically significant force constants. This introduces some complications in the handling of the linear terms of V since, whenever a dependent set of coordinates is used, the equilibrium conditions cannot be treated in the simple form of equations (8.10.1).

Before discussing this problem it is necessary to review briefly the nature of the coordinates used in the construction of the **G**- and **F**-matrices. There has been some confusion in the recent literature[147,148] because of insufficient attention to the definition of the coordinate system. As discussed in Chapter four, the vibrational problem is most conveniently set up in terms of internal coordinates. Although this is seldom explicitly stated, there are actually two possible types of internal coordinates that can be used for this purpose. The choice of one or other internal coordinate space implies a different handling of the redundancies in the potential function and involves a different definition of the harmonic part of the potential.

In the usual GF procedure, widely utilized in normal coordinate analysis, it is customary to set up the vibrational secular equation in terms of a set of $3N - 6$ independent internal coordinates, defined by means of a linear transformation from the $3N$ Cartesian displacement coordinates

$$\begin{aligned} R_i &= \sum_k B_k^i \Delta x_k \qquad i = 1, 2, \ldots, 3N - 6 \\ 0 &= \sum_k \beta_{ik} \Delta x_k \qquad i = 3N - 5, \ldots, 3N \end{aligned} \tag{8.10.2}$$

The last six equations are the Sayvetz conditions and represent six linear redundancies among the displacement coordinates. Because of the linear nature of the equations (8.10.2) the R_i coordinates span a linear configuration space. In order to emphasize this aspect of (8.10.2) we shall call the R_i 'rectilinear' coordinates.

If a number $m > 3N - 6$ of internal coordinates R_i is used, an additional $m - (3N - 6)$ linear equations are added to (8.10.2). Since there are only $3N - 6$ internal degrees of freedom, $m - (3N - 6)$ linear relations of the type

$$\sum_i \varphi_i R_i = 0 \qquad i = 1, 2, \ldots, m \tag{8.10.3}$$

will exist among the coordinates R_i, the linearity of these relations being a direct consequence of the linearity of (8.10.2). The kinetic and potential energy take, in the space of the R_i coordinates, the form

$$T = \tfrac{1}{2} \sum_{ij} G_{ij}^{-1} \dot{R}_i \dot{R}_j \tag{8.10.4}$$

$$V = \sum_i F_i R_i + \tfrac{1}{2} \sum_{ij} F_{ij} R_i R_j + \tfrac{1}{6} \sum_{ijk} F_{ijk} R_i R_j R_k + \ldots \tag{8.10.5}$$

where linear terms have been included in V on the assumption that the coordinate system is redundant. Without loss of generality, we can assume that only one redundancy occurs among the coordinates R_i. In order to take care of the linear part of V we can use the method of the undetermined Lagrange multiplier, by subtracting from (8.10.5) equation (8.10.3) multiplied by a parameter λ. In this way we obtain

$$V = \sum_i (F_i - \lambda \varphi_i) R_i + \tfrac{1}{2} \sum_{ij} F_{ij} R_i R_j + \ldots \tag{8.10.6}$$

and the conditions for V being stationary become

$$F_i - \lambda \varphi_i = 0 \qquad i = 1, 2, \ldots, m \tag{8.10.7}$$

One of these equations can be regarded as determining λ from one of the linear constants. Whether or not we consider λ or one of the F_i as the more basic quantity is immaterial. The value of λ remains in any case undetermined since it does not enter the vibrational secular equation. Usually λ is called intramolecular tension since it represents the residual tension existing at equilibrium among the molecular springs because of the constraint. In the harmonic approximation the potential energy is thus

$$V = \tfrac{1}{2} \sum_{ij} F_{ij} R_i R_j \tag{8.10.8}$$

and the presence of redundant coordinates has only the effect of introducing additional zero roots in the secular equation, as discussed in Section 6.7. If one wishes to eliminate the redundant coordinates, the methods described in Section 6.7 can be used. In any case, redundancies in the linearized space of the R_i coordinates have no influence on the form of the harmonic potential function and the linear terms can be ignored, just as in the case of a non-redundant basis.

In many cases it is, however, more convenient to express the potential energy in a different basis of internal coordinates $\mathcal{R}_s$, defined through a non-linear transformation from the Cartesian displacement coordinates[147,148,150]

$$\mathcal{R}_s = \sum_k B^s_k \,\Delta x_k + \tfrac{1}{2}\sum_{k,l} B^s_{kl}\,\Delta x_k\,\Delta x_l + \dots \tag{8.10.9}$$

The coordinates $\mathcal{R}_s$ are truly generalized coordinates and are often called curvilinear, since they span a curvilinear configuration space. They are geometrically defined for any position of the nuclei and not only for infinitesimal displacements as the R_i coordinates.[148]

The choice of a non-linear internal coordinate space may be dictated either by the structure of the model force field chosen or by the desire to get more information on the linear terms, as will be seen shortly. This is a perfectly legitimate choice as long as, before solving the vibrational secular equation, the force field expressed in terms of the curvilinear coordinates $\mathcal{R}_s$

$$V = \sum_s f_s \mathcal{R}_s + \tfrac{1}{2}\sum_{st} f_{st}\mathcal{R}_s\mathcal{R}_t + \tfrac{1}{6}\sum_{str} f_{str}\mathcal{R}_s\mathcal{R}_t\mathcal{R}_r + \dots \tag{8.10.10}$$

is transformed to the space of the linearized R_i coordinates in which the **G**-matrix is set up in the usual **G**–**F** procedure. The transformation from the $\mathcal{R}_s$ to the R_i space is a non-linear transformation[148] of the type

$$\mathcal{R}_s = \sum_i t_{si}R_i + \tfrac{1}{2}\sum_{ij} t_{sij}R_iR_j + \dots \tag{8.10.11}$$

where the t_{si} coefficients are mass-independent whereas the non-linear coefficients t_{sij} are mass-dependent.

If the coordinates $\mathcal{R}_s$ are non-redundant, the potential function (8.10.10) contains no linear terms and by performing the transformation (8.10.11) becomes

$$V = \tfrac{1}{2}\sum_{ij}\left(\sum_{st} f_{st}t_{si}t_{tj}\right)R_iR_j + \text{higher-order terms} \tag{8.10.12}$$

which is identical, in the harmonic approximation, to equation (8.10.8) with

$$F_{ij} = \sum_{st} f_{st}t_{si}t_{tj} \tag{8.10.13}$$

In the absence of redundancies the non-linear transformation from the $\mathcal{R}_s$ to the R_i coordinates thus has no effect on the harmonic part of the potential, except for a redefinition of the force constants. In particular the curvilinear and rectilinear coordinates can be chosen to be identical to the first order so that the coefficients t_{si} are the elements of the identity matrix and thus the harmonic force constants in the two bases are also identical. This is the reason why the curvilinear nature of the coordinates is normally ignored in defining the potential energy.

Assume now that a redundancy occurs among the curvilinear coordinates $\mathcal{R}_s$. In this case, before performing the transformation to the linearized R_i space, it is necessary to eliminate the redundancy and to impose the condition that V is stationary. In the space of the curvilinear coordinates $\mathcal{R}_s$, the redundancy is also non-linear and can be written in the symmetrical form

$$\sum_s \varphi_s \mathscr{R}_s + \tfrac{1}{2}\sum_{s,t} \varphi_{st}\mathscr{R}_s\mathscr{R}_t + \ldots = 0 \tag{8.10.14}$$

or in the equivalent, unsymmetrical form

$$\mathscr{R}_w = \sum_s{}' A_s\mathscr{R}_s + \tfrac{1}{2}\sum_{s,t}{}' A_{st}\mathscr{R}_s\mathscr{R}_t + \ldots \tag{8.10.15}$$

where the prime indicates that the summations run over all indices except w, this index labelling the redundant coordinate. Equation (8.10.14) is normally used when the redundancy involves a set of equivalent coordinates, as for instance the six angle-deformations of methane. Equation (8.10.15) is used when one type of coordinate can be expressed in terms of others as for instance the variation of the non-bonded atom–atom distance in the UBFF (see Section 8.9).

Again we can use the method of the undetermined multiplier to eliminate the redundancy. By subtracting from (8.10.10) equation (8.10.14) multiplied by a parameter λ, we have

$$V = \sum_s (f_s - \lambda\varphi_s)\mathscr{R}_s + \tfrac{1}{2}\sum_{s,t}(f_{st} - \lambda\varphi_{st})\mathscr{R}_s\mathscr{R}_t + \text{higher terms} \tag{8.10.16}$$

The conditions for V being stationary are then

$$f_s - \lambda\varphi_s = 0 \qquad s = 1, 2, \ldots, 3N - 5 \tag{8.10.17}$$

which are equivalent to equations (8.10.7) except for the fact that the parameter λ now enters the quadratic part of the potential and can be determined as any other quadratic force constant if one has enough experimental data. The force field is then transformed to the space of the rectilinear coordinates R_i using (8.10.11). The harmonic force constants thus assume the form

$$F_{ij} = \sum_{s,t} t_{si}t_{tj}(f_{st} - \lambda\varphi_{st}) \tag{8.10.18}$$

and include the linear parameter λ multiplied by the geometrical factors φ_{st}. Again, if desired, the redundancy can be eliminated from the secular equation using the methods described in Section 6.7. This amounts to a transformation to a new set of linearized coordinates, usually symmetry coordinates, one of which is just the first-order redundancy. If the redundancy is of the type (8.10.15) it is more convenient to eliminate the redundant coordinate (since in this case there is no loss of symmetry) by direct substitution of (8.10.15) in (8.10.10). If this is done we obtain

$$V = \sum_s{}'(f_s + A_s f_w)\mathscr{R}_s + \tfrac{1}{2}\sum_{st}{}'(f_{st} + A_{st}f_w + A_sA_tf_{ww} + 2A_tf_{sw})\mathscr{R}_s\mathscr{R}_t \tag{8.10.19}$$

and requiring V to be stationary we have

$$f_s + A_s f_w = 0 \qquad s = 1, 2, \ldots, 3N - 6 \tag{8.10.20}$$

Again equations (8.10.20) are equivalent to (8.10.7) and define each linear term f_s through the single linear term f_w which is now evaluable, being included in the harmonic part of the potential.

Transformation to the linearized R_i space yields for the harmonic force constants

$$F_{ij} = \sum_{st}{}' t_{si}t_{tj}(f_{st} + A_{st}f_w + A_sA_tf_{ww} + 2A_tf_{sw}) \tag{8.10.21}$$

As pointed out before, the coefficients t_{si} of the transformation (8.10.11) can be chosen to be the elements of the identity matrix. Equations (8.10.18) and (8.10.21) then become

$$F_{ij} = f_{ij} - \lambda\varphi_{ij} \tag{8.10.22}$$

$$F_{ij} = f_{ij} + A_{ij}f_w + A_iA_jf_{ww} + 2A_jf_{iw} \tag{8.10.23}$$

where it must be understood that the coordinates i, j and w must be involved in the redundancy since only for such coordinates are the coefficients A_i, A_j and A_{ij} different from zero. The appearance of the intramolecular tension λ in (8.10.22) is entirely equivalent to the appearance of the linear term f_w in (8.10.23). The two expressions of the force constant are formally different only because in the first case a basis of $3N - 5$ internal coordinates is used whereas in the second case the basis is reduced to $3N - 6$ coordinates. In any case each redundancy among the curvilinear coordinates introduces an intramolecular tension in the harmonic force field[148,149] except in the particular cases in which the redundancy condition is exactly linear.[151]

It is obvious that it makes sense to express the harmonic force constants in the form (8.10.22) only if one has at one's disposal enough experimental data to afford the evaluation of one extra force constant for each redundancy existing among the coordinates. In other words, one can be content to obtain a numerical value for the force constant F_{ij} in the linearized R_i space rather than for the force constant f_{ij} in the curvilinear $\mathfrak{R}_i$ space. Whether one chooses the first or the second definition of force constants is simply a matter of convenience and of personal preference. This is essentially the reason for the confusion that occurs whenever redundant sets of coordinates are used, since it is difficult to understand a non-linear redundancy relation if one thinks only in terms of a linearized space.

In Section 8.9 we have already seen how to use equation (8.10.15) for the elimination of the redundant coordinate Δq (variation of the distance between non-bonded atoms) in the UBFF. The force constants calculated by means of equation (8.10.23) are given by equations (8.9.9) and the relations (8.10.20) between the linear terms are given by (8.9.7).

As an example of the use of equation (8.10.14) for a redundancy involving a set of symmetrical coordinates we shall now consider the classical example of a tetrahedral molecule.

Example 8.10.1

For a tetrahedral molecule like methane the redundancy condition in the symmetrical form of equation (8.10.14) takes the form (see equation 6.7.17)

$$\sum_{i<j} \Delta\alpha_{ij} + \frac{1}{2}\left[\frac{3}{\sqrt{8}}\sum_{i<j} \Delta\alpha_{ij}^2 + \sqrt{2}\sum_{\substack{i\neq j\\ j<k}} \Delta\alpha_{ij}\,\Delta\alpha_{ik}\right] = 0$$

where the labelling of the internal coordinates refers to Figure 6.7.1. Using equation (8.10.22) we obtain for the harmonic force constants

$$F_r = F_{r_1r_1} = F_{r_2r_2} = \text{etc.} = f_r \qquad F_\alpha = F_{\alpha_{12}\alpha_{12}} = F_{\alpha_{13}\alpha_{13}} = \text{etc.} = f_\alpha - 3\lambda/\sqrt{8}$$

$$F_{rr} = F_{r_1r_2} = F_{r_1r_3} = \text{etc.} = f_{rr} \qquad F_{\alpha\alpha} = F_{\alpha_{12}\alpha_{13}} = F_{\alpha_{12}\alpha_{14}} = \text{etc.} = f_{\alpha\alpha} - \lambda/\sqrt{2}$$

$$F_{r\alpha} = F_{r_1\alpha_{12}} = F_{r_1\alpha_{13}} = \text{etc.} = f_{r\alpha} \qquad F'_{\alpha\alpha} = F_{\alpha_{12}\alpha_{34}} = F_{\alpha_{14}\alpha_{23}} = \text{etc.} = f'_{\alpha\alpha}$$

$$F'_{r\alpha} = F_{r_1\alpha_{23}} = F_{r_1\alpha_{34}} = \text{etc.} = f'_{r\alpha}$$

8.11 *Ab initio* calculations of force constants

In the mechanical model used until now, the potential energy of a polyatomic molecule is expressed in terms of elastic springs holding together the nuclei. These springs simulate the binding action of the electrons and thus the vibrational problem is solved in terms of adjustable parameters (the force constants) without any knowledge of the electron distribution in the molecule. If, however, the electronic wavefunctions and energies are well known, it is possible to calculate the force constants '*a priori*' and thus to establish a valuable connection between the mechanical model and the quantum-mechanical description of chemical bonds.

The calculation of force constants from the electronic distribution is not a simple problem and requires the knowledge of very accurate electronic wavefunctions. Although calculations of this type have been made for many years,[152–158] only very recently has a sufficient degree of accuracy been reached to make them valuable for the interpretation of the forces acting in the vibrating molecules. In this section we shall therefore give a general description of the methods used in the calculations and a short survey of the recent work in this field. For a deeper insight into the problem the reading of References 159, 160, 164 is strongly recommended.

In the limits of the Born–Oppenheimer approximation, the electronic Hamiltonian H_e is defined (in atomic units) for a molecule with N nuclei and n electrons as

$$\mathscr{H}_e = -\tfrac{1}{2}\sum_{\mu} \nabla_\mu^2 - \sum_{\mu}\sum_{i} z_i r_{\mu i}^{-1} + \sum_{\mu}\sum_{\nu} r_{\mu\nu}^{-1} + \sum_{i}\sum_{j} z_i z_j R_{ij}^{-1} \tag{8.11.1}$$

where μ and ν are indices used to label the electrons and i, j indices used for the nuclei with the conditions $\mu > \nu$ and $i > j$, in order to ensure that each term is counted only once. Z represents the number of charges on the nuclei and $r_{\mu\nu}$, $r_{\mu i}$ and R_{ij} are electron–electron, electron–nucleus and nucleus–nucleus distances respectively.

Since the force constants are best calculated in Cartesian coordinates, we define a Cartesian system in space and measure the positions of the nuclei by means of position vectors $\mathbf{R}_i$ with components R_i^α ($\alpha = x, y, z$) and the positions of the electrons by means of position vectors $\mathbf{r}_\mu$ with components $\mathbf{r}_\mu^\alpha$. The distances occurring in (8.11.1) are thus the scalar quantities $r_{\mu i} = |\mathbf{r}_\mu - \mathbf{R}_i|$, $r_{\mu\nu} = |\mathbf{r}_\mu - \mathbf{r}_\nu|$ and $R_{ij} = |\mathbf{R}_i - \mathbf{R}_j|$.

By definition the Cartesian force constants $f_{ij}^{\alpha\beta}$ are given by

$$f_{ij}^{\alpha\beta} = \left(\frac{\partial^2 E_e}{\partial R_i^\alpha \, \partial R_j^\beta}\right)_0 = \frac{\partial^2}{\partial R_i^\alpha \, \partial R_j^\beta} \langle \psi_e | \mathscr{H}_e | \psi_e \rangle \tag{8.11.2}$$

where E_e and ψ_e are the eigenvalues and the eigenfunctions respectively of the electronic wave-equation

$$\mathscr{H}_e \psi_e = E_e \psi_e \tag{8.11.3}$$

Three possible procedures can be used for the evaluation of the force constants, differing in the way in which the second derivatives (8.11.2) are obtained. These derivatives can actually be evaluated either twice numerically or twice analytically or even first analytically and then numerically.

The numerical derivation of the force constants represents the simplest approach and has been used in the past by several authors.[165] It amounts to the calculation of the energy for several nuclear configurations in which the coordinates R_i^α and R_j^β have been changed, all other coordinates being kept at equilibrium. A surface is then fitted to the various energies and the force constants are obtained by numerical differentiation of the surface equation. Bishop and Randič[159] and later Pulay[160] have pointed out that this method is not economical because it requires the calculation of the wavefunctions and the energies for several configurations of the nuclei and that it is not accurate because of the two consecutive numerical differentiations.

In order to obtain the force constants analytically from equation (8.11.2), we need to evaluate the first derivative of the energy with respect to a nuclear coordinate R_i^α, i.e. the force acting on the ith nucleus when the αth coordinate is changed. We have then

$$-f_i^\alpha = \frac{\partial E_e}{\partial R_i^\alpha} = \left\langle \psi_e \middle| \frac{\partial \mathscr{H}_e}{\partial R_i^\alpha} \middle| \psi_e \right\rangle + \left[\left\langle \frac{\partial \psi_e}{\partial R_i^\alpha} \middle| \mathscr{H}_e \middle| \psi_e \right\rangle + \left\langle \psi_e \middle| \mathscr{H}_e \middle| \frac{\partial \psi_e}{\partial R_i^\alpha} \right\rangle \right] \tag{8.11.4}$$

As long as ψ_e represents the exact wavefunction, this expression can be simplified by means of the Hellmann–Feynman theorem[161,162] which states that only the first term in the right-hand side of (8.11.4) is different from zero. In other words the theorem says that the forces exerted by the electrons on the nuclei are correctly calculated by considering the electrostatic attraction of the equilibrium charge cloud of the electrons on the nuclei. These latter move, therefore, in the undeformed charge distribution of the electrons toward the region of higher charge density until the nuclear repulsions balance the attractive forces exactly.

We shall call the first term the 'Hellmann–Feynman force' and the second term the 'wavefunction force' in (8.11.4). For the exact wavefunction the wavefunction force vanishes and the first derivative of the energy is equal to the Hellmann–Feynman force, being thus given by

$$f_i^\alpha = \frac{\partial E_e}{\partial R_i^\alpha} = \left\langle \psi_e \middle| \frac{\partial \mathscr{H}_e}{\partial R_i^\alpha} \middle| \psi_e \right\rangle = -\left\langle \psi_e \middle| \sum_\mu z_i \frac{\partial r_{\mu i}^{-1}}{\partial R_i^\alpha} \middle| \psi_e \right\rangle + \sum_j z_i z_j \frac{\partial R_{ij}^{-1}}{\partial R_i^\alpha} \tag{8.11.5}$$

From (8.11.5) we obtain for the second derivative

$$\begin{aligned} f_{ij}^{\alpha\beta} = \frac{\partial^2 E_e}{\partial R_i^\alpha \partial R_j^\beta} &= \left\langle \psi_e \middle| \frac{\partial^2 \mathscr{H}_e}{\partial R_i^\alpha \partial R_j^\beta} \middle| \psi_e \right\rangle + \left[\left\langle \frac{\partial \psi_e}{\partial R_j^\beta} \middle| \frac{\partial \mathscr{H}_e}{\partial R_i^\alpha} \middle| \psi_e \right\rangle + \left\langle \psi_e \middle| \frac{\partial \mathscr{H}_e}{\partial R_i^\alpha} \middle| \frac{\partial \psi_e}{\partial R_j^\beta} \right\rangle \right] \\ &= -\left\langle \psi_e \middle| \sum_\mu z_i \frac{\partial^2 r_{\mu i}^{-1}}{\partial R_i^\alpha \partial R_j^\beta} \middle| \psi_e \right\rangle + z_i z_j \frac{\partial^2 R_{ij}^{-1}}{\partial R_i^\alpha \partial R_j^\beta} \\ &\quad + \left[\left\langle \frac{\partial \psi_e}{\partial R_j^\beta} \middle| \frac{\partial \mathscr{H}_e}{\partial R_i^\alpha} \middle| \psi_e \right\rangle + \left\langle \psi_e \middle| \frac{\partial \mathscr{H}_e}{\partial R_i^\alpha} \middle| \frac{\partial \psi_e}{\partial R_j^\beta} \right\rangle \right] \end{aligned} \tag{8.11.6}$$

It follows from (8.11.6) that the force constants are made up of three types of terms. The first term represents the contribution from the nuclear displacements in the equilibrium charge distribution of the electrons and will be called the Hellmann–Feynman term. The second term arises from the variation in the nuclear repulsion and will be called the nuclear term. The third term arises from the readjustment of the electronic charge distribution to the new nuclear configuration and will be called the orbital following or the relaxation term. The Hellmann–Feynman term contributes only to the diagonal force constants ($i = j$), since $r_{\mu i}$ is a function of R_i only and hence $\partial^2 r_{\mu i}^{-1}/\partial R_i^\alpha \partial R_j^\beta$ is zero unless $i = j$.

As pointed out before, the Hellmann–Feynman theorem holds only when ψ_e is the exact electronic wavefunction. It has been shown,[163] however, that Hartree–Fock wavefunctions also obey the Hellmann–Feynman theorem and this permits the application of the above equations to molecular systems for which good self-consistent molecular orbitals (SCF–MO) converging upon the Hartree–Fock limit have been obtained. A detailed discussion of the accuracy expected for force constants calculated with HF functions is given in the paper by Gerratt and Mills.[164]

The exact expression for the derivatives of the Hartree–Fock energy was first derived by Bratos.[154] Later Gerratt and Mills[164] and Pulay[160] also derived expressions for the derivatives of the SCF energy. These expressions will not be reported here, because this would require a rather long discussion of the HF method.

The use of the Hellmann–Feynman theorem in deriving force constants has been criticized by various authors and particularly by Pulay.[160] The main point of these criticisms is that the Hellmann–Feynman force is extremely sensitive to the wavefunctions used and, unless these are very accurate, it may differ considerably from the derivative (8.11.4) of the energy. The amount of labour required to produce good wavefunctions is, however, so high that it is in many respects convenient to use medium accuracy wavefunctions and the complete energy derivative (8.11.4) instead of very accurate wavefunctions and the Hellmann–Feynman force. For this reason Pulay[160] has proposed the so-called 'force method' in which the first derivative of the energy is made analytically according to equation (8.11.4) and the second derivative numerically. The force method has been applied with noticeable success to the calculation of the force constants of several small molecules[166–169] using SCF–MO wavefunctions.

Semi-empirical simplifications of the SCF–MO method, such as the CNDO/2 method, have been very popular in recent years for the calculation of molecular observables. A review of the CNDO and of related methods can be found in Reference 170. The CNDO/2 method has also been used for the calculation of force constants.[171,172] Pulay and Török have recently extended the force method to the semi-empirical CNDO/2 wavefunctions. The diagonal force constants obtained in this way are normally much higher than the experimental values, whereas off-diagonal force constants seem in most cases to be of the right order of magnitude.

References

1. S. Califano, *Pure and Appl. Chem.*, **18**, 353 (1969).
2. A. Fadini, *Z. Angew. Math. Mechan.*, **44**, 506 (1964); **45**, 29 (1965).
3. W. Sawodny, A. Fadini and K. Ballein, *Spectrochim. Acta*, **21A**, 995 (1965).
4. H. J. Becher and R. Mattes, *Spectrochim. Acta*, **23A**, 2449 (1967).
5. A. Alix and L. Bernard, *J. Mol. Structure*, **20**, 51 (1974).
6. G. Strey, *J. Mol. Spectrosc.*, **24**, 87 (1967).
7. F. Billes, *Acta Chim. Acad. Sci. Hung.*, **47**, 53 (1966).
8. P. Pulay, *Acta Chim. Acad. Sci. Hung.*, **52**, 49 (1967).
9. S. J. Cyvin, L. A. Kristiansen and J. Brunvoll, *Acta Chim. Acad. Sci. Hung.*, **51**, 217 (1967).
10. J. Herranz and F. Castano, *Spectrochim. Acta*, **22A**, 1965 (1966); *Annales Real Soc. Espan. Fis. Quim.* (*Madrid*), Ser. **A62**, 199 (1966).
11. D. E. Freeman, *J. Mol. Spectrosc.*, **27**, 27 (1968).
12. P. Pulay and F. Török, *Acta Chim. Acad. Sci. Hung.*, **44**, 287 (1965); **47**, 273 (1966).
13. R. L. Redington and A. L. K. Aljibury, *J. Mol. Spectrosc.*, **37**, 494 (1971).
14. H. C. Urey and C. A. Bradley, *Phys. Rev.*, **38**, 1969 (1931).
15. T. Shimanouchi, *J. Chem. Phys.*, **17**, 245, 743, 848 (1949); **26**, 594 (1957).
16. T. Shimanouchi, *Pure and Appl. Chem.*, **7**, 131 (1963).
17. T. Shimanouchi, in *Physical Chemistry. An Advanced Treatise* (Eds. H. Eyring, D. Henderson and W. Jost), Vol. IV, Academic Press, 1970.
18. S. Lifson and A. Warshel, *J. Chem. Phys.*, **49**, 5116 (1968).
19. A. Warshel and S. Lifson, *Chem. Phys. Letters*, **4**, 255 (1969).
20. A. Warshel and S. Lifson, *J. Chem. Phys.*, **53**, 582 (1970).
21. D. F. Heath and J. W. Linnett, *Trans. Faraday Soc.*, **44**, 556, 561, 873, 878, 884 (1948); **45**, 264 (1949).
22. I. M. Mills, *Spectrochim. Acta*, **19**, 1585 (1963).
23. J. R. Scherer and J. Overend, *Spectrochim. Acta*, **17**, 719 (1961).
24. G. Herzberg, *Infrared and Raman Spectra of Polyatomic Molecules*, D. Van Nostrand, 1945.
25. W. S. Benedict, N. Gailar and E. K. Plyler, *J. Chem. Phys.*, **24**, 1139 (1956).
26. R. D. Shelton, A. H. Nielsen and W. H. Fletcher, *J. Chem. Phys.*, **21**, 2178 (1953); **22**, 1791 (1954).
27. L. H. Jones and R. S. McDowell, *J. Mol. Spectrosc.*, **3**, 632 (1959).
28. J. Aldous and I. M. Mills, *Spectrochim. Acta*, **18**, 1073 (1962).
29. W. T. King, I. M. Mills and B. Crawford jr., *J. Chem. Phys.*, **27**, 455 (1957).
30. R. L. Arnett and B. Crawford jr., *J. Chem. Phys.*, **18**, 118 (1950).

31. Quoted by Angus *et al.*, *J. Chem. Soc.*, 971 (1936).
32. O. Redlich, *Z. Phys. Chem.*, **B28**, 371 (1935).
33. J. C. Decius and E. B. Wilson jr., *J. Chem. Phys.*, **19**, 1409 (1951).
34. L. M. Sverdlov, *Doklady Acad. Nauk. SSSR*, **78**, 1115 (1951).
35. J. Heicklen, *J. Chem. Phys.*, **36**, 721 (1962).
36. L. M. Sverdlov, *Optics and Spectroscopy*, **8**, 17 (1960).
37. S. Brodersen and A. Langseth, *Kgl. Danske Videnskab Selbskab, Mat. Fys. Skrifter*, **1**, No. 5 (1958).
38. S. Brodersen and A. Langseth, *Ibidem*, **1**, No. 7 (1958).
39. S. Brodersen and A. Langseth, *J. Mol. Spectrosc.*, **3**, 114 (1959).
40. S. Brodersen, *J. Mol. Spectrosc.*, **3**, 450 (1959).
41. W. A. Person and B. Crawford jr., *J. Chem. Phys.*, **26**, 1295 (1957).
42. D. C. McKean and J. L. Duncan, *Spectrochim. Acta*, **27A**, 1879 (1971).
43. E. B. Wilson jr., J. C. Decius and P. C. Cross, *Molecular Vibrations*, McGraw-Hill, 1955.
44. A. A. Chalmers and D. C. McKean, *Spectrochim. Acta*, **22**, 251 (1966).
45. D. C. McKean, *Spectrochim. Acta*, **22**, 269 (1966).
46. M. Tsuboi, *J. Mol. Spectrosc.*, **19**, 4 (1966).
47. H. A. Jahn, *Phys. Rev.*, **56**, 680 (1939).
48. W. H. J. Childs and H. A. Jahn, *Proc. Roy. Soc.* (*London*), **A169**, 451 (1939).
49. S. J. Cyvin, *Molecular Vibrations and Mean Square Amplitudes*, Elsevier, 1968.
50. D. R. J. Boyd and H. C. Longuet Higgins, *Proc. Roy. Soc.* (*London*), **A213**, 55 (1952).
51. R. C. Lord and R. E. Merrifield, *J. Chem. Phys.*, **20**, 1348 (1952).
52. I. M. Mills and J. L. Duncan, *J. Mol. Spectrosc.*, **9**, 244 (1962).
53. D. W. Lepard, *Canad. J. Phys.*, **44**, 461 (1966).
54. E. Teller, *Hand- und Jahrb. Chem. Phys.*, **9**, II, 43 (1934).
55. M. Johnston and D. M. Dennison, *Phys. Rev.*, **48**, 868 (1935).
56. R. S. McDowell, *J. Chem. Phys.*, **43**, 319 (1965).
57. T. Oka, *J. Mol. Spectrosc.*, **29**, 84 (1969).
58. L. Nemes, *J. Mol. Spectrosc.*, **30**, 123 (1969).
59. J. K. G. Watson, *J. Mol. Spectrosc.*, **39**, 364 (1971).
60. J. H. Meal and S. R. Polo, *J. Chem. Phys.*, 1119, 1126 (1956).
61. D. Kivelson and E. B. Wilson jr., *J. Chem. Phys.*, **20**, 1575 (1952); **21**, 1229 (1953).
62. W. H. Kirchhoff, *J. Mol. Spectrosc.*, **41**, 333 (1972).
63. P. Pulay and W. Savodny, *J. Mol. Spectrosc.*, **26**, 150 (1968).
64. G. O. Sørensen, G. Hagen and S. J. Cyvin, *J. Mol. Spectrosc.*, **35**, 489 (1970).
65. F. Bloch, *Z. Physik*, **74**, 295 (1932).
66. R. W. James, *Phys. Z.*, **33**, 737 (1932).
67. Y. Morino, K. Kuchitsu, A. Takahashi and K. Maeda, *J. Chem. Phys.*, **21**, 1927 (1953).
68. J. Overend and J. Scherer, *J. Chem. Phys.*, **32**, 1289 (1960).
69. J. Aldous and I. M. Mills, *Spectrochim. Acta*, **19**, 1567 (1963).
70. J. H. Schachtschneider and R. G. Snyder, *Spectrochim. Acta*, **19**, 117 (1963).
71. L. Nemes, *Spectrochim. Acta*, **24A**, 300 (1968).
72. M. J. Bruton and L. A. Woodward, *Spectrochim. Acta*, **23A**, 175 (1967).
73. N. Neto, *Gazz. Chim. It.*, **96**, 1094 (1966).
74. R. Fletcher and M. J. D. Powell, *Computer J.*, **6**, 163 (1963).
75. P. Gans, *Chem. Phys. Letters*, **6**, 561 (1970).
76. P. Gans, *J. Chem. Soc.* (*A*), 2017 (1971).
77. J. W. Nibler and G. C. Pimentel, *J. Mol. Spectrosc.*, **26**, 294 (1968).
78. P. N. Schatz, *J. Chem. Phys.*, **29**, 481 (1958).
79. J. W. Levin and S. Abramowitz, *J. Chem. Phys.*, **44**, 2562 (1966).
80. A. M. Mirri, *J. Chem. Phys.*, **47**, 2823 (1967).
81. W. Savodny, A. Ruoff, C. J. Peacock and A. Müller, *Mol. Phys.*, **14**, 433 (1968).
82. M. Otake, C. Matsumura and Y. Morino, *J. Mol. Spectrosc.*, **28**, 316 (1968).
83. M. Otake, E. Hirota and Y. Morino, *J. Mol. Spectrosc.*, **28**, 325 (1968).
84. J. M. Dowling, R. Gold and A. G. Meister, *J. Mol. Spectrosc.*, **1**, 265 (1957).
85. R. D'Cunha, *J. Mol. Spectrosc.*, **43**, 282 (1972).
86. G. R. Bird *et al.*, *J. Chem. Phys.*, **40**, 3378 (1964).
87. A. M. Mirri and E. Mazzariol, *Spectrochim. Acta*, **22**, 785 (1966).
88. R. R. Ryan and L. H. Jones, *J. Chem. Phys.*, **50**, 1492 (1969).

89. J. Laane *et al.*, *J. Mol. Spectrosc.*, **30**, 489 (1969).
90. Y. Morino, Y. Kikuchi, S. Saito and E. Hirota, *J. Mol. Spectrosc.*, **13**, 95 (1964).
91. H. Takeo, E. Hirota and Y. Morino, *J. Mol. Spectrosc.*, **41**, 420 (1972).
92. T. Tanaka and Y. Morino, *J. Mol. Spectrosc.*, **33**, 538 (1970).
93. Y. Morino and S. Saito, *J. Mol. Spectrosc.*, **19**, 435 (1966).
94. A. M. Mirri and A. Guarnieri, *Spectrochim. Acta*, **23A**, 2159 (1967).
95. A. Ruoff, *Spectrochim. Acta*, **26A**, 545 (1970).
96. E. Hirota and Y. Morino, *J. Mol. Spectrosc.*, **33**, 460 (1970).
97. B. P. Winnewisser and M. Winnewisser, *Z. Naturforsch.*, **23A**, 832 (1968).
98. N. Mohan and A. Müller, *J. Mol. Spectrosc.*, **42**, 203 (1972).
99. A. Ruoff, *Spectrochim. Acta*, **23A**, 2421 (1967).
100. A. M. Mirri, G. Cazzoli and L. Ferretti, *J. Chem. Phys.*, **49**, 2775 (1968).
101. I. Suzuki, M. A. Pariseau and J. Overend, *J. Chem. Phys.*, **44**, 3561 (1966).
102. I. Suzuki, *J. Mol. Spectrosc.*, **32**, 54 (1969).
103. Y. Morino and T. Nakagawa, *J. Mol. Spectrosc.*, **26**, 496 (1968).
104. I. Suzuki, *J. Mol. Spectrosc.*, **25**, 479 (1968).
105. K. Nakamoto, *Infrared Spectra of Inorg. and Coord. Compounds*, J. Wiley, 1963.
106. J. L. Duncan, *J. Mol. Spectrosc.*, **13**, 338 (1964).
107. A. Müller, F. Königer, K. Nakamoto and N. Ohkaku, *Spectrochim. Acta*, **28A**, 1933 (1972).
108. G. Strey and I. M. Mills, *Mol. Phys.*, **26**, 129 (1973).
109. A. J. Dorney, A. R. Hoy and I. M. Mills, *J. Mol. Spectrosc.*, **45**, 253 (1973).
110. J. Giguere, V. K. Wang, J. Overend and A. Cabana, *Spectrochim. Acta*, **29A**, 1197 (1973).
111. Y. Morino, K. Kuchitsu and S. Yamamoto, *Spectrochim. Acta*, **24A**, 335 (1968).
112. R. G. Snyder and J. H. Schachtschneider, *Spectrochim. Acta*, **21**, 169 (1965).
113. J. N. Gayles, Jr., W. T. King and J. H. Schachtschneider, *Spectrochim. Acta*, **23A**, 703 (1967).
114. R. G. Snyder and G. Zerbi, *Spectrochim. Acta*, **23A**, 391 (1967).
115. G. Zerbi, *Appl. Spectrosc. Rev.*, **2**, 193 (1969).
116. See Reference 117.
117. C. G. Opaskar and S. Krimm, *Spectrochim. Acta*, **23A**, 2261 (1967).
118. R. Yamadera and S. Krimm, *Spectrochim. Acta*, **24A**, 1677 (1968).
119. See Reference 120.
120. G. Zerbi and M. Gussoni, *Spectrochim. Acta*, **22**, 2111 (1966).
121. N. Neto, C. Di Lauro, E. Castellucci and S. Califano, *Spectrochim. Acta*, **23A**, 1763 (1967).
122. N. Neto, C. Di Lauro and S. Califano, *Spectrochim. Acta*, **26A**, 1489 (1970).
123. N. Neto and S. Califano, unpublished work.
124. N. Neto, C. Di Lauro and S. Califano, *Spectrochim. Acta*, **24A**, 385 (1968).
125. C. Di Lauro, N. Neto and S. Califano, *J. Mol. Struct.*, **3**, 219 (1969).
126. N. Neto and C. Di Lauro, *Eur. Polymer J.*, **3**, 645 (1967).
127. D. H. Whiffen, *Phil. Trans. Roy. Soc.*, **A248**, 131 (1955).
128. A. C. Albrecht, *J. Mol. Spectrosc.*, **5**, 238 (1960).
129. J. R. Scherer, *Spectrochim. Acta*, **20**, 345 (1964).
130. N. Neto, M. Scrocco and S. Califano, *Spectrochim. Acta*, **22**, 1981 (1966).
131. J. C. Duinker and I. M. Mills, *Spectrochim. Acta*, **24A**, 417 (1968).
132. M. A. Kovner, *Dok. AK. Nauk. Khim. Otd.*, **41**, 499 (1953).
133. M. A. Kovner, *Zhur. Exp. Teor. Fiz.*, **26**, 598 (1954); **26**, 704 (1954).
134. M. A. Kovner, *Dok. AK. Nauk. Fiz. Otd.*, **97**, 65 (1954).
135. J. R. Scherer, *Spectrochim. Acta*, **24A**, 747 (1968).
136. D. J. Evans and D. B. Scully, *Spectrochim. Acta*, **20**, 891 (1964).
137. Y. Kakiuti and T. Shimanouchi, *J. Chem. Phys.*, **25**, 1252 (1956).
138. G. Varsanyi, *Vibrational Spectra of Benzene Derivatives*, Academic Press, 1969.
139. V. Schettino, N. Neto and S. Califano, *J. Chem. Phys.*, **44**, 2724 (1967).
140. N. Neto and C. Di Lauro, *Spectrochim. Acta*, **26A**, 1175 (1970).
141. V. Schettino, *J. Mol. Spectrosc.*, **34**, 78 (1970).
142. S. Califano and B. L. Crawford jr., *Spectrochim. Acta*, **16**, 889 (1960).
143. J. Scherer, *J. Chem. Phys.*, **36**, 3308 (1962).
144. S. Califano, *Ric. Sci.*, **3**, 461 (1963).
145. G. Zerbi, B. Crawford jr. and J. Overend, *J. Chem. Phys.*, **38**, 127 (1963).
146. M. Scrocco, C. Di Lauro and S. Califano, *Spectrochim. Acta*, **21**, 571 (1965).
147. M. Gussoni and G. Zerbi, *Chem. Phys. Letters*, **2**, 145 (1968).

148. I. M. Mills, *Chem. Phys. Letters*, **3**, 267 (1969).
149. B. Crawford and J. Overend, *J. Mol. Spectrosc.*, **42**, 307 (1964).
150. M. A. Eliashevich and L. A. Gribov, *Dokl. Akad Nauk. SSSR*, **166**, 1080 (1966).
151. S. Califano and J. Heicklen, *Spectrochim. Acta*, **17**, 1900 (1961).
152. C. Coulson and H. C. Longuet Higgins, *Proc. Roy. Soc.*, **A193**, 447, 456 (1948).
153. H. C. Longuet Higgins, *J. Inorg. and Nucl. Chem.*, **1**, 60 (1955).
154. S. Bratos, *Collq. int. CNRS*, **82**, 287 (1958).
155. J. N. Murrell, *J. Mol. Spectrosc.*, **4**, 446 (1960).
156. S. Bratos, R. Daudel, M. Allavena and M. Roux, *Rev. Mod. Phys.*, **32**, 412 (1960).
157. S. Bratos and M. Allavena, *J. Chem. Phys.* **37**, 2138 (1962).
158. W. L. Clinton, *J. Chem. Phys.*, **33**, 1603 (1960); **36**, 556 (1962).
159. D. M. Bishop and M. Randič, *J. Chem. Phys.*, **44**, 2480 (1966).
160. P. Pulay, *Mol. Phys.*, **17**, 197 (1969).
161. R. P. Feynman, *Phys. Rev.*, **56**, 340 (1939).
162. R. McWeeny, *Quantum Mechanics. Methods and Basic Applications*, Pergamon Press, 1973, Vol. II, p. 89.
163. R. E. Stanton, *J. Chem. Phys.*, **36**, 1298 (1960).
164. J. Gerratt and I. M. Mills, *J. Chem. Phys.*, **49**, 1719 (1968).
165. R. G. Body, D. S. McClure and E. Clementi, *J. Chem. Phys.*, **49**, 4916 (1968).
166. P. Pulay, *Mol. Phys.*, **18**, 473 (1970); **21**, 329 (1971).
167. P. Pulay and W. Meyer, *J. Mol. Spectrosc.*, **40**, 59 (1971).
168. P. Pulay and W. Meyer, *Mol. Phys.*, **27**, 473 (1974).
169. W. Meyer and P. Pulay, *J. Chem. Phys.*, **56**, 2109 (1972); **57**, 3337 (1972).
170. J. A. Pople and D. L. Beveridge, *Approximate Molecular Orbital Theory*, McGraw-Hill, 1970.
171. G. Klopman B. O'Learly, *Topics Curr. Chemistry*, **15**, 447 (1970).
172. P. Pulay and F. Török, *Mol. Phys.*, **25**, 1153 (1973).
173. S. D. Ross, *Inorganic Infrared and Raman Spectra*, McGraw Hill, 1972.

Chapter 9
Anharmonicity

9.1 Generality

The harmonic treatment of molecular vibrations, discussed in the previous chapters, rests on the assumption of infinitesimal displacements of the nuclei from their equilibrium positions. In this approximation the solution of the vibrational problem is straightforward and can be accomplished with the standard GF method outlined in Chapters 4 and 8. However, the vibrational spectra of molecular systems show definite deviations from the predictions of the harmonic model. The most important ones are briefly summarized below.

(1) The selection rules for the harmonic oscillator predict that for infinitesimal displacements of the nuclei, overtones and combination bands should be forbidden in the infrared spectrum. In actual spectra such bands are very often observed, sometimes with appreciable intensity.

(2) The harmonic model predicts that the vibrational levels associated with a given vibrational mode should have a constant spacing, i.e. that overtones corresponding to quantum jumps $\Delta v = 2, 3$, etc., should occur at frequencies which are exactly twice, three times, etc., the frequency of the fundamental mode and that hot bands should occur at the same frequency as the fundamental. In practice this is never observed. The frequency of overtones is normally lower than the corresponding multiple of the fundamental frequency. The same discrepancy is observed for combination and for hot bands.

(3) The Teller–Redlich product rule is never exactly satisfied by experimental frequencies of isotopic molecules. Differences up to some percent are normally observed.

In order to interpret these experimental results satisfactorily it is necessary to drop the assumption of infinitesimal displacement of the atoms and to use a more complex model in which terms higher than the second are retained in the potential function. This leads us to the anharmonic treatment of molecular vibrations which is discussed in the present chapter.

Although the quantum-mechanical solution of the anharmonic oscillator and the relationships between experimental data and higher terms in the potential function have been worked out for many years, only recently has the calculation of cubic and quartic potential energy terms received attention. Anharmonic force constants have been calculated for several triatomic molecules and iterative methods for their evaluation have been proposed. Unfortunately the extension of these calculations to larger molecules is severely limited by the enormous number of such terms which appear in the potential function of a polyatomic molecule.

Despite these limitations the anharmonic model is going to play a more important role in the near future, not only because of the interest in higher terms of the potential function of isolated molecules but also because the anharmonicity of the nuclear motions affects many properties of the solids.

The main difficulty associated with the anharmonic model is that it cannot be treated just as an extension of the harmonic one. First of all, when terms higher than the second are retained in the Taylor expansion of the potential, the Schrödinger equation cannot be solved exactly and approximate methods must be used. In addition, as soon as one abandons the conditions of infinitesimal amplitude of motion, new problems arise concerning the coordinate system used. In Section 8.10 we have already pointed out the difference between a linearized and a curvilinear coordinate system. The use of a curvilinear space was, however, presented more as a nuisance than as a genuine requirement of the physical problem. The reason for this is that in Section 8.10 we were only concerned with harmonic terms of the potential function. In dealing with anharmonic terms of the potential the use of a curvilinear space becomes mandatory in order to describe the finite nuclear displacements correctly.

In the next two sections we shall first present the quantum-mechanical treatment of the anharmonic oscillator and derive the relationships between the higher terms of the potential function and the vibro-rotational observables. In the following sections we shall then discuss the calculation of anharmonic force constants and analyse in some detail the transformations between the different spaces involved in actual calculations.

9.2 The anharmonic oscillator

When the assumption of infinitesimal displacements of the nuclei is abandoned, the quantum-mechanical treatment of a vibrating molecule becomes more complex than in Chapter 3, since the higher terms of the potential function must now be taken into account. In order to simplify the treatment, in this section we shall concentrate our attention on the vibrational part of the Hamiltonian and ignore the coupling between vibrations and rotations. In this way we shall derive expressions for the energy levels of the anharmonic oscillator and relationships between the anharmonicity coefficients and the cubic and quartic terms of the potential. In the next section we shall consider the complete Hamiltonian and derive more general relationships which include the vibro-rotational coupling terms.

The theory of the quantum-mechanical anharmonic oscillator has been derived in a general form by Nielsen[1,2] using the perturbation method. In his notation the potential function for a polyatomic molecule is written in the most general case in the form

$$V = \tfrac{1}{2}\sum_{i\sigma_i} \lambda_i Q_{i\sigma_i}^2 + \sum_{i\sigma_i}^{*}\sum_{j\sigma_j}^{*}\sum_{k\sigma_k}^{*} k'_{ijk} Q_{i\sigma_i} Q_{j\sigma_j} Q_{k\sigma_k} + \sum_{i\sigma_i}^{*}\sum_{j\sigma_j}^{*}\sum_{k\sigma_k}^{*}\sum_{l\sigma_l}^{*} k'_{ijkl} Q_{i\sigma_i} Q_{j\sigma_j} Q_{k\sigma_k} Q_{l\sigma_l} + \dots \quad (9.2.1)$$

where $\sum^*$ means $i \leqslant j \leqslant k \leqslant l$ and $\lambda_i = (2\pi c\omega_i)^2$, ω_i being the harmonic frequencies of vibration in cm^{-1}. We notice that in this notation the indices i, j, k and l are used to label distinct normal vibrations whereas the indices $\sigma_i, \sigma_j, \sigma_k$ and σ_l are used to label the different partners of degenerate vibrations. Restricted summations over the indices are used in (9.2.1) so that each cross-term appears only once.

In order to express all anharmonic force constants in the same units as the harmonic term, it is convenient to rewrite (9.2.1) in terms of dimensionless normal coordinates q_i, defined by the relation

$$Q_i = \left(\frac{\hbar^2}{\lambda_i}\right)^{\frac{1}{4}} q_i \quad (9.2.2)$$

In this way we obtain

$$V = \tfrac{1}{2}hc \sum_{i,\sigma_i} \omega_i q_i^2 + hc \sum_{i\sigma_i}^{*}\sum_{j\sigma_j}^{*}\sum_{k\sigma_k}^{*} k_{ijk} q_{i\sigma_i} q_{j\sigma_j} q_{k\sigma_k} + hc \sum_{i\sigma_i}^{*}\sum_{j\sigma_j}^{*}\sum_{k\sigma_k}^{*}\sum_{l\sigma_l}^{*} k_{ijkl} q_{i\sigma_i} q_{j\sigma_j} q_{k\sigma_k} q_{l\sigma_l} + \dots \quad (9.2.3)$$

The molecule symmetry puts some restrictions on the general expression (9.2.3). The invariance of the potential energy under all symmetry operations of the group requires that all cubic and quartic terms of the potential function vanish unless the direct product of the representations carried by the normal coordinates belongs to the totally symmetric representation or contains it in its structure.

If the potential function (9.2.3) is introduced into the Schrödinger equation, an exact solution of the type obtained for the harmonic oscillator cannot be found. However, one can take advantage of the fact that for finite but small displacements of the nuclei the quadratic part of V is much larger than the cubic, which is in turn larger than the quartic part. On this basis the vibrational Hamiltonian may be divided into orders of magnitude and perturbation theory used to calculate the corrections to the vibrational energy due to the various terms. For molecular systems with non-degenerate vibrations the first- and second-order corrections to the energy can be evaluated with relatively little difficulty from standard perturbation theory. For degenerate systems, however, the calculations become extremely involved in this way and it is then more convenient to use a modified form of the perturbation theory which involves a transformation of the Hamiltonian.

In this section, in order to make the reader acquainted with the various difficulties, we shall use the standard form of perturbation theory to discuss the anharmonic motion of molecules with non-degenerate states only. Degenerate systems will be treated in a more general and less detailed way in the next section where we shall consider the complete vibro-rotational Hamiltonian. Although the non-degenerate case discussed here is automatically included in the general treatment of the next section, we consider it more instructive for the reader to understand a simpler case in detail first and then to deal with the more complex one.

When we limit ourselves to molecules with non-degenerate vibrations we can drop the indices σ in the potential function (9.2.3). If we write the potential and the energy in orders of magnitude in the form

$$\begin{aligned} V &= V^{(0)} + \lambda V^{(1)} + \lambda^2 V^{(2)} && \text{(a)} \\ E &= \sum_i hc\omega_i(v_i + \tfrac{1}{2}) + \lambda E^{(1)} + \lambda^2 E^{(2)} && \text{(b)} \end{aligned} \tag{9.2.4}$$

where

$$V^{(0)} = \tfrac{1}{2}hc \sum_i \omega_i q_i^2 \tag{9.2.5}$$

$$V^{(1)} = hc \sum_{ijk}{}^{*} k_{ijk} q_i q_j q_k \tag{9.2.6}$$

$$V^{(2)} = hc \sum_{ijkl}{}^{*} k_{ijkl} q_i q_j q_k q_l \tag{9.2.7}$$

and λ is a parameter defining the order of magnitude of the various terms, we obtain from perturbation theory that the corrections to the energy to first and second order are given by

$$\text{First order } E^{(1)} = \sum_m \langle m|V^{(1)}|m\rangle \tag{9.2.8}$$

$$\text{Second order } E^{(2)} = \sum_m \left[\langle m|V^{(2)}|m\rangle + \sum_{n \neq m} \frac{\langle m|V^{(1)}|n\rangle\langle n|V^{(1)}|m\rangle}{E_m^0 - E_n^0} \right] \tag{9.2.9}$$

In these expressions $\langle m|$ and $\langle n|$ indicate the complete zero-order vibrational wavefunctions corresponding to the vibrational zero-order energies E_m^0 and E_n^0 respectively. These functions are, according to Chapter 3, the product of $3N - 6$ harmonic oscillator wavefunctions, each depending upon one single normal coordinate q_i and defined by one single quantum number v_i. For simplicity we shall indicate in what follows the harmonic oscillator wave-functions by their quantum number in expressions like (9.2.8) and (9.2.9).

By use of (9.2.6) we obtain that the first-order correction to the energy is given by

$$E^{(1)} = hc \sum_{ijk}{}^{*} k_{ijk} \langle v_i, v_j, v_k | q_i q_j q_k | v_i, v_j, v_k \rangle$$
$$= hc \sum_{ijk}{}^{*} k_{ijk} \langle v_i | q_i | v_i \rangle \langle v_j | q_j | v_j \rangle \langle v_k | q_k | v_k \rangle \tag{9.2.10}$$

where the symbol $\sum^*$ means restricted summation over the indices and $\langle v_i, v_j, v_k|$ represent the part of the product of harmonic functions on which the operators $q_i q_j q_k$ will operate.

Since, as discussed in Chapter 3, Section 2, the matrix elements of the type

$$\langle v_i | q_i | v_i \rangle \tag{9.2.11}$$

are always equal to zero, we conclude that the first-order correction to the energy, which involves only diagonal matrix elements of the type (9.2.11), is equal to zero.

The second-order correction to the energy (9.2.9) is made up of two parts, the first involving the diagonal matrix elements of the second-order perturbation terms $V^{(2)}$, and the second involving off-diagonal matrix elements of the first-order perturbation term $V^{(1)}$.

As far as the first part is concerned, it is easily seen that all terms of $V^{(2)}$ of the type

$$k_{ijkl} q_i q_j q_k q_l \qquad k_{iijk} q_i^2 q_j q_k \qquad k_{iiij} q_i^3 q_j$$

will give no contribution to $E^{(2)}$ since they always lead to at least one matrix element of the type (9.2.11). The first part of (9.2.9) is then

$$hc \sum_i k_{iiii} \langle v_i | q_i^4 | v_i \rangle + hc \sum_{ij}{}^{*} k_{iijj} \langle v_i | q_i^2 | v_i \rangle \langle v_j | q_j^2 | v_j \rangle \tag{9.2.12}$$

The matrix elements of the operators q^4 and q^2 are easily obtained by repeated application of the recursion formula (3.2.19) for Hermite polynomials, using the procedure described in Section 3.2. Table 9.2.1 collects the non-zero matrix elements of the operator q and of its powers that are needed for (9.2.12) as well as for successive terms.

Table 9.2.1 Matrix elements of the operators q, q^2, q^3 and q^4

$\langle v \| q \| v+1 \rangle = \left(\frac{v+1}{2}\right)^{\frac{1}{2}}$	$\langle v \| q^3 \| v-1 \rangle = \frac{3}{2}\left[\frac{v^3}{2}\right]^{\frac{1}{2}}$
$\langle v \| q \| v-1 \rangle = \left(\frac{v}{2}\right)^{\frac{1}{2}}$	$\langle v \| q^3 \| v-3 \rangle = \frac{1}{2}\left[\frac{v(v-1)(v-2)}{2}\right]^{\frac{1}{2}}$
$\langle v \| q^2 \| v+2 \rangle = \frac{1}{2}[(v+1)(v+2)]^{\frac{1}{2}}$	$\langle v \| q^4 \| v+4 \rangle = \frac{1}{4}[(v+1)(v+2)(v+3)(v+4)]^{\frac{1}{2}}$
$\langle v \| q^2 \| v \rangle = (v+\frac{1}{2})$	$\langle v \| q^4 \| v+2 \rangle = \frac{1}{2}(2v+3)[(v+1)(v+2)]^{\frac{1}{2}}$
$\langle v \| q^2 \| v-2 \rangle = \frac{1}{2}[v(v-1)]^{\frac{1}{2}}$	$\langle v \| q^4 \| v \rangle = \frac{3}{2}[(v+\frac{1}{2})^2+\frac{1}{4}]$
$\langle v \| q^3 \| v+3 \rangle = \frac{1}{2}\left[\frac{(v+1)(v+2)(v+3)}{2}\right]^{\frac{1}{2}}$	$\langle v \| q^4 \| v-2 \rangle = \frac{1}{2}(2v-1)[v(v-1)]^{\frac{1}{2}}$
$\langle v \| q^3 \| v+1 \rangle = \frac{3}{2}\left[\left(\frac{v+1}{2}\right)^3\right]^{\frac{1}{2}}$	$\langle v \| q^4 \| v-4 \rangle = \frac{1}{4}[v(v-1)(v-2)(v-3)]^{\frac{1}{2}}$

Expression (9.2.12) then becomes, using Table 9.2.1,

$$\tfrac{3}{2}hc\sum_i k_{iiii}[(v_i+\tfrac{1}{2})^2+\tfrac{1}{4}] + hc\sum_{ij}{}^{*} k_{iijj}(v_i+\tfrac{1}{2})(v_j+\tfrac{1}{2}) \tag{9.2.13}$$

The second part of (9.2.9) is given by

$$h^2c^2\sum_m\sum_{n\neq m}\frac{\left\langle m\left|\sum_{ijk}^{*}k_{ijk}q_iq_jq_k\right|n\right\rangle\left\langle n\left|\sum_{i'j'k'}^{*}k_{i'j'k'}q_{i'}q_{j'}q_{k'}\right|m\right\rangle}{E_m^0-E_n^0} \tag{9.2.14}$$

and is more complex to evaluate. From Table 9.2.1 it is seen that only terms of the type

$$k_{ijk}^2 \qquad k_{iij}^2 \qquad k_{iii}^2 \qquad k_{iii}k_{ijj} \qquad k_{iik}k_{jjk}$$

will be associated with non-zero matrix elements in (9.2.14). By writing down all non-zero terms, (9.2.14) becomes

$$\begin{aligned}
&-hc\sum_{ijk}^{\neq}k_{ijk}^2\left[\frac{\langle v_i|q_i|v_i+1\rangle^2\langle v_j|q_j|v_j+1\rangle^2\langle v_k|q_k|v_k+1\rangle^2-\langle v_i|q_i|v_i-1\rangle^2\langle v_j|q_j|v_j-1\rangle^2\langle v_k|q_k|v_k-1\rangle^2}{\omega_i+\omega_j+\omega_k}\right.\\
&\qquad+\frac{\langle v_i|q_i|v_i+1\rangle^2\langle v_j|q_j|v_j-1\rangle^2\langle v_k|q_k|v_k+1\rangle^2-\langle v_i|q_i|v_i-1\rangle^2\langle v_j|q_j|v_j+1\rangle^2\langle v_k|q_k|v_k-1\rangle^2}{\omega_i-\omega_j+\omega_k}\\
&\qquad+\frac{\langle v_i|q_i|v_i+1\rangle^2\langle v_j|q_j|v_j+1\rangle^2\langle v_k|q_k|v_k-1\rangle^2-\langle v_i|q_i|v_i-1\rangle^2\langle v_j|q_j|v_j-1\rangle^2\langle v_k|q_k|v_k+1\rangle^2}{\omega_i+\omega_j-\omega_k}\\
&\qquad\left.+\frac{\langle v_i|q_i|v_i-1\rangle^2\langle v_j|q_j|v_j+1\rangle^2\langle v_k|q_k|v_k+1\rangle^2-\langle v_i|q_i|v_i+1\rangle^2\langle v_j|q_j|v_j-1\rangle^2\langle v_k|q_k|v_k-1\rangle^2}{-\omega_i+\omega_j+\omega_k}\right]\\
&-hc\sum_{ij}^{\neq}k_{iij}^2\left[\frac{\langle v_i|q_i^2|v_i+2\rangle^2\langle v_j|q_j|v_j+1\rangle^2-\langle v_i|q_i^2|v_i-2\rangle^2\langle v_j|q_j|v_j-1\rangle^2}{2\omega_i+\omega_j}\right.\\
&\qquad+\frac{\langle v_i|q_i^2|v_i+2\rangle^2\langle v_j|q_j|v_j-1\rangle^2-\langle v_i|q_i^2|v_i-2\rangle^2\langle v_j|q_j|v_j+1\rangle^2}{2\omega_i-\omega_j}\\
&\qquad\left.+\frac{\langle v_i|q_i^2|v_i\rangle^2\langle v_j|q_j|v_j+1\rangle^2-\langle v_i|q_i^2|v_i\rangle\langle v_j|q_j|v_j-1\rangle^2}{\omega_j}\right]\\
&\qquad-\sum_i k_{iii}^2\left[\frac{\langle v_i|q_i^3|v_i+3\rangle^2-\langle v_i|q_i^3|v_i-3\rangle^2}{3\omega_i}+\frac{\langle v_i|q_i^3|v_i+1\rangle^2-\langle v_i|q_i^3|v_i-1\rangle^2}{\omega_i}\right]\\
&-2hc\sum_{ij}^{\neq}k_{iii}k_{ijj}\left[\frac{\langle v_i,v_j|q_i^3|v_i+1,v_j\rangle\langle v_i,v_j|q_iq_j^2|v_i+1,v_j\rangle\langle v_i,v_j|q_i^3|v_i-1,v_j\rangle\langle v_i,v_j|q_iq_j^2|v_i-1,v_j\rangle}{\omega_i}\right]\\
&-2hc\sum_{ijk}^{\neq}k_{iik}k_{jjk}\left[\frac{\begin{array}{l}\langle v_i,v_j,v_k|q_i^2q_k|v_i,v_j,v_{k+1}\rangle\langle v_i,v_j,v_k|q_j^2q_k|v_i,v_j,v_{k+1}\rangle\\ \qquad-\langle v_i,v_j,v_k|q_i^2q_k|v_i,v_j,v_k-1\rangle\langle v_i,v_j,v_k|q_j^2q_k|v_i,v_j,v_k-1\rangle\end{array}}{\omega_k}\right]
\end{aligned} \tag{9.2.15}$$

where $\sum_{ijk}^{\neq}$ means $i\neq j\neq k$.

By use of Table 9.2.1 this expression becomes

$$\begin{aligned}
-hc\sum_{ijk}^{\neq}\frac{k_{ijk}^2}{2N_{ijk}}[&\omega_k(\omega_k^2-\omega_j^2-\omega_i^2)(v_i+\tfrac{1}{2})(v_j+\tfrac{1}{2})+\omega_j(\omega_j^2-\omega_i^2-\omega_k^2)(v_i+\tfrac{1}{2})(v_k+\tfrac{1}{2})\\
&+\omega_i(\omega_i^2-\omega_j^2-\omega_k^2)(v_j+\tfrac{1}{2})(v_k+\tfrac{1}{2})+C]
\end{aligned}$$

$$-hc\sum_{ij}^{\neq} k_{iij}^2\left[\frac{2\omega_i}{(4\omega_i^2-\omega_j^2)}(v_i+\tfrac{1}{2})(v_j+\tfrac{1}{2})+\frac{(8\omega_i^2-3\omega_j^2)}{4\omega_j(4\omega_i^2-\omega_j^2)}(v_i+\tfrac{1}{2})^2-\frac{3}{16}\frac{\omega_j}{(4\omega_i^2-\omega_j^2)}\right] \tag{9.2.16}$$

$$-hc\sum_{ij}^{\neq} k_{iii}k_{ijj}\frac{3}{\omega_i}(v_i+\tfrac{1}{2})(v_j+\tfrac{1}{2})-hc\sum_{ijk}^{\neq} k_{iik}k_{jjk}\frac{1}{\omega_k}(v_i+\tfrac{1}{2})(v_j+\tfrac{1}{2})-hc\sum_i k_{iii}^2\frac{15}{4\omega_i}\left[(v_i+\tfrac{1}{2})^2+\frac{7}{60}\right]$$

where

$$\begin{aligned} N_{ijk} &= [(\omega_i+\omega_j+\omega_k)(\omega_i-\omega_j-\omega_k)(\omega_i-\omega_j+\omega_k)(\omega_i+\omega_j-\omega_k)] \\ C &= \tfrac{1}{4}[\omega_i^2(-\omega_i+\omega_j+\omega_k)+\omega_j^2(\omega_i-\omega_j+\omega_k)+\omega_k^2(\omega_i+\omega_j-\omega_k)-2\omega_i\omega_j\omega_k] \end{aligned} \tag{9.2.17}$$

The second-order correction to the energy $E^{(2)}$ is then the sum of (9.2.13) and (9.2.16). By collecting together similar terms we can rewrite equation (9.2.4b) in the form

$$E = E_0 + hc\sum_i \omega_i(v_i+\tfrac{1}{2}) + hc\sum_{ij}{}^{*} x_{ij}(v_i+\tfrac{1}{2})(v_j+\tfrac{1}{2}) + \ldots \tag{9.2.18}$$

where

$$x_{ii} = \frac{1}{4}\left[6k_{iiii} - \frac{15}{\omega_i}k_{iii}^2 - \sum_j^{j\neq i} k_{iij}^2\frac{(8\omega_i^2-3\omega_j^2)}{\omega_j(4\omega_i^2-\omega_j^2)}\right] \tag{9.2.19}$$

$$\begin{aligned} x_{ij} = \frac{1}{2}\Bigg\{&2k_{iijj} - \frac{6}{\omega_i}k_{iii}k_{ijj} - 4k_{iij}^2\frac{\omega_i}{(4\omega_i^2-\omega_j^2)} - \frac{6}{\omega_j}k_{jjj}k_{jii} - 4k_{jji}^2\frac{\omega_j}{(4\omega_j^2-\omega_i^2)} \\ &- \sum_k^{k\neq ij}\left[2k_{iik}k_{jjk}\frac{1}{\omega_k} + k_{ijk}^2\frac{\omega_k(\omega_k^2-\omega_i^2-\omega_j^2)}{N_{ijk}}\right]\Bigg\} \end{aligned} \tag{9.2.20}$$

Equations (9.2.19) and (9.2.20) are very useful since they give the relations between the cubic and quartic terms of the potential and the anharmonicity coefficients. These coefficients can be determined experimentally from overtones and combination bands in the infrared spectrum and can be used for the determination of the anharmonic terms of V. Since the force constants in (9.2.19) and (9.2.20) are given in cm^{-1}, the anharmonicity coefficients are expressed in the same units. We notice that in (9.2.13) and (9.2.16) there are some constant terms which contribute to the zero-point energy only. This correction is very small and is normally ignored. Its importance for linear triatomic molecules has been discussed by Bron.[3]

9.3 The contact transformation

If a molecule possesses two degenerate normal coordinates Q_{i1} and Q_{i2} associated with the same frequency ω_i, the zero-order harmonic wavefunctions describing the degenerate state E_i can be written, according to Section 3.2, in the form

$$\Psi_{v_i} = \left[\prod_{j=1}^{3N-8} \psi_{v_j}^0\right]\psi_{v_i1}(Q_{i1})\psi_{v_i2}(Q_{i2}) \tag{9.3.1}$$

and will thus depend upon two quantum numbers v_{i1} and v_{i2}. Correspondingly the energy E_i will be given by

$$E_i = hc\omega_i[(v_{i1}+\tfrac{1}{2})+(v_{i2}+\tfrac{1}{2})] = hc\omega_i(v_i+1) \tag{9.3.2}$$

where

$$v_i = v_{i1} + v_{i2} \tag{9.3.3}$$

It is obvious that to a given value of v_i correspond different combinations of v_{i1} and v_{i2}

and different wavefunctions Ψ_{v_i}. For instance if $v_i = 2$ we have three possibilities: $v_{i1} = 2$, $v_{i2} = 0$; $v_{i1} = 1$, $v_{i2} = 1$; $v_{i1} = 0$, $v_{i2} = 2$.

In Section 2.5 we have seen that, when a two-fold degenerate vibration is excited, the nuclei do not move in the most general case along straight lines but describe ellipses around the equilibrium positions. Associated with this motion there is therefore a vibrational angular momentum which is not adequately described by wavefunctions of the type (9.3.1), as discussed in Section 3.2(b).

The vibrational angular momentum arising from degenerate motions is conveniently handled by converting the degenerate normal coordinates Q_{i1} and Q_{i2} to polar coordinates ρ_i and φ_i and solving the Schrödinger equation in these new coordinates. The solutions are wavefunctions $\psi_{v_i l_i}$ which involve the associated Laguerre polynomials rather than the simpler Hermite polynomials occurring in (9.3.1). The Laguerre polynomials depend upon two quantum numbers v_i and l_i which can take the values

$$\begin{aligned} v_i &= 0, 1, 2, \ldots \\ l_i &= \pm v_i, \pm(v_i - 2), \pm(v_i - 4), \ldots, \pm 1 \quad \text{or} \quad 0 \end{aligned} \tag{9.3.4}$$

and which quantize the energy and the vibrational angular momentum arising from the degeneracy, respectively. In the case of three-fold degeneracy the problem is even more complex since the wavefunctions will depend upon three quantum numbers. The calculation of the energy corrections for such molecules is so involved and laborious that it is not convenient to treat them in general but rather as individual cases. In what follows therefore we shall limit ourselves to two-fold degeneracy. The reader interested in the anharmonicity correction for molecules of very high symmetry with three-fold degenerate motions is referred to the Nielsen article in the *Handbuch der Physik*.[2]

A significant simplification of the standard perturbation procedure is achieved by means of a contact transformation which changes the form of the perturbation Hamiltonian in such a way that the second order is reduced to a first-order calculation.

The most laborious and involved part of the calculation of $E^{(2)}$ is actually due to the large number of non-zero off-diagonal matrix elements of the first-order perturbation Hamiltonian which occur in the expression of $E^{(2)}$. If these terms were all equal to zero, only the diagonal matrix elements of $\mathscr{H}^{(2)}$ would be evaluated and the problem would then be formally reduced to a first-order calculation. These considerations show the convenience of transforming the Hamiltonian $\mathscr{H}$ into a new Hamiltonian $\mathscr{H}'$ through a similarity transformation with a unitary function T

$$\mathscr{H}' = T\mathscr{H}T^{-1} \tag{9.3.5}$$

such that $\mathscr{H}^{(1)\prime}$ is equal to zero and $\mathscr{H}'$ has the same eigenvalues as $\mathscr{H}$.

If we write down the Hamiltonians $\mathscr{H}$ and $\mathscr{H}'$ in order of magnitude

$$\mathscr{H} = \mathscr{H}^{(0)} + \lambda\mathscr{H}^{(1)} + \lambda^2\mathscr{H}^{(2)} \quad \text{(a)}$$

$$\mathscr{H}' = \mathscr{H}^{(0)\prime} + \lambda\mathscr{H}^{(1)\prime} + \lambda^2\mathscr{H}^{(2)\prime} \quad \text{(b)} \tag{9.3.6}$$

we can choose for T the form

$$T = \mathrm{e}^{\mathrm{i}\lambda S} \tag{9.3.7}$$

which ensures that the transformation is unitary since

$$T^{-1} = \mathrm{e}^{-\mathrm{i}\lambda S} \tag{9.3.8}$$

By developing T in powers of λ we have

$$T = 1 + i\lambda S + \frac{(i\lambda S)^2}{2!} + \ldots \qquad \text{(a)}$$

$$T^{-1} = 1 - i\lambda S + \frac{(-i\lambda S)^2}{2!} + \ldots \qquad \text{(b)} \qquad (9.3.9)$$

and thus (9.3.5) becomes

$$\mathscr{H}' = \left(1 + i\lambda S - \frac{\lambda^2 S^2}{2} + \ldots\right)(\mathscr{H}^{(0)} + \lambda\mathscr{H}^{(1)} + \lambda^2\mathscr{H}^{(2)} + \ldots)\left(1 - i\lambda S - \frac{\lambda^2 S^2}{2} + \ldots\right) \qquad (9.3.10)$$

By developing (9.3.10) and by collecting together terms with the same power of λ we obtain

$$\mathscr{H}^{(0)\prime} = \mathscr{H}^{(0)}$$

$$\mathscr{H}^{(1)\prime} = \mathscr{H}^{(1)} + i(S\mathscr{H}^{(0)} - \mathscr{H}^{(0)}S)$$

$$\mathscr{H}^{(2)\prime} = \mathscr{H}^{(2)} + i(S\mathscr{H}^{(1)} - \mathscr{H}^{(1)}S) - \tfrac{1}{2}(\mathscr{H}^{(0)}S^2 - 2S\mathscr{H}^{(0)}S + S^2\mathscr{H}^{(0)})$$

which can be rewritten in the form

$$\mathscr{H}^{(0)\prime} = \mathscr{H}^{(0)} \qquad \text{(a)}$$

$$\mathscr{H}^{(1)\prime} = \mathscr{H}^{(1)} + i(S\mathscr{H}^{(0)} - \mathscr{H}^{(0)}S) \qquad \text{(b)} \quad (9.3.11)$$

$$\mathscr{H}^{(2)\prime} = \mathscr{H}^{(2)} + \frac{i}{2}[S(\mathscr{H}^{(1)} + \mathscr{H}^{(1)\prime}) - (\mathscr{H}^{(1)} + \mathscr{H}^{(1)\prime})S] \qquad \text{(c)}$$

We can then choose the functions S in such a way that $\mathscr{H}^{(1)\prime} = 0$, i.e. from (9.3.11b) that

$$\mathscr{H}^{(1)} = i(\mathscr{H}^{(0)}S - S\mathscr{H}^{(0)}) \qquad (9.3.12)$$

Before discussing in detail the form of S it is necessary to indicate which form we choose for the Hamiltonian. The contact transformation (9.3.5) is very general and puts no restrictions on the form of $\mathscr{H}$. Thus in principle $\mathscr{H}$ can be the complete vibro-rotational Hamiltonian.[2,6]

Although in the non-degenerate case of the previous section we have limited the discussion to the pure vibrational part of $\mathscr{H}$, in this section we shall apply the contact transformation to the complete vibro-rotational Hamiltonian in order to derive the relationship between the anharmonic terms of the potential and both the anharmonicity coefficients and the vibro-rotational constants. It is obvious that the case discussed in the previous section is included in the more general treatment given here.

The actual form of a transformation function S which satisfies equation (9.3.12) has been discussed by Nielsen and coworkers[1] and later in considerable detail by Herman and Shaffer.[4]

The first-order perturbation Hamiltonian consists of a sum of terms[2,6] each of which is a function of q_i or of the conjugate momenta p_i, or of both, multiplied by a coefficient which is usually constant but in some cases is a function of the total angular momentum operators P_α. Therefore the transformation functions S will consist of a sum of terms

$$S = \sum_\rho S_\rho \qquad (9.3.13)$$

each S_ρ being chosen so that $i(\mathscr{H}^{(0)}S_\rho - S_\rho\mathscr{H}^{(0)})$ is equal to a single term $\mathscr{H}^{(1)}_\rho$ of $\mathscr{H}^{(1)}$. Herman and Shaffer have determined the various functions S_ρ to be used to remove the different terms of $\mathscr{H}^{(1)}$ from $\mathscr{H}^{(1)\prime}$. These functions are given in Table 9.3.1 together with

Table 9.3.1

ρ	S_ρ	$\mathscr{H}_\rho^{(1)}$
1	$-(a_1/hc\omega_i)(P_i/\hbar)$	$a_1 q_{i\sigma_i}$
2	$(a_2/2\pi c\omega_i)q_{i\sigma_i}$	$a_2 P_{i\sigma_i}$
3	$-(2a_3/hc\omega_i)[\frac{1}{3}(P_{i\sigma_i}/\hbar)^3 + \frac{1}{2}q_{i\sigma_i}(P_{i\sigma_i}/\hbar)q_{i\sigma_i}]$	$a_3 q_{i\sigma_i}^3$
4	$-[(a_4/2\pi c)/(\omega_i^2 - \omega_j^2)][\omega_j q_{i\sigma_i} q_{j\sigma_j} + \omega_i(P_{j\sigma_j}/\hbar)(P_{i\sigma_i}/\hbar)]$	$a_4 q_{i\sigma_i} P_{j\sigma_j}$
5	$-\left[\frac{a_5}{hc\omega_j(4\omega_i^2 - \omega_j^2)}\right]\left[(2\omega_i^2 - \omega_j^2)q_{i\sigma_i}^2\left(\frac{P_{j\sigma_j}}{\hbar}\right) + \omega_i\omega_j(P_{i\sigma_i}q_{i\sigma_i}\right.$	
	$\left. + q_{i\sigma_i}P_{i\sigma_i})\left(\frac{q_{j\sigma_j}}{\hbar}\right) + 2\omega_i^2\frac{P_{i\sigma_i}^2 P_{j\sigma_j}}{\hbar^3}\right]$	$a_5 q_{i\sigma_i}^2 q_{j\sigma_j}$
6	$-\left(\frac{a_6}{2\pi c}\right)[(\omega_i\omega_j)^{\frac{1}{2}}(\omega_i^2 - \omega_j^2)]^{-1}[(\omega_i^2 + \omega_j^2)q_{i\sigma_i}q_{j\sigma_j} + 2\omega_i\omega_j P_{i\sigma_i}P_{j\sigma_j}/\hbar^2]$	$a_6[(\omega_j/\omega_i)^{\frac{1}{2}} q_{i\sigma_i} P_{j\sigma_j}$
		$-(\omega_i/\omega_j)^{\frac{1}{2}} q_{j\sigma_j} P_{i\sigma_i}$
7	$-\left(\frac{a_7 D_{ijk}}{hc}\right)\left[\omega_i(\omega_i^2 - \omega_j^2 - \omega_k^2)\left(\frac{P_{i\sigma_i}}{\hbar}\right)q_{j\sigma_j}q_{k\sigma_k} + \omega_j(\omega_j^2 - \omega_i^2 - \omega_k^2)q_{i\sigma_j}\left(\frac{P_{j\sigma_j}}{\hbar}\right)\right.$	
	$\left. \times q_{k\sigma_k} + \omega_k(\omega_k^2 - \omega_i^2 - \omega_j^2)q_{i\sigma_i}q_{j\sigma_j}\left(\frac{P_{k\sigma_k}}{\hbar}\right) - 2\omega_i\omega_j\omega_k(P_{i\sigma_i}P_{j\sigma_j}P_{k\sigma_k}/\hbar^3)\right]$	$a_7 q_{i\sigma_i} q_{j\sigma_j} q_{k\sigma_k}$

$D_{ijk} = [(\omega_i + \omega_j + \omega_k)(\omega_i - \omega_j - \omega_k)(\omega_i - \omega_j + \omega_k)(\omega_i + \omega_j - \omega_k)]^{-1}$

the corresponding terms of $\mathscr{H}^{(1)}$ that they eliminate from $\mathscr{H}^{(1)\prime}$. It must be noticed that the S_ρ functions of the form given in Table 9.3.1 will remove from $\mathscr{H}^{(1)\prime}$ all terms except the Coriolis interactions which involve the two normal coordinates associated with a degenerate frequency.

Degeneracy due to molecular symmetry is taken into account in Table 9.3.1 by the indices σ_i, σ_j and σ_k which assume only one value (1) for non-degenerate, but two values (1 and 2) for two-fold degenerate modes. It may happen however that two frequencies ω_i and ω_j, or $2\omega_i$ and ω_j or even $(\omega_i + \omega_j)$ and ω_k fall very close together. In such cases the terms $(\omega_i^2 - \omega_j^2)$ or $(4\omega_i^2 - \omega_j^2)$ or $(\omega_i + \omega_j - \omega_k)$ which appear at the denominator in some of the S_ρ functions of Table 9.3.1 will approach zero. The contribution to the energy of the corresponding terms of the Hamiltonian will then become very large and of first-order importance.

The perturbation method on which the contact transformation is based is not valid in such cases. We shall discuss later in more detail how accidental degeneracy is taken into account. It might be anticipated, however, that the transformation method may still be employed although different S-functions must be used.

Once the S-function has been obtained, it can be used for the calculation of the second-order contributions to the energy $E^{(2)}$. These are given by the diagonal matrix elements

$$\langle m|\mathscr{H}^{(2)\prime}|m\rangle \qquad (9.3.14)$$

of the second-order transformed Hamiltonian $\mathscr{H}^{(2)\prime}$. It is important to notice that $\langle m|$

represents here the complete vibro-rotational zero-order wavefunction which is a product of a vibrational wavefunction and of a rotational wavefunction.

From equation (9.3.11c) we know that in the Hamiltonian $\mathscr{H}^{(2)\prime}$ will appear, in addition to the elements of the original Hamiltonian $\mathscr{H}^{(2)}$, terms of the type (i/2) $[S\mathscr{H}^{(1)} - \mathscr{H}^{(1)}S]$ which originate from the first-order perturbation Hamiltonian $\mathscr{H}^{(1)}$. This latter is made up, as discussed before, of a sum of terms and we shall indicate it by writing

$$\mathscr{H}^{(1)} = \sum_{\tau} \mathscr{H}_{\tau}^{(1)} \tag{9.3.15}$$

From (9.3.11c) we obtain then, remembering that $\mathscr{H}^{(1)\prime}$ is equal to zero and using (9.3.13) and (9.3.15)

$$\mathscr{H}^{(2)\prime} = \sum_{k} \mathscr{H}_{k}^{(2)} + \sum_{\tau\rho}{}^{\dagger} \frac{\mathrm{i}}{2}[S_{\rho}\mathscr{H}_{\tau}^{(1)} - \mathscr{H}_{\tau}^{(1)}S_{\rho}] \tag{9.3.16}$$

where the symbol $\sum^{\dagger}$ means that in the summation we include only terms which are of second-order importance. The elements $\mathscr{H}^{(2)\prime}$ which have non-zero matrix elements of the type (9.3.14) have been computed by Herman and Shaffer and are given in tables reported in the original paper by these authors[4] and reproduced also in Nielsen article.[2] The calculations are very long and tedious and will not be reported here. A very detailed exposition of how these matrix elements can be evaluated is given in the book by Barchewitz[5] *Spectroscopie Infrarouge* as well as in the original paper.

The matrix elements are evaluated first with respect to the vibrational and then with respect to the rotational wavefunctions, the sum of all contributions of the type (9.3.14) giving the total second-order energy correction. By collecting together terms with like powers of vibrational quantum numbers and no powers of the rotational quantum numbers, we obtain that the vibrational part of the energy is given by

$$\begin{aligned} E_v = E_0 &+ hc\left[\sum_i \omega_i(v_i + \tfrac{1}{2}) + \sum_t \omega_t(v_t + 1)\right] + hc\left[\sum_{ij}{}^{*} x_{ij}(v_i + \tfrac{1}{2})(v_j + \tfrac{1}{2})\right. \\ &\left. + \sum_{it} x_{it}(v_i + \tfrac{1}{2})(v_t + 1)\right] + hc\sum_{tt'}{}^{*} x_{tt'}(v_t + 1)(v_{t'} + 1) + hc\sum_{tt'}{}^{*} x_{l_t l_{t'}} l_t l_{t'} \end{aligned} \tag{9.3.17}$$

where i and j count non-degenerate and t, t' count two-fold degenerate vibrations. The anharmonicity coefficients in (9.3.17) are given by

$$x_{ii} = \frac{3}{2}k_{iiii} - \frac{15}{4\omega_i}k_{iii}^2 - \sum_{j}^{j\neq i} \frac{(8\omega_i^2 - 3\omega_j^2)}{4\omega_j(4\omega_i^2 - \omega_j^2)}k_{iij}^2 \tag{a}$$

$$x_{tt} = \frac{3}{2}k_{tttt} - \frac{15}{4\omega_t}k_{ttt}^2 - \sum_{j} \frac{(8\omega_t^2 - 3\omega_j^2)}{4\omega_j(4\omega_t^2 - \omega_j^2)}k_{ttj}^2 - \sum_{t'}^{t'\neq t} \frac{(8\omega_t^2 - 3\omega_{t'}^2)}{4\omega_{t'}^2(4\omega_t^2 - \omega_{t'}^2)}k_{ttt'}^2 \tag{b}$$

$$\begin{aligned} x_{ij} = k_{iijj} &- \frac{3}{\omega_i}k_{iii}k_{ijj} - \frac{3}{\omega_j}k_{jjj}k_{iij} - \frac{2\omega_i}{(4\omega_i^2 - \omega_j^2)}k_{iij}^2 - \frac{2\omega_j}{(4\omega_j^2 - \omega_i^2)}k_{jji}^2 \\ &- \sum_{k}^{k\neq i,j}\left[\omega_k\frac{(\omega_k^2 - \omega_i^2 - \omega_j^2)}{2N_{ijk}}k_{ijk}^2 + \frac{1}{\omega_k}k_{iik}k_{jjk}\right] + B_e^{zz}(\zeta_{ij}^z)^2\left(\frac{\omega_i}{\omega_j} + \frac{\omega_j}{\omega_i}\right) \end{aligned} \tag{c}$$

$$\begin{aligned} x_{it} = k_{iitt} &- \frac{3}{\omega_i}k_{iii}k_{itt} - \frac{2\omega_i}{(4\omega_t^2 - \omega_i^2)}k_{tti}^2 - \sum_{k}^{k\neq i}\frac{1}{\omega_k}k_{iik}k_{ttk} \\ &- \sum_{t'}^{t'\neq t}\omega_{t'}\frac{(\omega_{t'}^2 - \omega_i^2 - \omega_t^2)}{2N_{itt'}}k_{itt'}^2 + B_e^{xx}[(\zeta_{i,t_1}^x)^2 + (\zeta_{i,t_1}^y)^2]\left[\frac{\omega_i}{\omega_t} + \frac{\omega_t}{\omega_i}\right] \end{aligned} \tag{d}$$

(9.3.18)

$$x_{tt'} = k_{ttt't'} - \sum_i \left[\frac{1}{\omega_i} k_{itt} k_{it't'} + \omega_i \frac{(\omega_i^2 - \omega_t^2 - \omega_{t'}^2)}{2N_{itt'}} k_{itt'}^2 \right]$$

$$- \sum_{t''}^{t''\neq t,t'} \omega_{t''} \frac{(\omega_{t''}^2 - \omega_t^2 - \omega_{t'}^2)}{2N_{tt't''}} k_{tt't''}^2 - \frac{2\omega_t}{(4\omega_t^2 - \omega_{t'}^2)} k_{ttt'}^2 - \frac{2\omega_{t'}}{(4\omega_{t'}^2 - \omega_t^2)} k_{tt't'}^2$$

$$+ [\tfrac{1}{2} B_e^{zz} (\zeta_{t1,t'2}^z)^2 + B_e^{xx} (\zeta_{t1,t'1}^y)^2] \left[\frac{\omega_t}{\omega_{t'}} + \frac{\omega_{t'}}{\omega_t} \right] \qquad \text{(e)}$$

$$x_{l_t l_t} = -\tfrac{1}{2} k_{tttt} - \sum_i \frac{\omega_i}{(4\omega_t^2 - \omega_i^2)} k_{itt}^2 + \sum_{t'}^{t'\neq t} \frac{(8\omega_t^2 - 3\omega_{t'}^2)}{\omega_{t'}(4\omega_t^2 - \omega_{t'}^2)} k_{ttt'}^2 + \frac{15}{4\omega_t} k_{ttt}^2 + B_e^{zz} (\zeta_{t1,t2}^z)^2 \qquad \text{(f)}$$

$$x_{l_t l_{t'}} = \frac{1}{2} \sum_i \frac{\omega_i \omega_t \omega_{t'}}{N_{itt'}} k_{itt'}^2 - \sum_{t''}^{t''\neq t,t'} \frac{\omega_t \omega_{t'} \omega_{t''}}{N_{tt't''}} k_{tt't''}^2 + \frac{4\omega_t^2}{\omega_{t'}(4\omega_t^2 - \omega_{t'}^2)} k_{ttt'}^2$$

$$+ \frac{4\omega_{t'}^2}{\omega_t(4\omega_{t'}^2 - \omega_t^2)} k_{tt't'}^2 - 2B_e^{xx} (\zeta_{t1,t'1}^y)^2 + B_e^{zz} (\zeta_{t1,t'2}^z)^2 + 2B_e^{zz} \zeta_{t1,t2}^z \zeta_{t'1,t'2}^z \qquad \text{(g)}$$

Comparison of these expressions with the corresponding ones obtained in Section 9.2 shows that they differ only in some additional terms involving the Coriolis coupling constants. We notice that terms which are zero by symmetry, such as those involving cubic force constants of the type k_{tii} or k_{tij}, have been omitted in the expressions of the anharmonicity coefficients. We notice also that $t1$ and $t2$ label the two components of a degenerate normal coordinate.

In much the same way as done for the vibrational energy, one can collect together terms with the same powers of the rotational quantum numbers and obtain the second-order correction to the energy. Terms in $J(J+1)$, $[J(J+1)]^2$, $J(J+1)K^2$, K^2 and K^4 appear with very complex coefficients. A discussion of these terms is beyond the scope of this book and the reader interested in them is referred to the recent article by Mills[6] and to references contained therein.

As far as the calculation of force constants is concerned, the only terms employed so far are the effective rotational constants $B_v^{\alpha\alpha}$ which appear in the coefficients of $J(J+1)$. For a rigid rotor these constants depend only upon the moments of inertia whereas for the anharmonic vibro-rotor they became functions of the vibrational quantum numbers and are given by

$$B_v^{\alpha\alpha} = B_e^{\alpha\alpha} - \sum_i \alpha_i^{\alpha}(v_i + d_i/2) \qquad \alpha = x, y, z \qquad (9.3.19)$$

where d_i is the degeneracy of the ith mode and

$$B_e^{\alpha\alpha} = \hbar/4\pi I_{\alpha\alpha}^e c \qquad (9.3.20)$$

The vibro-rotational constants $\alpha^{\alpha\alpha}$ are given by the expressions

$$-\alpha_i^{\alpha\alpha} = \frac{2(B_e^{\alpha\alpha})^2}{\omega_i} \left[\sum_\gamma \frac{3(a_i^{\alpha\gamma})^2}{4I_{\gamma\gamma}^e} + \sum_j^{j\neq l} (\zeta_{ij}^{\alpha})^2 \frac{3\omega_i^2 + \omega_j^2}{\omega_i^2 - \omega_j^2} + 2\pi \left(\frac{c}{h}\right)^{\frac{1}{2}} \sum_j^{j\neq i} k_{iij} a_j^{\alpha\alpha} \left(\frac{\omega_i}{\omega_j^{\frac{3}{2}}}\right) + 6\pi \left(\frac{c}{h}\right)^{\frac{1}{2}} k_{iii} \frac{\alpha_i^{\alpha\alpha}}{\omega_i^{\frac{3}{2}}} \right] \qquad (9.3.21)$$

for non-degenerate molecules and by the expressions

$$-\alpha_i^{zz} = \frac{2(B_e^{zz})^2}{\omega_i} \left[\frac{3(a_i^{zz})^2}{4I_{zz}^e} + \sum_j^{j\neq i} (\zeta_{ij}^z)^2 \frac{3\omega_i^2 + \omega_j^2}{\omega_i^2 - \omega_j^2} \right.$$

$$\left. + 2\pi \left(\frac{c}{h}\right)^{\frac{1}{2}} \sum_j^{j\neq i} k_{iij} a_j^{zz} \left(\frac{\omega_i}{\omega_j^{\frac{3}{2}}}\right) + 6\pi \left(\frac{c}{h}\right)^{\frac{1}{2}} k_{iii} \frac{\alpha_i^{zz}}{\omega_i^{\frac{1}{2}}} \right] \qquad \text{(a)}$$

$$-\alpha_t^{zz} = \frac{2(B_e^{zz})^2}{\omega_t}\left[\frac{3(a_{t1}^{xz})^2}{4I_{xx}^{\mathrm{e}}} + \sum_{t'}^{t' \neq t} (\zeta_{t1,t'2}^{z})^2 \frac{3\omega_t^2 + \omega_{t'}^2}{\omega_t^2 - \omega_{t'}^2} + 2\pi\left(\frac{c}{h}\right)^{\frac{1}{2}} \sum_j k_{j,t1,t1} a_j^{zz} \left(\frac{\omega_t}{\omega_j^{\frac{3}{2}}}\right)\right] \tag{b}$$

$$-\alpha_i^{xx} = \frac{2(B_e^{xx})^2}{\omega_i}\left[\frac{3(a_i^{xx})^2 + 3(a_i^{xy})^2}{4I_{xx}^{\mathrm{e}}} + \sum_t [(\zeta_{i,t1}^{y})^2 + (\zeta_{i,t1}^{x})^2] \frac{3\omega_i^2 + \omega_t^2}{\omega_i^2 - \omega_t^2}\right. \tag{9.3.22}$$

$$\left. + 2\pi\left(\frac{c}{h}\right)^{\frac{1}{2}} \sum_j^{j \neq i} k_{iij} a_j^{xx} \frac{\omega_i}{\omega_j^{\frac{3}{2}}} + 6\pi\left(\frac{c}{h}\right)^{\frac{1}{2}} k_{iii} \frac{\alpha_i^{xx}}{\omega_i^{\frac{1}{2}}}\right] \tag{c}$$

$$-\alpha_t^{xx} = \frac{2(B_e^{xx})^2}{\omega_t}\left[\frac{3(a_{t1}^{xz})^2}{8I_{zz}^{\mathrm{e}}} + \frac{3(a_{t1}^{xx})^2}{4I_{xx}^{\mathrm{e}}} + \frac{1}{2}\sum_i [(\zeta_{i,t1}^{y})^2 + (\zeta_{i,t1}^{x})^2] \frac{3\omega_t^2 + \omega_i^2}{\omega_t^2 - \omega_i^2}\right.$$

$$\left. + \sum_{t'}^{t' \neq t} [(\zeta_{t1,t'1}^{y})^2 + (\zeta_{t1,t'1}^{x})^2] \frac{3\omega_t^2 + \omega_{t'}^2}{\omega_t^2 - \omega_{t'}^2} + 2\pi\left(\frac{c}{h}\right)^{\frac{1}{2}} \sum_j k_{j,t1,t1} a_j^{xx} \frac{\omega_t}{\omega_j^{\frac{3}{2}}}\right] \tag{d}$$

for molecules with two-fold degenerate vibrations.

The quantities $a_i^{\alpha\beta}$ occurring in these expressions are the derivatives

$$a_i^{\alpha\beta} = (\partial I_{\alpha\beta}/\partial Q_i)_e \tag{9.3.23}$$

of the inertia tensor components with respect to the normal coordinates.

9.4 Anharmonic force constants

The quantum-mechanical treatment of the anharmonic vibrations of a molecule, discussed in the previous sections, is based on the expansion of the potential function in a Taylor series of normal coordinates with coefficients that were generically called 'anharmonic force constants', even if not directly related to a simple physical picture of forces acting between the atoms when displaced from the equilibrium positions.

The normal coordinates are, however, not 'working' coordinates in the sense that they are known only '*a posteriori*' when the harmonic secular equation has been already solved using any working set of coordinates which correctly describes the instantaneous positions of the atoms. For infinitesimal displacements of the nuclei the transformation relating the working coordinates with the normal coordinates can be taken as linear. For finite displacements of the nuclei this is no longer true and the transformation will be in general non-linear.† Even if one wishes to leave this transformation unspecified and is satisfied with calculating the higher terms of the potential expansion in normal coordinates, one has to face further difficulties. In principle these higher terms can be evaluated from the experimental values of the anharmonicity coefficients and of the effective rotational constants possibly for two or more isotopic species. In practice this procedure is limited to a very few triatomic molecules, the reason being that the number of cubic and quartic coefficients of the potential expansion is always much larger for a polyatomic molecule than the number of experimental data.

The only way to overcome this difficulty is to build up a model potential either by introducing constraints, so that some higher-order force constants become functions of others, or by assuming a specific analytical form of the potential (or of part of it) and obtaining the force constants as functions of a limited number of parameters. In order to do this it is again necessary to use a working set of coordinates to build up a physically meaningful model potential. The choice of the working set of coordinates is, however, not unique and additional complications arise at this level.

† The transformation is linear only between normal and Cartesian coordinates.

The natural coordinate system for setting up the kinetic energy of a polyatomic molecule is a Cartesian displacement coordinate system, since in this basis the kinetic energy assumes a simple diagonal form. The vibrational potential is however best described in terms of internal coordinates (variation of bond lengths and bond angles). In terms of these co-ordinates the various coefficients of the potential expansion assume a concrete meaning directly related to our picture of the chemical bonds and become true 'force constants' associated with variations of bond distances and changes in bond angles.

In Chapter 8 we have already seen that there are two possible sets of internal coordinates one can use to describe the atomic motions. The first is a set of truly generalized internal coordinates $\mathcal{R}_k$, geometrically defined for any instantaneous position of the nuclei and related to the Cartesian displacement coordinates by a non-linear relation of the type

$$\mathcal{R}_s = \sum_m B^s_m x_m + \tfrac{1}{2}\sum_{mn} B^s_{mn} x_m x_n + \tfrac{1}{6}\sum_{mnp} B^s_{mnp} x_m x_n x_p + \ldots \tag{9.4.1}$$

The second set of internal coordinates R_s is defined by a linear relation with the Cartesian displacement coordinates of the type

$$R_s = \sum_m B^s_m x_m \tag{9.4.2}$$

which is simply obtained by dropping all terms in (9.4.1) higher than the first and assuming that the linear relation holds for any value of the x_m. For infinitesimal displacements of the nuclei the two sets of coordinates yield the same description of the molecular motions since the higher terms of (9.4.1) are vanishingly small. For this reason the whole treatment of harmonic vibrations given in the previous chapters was based on the linearized co-ordinates R_s which are obviously easier to handle. For finite displacements of the nuclei the two sets yield, however, a different description of the molecular motions and therefore imply a different definition of the force constants of order higher than the second.

This is easily seen by considering the schematic drawing of the bending motion in a bent triatomic molecule shown in Figure 9.4.1(a) in terms of a linearized bending coordinate R and in Figure 9.4.1(b) in terms of a curvilinear bending coordinate $\mathcal{R}$. Since the linearized

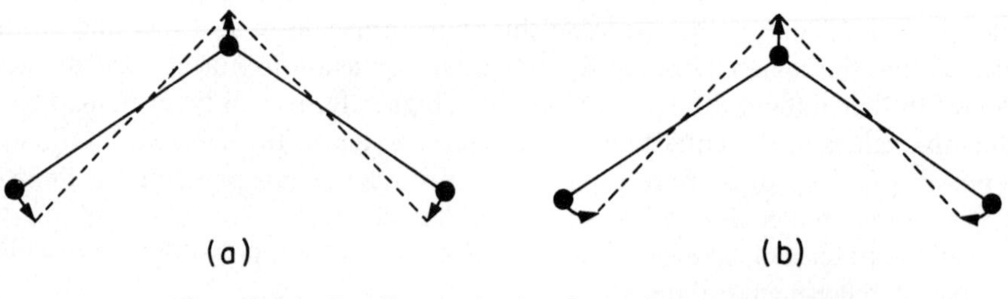

Figure 9.4.1 Bending motion in a bent triatomic molecule

coordinate R is defined through the linear equation (9.4.2), the atoms are constrained to move along straight lines from the equilibrium position. As a consequence, a finite variation of the bond angle will always involve some stretching of the bonds, the amount of stretching clearly being a function of the angle variation. The curvilinear coordinate $\mathcal{R}$ is defined through the non-linear relation (9.4.1) and the atoms therefore are no longer forced to follow a linear displacement from equilibrium. They can in general move on a curved path (Figure 9.4.1b) and thus preserved the equilibrium bond length in the distorted configuration. The higher force constants will obviously reflect this situation and will be different in the two cases. If one of the atoms is now substituted with an isotopic species the displacement of that atom in the bending coordinate will be different and thus, in the case of the linearized coordinate R, there will be a different contribution of bond stretching at higher order. The direct consequence of this fact is that the higher-order force constants are in general different for two isotopic species when a linearized coordinate basis is used.

We conclude therefore that, as far as the quadratic terms of the potential are concerned, the expansion in the $\mathcal{R}_s$ coordinates

$$V = \sum_{r \leqslant s} K_{rs}\mathcal{R}_r\mathcal{R}_s + \sum_{r \leqslant s \leqslant t} K_{rst}\mathcal{R}_r\mathcal{R}_s\mathcal{R}_t + \sum_{r \leqslant s \leqslant t \leqslant u} K_{rstu}\mathcal{R}_r\mathcal{R}_s\mathcal{R}_t\mathcal{R}_u + \ldots \tag{9.4.3}$$

and the expansion in the R_s coordinates

$$V = \sum_{r \leqslant s} \bar{K}_{rs}R_rR_s + \sum_{r \leqslant s \leqslant t} \bar{K}_{rst}R_rR_sR_t + \sum_{r \leqslant s \leqslant t \leqslant u} \bar{K}_{rstu}R_rR_sR_tR_u + \ldots \tag{9.4.4}$$

are equivalent in the sense that the coefficients K_{rs} are identical to the coefficients $\bar{K}_{rs}$. Higher coefficients of the expansion are, however, different, those expressed in the linearized basis R being mass-dependent[7] and thus different for different isotopic species of the same molecule. For this reason as well as because the R coordinates give an unfaithful description of the nuclear motions, with deviations which increase with the magnitude of the displacements, the expression (9.4.3) is normally used for the potential. In what follow, therefore, we shall always refer to equation (9.4.3) when dealing with the potential expanded in internal coordinates. In this respect we notice that an alternative way of writing the potential expansion (9.4.3) is through the use of unrestricted summations over the indices, which gives

$$V = \tfrac{1}{2}\sum_{rs} f_{rs}\mathcal{R}_r\mathcal{R}_s + \tfrac{1}{6}\sum_{rst} f_{rst}\mathcal{R}_r\mathcal{R}_s\mathcal{R}_t + \tfrac{1}{24}\sum_{rstu} f_{rstu}\mathcal{R}_r\mathcal{R}_s\mathcal{R}_t\mathcal{R}_u + \ldots \tag{9.4.5}$$

The force constants f are true derivatives of V with respect to the internal coordinates, whereas the force constants K include numerical factors. The relationships between these two types of force constant are rather simple and are given below.

$$\begin{aligned}
f_{rr} &= \left(\frac{\partial^2 V}{\partial \mathcal{R}_r^2}\right)_0 = 2K_{rr} & f_{rrrr} &= \left(\frac{\partial^4 V}{\partial \mathcal{R}_r^4}\right)_0 = 24K_{rrrr} \\
f_{rs} &= \left(\frac{\partial^2 V}{\partial \mathcal{R}_r \, \partial \mathcal{R}_s}\right)_0 = K_{rs} & f_{rrrs} &= \left(\frac{\partial^4 V}{\partial \mathcal{R}_r^3 \, \partial \mathcal{R}_s}\right)_0 = 6K_{rrrs} \\
f_{rrr} &= \left(\frac{\partial^3 V}{\partial \mathcal{R}_r^3}\right)_0 = 6K_{rrr} & f_{rrss} &= \left(\frac{\partial^4 V}{\partial \mathcal{R}_r^2 \, \partial \mathcal{R}_s^2}\right)_0 = 4K_{rrss} \\
f_{rrs} &= \left(\frac{\partial^3 V}{\partial \mathcal{R}_r^2 \, \partial \mathcal{R}_s}\right)_0 = 2K_{rrs} & f_{rrst} &= \left(\frac{\partial^4 V}{\partial \mathcal{R}_r^2 \, \partial \mathcal{R}_s \, \partial \mathcal{R}_t}\right)_0 = 2K_{rrst} \\
f_{rst} &= \left(\frac{\partial^3 V}{\partial \mathcal{R}_r \, \partial \mathcal{R}_s \, \partial \mathcal{R}_t}\right)_0 = K_{rst} & f_{rstu} &= \left(\frac{\partial^4 V}{\partial \mathcal{R}_r \, \partial \mathcal{R}_s \, \partial \mathcal{R}_t \, \partial \mathcal{R}_u}\right)_0 = K_{rstu}
\end{aligned} \tag{9.4.6}$$

Both types of definition of the anharmonic force constants have been utilized in recent literature and attention must thus be paid to the kind of potential expansion used, before comparing data from different authors. We shall utilize the f force constants defined in (9.4.5) in the following discussion because they are simpler to handle in non-linear transformations than the K force constants of (9.4.3). For this we shall rewrite equation (9.2.3) in a form consistent with (9.4.5)

$$V = \tfrac{1}{2}hc\sum_i \omega_i q_i^2 + \tfrac{1}{6}hc\sum_{ijk}\phi_{ijk}q_iq_jq_k + \tfrac{1}{24}hc\sum_{ijkl}\phi_{ijkl}q_iq_jq_kq_l + \ldots \tag{9.4.7}$$

with a similar expression for the Cartesian space

$$V = \tfrac{1}{2}\sum_{m,n} F_{mn}x_mx_n + \tfrac{1}{6}\sum_{mnp} F_{mnp}x_mx_nx_p + \tfrac{1}{24}\sum_{mnpq} F_{mnpq}x_mx_nx_px_q + \ldots \tag{9.4.8}$$

and investigate the relationships between the force constants in these different spaces.

The transformation from Cartesian to internal coordinates has already been given in equation (9.4.1) as a non-linear relation with coefficients B^s_m, B^s_{mn} and B^s_{mnp} which are the first, second and third derivatives respectively of the instantaneous internal coordinates with respect to the Cartesian displacements. The calculation of these coefficients from the basic expressions of the internal coordinates is illustrated in the next section.

The transformation from dimensionless normal to Cartesian displacement coordinates is linear by definition and according to Section 2.4 has the form

$$x_m = \sum_i \left(\frac{\hbar^2}{\lambda_i}\right)^{\frac{1}{4}} m_m^{-\frac{1}{2}} l_{mi} q_i \tag{9.4.9}$$

where l_{mi} are the elements of the $\mathscr{L}$-matrix defined in Chapter 2. These elements are obtained either by direct solution of the harmonic secular equation in mass-weighted Cartesian coordinates or by solution in linearized internal coordinates and successive transformation by means of the relation

$$\mathscr{L} = \mathbf{M}^{-\frac{1}{2}}\tilde{\mathbf{B}}\tilde{\mathbf{L}}^{-1} \tag{9.4.10}$$

The direct transformation from dimensionless normal to curvilinear internal coordinates is the combination of (9.4.1) and (9.4.9) and is therefore a non-linear relation of the type

$$\mathfrak{R}_s = \sum_i \gamma_i^{-\frac{1}{2}} L^s_i q_i + \tfrac{1}{2}\sum_{ij}(\gamma_i\gamma_j)^{-\frac{1}{2}} L^s_{ij} q_i q_j + \tfrac{1}{6}\sum_{ijk}(\gamma_i\gamma_j\gamma_k)^{-\frac{1}{2}} L^s_{ijk} q_i q_j q_k + \cdots \tag{9.4.11}$$

where

$$\gamma_i = \left(\frac{\lambda_i}{\hbar^2}\right)^{\frac{1}{2}} \tag{9.4.12}$$

By combining together (9.4.1) and (9.4.9) we obtain for the coefficients of (9.4.11) the expressions

$$\begin{aligned} L^s_i &= \sum_m B^s_m (m_m)^{-\frac{1}{2}} l_{mi} \\ L^s_{ij} &= \sum_{mn} B^s_{mn} (m_m m_n)^{-\frac{1}{2}} l_{mi} l_{nj} \\ L^s_{ijk} &= \sum_{mnp} B^s_{mnp} (m_m m_n m_p)^{-\frac{1}{2}} l_{mi} l_{nj} l_{pk} \end{aligned} \tag{9.4.13}$$

Many authors have preferred in the past to construct the coefficients of the transformation (9.4.11) following the two-step method described above, i.e. first obtaining the *B*-coefficients of (9.4.1) and then transforming them into the *L*-coefficients by means of (9.4.13). Recently Hoy *et al.*[7] and Tsaune *et al.*[8] have proposed a different method which furnishes the *L*-coefficients directly and avoids the long and tedious computation of the *B*-coefficients. The construction of the *L*-coefficients in this way is discussed in the next section.

Once the coefficients of the transformation (9.4.11) have been obtained it is possible to transform the force constants f_{rs}, f_{rst} and f_{rstu} defined in the internal curvilinear space into the force constants ω_i, ϕ_{ijk} and ϕ_{ijkl} defined in the dimensionless normal coordinate space. We notice at this point that, owing to the non-linear nature of (9.4.5) and (9.4.7), it is sufficient to include terms up to the third order in the coordinate transformation to obtain the anharmonic force-constant relations up to the fourth order.

The relationships between force constants in the Я- and q-spaces are easily obtained by partial differentiation of V with respect to the normal coordinates. From the general relation

$$\left(\frac{\partial V}{\partial q_i}\right) = \sum_s \left(\frac{\partial V}{\partial \text{Я}_s}\right)\left(\frac{\partial \text{Я}_s}{\partial q_i}\right) \tag{9.4.14}$$

we obtain

$$\begin{aligned}
\left(\frac{\partial^2 V}{\partial q_i \partial q_j}\right) &= \sum_{rs} \left(\frac{\partial^2 V}{\partial \text{Я}_r \partial \text{Я}_s}\right)\left(\frac{\partial \text{Я}_r}{\partial q_i}\frac{\partial \text{Я}_s}{\partial q_j}\right) + \sum_s \left(\frac{\partial V}{\partial \text{Я}_s}\frac{\partial^2 \text{Я}_s}{\partial q_i \partial q_j}\right) \\
\left(\frac{\partial^3 V}{\partial q_i \partial q_j \partial q_k}\right) &= \sum_{rst} \left(\frac{\partial^3 V}{\partial \text{Я}_r \partial \text{Я}_s \partial \text{Я}_t}\right)\left(\frac{\partial \text{Я}_r}{\partial q_i}\frac{\partial \text{Я}_s}{\partial q_j}\frac{\partial \text{Я}_t}{\partial q_k}\right) \\
&\quad + \sum_{rs} \left(\frac{\partial^2 V}{\partial \text{Я}_r \partial \text{Я}_s}\right)\left(\frac{\partial^2 \text{Я}_r}{\partial q_i \partial q_j}\frac{\partial \text{Я}_s}{\partial q_k} + \frac{\partial^2 \text{Я}_r}{\partial q_i \partial q_k}\frac{\partial \text{Я}_s}{\partial q_j} + \frac{\partial^2 \text{Я}_r}{\partial q_j \partial q_k}\frac{\partial \text{Я}_s}{\partial q_i}\right) + \sum_s \left(\frac{\partial V}{\partial \text{Я}_s}\right)\left(\frac{\partial^3 \text{Я}_s}{\partial q_i \partial q_j \partial q_k}\right)
\end{aligned} \tag{9.4.15}$$

and similar expressions for the fourth derivatives.

By use of (9.4.11) and by evaluating the above derivatives at equilibrium, we have

$$\begin{aligned}
\lambda_i &= \sum_{rs} f_{rs} L_i^s L_i^r \\
\phi_{ijk} &= \left[\sum_{rst} f_{rst} L_i^r L_j^s L_k^t + \sum_{rs} f_{rs}(L_{ij}^r L_k^s + L_{ik}^r L_j^s + L_{jk}^r L_i^s)\right](h^2c^2\gamma_i\gamma_j\gamma_k)^{-\frac{1}{2}} \\
\phi_{ijkl} &= \left[\sum_{rstu} f_{rstu} L_i^r L_j^s L_k^t L_l^u + \sum_{rst} f_{rst}(L_{ij}^r L_k^s L_l^t + L_{ik}^r L_j^s L_l^t + L_{il}^r L_j^s L_k^t\right. \\
&\quad + L_{jk}^r L_i^s L_l^t + L_{jl}^r L_i^s L_k^t + L_{kl}^r L_i^s L_j^t) + \sum_{rs} f_{rs}(L_{ij}^r L_{kl}^s + L_{ik}^r L_{jl}^s \\
&\quad \left. + L_{il}^r L_{jk}^s + L_{ijk}^r L_l^s + L_{ijl}^r L_k^s + L_{ikl}^r L_j^s + L_{jkl}^r L_i^s)\right](h^2c^2\gamma_i\gamma_j\gamma_k\gamma_l)^{-\frac{1}{2}}
\end{aligned} \tag{9.4.16}$$

Alternatively we can transform the force constants from the internal to the Cartesian space using (9.4.1) and then from the Cartesian to the dimensionless normal coordinate space using (9.4.9). The first transformation leads to expressions which are identical in form to (9.4.16) with only the difference that the coefficients L must be replaced by the corresponding coefficients B of (9.4.1) and the force constants ϕ by the corresponding force constants F defined through (9.4.8). The second transformation is linear and yields the relations

$$\begin{aligned}
\omega_i &= \sum_{mn} F_{mn}(m_m m_n)^{-\frac{1}{2}} l_{mi} l_{ni} (hc\gamma_i)^{-1} \\
\phi_{ijk} &= \left[\sum_{mnp} F_{mnp}(m_m m_n m_p)^{-\frac{1}{2}} l_{mi} l_{nj} l_{pk}\right](h^2c^2\gamma_i\gamma_j\gamma_k)^{-\frac{1}{2}} \\
\phi_{ijkl} &= \left[\sum_{mnpq} F_{mnpq}(m_m m_n m_p m_q)^{-\frac{1}{2}} l_{mi} l_{nj} l_{pk} l_{ql}\right](h^2c^2\gamma_i\gamma_j\gamma_k\gamma_l)^{-\frac{1}{2}}
\end{aligned} \tag{9.4.17}$$

An important consequence of the non-linear coordinate transformation (9.4.11) is that the cubic force constants in the q-space depend on both quadratic and cubic force constants in internal coordinates whereas the quartic force constants depend on quadratic, cubic and quartic terms.

The calculation of anharmonic force constants proceeds as follows. An initial set of force constants in the Я-space is transformed to the q-space by means of (9.4.16), the L_i^s elements being obtained by solution of the harmonic secular equation and the L_{ij}^s and L_{ijk}^s elements being calculated with the methods of next section. The force constants ϕ_{ijk} and ϕ_{ijkl} are then used to produce the force constants k_{ijk} and k_{ijkl} of (9.2.3) by means of (9.4.6) and these force constants are utilized for the calculation of the anharmonic coefficients x_{ij} and of the vibro-rotational interaction constants $\alpha^{\alpha\alpha}$ of the previous section. These calculated quantities are then compared with experimental values if available. A Jacobian matrix with elements given by the partial derivatives of the anharmonic constants with respect to the force constants is then produced and a least-squares refinement is made to obtain the best fit of the data.

Computer programs for this kind of calculation have been produced by various authors, including Overend,[9] Morino,[11] Mills[7] and their coworkers. The programs differ essentially in the transformation from internal to normal coordinates which is made in one step or in two steps (via the Cartesian space) by different authors. Most of the work accumulated until now on anharmonic force constants is limited to very small molecules. Morino and coworkers[10–17] have investigated several bent XY_2 triatomic molecules and have determined first the cubic terms of the potential using the vibro-rotational interaction constants and then the quartic terms using the anharmonicity coefficients. Overend and coworkers have calculated the anharmonic force field of some linear triatomic molecules including CO_2,[9,20] CS_2[18,19] and HCN.[21] More recently calculations of the anharmonic force field of NH_3[7] and SO_3[22] have been reported and some additional work on the triatomic molecules has been published.[23,24,25]

In order to make the reader acquainted with the use of the various equations discussed before, we shall discuss in some detail the specific case of the bent XY_2 molecules, following the treatment of Morino and coworkers.[10,13] Molecules of this type (C_{2v} point group) possess three non-degenerate vibrations, two of A_1 species (ω_1, ω_2) and one of B_1 species (ω_3) and are asymmetric tops ($I_{xx} \neq I_{yy} \neq I_{zz}$). There are then three different rotational constants B_v^{xx}, B_v^{yy} and B_v^{zz} in the expression of the rotational energy and thus nine different $\alpha_i^{\alpha\alpha}$ constants in (9.3.19).

Darling and Dennison[26] and later Nielsen[1] have worked out the analytical form of these interaction constants and of the $a_i^{\alpha\alpha}$ and ζ_{ij}^{α} constants appearing in equation (9.3.21). By assuming the x- and y-axes in the molecular plane, the following expressions are found for the constants having non-zero values

$$
\begin{aligned}
\zeta_{13}^{z} &= [(I_{xx}^{e})^{\frac{1}{2}} \cos\gamma - (I_{yy}^{e})^{\frac{1}{2}} \sin\gamma]/(I_{zz}^{e})^{\frac{1}{2}} \\
\zeta_{23}^{z} &= [(I_{xx}^{e})^{\frac{1}{2}} \sin\gamma + (I_{yy}^{e})^{\frac{1}{2}} \cos\gamma]/(I_{zz}^{e})^{\frac{1}{2}} \\
a_1^{xx} &= 2(I_{xx}^{e})^{\frac{1}{2}} \sin\gamma \qquad a_2^{xx} = 2(I_{xx}^{e})^{\frac{1}{2}} \cos\gamma \\
a_1^{yy} &= 2(I_{yy}^{e})^{\frac{1}{2}} \cos\gamma \qquad a_2^{yy} = -2(I_{yy}^{e})^{\frac{1}{2}} \sin\gamma \\
a_3^{xy} &= 2(I_{xx}^{e} I_{yy}^{e}/I_{zz}^{e})^{\frac{1}{2}} \qquad a_s^{zz} = a_s^{xx} + a_s^{yy}
\end{aligned}
\tag{9.4.18}
$$

where the superscript e indicates equilibrium value and $\sin\gamma$ and $\cos\gamma$ are functions of the harmonic force constants, of the equilibrium geometry and of the masses. By use of

(9.4.18), of (9.3.21) and of the ζ sum rule $(\zeta_{13}^z)^2 + (\zeta_{23}^z)^2 = 1$ we obtain

$$\alpha_1^{xx} = -\frac{6A^2}{\omega_1}\sin^2\gamma - 6\frac{k_{111}}{\omega_1}\left(\frac{2A^3}{\omega_1}\right)^{\frac{1}{2}}\sin\gamma + 2\frac{k_{112}}{\omega_2}\left(\frac{2A^3}{\omega_2}\right)^{\frac{1}{2}}\cos\gamma$$

$$\alpha_1^{yy} = -\frac{6B^2}{\omega_1}\cos^2\gamma - 6\frac{k_{111}}{\omega_1}\left(\frac{2B^3}{\omega_1}\right)^{\frac{1}{2}}\cos\gamma - 2\frac{k_{112}}{\omega_2}\left(\frac{2B^3}{\omega_2}\right)^{\frac{1}{2}}\sin\gamma$$

$$\alpha_1^{zz} = \frac{C^2}{A^2}\alpha_1^{xx} + \frac{C^2}{B^2}\alpha_1^{yy} + \frac{8C^2\omega_3^2}{\omega_1(\omega_3^2 - \omega_1^2)}(\zeta_{13}^z)^2$$

$$\alpha_2^{xx} = -\frac{6A^2}{\omega_2}\cos^2\gamma + 6\frac{k_{222}}{\omega_2}\left(\frac{2A^3}{\omega_2}\right)^{\frac{1}{2}}\cos\gamma - 2\frac{k_{122}}{\omega_1}\left(\frac{2A^3}{\omega_1}\right)^{\frac{1}{2}}\sin\gamma$$

$$\alpha_2^{yy} = -\frac{6B^2}{\omega_2}\sin^2\gamma - 6\frac{k_{222}}{\omega_2}\left(\frac{2B^3}{\omega_2}\right)^{\frac{1}{2}}\sin\gamma - 2\frac{k_{122}}{\omega_1}\left(\frac{2B^3}{\omega_1}\right)^{\frac{1}{2}}\cos\gamma \qquad (9.4.19)$$

$$\alpha_2^{zz} = \frac{C^2}{A^2}\alpha_2^{xx} + \frac{C^2}{B^2}\alpha_2^{yy} + \frac{8C^2\omega_3^2}{\omega_2(\omega_3^2 - \omega_2^2)}(\zeta_{23}^z)^2$$

$$\alpha_3^{xx} = -\frac{6CA}{\omega_3} - 2\frac{k_{133}}{\omega_1}\left(\frac{2A^3}{\omega_1}\right)^{\frac{1}{2}}\sin\gamma + 2\frac{k_{233}}{\omega_2}\left(\frac{2A^3}{\omega_2}\right)^{\frac{1}{2}}\cos\gamma$$

$$\alpha_3^{yy} = -\frac{6CB}{\omega_3} - 2\frac{k_{133}}{\omega_1}\left(\frac{2B^3}{\omega_1}\right)^{\frac{1}{2}}\cos\gamma - 2\frac{k_{233}}{\omega_2}\left(\frac{2B^3}{\omega_2}\right)^{\frac{1}{2}}\sin\gamma$$

$$\alpha_3^{zz} = \frac{C^2}{A^2}\alpha_3^{xx} + \frac{C^2}{B^2}\alpha_3^{yy} - \frac{8C^2\omega_1^2}{\omega_3(\omega_3^2 - \omega_1^2)}(\zeta_{13}^z)^2 - \frac{8C^2\omega_2^2}{\omega_3(\omega_3^2 - \omega_2^2)}(\zeta_{23}^z)^2$$

where

$$A = B_e^{xx} \qquad B = B_e^{yy} \qquad C = B_e^{zz}$$

and where the potential function, expressed in terms of the dimensionless normal coordinates (q_1 and q_2 of species A_1, q_3 of species B_1), is given by

$$\begin{aligned} V/hc = {} & \tfrac{1}{2}(\omega_1 q_1^2 + \omega_2 q_2^2 + \omega_3 q_3^2) + [k_{111}q_1^3 + k_{112}q_1^2 q_2 + k_{122}q_1 q_2^2 + k_{222}q_2^3 \\ & + k_{133}q_1 q_3^2 + k_{233}q_2 q_3^2] + [k_{1111}q_1^4 + k_{1112}q_1^3 q_2 + k_{1122}q_1^2 q_2^2 + k_{1133}q_1^2 q_3^2 \\ & + k_{1222}q_1 q_2^3 + k_{1233}q_1 q_2 q_3^2 + k_{2222}q_2^4 + k_{2233}q_2^2 q_3^2 + k_{3333}q_3^4] \end{aligned} \qquad (9.4.20)$$

since all terms in q_3 and q_3^3 are zero by symmetry.

The vibro-rotational interaction constants are obtained experimentally from the analysis of the rotational spectra. By using these data the six cubic force constants are then evaluated by a best fit of equations (9.4.19). Once the cubic force constants k in the normal coordinate space have been obtained, they are transformed into the f force constants in the curvilinear internal coordinate as true derivatives of the potential. According to equation (9.4.5) the potential function assumes the form

$$\begin{aligned} V = {} & [\tfrac{1}{2}f_{rr}(\Delta r_1^2 + \Delta r_2^2) + f_{rr'}\Delta r_1 \Delta r_2 + f_{r\alpha}(\Delta r_1 \Delta\alpha + \Delta r_2 \Delta\alpha) + \tfrac{1}{2}f_{\alpha\alpha}\Delta\alpha^2] \\ & + [\tfrac{1}{6}f_{rrr}(\Delta r_1^3 + \Delta r_2^3) + \tfrac{1}{2}f_{rrr'}(\Delta r_1^2 \Delta r_2 + \Delta r_2^2 \Delta r_1) + \tfrac{1}{2}f_{rr\alpha}(\Delta r_1^2 \Delta\alpha + \Delta r_2^2 \Delta\alpha) \\ & + f_{rr'\alpha}\Delta r_1 \Delta r_2 \Delta\alpha + \tfrac{1}{2}f_{r\alpha\alpha}(\Delta r_1 \Delta\alpha^2 + \Delta r_2 \Delta\alpha^2) + \tfrac{1}{6}f_{\alpha\alpha\alpha}\Delta\alpha^3] + [\tfrac{1}{24}f_{rrrr}(\Delta r_1^4 + \Delta r_2^4) \\ & + \tfrac{1}{6}f_{rrrr'}(\Delta r_1^3 \Delta r_2 + \Delta r_2^3 \Delta r_1) + \tfrac{1}{4}f_{rrr'r'}\Delta r_1^2 \Delta r_2^2 + \tfrac{1}{6}f_{rrr\alpha}(\Delta r_1^3 \Delta\alpha + \Delta r_2^3 \Delta\alpha) + \tfrac{1}{4}f_{rr\alpha\alpha}(\Delta r_1^2 \Delta\alpha^2 \\ & + \Delta r_2^2 \Delta\alpha^2) + \tfrac{1}{2}f_{rrr'\alpha}(\Delta r_1^2 \Delta r_2 \Delta\alpha + \Delta r_2^2 \Delta r_1 \Delta\alpha) + \tfrac{1}{2}f_{rr'\alpha\alpha}\Delta r_1 \Delta r_2 \Delta\alpha^2 + \tfrac{1}{6}f_{r\alpha\alpha\alpha}(\Delta r_1 \Delta\alpha^3 \\ & + \Delta r_2 \Delta\alpha^3) + \tfrac{1}{24}f_{\alpha\alpha\alpha\alpha}\Delta\alpha^4] \end{aligned} \qquad (9.4.21)$$

There are therefore six independent cubic f force constants which can be uniquely determined from the six K force constants in normal coordinates. The values obtained in this way for some triatomic molecules are reported in Table 9.4.1.

Table 9.4.1 Cubic anharmonic force constants of some bent triatomic molecules

	Units	O_3	OF_2	SO_2	SeO_2
f_{rrr}	mdyn/Å^2	−54·2	−27·2	−71·1	−44·8
$f_{rrr'}$	mdyn/Å^2	−2·38	−1·64	−1·15	−0·58
$f_{rr\alpha}$	mdyn/Å	−3·19	−1·34	−1·36	−0·60
$f_{rr'\alpha}$	mdyn/Å	−1·16	−0·41	−0·76	−0·20
$f_{r\alpha\alpha}$	mdyn	−3·85	−2·61	−1·80	−1·23
$f_{\alpha\alpha\alpha}$	mdyn Å	−3·92	−3·30	−2·50	−1·91
Reference		15	15	15	17

Some useful information can be obtained from these numerical values. First, all force constants are negative if the plus direction of the displacements is taken toward the increase of bond lengths and bond angles. Second, f_{rrr} is much larger than any other cubic term, thus revealing that the anharmonicity of bond stretchings is by far the most important factor. Using the experimental values of the anharmonicity constants x_{ij} and the cubic terms calculated above, it is then possible to evaluate the quartic force constants. From equations (9.3.18) we obtain

$$
\begin{aligned}
x_{11} &= \tfrac{3}{2}k_{1111} - \frac{15k_{111}^2}{4\omega_1} - \frac{k_{112}^2(8\omega_1^2 - 3\omega_2^2)}{4\omega_2(4\omega_1^2 - \omega_2^2)} \\
x_{22} &= \tfrac{3}{2}k_{2222} - \frac{15k_{222}^2}{4\omega_2} - \frac{k_{122}^2(8\omega_2^2 - 3\omega_1^2)}{4\omega_1(4\omega_2^2 - \omega_1^2)} \\
x_{33} &= \tfrac{3}{2}k_{3333} - \frac{k_{133}^2(8\omega_3^2 - 3\omega_1^2)}{4\omega_1(4\omega_3^2 - \omega_1^2)} - \frac{k_{233}^2(8\omega_3^2 - 3\omega_2^2)}{4\omega_2(4\omega_3^2 - \omega_2^2)} \\
x_{12} &= k_{1122} - \frac{3}{\omega_1}k_{111}k_{122} - \frac{3}{\omega_2}k_{222}k_{112} - \frac{2\omega_1}{4\omega_1^2 - \omega_2^2}k_{112}^2 - \frac{2\omega_2}{4\omega_2^2 - \omega_1^2}k_{122}^2 \\
x_{13} &= k_{1133} - \frac{3}{\omega_1}k_{111}k_{133} - \frac{2\omega_3}{4\omega_3^2 - \omega_1^2}k_{133}^2 - \frac{1}{\omega_2}k_{112}k_{233} + B_e^{zz}\left(\frac{\omega_3}{\omega_1} + \frac{\omega_1}{\omega_3}\right)(\zeta_{13}^z)^2 \\
x_{23} &= k_{2233} - \frac{3}{\omega_2}k_{222}k_{233} - \frac{2\omega_3}{4\omega_3^2 - \omega_2^2}k_{233}^2 - \frac{1}{\omega_1}k_{122}k_{133} + B_e^{zz}\left(\frac{\omega_3}{\omega_2} + \frac{\omega_2}{\omega_3}\right)(\zeta_{23}^z)^2
\end{aligned}
\qquad (9.4.22)
$$

In the expansion of the potential (9.4.20) there are actually nine independent quartic force constants whereas only six appear in the expressions of the anharmonicity coefficients given above. The three force constants k_{1112}, k_{1222} and k_{1333} therefore remain undetermined. The other six quartic terms can be determined from equations (9.4.22) and transformed then into the f force constants of expansion (9.4.21) in the usual way. Since there are also nine quartic force constants in (9.4.21), three of them will remain unknown. Kuchitsu and Morino[10] have assumed a value of zero for $f_{rrr\alpha}$, $f_{rrr'\alpha}$ and $f_{r\alpha\alpha\alpha}$ and on this basis have determined the remaining six force constants for a series of molecules; Owing to the indeterminacy in the experimental values of the x_{ij} and to error accumulation in the calculations, the indeterminacy on the quartic terms is rather large. The results

Table 9.4.2 Higher-order potential constants of some bent triatomic molecules in mdyne/Å

Molecule	f_{rrr}	$f_{rrr'}$	$f_{rr\alpha}$	$f_{rr'\alpha}$	$f_{r\alpha\alpha}$	$f_{\alpha\alpha\alpha}$
H_2O	$-38{\cdot}20 \pm 0{\cdot}24$	$-0{\cdot}64 \pm 0{\cdot}32$	$0{\cdot}32 \pm 0{\cdot}06$	$-0{\cdot}66 \pm 0{\cdot}01$	$0{\cdot}30 \pm 0{\cdot}40$	$-0{\cdot}84 \pm 0{\cdot}06$
D_2O	$-40{\cdot}72 \pm 1{\cdot}80$	$0{\cdot}96 \pm 0{\cdot}94$	$0{\cdot}44 \pm 0{\cdot}12$	$-0{\cdot}46 \pm 0{\cdot}04$	$0{\cdot}62 \pm 0{\cdot}52$	$-0{\cdot}78 \pm 0{\cdot}12$
H_2S	$-20{\cdot}64 \pm 0{\cdot}24$	$0{\cdot}06 \pm 0{\cdot}30$	$-0{\cdot}80 \pm 0{\cdot}06$	$-0{\cdot}15 \pm 0{\cdot}02$	$-0{\cdot}08 \pm 0{\cdot}16$	$-0{\cdot}06 \pm 0{\cdot}06$
SO_2	$-67{\cdot}80 \pm 0{\cdot}32$	$-1{\cdot}64 \pm 0{\cdot}46$	$-1{\cdot}36 \pm 0{\cdot}10$	$-0{\cdot}76 \pm 0{\cdot}03$	$-1{\cdot}26 \pm 0{\cdot}10$	$-1{\cdot}20 \pm 0{\cdot}06$
	f_{rrrr}	$f_{rrrr'}$	$f_{rrr'r'}$	$f_{rr\alpha\alpha}$	$f_{rr'\alpha\alpha}$	$f_{\alpha\alpha\alpha\alpha}$
H_2O	$369{\cdot}6 \pm 7{\cdot}2$	$4{\cdot}8 \pm 3{\cdot}6$	$5{\cdot}2 \pm 4{\cdot}4$	$-6{\cdot}8 \pm 3{\cdot}2$	$-1{\cdot}0 \pm 3{\cdot}4$	0
D_2O	$444{\cdot}0 \pm 36{\cdot}0$	$-16{\cdot}8 \pm 12{\cdot}0$	$0{\cdot}4 \pm 16{\cdot}0$	$-10{\cdot}8 \pm 5{\cdot}6$	$-3{\cdot}0 \pm 5{\cdot}6$	0
H_2S	$204{\cdot}0 \pm 7{\cdot}2$	$-1{\cdot}2 \pm 3{\cdot}6$	$-1{\cdot}2 \pm 4{\cdot}4$	$-2{\cdot}8 \pm 1{\cdot}6$	$-0{\cdot}6 \pm 1{\cdot}6$	0
SO_2	$811{\cdot}2 \pm 28{\cdot}8$	$-13{\cdot}2 \pm 15{\cdot}6$	$-1{\cdot}2 \pm 17{\cdot}2$	$2{\cdot}8 \pm 12{\cdot}8$	$5{\cdot}6 \pm 12{\cdot}4$	$-7{\cdot}2 \pm 2{\cdot}4$

obtained by Kuchitsu and Morino for a series of triatomic molecules are shown in Table 9.4.2. In order to show the relative magnitude, the cubic force constants are also reported in the table and all force constants are given in the same units (mdyn/Å). From the table it is evident that f_{rrrr} is much larger than any other quartic force constant and is the only one which is not indeterminate. The anharmonic force field of H_2O has been recently recalculated by Hoy, Mills and Strey.[7] The numerical values of force constants obtained by these authors by refining the α and x constants for both isotopic species show some noticeable differences from those given in the table. The anharmonic force field of O_3 has been also revised recently by Foss Smith[27] and the quartic force constants evaluated by a complete refinement procedure. The cubic part of the force field is not very different from that shown in Table 9.4.1.

As discussed before, the determination of the complete anharmonic force field of a polyatomic molecule is severely limited by the enormous number of anharmonic force constants that appear in the Taylor expansion. For very small molecules, such as those discussed before and a few others of comparable size, the rotational spectra furnish a great deal of information through the vibro-rotational interaction constants $\alpha_i^{\alpha\alpha}$ and related quantities such as the l-doubling constants not discussed here. For larger molecules, as the molecular size increases, these data become more and more difficult to measure and in any case are completely insufficient to determine the anharmonic constants, even if several isotopic species are used. The only way to extend this type of calculation to large molecules is by the use of a simplified model of the anharmonic part of the potential. We have already seen that force constants of the type f_{rrr} and f_{rrrr} are much larger in value than all other terms of the anharmonic potential. On this basis many authors have thus tried to develop model force fields which give particular emphasis to the bond-stretching anharmonic force constants.

The first attempt to derive anharmonic force constants from empirical potentials in analytical form was made by Redlich[28] who used for diatomic molecules the well known Morse function to represent the bond-stretching potential. Later Pliva[29] utilized for the bond-stretching potential in polyatomic molecules an extension of the function proposed by Lippincott and Schroeder,[30] in the form

$$V_i^{\text{str}} = D_i\{1 - \exp(-\tfrac{1}{2}q_i^2 n_i/r_i)[1 - \alpha_i q_i (n_i/r_i)^{\frac{1}{2}} + \beta_i q_i^2 n_i/r_i]\} \qquad (9.4.23)$$

where D_i is the dissociation energy of the ith bond, $q_i = r_i - r_i^e$ is the stretching coordinate, r_i and r_i^e the instantaneous and equilibrium bond lengths and n_i, α_i, β_i parameters related to the molecular structure. Furthermore Pliva proposed for the angle deformation potential a function of the type

$$V_m^{\text{def}} = D_m^d(\sin^2 \gamma_m/\sin \tfrac{1}{2}\varphi_m)[1 + b_m(\sin^2 \gamma_m/\sin \tfrac{1}{2}\varphi_m)] \tag{9.4.24}$$

where γ_m is the mth angle deformation coordinate, φ_m is the instantaneous value of the angle, D_m^d a parameter related to the height of the barrier and b_m an adjustable parameter.

More recently Morino and coworkers[10–17] have extensively investigated the use of non-bonded atom–atom potentials of Lennard-Jones or Buckingham type to derive cubic interaction constants. In their method the potential for a bent XY_2 molecule is written in the form

$$V = V_{\text{harm}} + (1/r_e)f_{rrr}(\Delta r_1^3 + \Delta r_2^3) + V(\Delta q) \tag{9.4.25}$$

where Δq is the variation of the non-bonded distance (see Section 8.9) and

$$V(\Delta q) = F'q_e\,\Delta q + \tfrac{1}{2}F\,\Delta q^2 + \frac{1}{6q_e}F_3\,\Delta q^3 \tag{9.4.26}$$

The non-bonded force constants $F' = (1/q_e)(dV/dq)_e$, $F = (d^2V/dq^2)_e$ and $F_3 = (d^3V/dq^3)_e$ are obtained by differentiation of the analytical expression of the non-bonded potential. By substitution into (9.4.26) of the expression of Δq expanded to the third order in terms of the internal coordinates, the cubic force constants are then obtained as functions of the non-bonded force constants F', F and F_3. Furthermore the f_{rrr} anharmonic force constant is expressed by these authors in terms of the corresponding quadratic force constant by means of a parameter obtained from analytical forms of the potential such as (9.4.23).

Machida and Overend[31] have discussed the cubic and quartic stretching interaction constants in triatomic molecules starting from the idea that, when a bond is stretched, the effective potential function of the second bond is displaced in order to approach that of a diatomic molecule when the first bond is broken. This produces a variation of the quadratic force constant and of the equilibrium bond length of the second bond, variation that is related to the interaction constants. On this basis these authors were able to predict the anharmonic stretching–stretching force constants of CO_2 from the quadratic force constant. Later the same authors[32] extended this treatment to polyatomic molecules and derived useful parametric relations for the anharmonic force constants with two indices like f_{iij} and f_{iijj} as well as for those with three indices like f_{ijk} and f_{iijk}.

Foss Smith and Overend[23,33] have also investigated model potentials for the stretching–bending and for the bending interaction constants. Using a double-minimum function symmetric about the linear configuration, these authors were able to express the cubic and quartic terms of the potential in triatomic molecules as a function of a limited number of parameters.

9.5 Curvilinear coordinates

In the previous section we have seen that a necessary step in the calculation of the anharmonic force constants is the determination of the coefficients of the non-linear transformation (9.4.11) that we shall rewrite here in a slightly different form

$$\mathfrak{R}_s = \sum_i L_i^s Q_i + \tfrac{1}{2}\sum_{ij} L_{ij}^s Q_i Q_j + \tfrac{1}{6}\sum_{ijk} L_{ijk}^s Q_i Q_j Q_k + \ldots \tag{9.5.1}$$

using normal coordinates $Q_i = \gamma_i^{-\frac{1}{4}} q_i$. The coefficients are the 1st, 2nd and 3rd derivatives of the coordinate $\mathfrak{R}_s$ with respect to the normal coordinates, taken at equilibrium. The

1st derivatives are simply the elements of the well-known **L**-matrix obtained by diagonalization of the harmonic secular equation. Higher coefficients can be obtained in essentially two ways, either by direct differentiation with respect to the Qs of the basic equations defining the internal coordinates or by calculating first the coefficients of the transformation (9.4.1)

$$\text{Я}_s = \sum_m B^s_m x_m + \tfrac{1}{2}\sum_{mn} B^s_{mn} x_m x_n + \tfrac{1}{6}\sum_{mnp} B^s_{mnp} x_m x_n x_p + \cdots \qquad (9.5.2)$$

and then the L-coefficients by means of equations (9.4.13). In this second case the problem is reduced to the calculation of the B-coefficients, which are the 1st, 2nd and 3rd derivatives of the Я_s coordinates with respect to Cartesian displacements, taken at equilibrium.

Various authors have used the last method in performing the non-linear transformation. In a series of papers[29,34,35,36] Pliva and coworkers have discussed the construction of the B-coefficients for the usual type of internal coordinates. Later Parisean, Suzuki and Overend[9] have derived general expressions for the stretching and bending coordinates and Morino, Kuchitsu and Yamamoto[37] have presented a compact method for the evaluation of the same coefficients. The calculation of the B-coefficients by differentiation of the basic equations defining the instantaneous values of the Я-coordinates in terms of the Cartesian displacements of the atoms is straightforward, although long and tedious. The process is simply an extension of the method described in Chapter 4 for the evaluation of the 1st derivatives. Parisean, Suzuki and Overend give explicit formulas for the stretching and in-plane bending coordinates only but the same procedure can be applied to more complex internal coordinates.

Recently Hoy, Mills and Strey[7] have shown how to compute the L-coefficients without going through the transformation (9.5.2). In their direct method the higher coefficients L^s_{ij} and L^s_{ijk} are expressed in terms of the first-order derivatives L^s_i, i.e. in terms of the eigenvectors of the secular equation in internal coordinates. For the construction of the L- or B-coefficients it is necessary to express the variation of the bond lengths or angles in terms of the instantaneous Cartesian coordinates of the atoms. In doing this it is very convenient to use *Cartesian difference* coordinates relative to the bonds since in this way more compact expressions are obtained. Therefore we shall associate with the bond s between two atoms a and b a vector $\mathbf{r}_s = \mathbf{r}_{ab}$, which is the difference of two 'atom vectors'

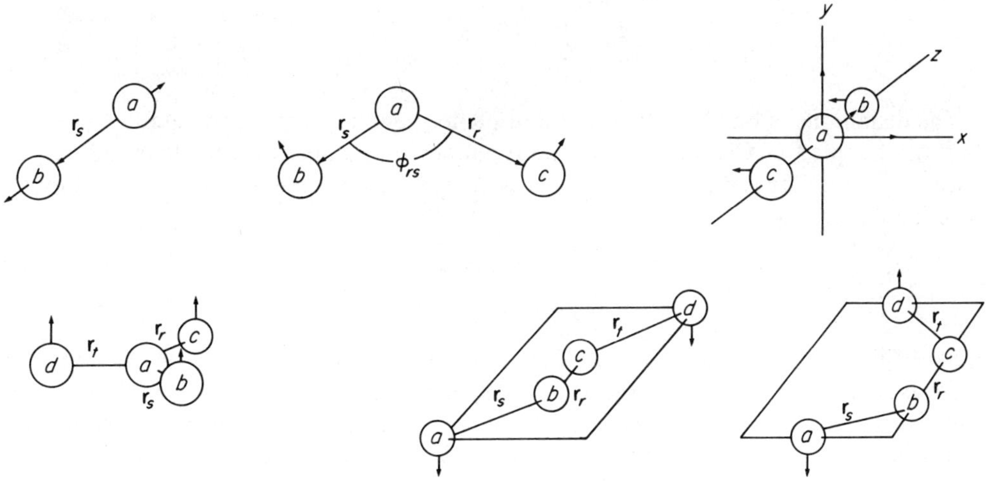

Figure 9.5.1 Labelling of atoms, bonds and angles for the definition of internal coordinates

whose components are the Cartesian coordinates of the atoms (see Figure 9.5.1). We then have the following obvious relations

$$\mathbf{r}_a = x_a\mathbf{e}_x + y_a\mathbf{e}_y + z_a\mathbf{e}_z \quad \text{(a)}$$

$$\mathbf{r}_b = x_b\mathbf{e}_x + y_b\mathbf{e}_y + z_b\mathbf{e}_z \quad \text{(b)} \qquad (9.5.3)$$

$$\mathbf{r}_s = \mathbf{r}_{ab} = \mathbf{r}_b - \mathbf{r}_a = (x_b - x_a)\mathbf{e}_x + (y_b - y_a)\mathbf{e}_y + (z_b - z_a)\mathbf{e}_z \quad \text{(c)}$$

We shall define Cartesian difference coordinates relative to the bond by the relation

$$\alpha_s = \alpha_{ab} = \alpha_b - \alpha_a \qquad \alpha = x, y, z \qquad (9.5.4)$$

and rewrite (9.5.3c) in the form

$$\mathbf{r}_s = x_{ab}\mathbf{e}_x + y_{ab}\mathbf{e}_y + z_{ab}\mathbf{e}_z \qquad (9.5.5)$$

In these expressions $\mathbf{e}_x$, $\mathbf{e}_y$ and $\mathbf{e}_z$ are unit vectors directed along the axes of the reference system. Using a superscript e to indicate the equilibrium values, the five standard types of internal coordinates are then defined as follows (see Figure 9.5.1):

(i) Bond stretching

$$\mathcal{R}_s = [\mathbf{r}_s \cdot \mathbf{r}_s]^{\frac{1}{2}} - r_s^e = [\sum_\alpha \alpha_{ab}^2]^{\frac{1}{2}} - r_s^e \qquad (9.5.6)$$

(ii) Angle bending

$$\mathcal{R}_\phi = \arccos\,[\mathbf{r}_s \cdot \mathbf{r}_r/r_s r_r] - \phi_{rs}^e = \arccos\,[\sum_\alpha \alpha_{ab}\alpha_{ac}/r_r r_s] - \phi_{rs}^e \qquad (9.5.7)$$

(iii) Angle bending (linear)

$$\mathcal{R}_\phi^{+y} = \mathbf{e}_y \cdot \mathbf{r}_s \times \mathbf{r}_r/r_s r_r = (z_{ab}x_{ac} - x_{ab}z_{ac})/r_s r_r$$

$$\mathcal{R}_\phi^{-x} = -\mathbf{e}_x \cdot \mathbf{r}_s \times \mathbf{r}_r/r_s r_r = (z_{ab}y_{ac} - y_{ab}z_{ac})/r_s r_r \qquad (9.5.8)$$

(iv) Out-of-plane bending

$$\mathcal{R}_\gamma = \mathbf{r}_s \cdot \mathbf{r}_r \times \mathbf{r}_t/r_s r_r r_t \qquad (9.5.9)$$

(v) Torsion (planar molecules only)

$$\mathcal{R}_\tau = \pm \arcsin\,[\mathbf{r}_s \cdot \mathbf{r}_r \times \mathbf{r}_t/r_s r_r r_t \sin\phi_{sr} \sin\phi_{rt}] \qquad \begin{pmatrix} +cis \\ -trans \end{pmatrix} \qquad (9.5.10)$$

The definitions of bond-stretching and in-plane angle-bending coordinates are the same as those adopted in Chapter 4 and require no special comments. The definition for the linear angle bending is that proposed by Hoy, Mills and Strey and differs from that adopted by other authors, given by

$$\beta^+ = \arcsin\,(\mathcal{R}_\phi^{+y}) \quad \text{(a)}$$

$$\beta^- = \arcsin\,(-\mathcal{R}_\phi^{-x}) \quad \text{(b)} \qquad (9.5.11)$$

Hoy, Mills and Strey have pointed out that the pair of coordinates (9.5.8) transform according to the Π-species of the $C_{\infty v}$ point group under a rotation about the z-axis, whereas the pair of coordinates (9.5.11) do not, as a result of the non-linear arc sin function.

The definition of the out-of-plane bending coordinate is different from that given in Chapter 4. It has been chosen in the form (9.5.9) by Hoy *et al.* because, being symmetrical in the indices s, r and t, it gives rise to relatively simple expressions for the L-coefficients.

The definition used in Chapter 4 can be expressed in terms of (9.5.9) by means of the equation

$$\theta = \arccos(Я_y/\sin\phi_{rt}) \tag{9.5.12}$$

and accordingly the L-coefficient obtained using (9.5.9) can be transformed[7] easily to conform to definition (9.5.12).

Finally, the definition used here for the torsion angle τ is also different from that used in Chapter 4, again because it leads to simpler expressions for the L-coefficients. Furthermore, the definition is limited to the planar *cis* ($\tau^e = 0$) and *trans* ($\tau^e = \pi$) configurations since for non-planar structures very complicated L-coefficients are obtained (see Reference 7).

Hoy, Mills and Strey[7] have calculated the L-coefficients for all five types of coordinates defined above, whereas Parisean, Suzuki and Overend[9] have calculated the B-coefficient for the first three types of coordinates only. For the linear angle bending these authors have used definition (9.5.11) which leads to a different definition of the quartic force constands as pointed out by Hoy *et al.* The force constants obtained using (9.5.8) are related to those obtained using (9.5.11) by the relation

$$f_{ЯЯЯЯ} = f_{\beta\beta\beta\beta} + 4f_{\beta\beta} \tag{9.5.13}$$

Since, as discussed before, anharmonic force constants are currently calculated using both the direct transformation (9.5.1) and the two-step transformation (9.5.2) and (9.4.9), we consider it worthwhile to present here expressions for the L- and B-coefficients. For the L-coefficients we shall give the expressions obtained by Hoy *et al.* for the five types of coordinates defined before. For the B-coefficients we shall give the expressions obtained by Parisean *et al.* for the stretching and in-plane bending coordinates as well as for the linear bending defined according to (9.5.11).

(A) *Calculation of the L-coefficients*

Since the definitions of the basic internal coordinates are given in terms of difference Cartesian coordinates, it is convenient to define, as a direct consequence of (9.4.9), the linear transformation from normal to difference Cartesian coordinates in the form

$$\alpha_s = \alpha_b - \alpha_a = \sum_i P^s_{\alpha i} Q_i \tag{9.5.14}$$

where

$$P^s_{\alpha i} = \frac{\partial \alpha_s}{\partial Q_i} = \frac{\partial(\alpha_b - \alpha_a)}{\partial Q_i} = l^b_{\alpha i} m_b^{-\frac{1}{2}} - l^a_{\alpha i} m_a^{-\frac{1}{2}} \tag{9.5.15}$$

The derivatives $P^s_{\alpha i}$ relative to the same internal coordinate can be arranged in groups of three to define a vector

$$\mathbf{P}^s_i = P^s_{xi}\mathbf{e}_x + P^s_{yi}\mathbf{e}_y + P^s_{zi}\mathbf{e}_z = \frac{\partial \mathbf{r}_s}{\partial Q_i} \tag{9.5.16}$$

in much the same way as the elements of the **B**-matrix are grouped together in the **s**-vector method of Wilson. Using this notation the L-coefficients are easily calculated by forming the derivatives with respect to the normal coordinates of the relations (9.5.6) to (9.5.10). The formulae obtained for these coefficients are given below.†

† In the original paper by Hoy, Mills and Strey a different system of indices is used.

(i) Bond stretching $\mathcal{R}_s$

$$
\begin{aligned}
L_i^s &= \left(\frac{\partial \mathcal{R}_s}{\partial Q_i}\right)_0 = \mathbf{r}_s \cdot \mathbf{P}_i^s / r_s \\
L_{ij}^s &= \left(\frac{\partial^2 \mathcal{R}_s}{\partial Q_i \, \partial Q_j}\right)_0 = \frac{1}{r_s}(\mathbf{P}_i^s \cdot \mathbf{P}_j^s - L_i^s L_j^s) \\
L_{ijk}^s &= \left(\frac{\partial^3 \mathcal{R}_s}{\partial Q_i \, \partial Q_j \, \partial Q_k}\right)_0 = -\frac{1}{r_s}(L_i^s L_{jk}^s + L_j^s L_{ik}^s + L_k^s L_{ij}^s)
\end{aligned}
\tag{9.5.17}
$$

(ii) Angle bending $\mathcal{R}_\phi$

$$
\begin{aligned}
L_i^\phi &= \left(\frac{\partial \mathcal{R}_\phi}{\partial Q_i}\right)_0 = \left(\frac{L_i^s}{r_s} - \frac{L_i^r}{r_r}\right)\cot\phi - (\mathbf{P}_i^s \cdot \mathbf{r}_r + \mathbf{r}_s \cdot \mathbf{P}_i^r)/r_s r_r \sin\phi \\
L_{ij}^\phi &= \left(\frac{\partial^2 \mathcal{R}_\phi}{\partial Q_i \, \partial Q_j}\right)_0 = -L_i^\phi L_j^\phi \cot\phi - L_i^\phi(L_j^s/r_s + L_j^r/r_r) - L_j^\phi(L_i^s/r_s + L_i^r/r_r) \\
&\quad + (L_{ij}^s/r_s + L_{ij}^r/r_r)\cot\phi + (L_i^s L_j^r + L_i^r L_j^s)\cot\phi/r_s r_r \\
&\quad - (\mathbf{P}_i^s \cdot \mathbf{P}_j^r + \mathbf{P}_j^s \cdot \mathbf{P}_i^r)/r_s r_r \sin\phi \\
L_{ijk}^\phi &= \left(\frac{\partial^3 \mathcal{R}_\phi}{\partial Q_i \, \partial Q_j \, \partial Q_k}\right)_0 = L_i^\phi L_j^\phi L_k^\phi - \sum_{\substack{\text{perm} \\ ijk}} \{(L_i^s/r_s + L_i^r/r_r)(L_{jk}^\phi + L_j^\phi L_k^\phi \cot\phi) \\
&\quad + L_i^\phi[L_{jk}^\phi \cot\phi + (L_j^s L_k^r + L_k^s L_j^r)/r_s r_r + L_{jk}^s/r_s + L_{jk}^r/r_r] \\
&\quad + (L_i^s/r_s - L_i^r/r_r)(L_{jk}^s/r_s - L_{jk}^r/r_r)\cot\phi\}
\end{aligned}
\tag{9.5.18}
$$

(iii) Angle bending (linear) $\mathcal{R}_\phi^A$ $(A = (+y), (-x))$

$$
\begin{aligned}
L_i^{\phi A} &= \left(\frac{\partial R_\phi^A}{\partial Q_i}\right)_0 = \mathbf{e}_A \cdot (\mathbf{P}_i^s \times \mathbf{r}_r + \mathbf{r}_s \times \mathbf{P}_i^r)/r_s r_r \\
L_{ij}^{\phi A} &= \left(\frac{\partial^2 \mathcal{R}_\phi^A}{\partial Q_i \, \partial Q_j}\right)_0 = -L_i^{\phi A}(L_j^s/r_s + L_j^r/r_r) - L_j^{\phi A}(L_i^s/r_s + L_i^r/r_r) + \mathbf{e}_A \cdot (\mathbf{P}_i^s \times \mathbf{P}_j^r + \mathbf{P}_j^s \times \mathbf{P}_i^r)/r_s r_r \\
L_{ijk}^{\phi A} &= \left(\frac{\partial^3 \mathcal{R}_\phi^A}{\partial Q_i \, \partial Q_j \, \partial Q_k}\right)_0 = -\sum_{\substack{\text{perm} \\ i,j,k}} \{(L_i^s/r_s + L_i^r/r_r)L_{jk}^{\phi A} + L_i^{\phi A}[(L_j^s L_k^r + L_k^s L_j^r)/r_s r_r + L_{jk}^s/r_s + L_{jk}^r/r_r]\}
\end{aligned}
\tag{9.5.19}
$$

If the definition (9.5.11) is used for the linear angle bending the L_i^β and L_{ij}^β coefficients are given by expressions identical to $L_i^{\phi A}$ and $L_{ij}^{\phi A}$. The coefficients L_{ijk}^β will, however, contain an extra term $L_i^\beta L_j^\beta L_k^\beta$.

(iv) Out-of-plane bending $\mathcal{R}_\gamma$

$$
\begin{aligned}
L_i^\gamma &= \left(\frac{\partial \mathcal{R}_\gamma}{\partial Q_i}\right)_0 = (\mathbf{P}_i^s \cdot \mathbf{r}_r \times \mathbf{r}_t + \mathbf{r}_s \cdot \mathbf{P}_i^r \times \mathbf{r}_t + \mathbf{r}_s \cdot \mathbf{r}_r \times \mathbf{P}_i^t)/r_s r_r r_t \\
L_{ij}^\gamma &= \left(\frac{\partial^2 \mathcal{R}_\gamma}{\partial Q_i \, \partial Q_j}\right)_0 = -L_i^\gamma(L_j^s/r_s + L_j^r/r_r + L_j^t/r_t) - L_j^\gamma(L_i^s/r_s + L_i^r/r_r + L_i^t/r_t) \\
&\quad + [\mathbf{r}_s \cdot (\mathbf{P}_i^r \times \mathbf{P}_j^t + \mathbf{P}_j^r \times \mathbf{P}_i^t) + \mathbf{r}_r \cdot (\mathbf{P}_i^t \times \mathbf{P}_j^s + \mathbf{P}_j^t \times \mathbf{P}_i^s) \\
&\quad + \mathbf{r}_t \cdot (\mathbf{P}_i^s \times \mathbf{P}_j^r + \mathbf{P}_j^s \times \mathbf{P}_i^r)]/r_s r_r r_t
\end{aligned}
\tag{9.5.20}
$$

$$L_{ijk}^{\gamma} = \left(\frac{\partial^3 \text{Я}_\gamma}{\partial Q_i\,\partial Q_j\,\partial Q_k}\right)_0 = \sum_{\substack{\text{perm}\\ i,j,k}} \{\mathbf{P}_i^s\,.\,(\mathbf{P}_j^r \times \mathbf{P}_k^t + \mathbf{P}_k^r \times \mathbf{P}_j^t)/r_s r_r r_t - L_i^{\gamma}[L_{jk}^s/r_s + L_{jk}^r/r_r$$

$$+ L_{jk}^t/r_t - L_j^s L_k^s/r_s^2 - L_j^r L_k^r/r_r^2 - L_j^t L_k^t/r_t^2 + (L_j^s/r_s + L_j^r/r_r +$$

$$+ L_j^t/r_t)(L_k^s/r_s + L_k^r/r_r + L_k^t/r_t)] - L_{ij}^{\gamma}(L_k^s/r_s + L_k^r/r_r + L_k^t/r_t)\}$$

(v) Torsion Я_τ

$$L_i^{\tau} = \left(\frac{\partial \text{Я}_\tau}{\partial Q_i}\right)_0 = \pm L_i^{\gamma}/\sin\phi_{rs}\sin\phi_{rt}$$

$$L_{ij}^{\tau} = \left(\frac{\partial^2 \text{Я}_\tau}{\partial Q_i\,\partial Q_j}\right)_0 = \pm\{L_{ij}^{\gamma} - L_i^{\gamma}(\cot\phi_{sr}L_j^{\phi_{sr}} + \cot\phi_{rt}L_j^{\phi_{rt}}) - L_j^{\gamma}(\cot\phi_{rs}L_i^{\phi_{rs}}$$

$$+ \cot\phi_{rt}L_i^{\phi_{rt}})\}/\sin\phi_{sr}\sin\phi_{rt} \tag{9.5.21}$$

$$L_{ijk}^{\tau} = \left(\frac{\partial^3 \text{Я}_\tau}{\partial Q_i\,\partial Q_j\,\partial Q_k}\right) = \pm\Bigg\{L_{ijk}^{\gamma} + L_i^{\gamma}L_j^{\gamma}L_k^{\gamma}/(\sin\phi_{sr}\sin\phi_{rt})^2 + \sum_{\substack{\text{perm}\\ i,j,k}} [L_i^{\gamma}(L_j^{\phi_{sr}}L_k^{\phi_{sr}}/\sin^2\phi_{sr}$$

$$+ L_j^{\phi_{rt}}L_k^{\phi_{rt}}/\sin^2\phi_{rt} - (\cot\phi_{sr})L_{jk}^{\phi_{sr}} - (\cot\phi_{rt})L_{jk}^{\phi_{rt}} + (L_j^{\phi_{sr}}\cot\phi_{sr}$$

$$+ L_j^{\phi_{rt}}\cot\phi_{rt})(L_k^{\phi_{sr}}\cot\phi_{sr} + L_k^{\phi_{rt}}\cot\phi_{rt}) - L_{ij}^{\gamma}(L_k^{\phi_{sr}}\cot\phi_{sr}$$

$$+ L_k^{\phi_{rt}}\cot\phi_{rt})]\Bigg\}\Bigg/\sin\phi_{sr}\sin\phi_{rt}$$

In all these expressions the symbol

$$\sum_{\substack{\text{perm}\\ i,j,k}}$$

means a sum over three similar terms obtained by cyclic permutation of the indices i, j and k. Furthermore, all bond lengths and angles are to be taken at equilibrium. We notice that the L-coefficients for the torsion coordinate are expressed in terms of the coefficients of the out-of-plane bending owing to the fact that the definition of the torsion can be expressed in the form

$$\text{Я}_\tau(\text{planar}) = \pm\arcsin[\text{Я}_\gamma/\sin\phi_{sr}\sin\phi_{rt}] \tag{9.5.22}$$

One important feature of the formulation of the L-coefficients given above is that all higher-order derivatives are expressed in terms of the first-order derivative elements, i.e. by the elements of the **L**-matrix. The complete transformation is thus completely determined by the solution of the harmonic secular equation.

(B) *Calculation of the B-coefficients*

The expressions for the B-coefficients are conveniently given in terms of difference Cartesian coordinates α_s since more compact formulae are obtained. The transformation to Cartesian coordinates is very simple since, from (9.5.4), $\partial\alpha_{ab}/\partial\alpha_b = 1$ and $\partial\alpha_{ab}/\partial\alpha_a = -1$, all other derivatives being equal to zero. The derivatives of the basic definitions (9.5.6), (9.5.7) and (9.5.11) with respect to difference Cartesian coordinates are straightforward. By using the abbreviated notation

$$A_s, B_s, C_s = \left(\frac{\partial r_s}{\partial \alpha_s}\right)_0 = \frac{\alpha_s}{r_s} \qquad A_r, B_r, C_r = \left(\frac{\partial r_r}{\partial \alpha_r}\right)_0 = \frac{\alpha_r}{r_r} \qquad \alpha, \beta, \gamma = x, y, z \tag{9.5.23}$$

we easily obtain the elements given below.

(i) Bond stretching Я_s

$$
\begin{aligned}
B^s_{\alpha_s} &= \left(\frac{\partial \text{Я}_s}{\partial \alpha_s}\right)_0 = A_s \\
B^s_{\alpha_s\alpha_s} &= \left(\frac{\partial^2 R_s}{\partial \alpha_s^2}\right)_0 = (1 - A_s^2)/r_s \\
B^s_{\alpha_s\beta_s} &= \left(\frac{\partial^2 \text{Я}_s}{\partial \alpha_s\, \partial \beta_s}\right)_0 = -A_sB_s/r_s \\
B^s_{\alpha_s\alpha_s\alpha_s} &= \left(\frac{\partial^3 \text{Я}_s}{\partial \alpha_s^3}\right)_0 = 3A_s(A_s^2 - 1)/r_s^2 \\
B^s_{\alpha_s\alpha_s\beta_s} &= \left(\frac{\partial^3 \text{Я}_s}{\partial \alpha_s^2\, \partial \beta_s}\right)_0 = B_s(3A_s^2 - 1)/r_s^2 \\
B^s_{x_sy_sz_s} &= \left(\frac{\partial^3 \text{Я}_s}{\partial x_s\, \partial y_s\, \partial z_s}\right)_0 = 3A_sB_sC_s/r_s^2
\end{aligned}
\qquad (9.5.24)
$$

(ii) Angle bending Я_ϕ

$$B^\phi_{\alpha_s} = \left(\frac{\partial \text{Я}_\phi}{\partial \alpha_s}\right)_0 = (A_s \cos\phi - A_r)/r_s \sin\phi$$

$$B^\phi_{\alpha_r} = \left(\frac{\partial \text{Я}_\phi}{\partial \alpha_r}\right)_0 = (A_r \cos\phi - A_s)/r_r \sin\phi$$

$$B^\phi_{\alpha_s\alpha_s} = \left(\frac{\partial^2 \text{Я}_\phi}{\partial \alpha_s^2}\right)_0 = (2A_sA_r/r_s^2 \sin\phi) + \cot\phi[(1 - 3A_s^2) - (B^\phi_{\alpha_s})^2]$$

$$B^\phi_{\alpha_s\beta_s} = \left(\frac{\partial^2 \text{Я}_\phi}{\partial \alpha_s\, \partial \beta_s}\right)_0 = (A_sB_r + A_rB_s - 3A_sB_s\cos\phi)/r_s^2 \sin\phi - \cot\phi(B^\phi_{\alpha_s}B^\phi_{\beta_s})$$

$$B^\phi_{\alpha_s\alpha_r} = \left(\frac{\partial^2 \text{Я}_\phi}{\partial \alpha_s\, \partial \alpha_r}\right)_0 = (A_s^2 + A_r^2 - A_sA_r\cos\phi - 1)/r_sr_r \sin\phi - \cot\phi(B^\phi_{\alpha_s}B^\phi_{\alpha_r})$$

$$B^\phi_{\alpha_s\beta_r} = \left(\frac{\partial^2 \text{Я}_\phi}{\partial \alpha_s\, \partial \beta_r}\right)_0 = (A_sB_s + A_rB_r - A_sB_r\cos\phi)/r_sr_r \sin\phi - \cot\phi(B^\phi_{\alpha_s}B^\phi_{\beta_r})$$

$$
\begin{aligned}
B^\phi_{\alpha_s\alpha_s\alpha_s} = \left(\frac{\partial^3 \text{Я}_\phi}{\partial \alpha_s^3}\right)_0 &= 3[A_r - 3A_s(\cos\phi + A_rA_s) - 5A_s^3\cos\phi]/r_s^3 \sin\phi + (B^\phi_{\alpha_s})^3 \\
&\quad - 3\cot\phi B^\phi_{\alpha_s}B^\phi_{\alpha_s\alpha_s}
\end{aligned}
$$

$$
\begin{aligned}
B^\phi_{\alpha_s\alpha_s\beta_s} = \left(\frac{\partial^3 \text{Я}_\phi}{\partial \alpha_s^2\, \partial \beta_s}\right)_0 &= \{B_r - 3[B_s\cos\phi + 2A_sA_rB_s + A_s^2(B_r - 5B_s\cos\phi)]\}/r_s^3 \sin\phi \\
&\quad - \cot\phi(2B^\phi_{\alpha_s}B^\phi_{\alpha_s\beta_s} + B^\phi_{\beta_s}B^\phi_{\alpha_s\alpha_s}) + (B^\phi_{\alpha_s})^2B^\phi_{\beta_s}
\end{aligned}
$$

$$
\begin{aligned}
B^{\phi}_{x_s y_s z_s} &= \left(\frac{\partial^3 \mathcal{R}_{\phi}}{\partial x_s\, \partial y_s\, \partial z_s}\right)_0 = 3[5A_sB_sC_s\cos\phi - (A_rB_sC_s + A_sB_rC_s + A_sB_sC_r)]/r_s^3 \sin\phi \\
&\qquad - \cot\phi(B^{\phi}_{x_s}B^{\phi}_{y_s z_s} + B^{\phi}_{y_s}B^{\phi}_{x_s z_s} + B^{\phi}_{z_s}B^{\phi}_{x_s y_s}) + B^{\phi}_{x_s}B^{\phi}_{y_s}B^{\phi}_{z_s} \\
B^{\phi}_{\alpha_s\alpha_s\alpha_r} &= \left(\frac{\partial^3 \mathcal{R}_{\phi}}{\partial \alpha_s^2\, \partial \alpha_r}\right)_0 = [3A_s(1 - A_s^2 + A_rA_s\cos\phi) - 2A_r^2A_s - A_r\cos\phi]/r_s^2 r_r \sin\phi \\
&\qquad - \cot\phi(2B^{\phi}_{\alpha_s}B^{\phi}_{\alpha_s\alpha_r} + B^{\phi}_{\alpha_r}B^{\phi}_{\alpha_s\alpha_s}) + (B^{\phi}_{\alpha_s})^2 B^{\phi}_{\alpha_r} \\
B^{\phi}_{\alpha_s\alpha_s\beta_r} &= \left(\frac{\partial^3 \mathcal{R}_{\phi}}{\partial \alpha_s^2\, \partial \beta_r}\right)_0 = [B_s - 2A_sA_rB_r - B_r\cos\phi + 3A_s^2(B_r\cos\phi - B_s)]/r_s^2 r_r \sin\phi \\
&\qquad - \cot\phi(2B^{\phi}_{\alpha_s}B^{\phi}_{\alpha_s\beta_r} + B^{\phi}_{\beta_r}B^{\phi}_{\alpha_s\alpha_s}) + (B^{\phi}_{\alpha_s})^2 B^{\phi}_{\beta_r} \\
B^{\phi}_{\alpha_s\beta_s\alpha_r} &= \left(\frac{\partial^3 \mathcal{R}_{\phi}}{\partial \alpha_s\, \partial \beta_s\, \partial \alpha_r}\right)_0 = [B_s(1 - 3A_s^2 + 3A_sA_r\cos\phi) - A_r(A_rB_s + A_sB_r)]/r_s^2 r_r \sin\phi \\
&\qquad - \cot\phi(B^{\phi}_{\alpha_s}B^{\phi}_{\beta_s\alpha_r} + B^{\phi}_{\beta_s}B^{\phi}_{\alpha_s\alpha_r} + B^{\phi}_{\alpha_r}B^{\phi}_{\alpha_s\beta_s}) + B^{\phi}_{\alpha_s}B^{\phi}_{\beta_s}B^{\phi}_{\alpha_r} \\
B^{\phi}_{\alpha_s\beta_s\gamma_r} &= \left(\frac{\partial^3 \mathcal{R}_{\phi}}{\partial \alpha_s\, \partial \beta_s\, \partial \gamma_r}\right)_0 = -[C_r(A_rB_s + B_rA_s) + 3A_sB_s(C_s - C_r\cos\phi)]/r_s^2 r_r \sin\phi \\
&\qquad - \cot\phi(B^{\phi}_{\alpha_s}B^{\phi}_{\beta_s\gamma_r} + B^{\phi}_{\beta_s}B^{\phi}_{\alpha_s\gamma_r} + B^{\phi}_{\gamma_r}B^{\phi}_{\alpha_s\beta_s}) + B^{\phi}_{\alpha_s}B^{\phi}_{\beta_s}B^{\phi}_{\gamma_r}
\end{aligned}
\tag{9.5.25}
$$

We notice that in (9.5.25) the B^{ϕ} coefficients relative to the Cartesian coordinates of the atom a in common between the two bonds s and r are linear combinations of those relative to the coordinates of the end atoms b and c since from (9.5.7) it follows that

$$
\frac{\partial \mathcal{R}_{\phi}}{\partial \alpha_a} = -\left(\frac{\partial R_{\phi}}{\partial \alpha_b} + \frac{\partial \mathcal{R}_{\phi}}{\partial \alpha_c}\right) \tag{9.5.26}
$$

Attention must therefore be paid to the fact that, in the compact expressions in difference Cartesian coordinates, a B-element involving the coordinates of atom a will appear in different elements of (9.5.25) that must thus be added together. For the convenience of the reader we shall give below the expressions for the B-coefficients relative to the coordinates x_a, y_a and z_a of atom a in terms of those of the end atoms.

$$
\begin{aligned}
&B^{\phi}_{\alpha_a} = -(B^{\phi}_{\alpha_b} + B^{\phi}_{\alpha_c}) \qquad B^{\phi}_{\alpha_a\alpha_a} = (B^{\phi}_{\alpha_b\alpha_b} + B^{\phi}_{\alpha_c\alpha_c} + 2B^{\phi}_{\alpha_b\alpha_c}) \qquad B^{\phi}_{\alpha_a\alpha_b} = -(B^{\phi}_{\alpha_b\alpha_b} + B^{\phi}_{\alpha_b\alpha_c}) \\
&B^{\phi}_{\alpha_a\beta_a} = (B^{\phi}_{\alpha_b\beta_b} + B^{\phi}_{\alpha_b\beta_c} + B^{\phi}_{\alpha_c\beta_b} + B^{\phi}_{\alpha_c\beta_c}) \qquad B^{\phi}_{\alpha_a\alpha_a\alpha_a} = -(B^{\phi}_{\alpha_b\alpha_b\alpha_b} + 3B^{\phi}_{\alpha_b\alpha_b\alpha_c} + 3B^{\phi}_{\alpha_b\alpha_c\alpha_c} + B^{\phi}_{\alpha_c\alpha_c\alpha_c}) \\
&B^{\phi}_{\alpha_b\alpha_b\alpha_a} = -(B^{\phi}_{\alpha_b\alpha_b\alpha_b} + B^{\phi}_{\alpha_b\alpha_b\alpha_c}) \qquad B^{\phi}_{\alpha_b\alpha_a\alpha_a} = (B^{\phi}_{\alpha_b\alpha_b\alpha_b} + 2B^{\phi}_{\alpha_b\alpha_b\alpha_c} + B^{\phi}_{\alpha_b\alpha_c\alpha_c}) \\
&B^{\phi}_{\alpha_a\alpha_b\alpha_c} = -(B^{\phi}_{\alpha_b\alpha_b\alpha_c} + B^{\phi}_{\alpha_b\alpha_c\alpha_c}) \qquad B^{\phi}_{\alpha_b\alpha_b\beta_a} = -(B^{\phi}_{\alpha_b\alpha_b\beta_b} + B^{\phi}_{\alpha_b\alpha_b\beta_c}) \\
&B^{\phi}_{\alpha_b\alpha_a\beta_b} = -(B^{\phi}_{\alpha_b\alpha_b\beta_b} + B^{\phi}_{\alpha_b\alpha_c\beta_b}) \qquad B^{\phi}_{\alpha_b\alpha_a\beta_a} = (B^{\phi}_{\alpha_b\alpha_b\beta_b} + B^{\phi}_{\alpha_b\alpha_b\beta_c} + B^{\phi}_{\alpha_b\alpha_c\beta_b} + B^{\phi}_{\alpha_b\alpha_c\beta_c}) \\
&B^{\phi}_{\alpha_a\alpha_a\beta_b} = (B^{\phi}_{\alpha_b\alpha_b\beta_b} + 2B^{\phi}_{\alpha_b\alpha_c\beta_b} + B^{\phi}_{\alpha_c\alpha_c\beta_b}) \qquad B^{\phi}_{\alpha_a\alpha_a\beta_a} = -(B^{\phi}_{\alpha_b\alpha_b\beta_b} + 2B^{\phi}_{\alpha_b\alpha_c\beta_b} + B^{\phi}_{\alpha_c\alpha_c\beta_b} \\
&\qquad\qquad + B^{\phi}_{\alpha_b\alpha_b\beta_c} + 2B^{\phi}_{\alpha_b\alpha_c\beta_c} + B^{\phi}_{\alpha_c\alpha_c\beta_c}) \\
&B^{\phi}_{\alpha_b\beta_b\gamma_a} = -(B^{\phi}_{\alpha_b\beta_b\gamma_b} + B^{\phi}_{\alpha_b\beta_b\gamma_c}) \qquad B^{\phi}_{\alpha_b\beta_a\gamma_a} = (B^{\phi}_{\alpha_b\beta_b\gamma_b} + B^{\phi}_{\alpha_b\beta_c\gamma_b} + B^{\phi}_{\alpha_b\beta_b\gamma_c} + B^{\phi}_{\alpha_b\beta_c\gamma_c}) \\
&B^{\phi}_{x_a y_a z_a} = -(B^{\phi}_{x_b y_a z_a} + B^{\phi}_{x_c y_a z_a})
\end{aligned}
\tag{9.5.27}
$$

(iii) Linear angle bending β (definition 9.5.11)

$$B^{\beta^+}_{x_{ab}} = \left(\frac{\partial\beta^+}{\partial x_{ab}}\right) = \frac{1}{r_s} \qquad B^{\beta^-}_{y_{ab}} = \left(\frac{\partial\beta^-}{\partial y_{ab}}\right) = -\frac{1}{r_s}$$

$$B^{\beta^+}_{x_{ac}} = \left(\frac{\partial\beta^+}{\partial x_{ac}}\right) = \frac{1}{r_r} \qquad B^{\beta^-}_{y_{ac}} = \left(\frac{\partial\beta^-}{\partial y_{ac}}\right) = -\frac{1}{r_r}$$

$$B^{\beta^+}_{x_{ab}z_{ab}} = \left(\frac{\partial^2\beta^+}{\partial x_{ab}\,\partial z_{ab}}\right) = -\frac{1}{r_s^2} \qquad B^{\beta^-}_{y_{ab}z_{ab}} = \left(\frac{\partial^2\beta^-}{\partial y_{ab}\,\partial z_{ab}}\right) = \frac{1}{r_s^2}$$

$$B^{\beta^+}_{x_{ac}z_{ac}} = \left(\frac{\partial^2\beta^+}{\partial x_{ac}\,\partial z_{ac}}\right) = \frac{1}{r_r^2} \qquad B^{\beta^-}_{y_{ac}z_{ac}} = \left(\frac{\partial^2\beta^-}{\partial y_{ac}\,\partial z_{ac}}\right) = -\frac{1}{r_r^2} \tag{9.5.28}$$

$$B^{\beta^+}_{x_{ab}x_{ab}x_{ab}} = \left(\frac{\partial^3\beta^+}{\partial x_{ab}^3}\right) = -\frac{2}{r_s^3} \qquad B^{\beta^+}_{x_{ab}z_{ab}z_{ab}} = \left(\frac{\partial^3\beta^+}{\partial x_{ab}\,\partial z_{ab}^2}\right) = \frac{2}{r_s^3} \qquad B^{\beta^-}_{y_{ab}y_{ab}y_{ab}} = \left(\frac{\partial^3\beta^-}{\partial y_{ab}^3}\right) = \frac{2}{r_s^3}$$

$$B^{\beta^+}_{x_{ac}x_{ac}x_{ac}} = \left(\frac{\partial^3\beta^+}{\partial x_{ac}^3}\right) = -\frac{2}{r_r^3} \qquad B^{\beta^+}_{x_{ac}z_{ac}z_{ac}} = \left(\frac{\partial^3\beta^+}{\partial x_{ac}\,\partial z_{ac}^2}\right) = \frac{2}{r_r^3} \qquad B^{\beta^-}_{y_{ac}y_{ac}y_{ac}} = \left(\frac{\partial^3\beta^-}{\partial y_{ac}^3}\right) = \frac{2}{r_r^3}$$

$$B^{\beta^-}_{y_{ab}z_{ab}z_{ab}} = \left(\frac{\partial^3\beta^-}{\partial y_{ab}\,\partial z_{ab}^2}\right) = -\frac{2}{r_s^3} \qquad B^{\beta^-}_{y_{ac}z_{ac}z_{ac}} = \left(\frac{\partial^3\beta^-}{\partial y_{ac}\,\partial z_{ac}^2}\right) = -\frac{2}{r_r^3}$$

In these expressions the positive z-direction is taken along the $\mathbf{v}_s$ vector. Again the coefficients of the coordinates of the common atom a are linear combinations of those relative to the end atoms. The expressions given before can also be used in this case.

9.6 Fermi resonance

In polyatomic molecules with several vibrational modes it may happen that the level associated with a given vibration (or combination of vibrations) has the same energy as another level associated with a different vibration (or combination of vibrations). In such cases, if some symmetry conditions are fulfilled, a mutual perturbation sets in and, as a result, the two levels will repel each other. The wavefunctions describing the perturbed states then become combinations of the original wavefunctions and the states assume a mixed character. The direct consequence of this mixing of states is a redistribution of the intensity between the transitions from the ground to the two perturbed levels.

This phenomenon was first recognized by Fermi[38] in the case of CO_2 where the overtone $2\omega_2$ has nearly the same energy as the fundamental ω_1 so that the two levels interact strongly. Phenomenologically this interaction is revealed by the occurrence of two strong Raman lines in a frequency region where only one is expected. Since then the interaction of accidentally degenerate levels has taken the name of *Fermi resonance* and is now a concept widely used by spectroscopists in the interpretation of vibrational spectra.

Fermi resonance is a purely vibrational perturbation in the sense that the vibrational Hamiltonian is responsible for the interaction between states which are accidentally degenerate. Both the standard perturbation theory and the equivalent method of the contact transformation can be used to evaluate the quantitative effects of the perturbation. In our opinion the perturbation theory is more convenient from a didactic point of view since in this approach it is easier to identify the terms of the vibrational Hamiltonian that are involved in the Fermi resonance. For this reason we shall emphasize the perturbation more than the contact transformation approach in the following discussion.

In Section 9.2 we have seen that the first-order perturbation Hamiltonian $\mathscr{H}^{(1)}$ normally has no first-order influence on the energy of the vibrational states, since the diagonal matrix elements of $\mathscr{H}^{(1)}$ are equal to zero. The off-diagonal matrix elements of $\mathscr{H}^{(1)}$, however, play a role in the second-order correction to the energy, with a contribution given by

$$\sum_{n \neq m} \frac{\langle m|\mathscr{H}^{(1)}|n\rangle \langle n|\mathscr{H}^{(1)}|m\rangle}{E_m^0 - E_n^0} \tag{9.6.1}$$

If the energies E_m^0 and E_n^0 are very different, the contribution to the energy of the terms (9.6.1) is small. If, however, one value of E_n^0, say E_p^0, is very close to E_m^0, so that the difference $E_m^0 - E_p^0$ is of the same order of magnitude as the non-diagonal elements of $\mathscr{H}^{(1)}$, the denominator of that particular term in (9.6.1) will blow up and the term will become of first-order importance. If we consider in more detail the various terms in (9.6.1), as given in (9.2.15), we realize that only two types of terms can have the 'resonant' denominator discussed before. The first type couples states $\langle v_a, v_b, v_c|$ with states $\langle v_a + 1, v_b + 1, v_c - 1|$ and has the form (see 9.2.15)

$$-hck_{abc}^2 \frac{\langle v_a, v_b, v_c|q_a q_b q_c|v_a + 1, v_b + 1, v_c - 1\rangle^2}{\omega_a + \omega_b - \omega_c} \tag{9.6.2}$$

and becomes resonant when $\omega_a + \omega_b \simeq \omega_c$. A typical example of this kind of resonance is that between a fundamental ω_c ($v_c = 1$, $v_a = v_b = 0$) and a combination $\omega_a + \omega_b$ ($v_a + 1 = v_b + 1 = 1$, $v_c = 0$). Since the potential function is totally symmetric, the cubic term $k_{abc}q_a q_b q_c$ will be different from zero only if the direct product of the representations carried by q_a, q_b and q_c is equal to the totally symmetric representation of the group (or contains the totally symmetric representation in case two modes are degenerate). In other words the levels $\omega_a + \omega_b$ and ω_c will interact only if they belong to the same symmetry species.

The second type of resonant term couples states $\langle v_a, v_b|$ and states $\langle v_a + 2, v_b - 1|$ and has the form

$$-hck_{aab}^2 \frac{\langle v_a, v_b|q_a^2 q_b|v_a + 2, v_b - 1\rangle^2}{2\omega_a - \omega_b} \tag{9.6.3}$$

It becomes of first-order importance only when $2\omega_a \simeq \omega_b$. An example of this kind of resonance is that between a fundamental ω_b ($v_a = 0$, $v_b = 1$) and an overtone $2\omega_a$ ($v_a + 2 = 2$, $v_b = 0$). The coupling term $k_{aab}q_a^2 q_b$ will be different from zero by symmetry only if the direct product of the representations carried by q_a^2 and by q_b is equal to the totally symmetric representation of the molecular group (or contains the totally symmetric representation if q_a is degenerate). This means that, when q_b is non-degenerate, Fermi resonance can occur between $2\omega_a$ and ω_b only if ω_b belongs to the totally symmetric representation of the group.

We reach, of course, the same conclusions using the contact transformation method. In this case the S_p functions of Table 9.3.1, used to remove the terms $hck_{abc}q_a q_b q_c$ and $hck_{aab}q_a^2 q_b$ from the first-order perturbation Hamiltonian, have denominators of the type $\omega_i + \omega_j - \omega_k$ and $4\omega_i^2 - \omega_j^2$ respectively and thus will produce second-order terms of the Hamiltonian H' with the same denominators. For the particular values a, b and c of i, j and k for which the resonance conditions $\omega_a + \omega_b \simeq \omega_c$ and $2\omega_a \simeq \omega_b$ hold, these terms will become of first-order importance.

In addition, there are also S_ρ functions in Table 9.3.1 with denominators $\omega_i^2 - \omega_j^2$ which will give rise to resonance if $\omega_i \simeq \omega_j$. These S_ρ functions ($\rho = 4$ and $\rho = 6$ in the table) remove vibro-rotational coupling terms from the first-order Hamiltonian and affect the rotational levels only. This kind of resonance takes the name of Coriolis resonance and will not be discussed here. It will be sufficient to notice that the vibro-rotational constant $\alpha_i^{\alpha\alpha}$ of equation (9.3.21) has a term with the denominator $\omega_i^2 - \omega_j^2$. This term is replaced by a different one when resonance occurs. A complete discussion of Coriolis resonance can be found in the articles by Nielsen.[1,2]

When accidental degeneracy occurs between two vibrational levels it is no longer legitimate to use the perturbation theory in the form of Section 9.2. The energies of the two perturbed levels are obtained by direct diagonalization of the corresponding part of the energy matrix corrected to the first order, i.e. by solution of the secular determinant

$$\begin{vmatrix} \langle m|\mathscr{H}^{(0)} + \mathscr{H}^{(1)}|m\rangle - E & \langle m|\mathscr{H}^{(1)}|p\rangle \\ \langle p|\mathscr{H}^{(1)}|m\rangle & \langle p|\mathscr{H}^{(0)} + \mathscr{H}^{(1)}|p\rangle - E \end{vmatrix} = 0 \tag{9.6.4}$$

which, the diagonal matrix elements of $\mathscr{H}^{(1)}$ being equal to zero, can be written

$$\begin{vmatrix} E_m^0 - E & \langle m|\mathscr{H}^{(1)}|p\rangle \\ \langle p|\mathscr{H}^{(1)}|m\rangle & E_p^0 - E \end{vmatrix} = 0 \tag{9.6.5}$$

yielding

$$E = \tfrac{1}{2}(E_m^0 + E_p^0) \pm [(E_m^0 - E_p^0)^2/4 + \langle m|\mathscr{H}^{(1)}|p\rangle^2]^{\frac{1}{2}} \tag{9.6.6}$$

We need, therefore, to evaluate the matrix elements of the first-order Hamiltonian coupling two different states. If the states m and p are non-degenerate this is simply done by use of Table 9.2.1. If one of the states is degenerate, the problem becomes more complex since, as discussed in Section 9.3, the wavefunctions for degenerate states depend upon the two quantum numbers v and l and involve the associate Laguerre polynomials (for two-fold degeneracy). The derivation of the matrix elements of the operators q_{i,σ_i} and q_{i,σ_i}^2 with the appropriate wavefunctions for degenerate states is discussed in detail in the book by Barchewitz.[5] Some of them, of specific interest in the discussion of typical Fermi resonance cases, are collected in Table 9.6.1.

Table 9.6.1

$\langle v, l\|q_{i,\sigma_i}\|v+1, l+1\rangle = -\frac{1}{2}\left(\frac{v+l+2}{2}\right)^{\frac{1}{2}}$	$\langle v, l\|q_{i,\sigma_i}^2\|v+2, l+2\rangle = \frac{1}{8}[(v+l+2)(v+l+4)]^{\frac{1}{2}}$
$\langle v, l\|q_{i,\sigma_i}\|v+1, l-1\rangle = \frac{1}{2}\left(\frac{v-l+2}{2}\right)^{\frac{1}{2}}$	$\langle v, l\|q_{i,\sigma_i}^2\|v+2, l\rangle = -\frac{1}{4}[(v+l+2)(v-l+2)]^{\frac{1}{2}}$
$\langle v, l\|q_{i,\sigma_i}\|v-1, l+1\rangle = \frac{1}{2}\left(\frac{v-l}{2}\right)^{\frac{1}{2}}$	$\langle v, l\|q_{i,\sigma_i}^2\|v+2, l-2\rangle = \frac{1}{8}[(v-l+2)(v-l+4)]^{\frac{1}{2}}$
$\langle v, l\|q_{i,\sigma_i}\|v-1, l-1\rangle = -\frac{1}{2}\left(\frac{v+l}{2}\right)^{\frac{1}{2}}$	

Let us first consider the Fermi resonance between a state $\langle v_a, v_b, v_c|$ and a state $\langle v_a + 1, v_b + 1, v_c - 1|$ occurring when $\omega_a + \omega_b \simeq \omega_c$. In discussing this case we shall assume that ω_a, ω_b and ω_c are all non-degenerate, as actually happens in the great majority of cases.

The zero-order energy relative to this triad of frequencies is given by

$$hc\omega_a(v_a + \tfrac{1}{2}) + hc\omega_b(v_b + \tfrac{1}{2}) + hc\omega_c(v_c + \tfrac{1}{2}) \tag{9.6.7}$$

which, by writing $\omega_c = \omega_a + \omega_b + \Delta_0$, where Δ_0 is a small frequency interval, becomes

$$hc\omega_a(v_a + v_c + 1) + hc\omega_b(v_b + v_c + 1) + hc\,\Delta_0(v_c + \tfrac{1}{2}) \tag{9.6.8}$$

The matrix elements for the coupling of the two states can be obtained directly from Table 9.2.1 and are given by

$$hck_{abc}\langle v_a, v_b, v_c|q_aq_bq_c|v_a + 1, v_b + 1, v_c - 1\rangle$$

$$= hck_{abc}\langle v_a + 1, v_b + 1, v_c - 1|q_aq_bq_c|v_a, v_b, v_c\rangle = \frac{hc}{2\sqrt{2}}k_{abc}[(v_a + 1)(v_b + 1)v_c]^{\frac{1}{2}} \tag{9.6.9}$$

By substitution of (9.6.8) and (9.6.9) in (9.6.5) and by solution of the secular equation the energies of the two perturbed states are then obtained.

Example 9.6.1

As an example consider the case in which a combination level $\omega_a + \omega_b$ is in Fermi resonance with a fundamental ω_c. The two interacting states are then the states $\langle 1, 1, 0|$ and $\langle 0, 0, 1|$. We thus obtain the secular equation

$$\begin{array}{c|cc|}
 & v_a = 1, v_b = 1, v_c = 0 & v_a = 0, v_b = 0, v_c = 1 \\
v_a = 1, v_b = 1, v_c = 0 & 2(\omega_a + \omega_b) + \frac{1}{2}\Delta_0 - E/hc & \frac{1}{2\sqrt{2}}k_{abc} \\
v_a = 0, v_b = 0, v_c = 1 & \frac{1}{2\sqrt{2}}k_{abc} & 2(\omega_a + \omega_b) + \frac{3}{2}\Delta_0 - E/hc
\end{array} = 0 \tag{9.6.10}$$

which gives, for the energy of the perturbed states,

$$E/hc = 2(\omega_a + \omega_b) + \Delta_0 \pm \Delta \tag{9.6.11}$$

where

$$\Delta = [\Delta_0^2 + k_{abc}^2/2]^{\frac{1}{2}} \tag{9.6.12}$$

By subtraction of the zero-point energy $(\omega_a + \omega_b + \Delta_0/2)$ from (9.6.11) we obtain the perturbed frequencies

$$\begin{aligned} \omega_1 &= \omega_a + \omega_b + \tfrac{1}{2}(\Delta_0 + \Delta) \\ \omega_2 &= \omega_c - \tfrac{1}{2}(\Delta_0 + \Delta) \end{aligned} \tag{9.6.13}$$

Owing to the fact that the term (9.6.2) is now included in the first-order energy, it will not enter the second-order terms as in the normal case discussed in Section 9.2. As a consequence the expression of x_{ij} will be different. It is easily seen that the coefficient of k_{ijk} in (9.2.20) then changes from $\omega_k(\omega_k^2 - \omega_i^2 - \omega_j^2)/2N_{ijk}$ to

$$\frac{1}{8}\left[\frac{1}{\omega_i + \omega_j + \omega_k} + \frac{1}{\omega_i - \omega_j + \omega_k} - \frac{1}{\omega_i - \omega_j - \omega_k}\right] \tag{9.6.14}$$

for the particular values $i = a, j = b$ and $k = c$.

Consider now the second type of Fermi resonance discussed before, namely the resonance involving the states $\langle v_a, v_b|$ and $\langle v_a + 2, v_b - 1|$. If ω_a and ω_b are both non-degenerate, the zero-order energy is given by

$$hc\omega_a(v_a + \tfrac{1}{2}) + hc\omega_b(v_b + \tfrac{1}{2}) \tag{9.6.15}$$

and, assuming $\omega_b = 2\omega_a + \Delta_0$, where Δ_0 is, as before, a small frequency interval, can be rewritten in the form

$$hc\omega_a(v_a + 2v_b + \tfrac{3}{2}) + hc\,\Delta_0(v_b + \tfrac{1}{2}) \tag{9.6.16}$$

The matrix elements for the coupling of the two non-degenerate states $\langle v_a, v_b|$ and $\langle v_a + 2, v_b - 1|$ are again obtained from Table 9.2.1 and are given by

$$\begin{aligned} &hck_{aab}\langle v_a, v_b|q_a^2 q_b|v_a + 2, v_b - 1\rangle \\ &\quad = hck_{aab}\langle v_a + 2, v_b - 1|q_a^2 q_b|v_a, v_b\rangle = \frac{1}{2\sqrt{2}}hck_{aab}[(v_a + 1)(v_a + 2)v_b]^{\frac{1}{2}} \end{aligned} \tag{9.6.17}$$

By substitution of (9.6.16) and (9.6.17) in (9.6.5) and by solution of the secular determinant we obtain the energies of the two perturbed states.

Example 9.6.2

In this example we discuss the Fermi resonance between the first overtone $2\omega_a$ of a non-degenerate vibration and a non-degenerate fundamental ω_b. The two interacting states are then the states $\langle 2, 0|$ and $\langle 0, 1|$ and thus the secular equation (9.6.5) assumes the form

$$\begin{array}{c|cc|c} & v_a = 2, v_b = 0 & v_a = 0, v_b = 1 & \\ v_a = 2, v_b = 0 & \tfrac{7}{2}\omega_a + \tfrac{1}{2}\Delta_0 - E/hc & \tfrac{1}{2}k_{aab} & = 0 \\ v_a = 0, v_b = 1 & \tfrac{1}{2}k_{aab} & \tfrac{7}{2}\omega_a + \tfrac{3}{2}\Delta_0 - E/hc & \end{array} \tag{9.6.18}$$

yielding

$$E/hc = \tfrac{7}{2}\omega_a + \Delta_0 \pm \tfrac{1}{2}\Delta' = \tfrac{3}{2}\omega_a + \omega_b \pm \tfrac{1}{2}\Delta' \tag{9.6.19}$$

where

$$\Delta' = [k_{aab}^2 + \Delta_0^2]^{\frac{1}{2}} \tag{9.6.20}$$

The perturbed frequencies are easily obtained from (9.6.19) by subtracting the zero-point energy $[\tfrac{3}{2}\omega_a + \tfrac{1}{2}\Delta_0]$ and are

$$\begin{aligned} \omega_1 &= 2\omega_a + \tfrac{1}{2}(\Delta_0 + \Delta') \\ \omega_2 &= \omega_b - \tfrac{1}{2}(\Delta_0 + \Delta') \end{aligned} \tag{9.6.21}$$

The exclusion of the term (9.6.3) from the second-order energy introduces some changes in x_{ii} and x_{ij}. It is easily verified that the coefficient of k_{iij} changes from

$$-\frac{k_{iij}^2}{4\omega_j}\frac{(8\omega_i^2 - 3\omega_j^2)}{(4\omega_i^2 - \omega_j^2)} \quad \text{to} \quad -k_{iij}^2\left[\frac{1}{2\omega_j} + \frac{1}{8(2\omega_i + \omega_j)}\right] \quad \text{in } x_{ii}$$

and from

$$-2k_{iij}^2\frac{\omega_i}{(4\omega_i^2 - \omega_j^2)} \quad \text{to} \quad -k_{iij}^2\frac{1}{2(2\omega_i + \omega_j)} \quad \text{in } x_{ij} \tag{9.6.22}$$

for the values $i = a$ and $j = b$.

In many practical cases ω_a is a two-fold degenerate frequency. If this is the situation, the zero-order energy is given by

$$hc\omega_a(v_a + 1) + hc\omega_b(v_b + \tfrac{1}{2}) \tag{9.6.23}$$

i.e., setting $\omega_b = 2\omega_a + \Delta_0$,

$$hc\omega_a(v_a + 2v_b + 2) + hc\,\Delta_0(v_b + \tfrac{1}{2}) \tag{9.6.24}$$

The term of the potential energy that now couples the two states is

$$hck_{aab}(q_{a1}^2 + q_{a2}^2)q_b \tag{9.6.25}$$

where q_{a1} and q_{a2} are the two components of the ath degenerate normal coordinate. There are now several degenerate wavefunctions associated with the same vibrational quantum number v_a because of the degeneracy. For instance, if $v_a = 2$ and $v_b = 0$ we have the following functions

$$\langle 2^{+2}, 0| \qquad \langle 2^0, 0| \qquad \langle 2^{-2}, 0|$$

which differ for the value of the quantum number l, shown as the superscript of v_a and taking the values $+2$, 0 and -2. These states are not exactly of the same energy as would appear from equation (9.3.17), since the degeneracy is removed to the second order by the l-doubling interaction.[2] Thus, even if only one of the states $\langle (v_a + 2)^l, v_b - 1|$ has the correct symmetry to interact with the state $\langle v_a^l, v_b|$, all of them are in principle involved in the interaction via the l-doubling. Since, however, the l-doubling is small compared to the Fermi resonance, we shall ignore this further complication in our discussion and consider only the direct interaction due to (9.6.25). The matrix elements of the operator (9.6.25) coupling the states $\langle (v_a + 2)^l, v_b - 1|$ and $\langle v_a^l, v_b|$ can be obtained from Tables 9.6.1 and 9.2.1 and are

$$hck_{aab}\langle v_a^l, v_b|(q_{a1}^2 + q_{a2}^2)q_b|(v_a + 2)^l, v_b - 1\rangle =$$

$$hck_{aab}\langle (v_a + 2)^l, v_b - 1|(q_{a1}^2 + q_{a2}^2)q_b|v_a^l, v_b\rangle = -hc\frac{k_{aab}}{2\sqrt{2}}[(v_a + 2)^2 - l^2]^{\frac{1}{2}}(v_b)^{\frac{1}{2}} \tag{9.6.26}$$

By substitution of (9.6.26) and (9.6.24) in (9.6.5) and by solution of the secular equation we obtain the perturbed energy levels.

Example 9.6.3

CO_2 is a linear triatomic molecule belonging to the $D_{\infty h}$ point group and possessing three fundamental modes, one of species Σ_g^+ (ω_1), one two-fold degenerate of species Π_u (ω_2) and one of species Σ_u^+ (ω_3). Following the usual nomenclature we shall indicate the energy levels of CO_2 using the symbol (v_1, v_2^l, v_3) and specifying the values of the three quantum numbers. It happens for this molecule that the levels

$$\begin{array}{ll} (1\;\; 0^0\;\; 0) & \Sigma_g^+ \\ (0\;\; 2^0\;\; 0) & \Sigma_g^+ \\ \left.\begin{array}{l}(0\;\; 2^{+2}\;\; 0) \\ (0\;\; 2^{-2}\;\; 0)\end{array}\right\} & \Delta_g \end{array}$$

have approximately the same energy and thus Fermi resonance takes place between the levels $(1\;\; 0^0\;\; 0)$ and $(0\;\; 2^0\;\; 0)$ belonging to the same symmetry species (consult Table 7.6.2 for finding the symmetry of the $(0\;\; 2^l\;\; 0)$ levels).

The matrix elements (9.6.26) assume, for $v_a = 0$, $l = 0$ and $v_b = 1$, the value $-hck_{aab}/\sqrt{2}$ whereas the zero-order energies of the two interacting states are from (9.6.24)

$hc[4\omega_a + \frac{1}{2}\Delta_0]$ and $hc[4\omega_a + \frac{3}{2}\Delta_0]$ respectively. The secular equation (9.6.5) is then (ignoring the l-doubling)

$$\begin{array}{c|cc|}
 & v_a = 2, l = 0, v_b = 0 & v_a = 0, l = 0, v_b = 1 \\
\begin{matrix} v_a = 2, l = 0 \\ v_b = 0 \end{matrix} & 4\omega_a + \frac{1}{2}\Delta_0 - E/hc & -\dfrac{k_{aab}}{\sqrt{2}} \\
\begin{matrix} v_a = 0, l = 0 \\ v_b = 1 \end{matrix} & -\dfrac{k_{aab}}{\sqrt{2}} & 4\omega_a + \frac{3}{2}\Delta_0 - E/hc
\end{array} = 0 \qquad (9.6.27)$$

and gives

$$E/hc = 4\omega_a + \Delta_0 \pm \Delta''/2 = 2\omega_a + \omega_b \pm \Delta''/2 \qquad (9.6.28)$$

where

$$\Delta'' = [\Delta_0^2 + 2k_{aab}^2]^{\frac{1}{2}} \qquad (9.6.29)$$

By subtraction of the zero-point energy $[2\omega_a + \frac{1}{2}\Delta_0]$ from (9.6.28) we obtain the perturbed frequencies

$$\begin{aligned} \omega_1 &= 2\omega_a + \tfrac{1}{2}\Delta_0 + \tfrac{1}{2}\Delta'' \\ \omega_2 &= 2\omega_a + \tfrac{1}{2}\Delta_0 - \tfrac{1}{2}\Delta'' \end{aligned} \qquad (9.6.30)$$

For systems with degenerate states the term (9.6.3) contributes to the anharmonicity coefficients† $x_{l_tl_t}$ (see 9.3.18). If this term is involved in the first-order Fermi resonance interaction, it will not enter the second-order correction to the energy and thus the coefficient of k_{ttj} will change from

$$-\frac{1}{4}\frac{\omega_j^2}{(4\omega_t^2 - \omega_j^2)} \quad \text{to} \quad \frac{1}{8}\frac{1}{2\omega_t + \omega_j} \quad \text{in } x_{l_tl_t} \qquad (9.6.31)$$

Let us now briefly consider the Fermi resonance problem in terms of the contact transformation. We recall from Section 9.3 that the contact transformation transforms the original Hamiltonian $\mathscr{H}$ into a new Hamiltonian $\mathscr{H}'$ such that the first-order Hamiltonian $\mathscr{H}^{(1)'}$ is equal to zero, the terms occurring in $\mathscr{H}^{(1)}$ being transferred to the second-order Hamiltonian $\mathscr{H}^{(2)'}$ by means of the transformation. If this is done using the S_ρ functions of Table 9.3.1, resonant denominators of the type $(\omega_i + \omega_j - \omega_k)$ or $(4\omega_i^2 - \omega_j^2)$ will appear, thus giving rise to second-order terms in $\mathscr{H}^{(2)'}$ which can reach first-order importance when the resonance conditions are met.

This suggests that in such cases, in place of the S function of Section 9.3, a new transformation S^* must be used to produce a new first-order Hamiltonian $\mathscr{H}^{(1)''}$, whose matrix elements are all zero, except those of the type (9.6.9) or (9.6.17) involving the terms $hck_{abc}q_aq_bq_c$ or $hck_{aab}q_a^2q_b$ which are left unchanged.

The S^* function to be used when $\omega_i + \omega_j \simeq \omega_k$ has been worked out by Nielsen[1,2] and has the form

$$\begin{aligned} S^* = \frac{hck_{ijk}}{4}&[(\lambda_i + \lambda_j + \lambda_k - 2\lambda_i^{\frac{1}{2}}\lambda_j^{\frac{1}{2}} + 2\lambda_i^{\frac{1}{2}}\lambda_k^{\frac{1}{2}} + 2\lambda_j^{\frac{1}{2}}\lambda_k^{\frac{1}{2}})(P_{i\sigma_i}P_{j\sigma_j}P_{k\sigma_k}/\hbar^4) \\ &-(\lambda_i + \lambda_j - 3\lambda_k - 2\lambda_i^{\frac{1}{2}}\lambda_j^{\frac{1}{2}} - 2\lambda_i^{\frac{1}{2}}\lambda_k^{\frac{1}{2}} - 2\lambda_j^{\frac{1}{2}}\lambda_k^{\frac{1}{2}})(q_{i\sigma_i}q_{j\sigma_j}P_{k\sigma_k}/\hbar^2) \\ &+(\lambda_i - 3\lambda_j + \lambda_k + 2\lambda_i^{\frac{1}{2}}\lambda_j^{\frac{1}{2}} + 2\lambda_i^{\frac{1}{2}}\lambda_k^{\frac{1}{2}} - 2\lambda_j^{\frac{1}{2}}\lambda_k^{\frac{1}{2}})(q_{i\sigma_i}P_{j\sigma_j}q_{k\sigma_k}/\hbar^2) \\ &+(-3\lambda_i + \lambda_j + \lambda_k + 2\lambda_i^{\frac{1}{2}}\lambda_j^{\frac{1}{2}} - 2\lambda_i^{\frac{1}{2}}\lambda_k^{\frac{1}{2}} + 2\lambda_j^{\frac{1}{2}}\lambda_k^{\frac{1}{2}})(P_{i\sigma_i}q_{j\sigma_j}q_{k\sigma_k}/\hbar^2)] \\ &\times[(\lambda_i + \lambda_j + \lambda_k)(\lambda_i^{\frac{1}{2}} - \lambda_j^{\frac{1}{2}} + \lambda_k^{\frac{1}{2}})(\lambda_i^{\frac{1}{2}} - \lambda_j^{\frac{1}{2}} - \lambda_k^{\frac{1}{2}})]^{-1} \end{aligned} \qquad (9.6.32)$$

† The coefficients x_{it} and $x_{tt'}$ are also changed as a consequence of the absence of the term (9.6.3). The change is given by equation (9.6.22).

The contact transformation then proceeds as in Section 9.3 and yields, instead of $\mathscr{H}^{(1)'} = 0$, a first-order Hamiltonian $\mathscr{H}^{(1)''}$

$$\mathscr{H}^{(1)''} = \frac{hck_{ijk}}{4}\{(q_{i\sigma_i}P_{j\sigma_j} + P_{i\sigma_i}q_{j\sigma_j})(P_{k\sigma_k}/\hbar^2) - [(P_{i\sigma_i}P_{j\sigma_j}/\hbar^2) - q_{i\sigma_i}q_{j\sigma_j}]q_{k\sigma_k}\} \tag{9.6.33}$$

which can be shown to have the same matrix element of the type (9.6.9) as the original term $hck_{ijk}q_{i\sigma_i}q_{j\sigma_j}q_{k\sigma_k}$.

The S^* function to be used when $2\omega_i \simeq \omega_k$ can be obtained directly from (9.6.32) by setting $i = j$ and preserving the order of the operators. In this case the new first-order Hamiltonian $\mathscr{H}^{(1)''}$ is given by

$$\mathscr{H}^{(1)''} = \frac{hck_{iik}}{4}[(P_{i\sigma_i}q_{i\sigma_i} + q_{i\sigma_i}P_{i\sigma_i})(P_{k\sigma_k}/\hbar^2) - (P^2_{i\sigma_i}/\hbar^2 - q^2_{i\sigma_i})q_{k\sigma_k}] \tag{9.6.34}$$

and has the same matrix element of the type (9.6.17) as the original term $hck_{iik}q^2_{i\sigma_i}q_{k\sigma_k}$. By diagonalization of the energy matrix to the first-order, i.e. by solving the secular determinant (9.6.4) with the operator $\mathscr{H}^{(0)} + \mathscr{H}^{(1)''}$ we then obtain the same results shown above using the perturbation method.

References

1. H. H. Nielsen, *Rev. Mod. Phys.*, **23**, 90 (1951).
2. H. H. Nielsen, in *Handbuch der Physik* (Ed. S. Flugge), Vol. 38, Springer Verlag, 1959.
3. J. Bron, *J. Chem. Soc. Faraday Trans. II*, **70**, 871 (1974).
4. R. C. Herman and W. H. Shaffer, *J. Chem. Phys.*, **16**, 453 (1948).
5. M. P. Barchewitz, *Spectroscopie Infrarouge*, Gauthier-Villars, Vol. I, 1961.
6. I. M. Mills in *Molecular Spectroscopy: Modern Research* (Ed. by K. Narahari Rao and C. Weldo Mathews), Academic Press, 1972.
7. A. R. Hoy, I. M. Mills and G. Strey, *Mol. Phys.*, **24**, 1265 (1972).
8. A. Ya. Tsaune, N. T. Storchai, L. V. Belyavskaya and V. P. Morozov, *Optics and Spectroscopy*, **26**, 502 (1969).
9. M. A. Parisean, I. Suzuki and J. Overend, *J. Chem. Phys.*, **42**, 233 (1965).
10. K. Kuchitsu and Y. Morino, *Bull. Chem. Soc. (Japan)*, **38**, 814 (1965).
11. K. Kuchitsu and Y. Morino, *Spectrochim. Acta*, **22**, 33 (1966).
12. Y. Morino, *Pure and Appl. Chem.*, **18**, 323 (1969).
13. Y. Morino, Y. Kikuchi, S. Saito and E. Hirota, *J. Mol. Spectrosc.*, **13**, 95 (1964).
14. Y. Morino and S. Saito, *J. Mol. Spectrosc.*, **19**, 435 (1966).
15. T. Tanaka and Y. Morino, *J. Mol. Spectrosc.*, **33**, 538 (1970).
16. H. Takeo, E. Hirota and Y. Morino, *J. Mol. Spectrosc.*, **34**, 370 (1970).
17. H. Takeo, E. Hirota and Y. Morino, *J. Mol. Spectrosc.*, **41**, 420 (1972).
18. D. F. Smith, Jr. and J. Overend, *J. Chem. Phys.*, **54**, 3632 (1971).
19. J. Giguere, V. K. Wang, J. Overend and A. Cabana, *Spectrochim. Acta*, **29A**, 1197 (1973).
20. I. Suzuki, *J. Mol. Spectrosc.*, **25**, 477 (1968).
21. I. Suzuki, M. A. Parisean and J. Overend, *J. Chem. Phys.*, **44**, 3561 (1966).
22. A. J. Dorney, A. R. Hoy and I. M. Mills, *J. Mol. Spectrosc.*, **45**, 253 (1973).
23. D. F. Smith jr. and J. Overend, *Spectrochim. Acta*, **28A**, 2387 (1972).
24. G. Strey and I. M. Mills, *Mol. Phys.*, **26**, 129 (1973).
25. D. R. Lide jr., *J. Mol. Spectrosc.*, **33**, 448 (1970).
26. B. T. Darling and D. M. Dennison, *Phys. Rev.*, **57**, 128 (1940).
27. D. F. Smith jr., *Spectrochim. Acta*, **29A**, 1517 (1973).
28. O. Redlich, *J. Chem. Phys.*, **9**, 298 (1941).
29. J. Pliva, *Coll. Czechoslov. Chem. Comm.*, **23**, 777 (1958).
30. E. R. Lippincott and R. Schroeder, *J. Chem. Phys.*, **23**, 1131 (1955).
31. K. Machida and J. Overend, *J. Chem. Phys.*, **50**, 4429 (1969).
32. K. Machida and J. Overend, *J. Chem. Phys.*, **51**, 2537 (1969).
33. D. F. Smith jr. and J. Overend, *Spectrochim. Acta*, **29A**, 1517 (1973).

34. J. Pliva, *Coll. Czechoslov. Chem. Comm.*, **23**, 1839 (1958).
35. J. Pliva, *Coll. Czechoslov. Chem. Comm.*, **23**, 1846 (1958).
36. Z. Cihla and J. Pliva, *Coll. Czechoslov. Chem. Comm.*, **28**, 1232 (1962).
37. Y. Morino, K. Kuchitsu and S. Yamamoto, *Spectrochim. Acta*, **24A**, 335 (1968).
38. E. Fermi, *Z. Physik*, **71**, 251 (1931).

Appendix I
Character Tables of the Point Groups

C_n groups

C_2	E	C_2		
A	1	1	$T_z; R_z$	$\alpha_{xx}, \alpha_{yy}, \alpha_{zz}, \alpha_{xy}$
B	1	-1	$T_x, T_y; R_x, R_y$	α_{yz}, α_{zx}

C_3	E	C_3	C_3^{-1}	$\varepsilon = \exp(2\pi i/3)$	
A	1	1	1	$T_z; R_z$	$\alpha_{xx} + \alpha_{yy}, \alpha_{zz}$
E	$\begin{cases} 1 \\ 1 \end{cases}$	$\begin{matrix} \varepsilon \\ \varepsilon^* \end{matrix}$	$\begin{rcases} \varepsilon^* \\ \varepsilon \end{rcases}$	$(T_x, T_y); (R_x, R_y)$	$(\alpha_{xx} - \alpha_{yy}, \alpha_{xy}); (\alpha_{yz}, \alpha_{zx})$

C_4	E	C_4	C_2	C_4^{-1}		
A	1	1	1	1	$T_z; R_z$	$\alpha_{xx} + \alpha_{yy}, \alpha_{zz}$
B	1	-1	1	-1		$\alpha_{xx} - \alpha_{yy}, \alpha_{xy}$
E	$\begin{cases} 1 \\ 1 \end{cases}$	$\begin{matrix} i \\ -i \end{matrix}$	$\begin{matrix} -1 \\ -1 \end{matrix}$	$\begin{rcases} -i \\ i \end{rcases}$	$(T_x, T_y); (R_x, R_y)$	$(\alpha_{yz}, \alpha_{zx})$

C_5	E	C_5	C_5^2	C_5^3	C_5^{-1}	$\varepsilon = \exp(2\pi i/5)$	
A	1	1	1	1	1	$T_z; R_z$	$\alpha_{xx} + \alpha_{yy}, \alpha_{zz}$
E_1	$\begin{cases} 1 \\ 1 \end{cases}$	$\begin{matrix} \varepsilon \\ \varepsilon^* \end{matrix}$	$\begin{matrix} \varepsilon^2 \\ \varepsilon^{2*} \end{matrix}$	$\begin{matrix} \varepsilon^{2*} \\ \varepsilon^2 \end{matrix}$	$\begin{rcases} \varepsilon^* \\ \varepsilon \end{rcases}$	$(T_x, T_y); (R_x, R_y)$	$(\alpha_{yz}, \alpha_{zx})$
E_2	$\begin{cases} 1 \\ 1 \end{cases}$	$\begin{matrix} \varepsilon^2 \\ \varepsilon^{2*} \end{matrix}$	$\begin{matrix} \varepsilon^* \\ \varepsilon \end{matrix}$	$\begin{matrix} \varepsilon \\ \varepsilon^* \end{matrix}$	$\begin{rcases} \varepsilon^{2*} \\ \varepsilon^2 \end{rcases}$		$(\alpha_{xx} - \alpha_{yy}, \alpha_{xy})$

C_6	E	C_6	C_3	C_2	C_3^{-1}	C_6^{-1}	$\varepsilon = \exp(2\pi i/6)$	
A	1	1	1	1	1	1	$T_z; R_z$	$\alpha_{xx}+\alpha_{yy}, \alpha_{zz}$
B	1	−1	1	−1	1	−1		
E_1	$\begin{Bmatrix}1\\1\end{Bmatrix}$	ε ε^*	$-\varepsilon^*$ $-\varepsilon$	−1 −1	$-\varepsilon$ $-\varepsilon^*$	ε^* ε	$(T_x, T_y); (R_x, R_y)$	$(\alpha_{yz}, \alpha_{zx})$
E_2	$\begin{Bmatrix}1\\1\end{Bmatrix}$	$-\varepsilon^*$ $-\varepsilon$	$-\varepsilon$ $-\varepsilon^*$	1 1	$-\varepsilon^*$ $-\varepsilon$	$-\varepsilon$ $-\varepsilon^*$		$(\alpha_{xx}-\alpha_{yy}, \alpha_{xy})$

S_n groups

$S_2 = C_i$	E	i		
A_g	1	1	R_x, R_y, R_z	$\alpha_{xx}, \alpha_{yy}, \alpha_{zz}, \alpha_{xy}, \alpha_{yz}, \alpha_{zx}$
A_u	1	−1	T_x, T_y, T_z	

S_4	E	S_4	C_2	S_4^{-1}		
A	1	1	1	1	R_z	$\alpha_{xx}+\alpha_{yy}, \alpha_{zz}$
B	1	−1	1	−1	T_z	$\alpha_{xx}-\alpha_{yy}, \alpha_{xy}$
E	$\begin{Bmatrix}1\\1\end{Bmatrix}$	i −i	−1 −1	−i i	$(T_x, T_y); (R_x, R_y)$	$(\alpha_{yz}, \alpha_{zx})$

S_6	E	S_6	C_3	i	C_3^{-1}	S_6^{-1}	$\varepsilon = \exp(2\pi i/6)$	
A_g	1	1	1	1	1	1	R_z	$\alpha_{xx}+\alpha_{yy}, \alpha_{zz}$
A_u	1	−1	1	−1	1	−1	T_z	
E_u	$\begin{Bmatrix}1\\1\end{Bmatrix}$	ε ε^*	$-\varepsilon^*$ $-\varepsilon$	−1 −1	$-\varepsilon$ $-\varepsilon^*$	ε^* ε	(T_x, T_y)	
E_g	$\begin{Bmatrix}1\\1\end{Bmatrix}$	$-\varepsilon$ $-\varepsilon^*$	$-\varepsilon^*$ $-\varepsilon$	1 1	$-\varepsilon$ $-\varepsilon^*$	$-\varepsilon^*$ $-\varepsilon$	(R_x, R_y)	$(\alpha_{xx}-\alpha_{yy}, \alpha_{xy}); (\alpha_{yz}, \alpha_{zx})$

C_{nv} groups

C_{2v}	E	C_2	$\sigma_v(zx)$	$\sigma_v(yz)$		
A_1	1	1	1	1	T_z	$\alpha_{xx}, \alpha_{yy}, \alpha_{zz}$
A_2	1	1	−1	−1	R_z	α_{xy}
B_1	1	−1	1	−1	$T_x; R_y$	α_{zx}
B_2	1	−1	−1	1	$T_y; R_x$	α_{yz}

C_{3v}	E	$2C_3$	$3\sigma_v$		
A_1	1	1	1	T_z	$\alpha_{xx}+\alpha_{yy}, \alpha_{zz}$
A_2	1	1	−1	R_z	
E	2	−1	0	$(T_x, T_y); (R_x, R_y)$	$(\alpha_{xx}-\alpha_{yy}, \alpha_{xy}); (\alpha_{yz}, \alpha_{zx})$

C_{4v}	E	$2C_4$	C_2	$2\sigma_v$	$2\sigma_d$		
A_1	1	1	1	1	1	T_z	$\alpha_{xx}+\alpha_{yy}, \alpha_{zz}$
A_2	1	1	1	−1	−1	R_z	
B_1	1	−1	1	1	−1		$\alpha_{xx}-\alpha_{yy}$
B_2	1	−1	1	−1	1		α_{xy}
E	2	0	−2	0	0	$(T_x, T_y); (R_x, R_y)$	$(\alpha_{yz}, \alpha_{zx})$

C_{5v}	E	$2C_5$	$2C_5^2$	$5\sigma_v$		
A_1	1	1	1	1	T_z	$\alpha_{xx}+\alpha_{yy}, \alpha_{zz}$
A_2	1	1	1	−1	R_z	
E_1	2	2 cos 72°	2 cos 144°	0	$(T_x, T_y); (R_x, R_y)$	$(\alpha_{yz}, \alpha_{zx})$
E_2	2	2 cos 144°	2 cos 72°	0		$(\alpha_{xx}-\alpha_{yy}, \alpha_{xy})$

C_{6v}	E	$2C_6$	$2C_3$	C_2	$3\sigma_v$	$3\sigma_d$		
A_1	1	1	1	1	1	1	T_z	$\alpha_{xx}+\alpha_{yy}, \alpha_{zz}$
A_2	1	1	1	1	−1	−1	R_z	
B_1	1	−1	1	−1	1	−1		
B_2	1	−1	1	−1	−1	1		
E_1	2	1	−1	−2	0	0	$(T_x, T_y); (R_x, R_y)$	$(\alpha_{yz}, \alpha_{zx})$
E_2	2	−1	−1	2	0	0		$(\alpha_{xx}-\alpha_{yy}, \alpha_{xy})$

C_{nh} groups

$C_{1h}=C_s$	E	σ_h		
A'	1	1	$T_x, T_y; R_z$	$\alpha_{xx}, \alpha_{yy}, \alpha_{zz}, \alpha_{xy}$
A''	1	−1	$T_z; R_x, R_y$	α_{yz}, α_{zx}

C_{2h}	E	C_2	i	σ_h		
A_g	1	1	1	1	R_z	$\alpha_{xx}, \alpha_{yy}, \alpha_{zz}, \alpha_{xy}$
B_g	1	-1	1	-1	R_x, R_y	α_{yz}, α_{zx}
A_u	1	1	-1	-1	T_z	
B_u	1	-1	-1	1	T_x, T_y	

C_{3h}	E	C_3	C_3^{-1}	σ_h	S_3	S_3^{-1}		
A'	1	1	1	1	1	1	R_z	$\alpha_{xx} + \alpha_{yy}, \alpha_{zz}$
E'	1	ε	ε^*	1	ε	ε^*	(T_x, T_y)	$(\alpha_{xx} - \alpha_{yy}, \alpha_{xy})$
	1	ε^*	ε	1	ε^*	ε		
A''	1	1	1	-1	-1	-1	T_z	
E''	1	ε	ε^*	-1	$-\varepsilon$	$-\varepsilon^*$	(R_x, R_y)	$(\alpha_{yz}, \alpha_{zx})$
	1	ε^*	ε	-1	$-\varepsilon^*$	$-\varepsilon$		

C_{4h}	E	C_4	C_2	C_4^{-1}	i	S_4^{-1}	σ_h	S_4		
A_g	1	1	1	1	1	1	1	1	R_z	$\alpha_{xx} + \alpha_{yy}, \alpha_{zz}$
B_g	1	-1	1	-1	1	-1	1	-1		$\alpha_{xx} - \alpha_{yy}, \alpha_{xy}$
E_g	1	i	-1	$-$i	1	i	-1	$-$i	(R_x, R_y)	$(\alpha_{yz}, \alpha_{zx})$
	1	$-$i	-1	i	1	$-$i	-1	i		
A_u	1	1	1	1	-1	-1	-1	-1	T_z	
B_u	1	-1	1	-1	-1	1	-1	1		
E_u	1	i	-1	$-$i	-1	$-$i	1	i	(T_x, T_y)	
	1	$-$i	-1	i	-1	i	1	$-$i		

C_{5h}	E	C_5	C_5^2	C_5^3	C_5^4	σ_h	S_5	S_5^7	S_5^3	S_5^9	$\varepsilon = \exp(2\pi i/5)$	
A'	1	1	1	1	1	1	1	1	1	1	R_z	$\alpha_{xx} + \alpha_{yy}, \alpha_{zz}$
E_1'	1	ε	ε^2	ε^{2*}	ε^*	1	ε	ε^2	ε^{2*}	ε^*	(T_x, T_y)	
	1	ε^*	ε^{2*}	ε^2	ε	1	ε^*	ε^{2*}	ε^2	ε		
E_2'	1	ε^2	ε^*	ε	ε^{2*}	1	ε^2	ε^*	ε	ε^{2*}		$(\alpha_{xx} - \alpha_{yy}, \alpha_{xy})$
	1	ε^{2*}	ε	ε^*	ε^2	1	ε^{2*}	ε	ε^*	ε^2		
A''	1	1	1	1	1	-1	-1	-1	-1	-1	T_z	
E_1''	1	ε	ε^2	ε^{2*}	ε^*	-1	$-\varepsilon$	$-\varepsilon^2$	$-\varepsilon^{2*}$	$-\varepsilon^*$	(R_x, R_y)	$(\alpha_{yz}, \alpha_{zx})$
	1	ε^*	ε^{2*}	ε^2	ε	-1	$-\varepsilon^*$	$-\varepsilon^{2*}$	$-\varepsilon^2$	$-\varepsilon$		
E_2''	1	ε^2	ε^*	ε	ε^{2*}	-1	$-\varepsilon^2$	$-\varepsilon^*$	$-\varepsilon$	$-\varepsilon^{2*}$		
	1	ε^{2*}	ε	ε^*	ε^2	-1	$-\varepsilon^{2*}$	$-\varepsilon$	$-\varepsilon^*$	$-\varepsilon^2$		

C_{6h}	E	C_6	C_3	C_2	C_3^{-1}	C_6^{-1}	i	S_3^5	S_6^5	σ_h	S_6	S_3	$\varepsilon = \exp(2\pi i/6)$	
A_g	1	1	1	1	1	1	1	1	1	1	1	1	R_z	$\alpha_{xx} + \alpha_{yy}, \alpha_{zz}$
B_g	1	−1	1	−1	1	−1	1	−1	1	−1	1	−1		
E_{1g}	1	ε	$-\varepsilon^*$	−1	$-\varepsilon$	ε^*	1	ε	$-\varepsilon^*$	−1	$-\varepsilon$	ε^*	(R_x, R_y)	$(\alpha_{yz}, \alpha_{zx})$
	1	ε^*	$-\varepsilon$	−1	$-\varepsilon^*$	ε	1	ε^*	$-\varepsilon$	−1	$-\varepsilon^*$	ε		
E_{2g}	1	$-\varepsilon^*$	$-\varepsilon$	1	$-\varepsilon^*$	$-\varepsilon$	1	$-\varepsilon^*$	$-\varepsilon$	1	$-\varepsilon^*$	$-\varepsilon$		$(\alpha_{xx} - \alpha_{yy}, \alpha_{xy})$
	1	$-\varepsilon$	$-\varepsilon^*$	1	$-\varepsilon$	$-\varepsilon^*$	1	$-\varepsilon$	$-\varepsilon^*$	1	$-\varepsilon$	$-\varepsilon^*$		
A_u	1	1	1	1	1	1	−1	−1	−1	−1	−1	−1	T_z	
B_u	1	−1	1	−1	1	−1	−1	1	−1	1	−1	1		
E_{1u}	1	ε	$-\varepsilon^*$	−1	$-\varepsilon$	ε^*	−1	$-\varepsilon$	ε^*	1	ε	$-\varepsilon^*$	(T_x, T_y)	
	1	ε^*	$-\varepsilon$	−1	$-\varepsilon^*$	ε	−1	$-\varepsilon^*$	ε	1	ε^*	$-\varepsilon$		
E_{2u}	1	$-\varepsilon^*$	$-\varepsilon$	1	$-\varepsilon^*$	$-\varepsilon$	−1	ε^*	ε	−1	ε^*	ε		
	1	$-\varepsilon$	$-\varepsilon^*$	1	$-\varepsilon$	$-\varepsilon^*$	−1	ε	ε^*	−1	ε	ε^*		

D_n groups

D_2	E	$C_2(z)$	$C_2(y)$	$C_2(x)$		
A	1	1	1	1		$\alpha_{xx}, \alpha_{yy}, \alpha_{zz}$
B_1	1	1	−1	−1	$T_z; R_z$	α_{xy}
B_2	1	−1	1	−1	$T_y; R_y$	α_{zx}
B_3	1	−1	−1	1	$T_x; R_x$	α_{yz}

D_3	E	$2C_3$	$3C_2$		
A_1	1	1	1		$\alpha_{xx} + \alpha_{yy}, \alpha_{zz}$
A_2	1	1	−1	$T_z; R_z$	
E	2	−1	0	$(T_x, T_y); (R_x, R_y)$	$(\alpha_{xx} - \alpha_{yy}, \alpha_{xy}); (\alpha_{yz}, \alpha_{zx})$

D_4	E	$2C_4$	C_2	$2C_2'$	$2C_2''$		
A_1	1	1	1	1	1		$\alpha_{xx} + \alpha_{yy}, \alpha_{zz}$
A_2	1	1	1	−1	−1	$T_z; R_z$	
B_1	1	−1	1	1	−1		$\alpha_{xx} - \alpha_{yy}$
B_2	1	−1	1	−1	1		α_{xy}
E	2	0	−2	0	0	$(T_x, T_y); (R_x, R_y)$	$(\alpha_{yz}, \alpha_{zx})$

D_5	E	$2C_5$	$2C_5^2$	$5C_2$		
A_1	1	1	1	1		$\alpha_{xx}+\alpha_{yy}, \alpha_{zz}$
A_2	1	1	1	−1	$T_z; R_z$	
E_1	2	2 cos 72°	2 cos 144°	0	$(T_x, T_y); (R_x, R_y)$	$(\alpha_{yz}, \alpha_{zx})$
E_2	2	2 cos 144°	2 cos 72°	0		$(\alpha_{xx}-\alpha_{yy}, \alpha_{xy})$

D_6	E	$2C_6$	$2C_3$	C_2	$3C_2'$	$3C_2''$		
A_1	1	1	1	1	1	1		$\alpha_{xx}+\alpha_{yy}, \alpha_{zz}$
A_2	1	1	1	1	−1	−1	$T_z; R_z$	
B_1	1	−1	1	−1	1	−1		
B_2	1	−1	1	−1	−1	1		
E_1	2	1	−1	−2	0	0	$(T_x, T_y); (R_x, R_y)$	$(\alpha_{yz}, \alpha_{zx})$
E_2	2	−1	−1	2	0	0		$(\alpha_{xx}-\alpha_{yy}, \alpha_{xy})$

D_{nd} groups

D_{2d}	E	$2S_4$	C_2	$2C_2'$	$2\sigma_d$		
A_1	1	1	1	1	1		$\alpha_{xx}+\alpha_{yy}, \alpha_{zz}$
A_2	1	1	1	−1	−1	R_z	
B_1	1	−1	1	1	−1		$\alpha_{xx}-\alpha_{yy}$
B_2	1	−1	1	−1	1	T_z	α_{xy}
E	2	0	−2	0	0	$(T_x, T_y); (R_x, R_y)$	$(\alpha_{yz}, \alpha_{zx})$

D_{3d}	E	$2C_3$	$3C_2$	i	$2S_6$	$3\sigma_d$		
A_{1g}	1	1	1	1	1	1		$\alpha_{xx}+\alpha_{yy}, \alpha_{zz}$
A_{2g}	1	1	−1	1	1	−1	R_z	
E_g	2	−1	0	2	−1	0	(R_x, R_y)	$(\alpha_{xx}-\alpha_{yy}, \alpha_{xy}); (\alpha_{yz}, \alpha_{zx})$
A_{1u}	1	1	1	−1	−1	−1		
A_{2u}	1	1	−1	−1	−1	1	T_z	
E_u	2	−1	0	−2	1	0	(T_x, T_y)	

D_{4d}	E	$2S_8$	$2C_4$	$2S_8^3$	C_2	$4C_2'$	$4\sigma_d$		
A_1	1	1	1	1	1	1	1		$\alpha_{xx}+\alpha_{yy}, \alpha_{zz}$
A_2	1	1	1	1	1	-1	-1	R_z	
B_1	1	-1	1	-1	1	1	-1		
B_2	1	-1	1	-1	1	-1	1	T_z	
E_1	2	$2^{\frac{1}{2}}$	0	$-2^{\frac{1}{2}}$	-2	0	0	(T_x, T_y)	
E_2	2	0	-2	0	2	0	0		$(\alpha_{xx}-\alpha_{yy}, \alpha_{xy})$
E_3	2	$-2^{\frac{1}{2}}$	0	$2^{\frac{1}{2}}$	-2	0	0	(R_x, R_y)	$(\alpha_{yz}, \alpha_{zx})$

D_{5d}	E	$2C_5$	$2C_5^2$	$5C_2$	i	$2S_{10}^3$	$2S_{10}$	$5\sigma_d$		
A_{1g}	1	1	1	1	1	1	1	1		$\alpha_{xx}+\alpha_{yy}, \alpha_{zz}$
A_{2g}	1	1	1	-1	1	1	1	-1	R_z	
E_{1g}	2	2 cos 72°	2 cos 144°	0	2	2 cos 72°	2 cos 144°	0	(R_x, R_y)	$(\alpha_{yz}, \alpha_{zx})$
E_{2g}	2	2 cos 144°	2 cos 72°	0	2	2 cos 144°	2 cos 72°	0		$(\alpha_{xx}-\alpha_{yy}, \alpha_{xy})$
A_{1u}	1	1	1	1	-1	-1	-1	-1		
A_{2u}	1	1	1	-1	-1	-1	-1	1	T_z	
E_{1u}	2	2 cos 72°	2 cos 144°	0	-2	-2 cos 72°	-2 cos 144°	0	(T_x, T_y)	
E_{2u}	2	2 cos 144°	2 cos 72°	0	-2	-2 cos 144°	-2 cos 72°	0		

D_{6d}	E	$2S_{12}$	$2C_6$	$2S_4$	$2C_3$	$2S_{12}^5$	C_2	$6C_2'$	$6\sigma_d$		
A_1	1	1	1	1	1	1	1	1	1		$\alpha_{xx}+\alpha_{yy}, \alpha_{zz}$
A_2	1	1	1	1	1	1	1	-1	-1	R_z	
B_1	1	-1	1	-1	1	-1	1	1	-1		
B_2	1	-1	1	-1	1	-1	1	-1	1	T_z	
E_1	2	$3^{\frac{1}{2}}$	1	0	-1	$-3^{\frac{1}{2}}$	-2	0	0	(T_x, T_y)	
E_2	2	1	-1	-2	-1	1	2	0	0		$(\alpha_{xx}-\alpha_{yy}, \alpha_{xy})$
E_3	2	0	-2	0	2	0	-2	0	0		
E_4	2	-1	-1	2	-1	-1	2	0	0		
E_5	2	$-3^{\frac{1}{2}}$	1	0	-1	$3^{\frac{1}{2}}$	-2	0	0	(R_x, R_y)	$(\alpha_{yz}, \alpha_{zx})$

D_{nh} groups

$D_{2h} = V_h$	E	$C_2(z)$	$C_2(y)$	$C_2(x)$	i	$\sigma(xy)$	$\sigma(zx)$	$\sigma(yz)$		
A_g	1	1	1	1	1	1	1	1		$\alpha_{xx}, \alpha_{yy}, \alpha_{zz}$
B_{1g}	1	1	−1	−1	1	1	−1	−1	R_z	α_{xy}
B_{2g}	1	−1	1	−1	1	−1	1	−1	R_y	α_{zx}
B_{3g}	1	−1	−1	1	1	−1	−1	1	R_x	α_{yz}
A_u	1	1	1	1	−1	−1	−1	−1		
B_{1u}	1	1	−1	−1	−1	−1	1	1	T_z	
B_{2u}	1	−1	1	−1	−1	1	−1	1	T_y	
B_{3u}	1	−1	−1	1	−1	1	1	−1	T_x	

D_{3h}	E	$2C_3$	$3C_2$	σ_h	$2S_3$	$3\sigma_v$		
A_1'	1	1	1	1	1	1		$\alpha_{xx} + \alpha_{yy}, \alpha_{zz}$
A_2'	1	1	−1	1	1	−1	R_z	
E'	2	−1	0	2	−1	0	(T_x, T_y)	$(\alpha_{xx} - \alpha_{yy}, \alpha_{xy})$
A_1''	1	1	1	−1	−1	−1		
A_2''	1	1	−1	−1	−1	1	T_z	
E''	2	−1	0	−2	1	0	(R_x, R_y)	$(\alpha_{yz}, \alpha_{zx})$

D_{4h}	E	$2C_4$	C_2	$2C_2'$	$2C_2''$	i	$2S_4$	σ_h	$2\sigma_v$	$2\sigma_d$		
A_{1g}	1	1	1	1	1	1	1	1	1	1		$\alpha_{xx} + \alpha_{yy}, \alpha_{zz}$
A_{2g}	1	1	1	−1	−1	1	1	1	−1	−1	R_z	
B_{1g}	1	−1	1	1	−1	1	−1	1	1	−1		$\alpha_{xx} - \alpha_{yy}$
B_{2g}	1	−1	1	−1	1	1	−1	1	−1	1		α_{xy}
E_g	2	0	−2	0	0	2	0	−2	0	0	(R_x, R_y)	$(\alpha_{yz}, \alpha_{zx})$
A_{1u}	1	1	1	1	1	−1	−1	−1	−1	−1		
A_{2u}	1	1	1	−1	−1	−1	−1	−1	1	1	T_z	
B_{1u}	1	−1	1	1	−1	−1	1	−1	−1	1		
B_{2u}	1	−1	1	−1	1	−1	1	−1	1	−1		
E_u	2	0	−2	0	0	−2	0	2	0	0	(T_x, T_y)	

D_{5h}	E	$2C_5$	$2C_5^2$	$5C_2$	σ_h	$2S_5$	$2S_5^3$	$5\sigma_v$		
A_1'	1	1	1	1	1	1	1	1		$\alpha_{xx}+\alpha_{yy}, \alpha_{zz}$
A_2'	1	1	1	-1	1	1	1	-1	R_z	
E_1'	2	2 cos 72°	2 cos 144°	0	2	2 cos 72°	2 cos 144°	0	(T_x, T_y)	
E_2'	2	2 cos 144°	2 cos 72°	0	2	2 cos 144°	2 cos 72°	0		$(\alpha_{xx}-\alpha_{yy}, \alpha_{xy})$
A_1''	1	1	1	1	-1	-1	-1	-1		
A_2''	1	1	1	-1	-1	-1	-1	1	T_z	
E_1''	2	2 cos 72°	2 cos 144°	0	-2	-2 cos 72°	-2 cos 144°	0	(R_x, R_y)	$(\alpha_{yz}, \alpha_{zx})$
E_2''	2	2 cos 144°	2 cos 72°	0	-2	-2 cos 144°	-2 cos 72°	0		

D_{6h}	E	$2C_6$	$2C_3$	C_2	$3C_2'$	$3C_2''$	i	$2S_3$	$2S_6$	σ_h	$3\sigma_d$	$3\sigma_v$		
A_{1g}	1	1	1	1	1	1	1	1	1	1	1	1		
A_{2g}	1	1	1	1	-1	-1	1	1	1	1	-1	-1	R_z	$\alpha_{xx}+\alpha_{yy}, \alpha_{zz}$
B_{1g}	1	-1	1	-1	1	-1	1	-1	1	-1	1	-1		
B_{2g}	1	-1	1	-1	-1	1	1	-1	1	-1	-1	1		
E_{1g}	2	1	-1	-2	0	0	2	1	-1	-2	0	0	(R_x, R_y)	$(\alpha_{yz}, \alpha_{zx})$
E_{2g}	2	-1	-1	2	0	0	2	-1	-1	2	0	0		$(\alpha_{xx}-\alpha_{yy}, \alpha_{xy})$
A_{1u}	1	1	1	1	1	1	-1	-1	-1	-1	-1	-1		
A_{2u}	1	1	1	1	-1	-1	-1	-1	-1	-1	1	1	T_z	
B_{1u}	1	-1	1	-1	1	-1	-1	1	-1	1	-1	1		
B_{2u}	1	-1	1	-1	-1	1	-1	1	-1	1	1	-1		
E_{1u}	2	1	-1	-2	0	0	-2	-1	1	2	0	0	(T_x, T_y)	
E_{2u}	2	-1	-1	2	0	0	-2	1	1	-2	0	0		

Cubic groups

T	E	$4C_3$	$4C_3^2$	$3C_2$		
A	1	1	1	1		$\alpha_{xx}+\alpha_{yy}+\alpha_{zz}$
E	$\begin{Bmatrix}1 & \varepsilon & \varepsilon^* & 1\\ 1 & \varepsilon^* & \varepsilon & 1\end{Bmatrix}$					$(\alpha_{xx}+\alpha_{yy}-2\alpha_{zz}, \alpha_{xx}-\alpha_{yy})$
F	3	0	0	-1	T, R	$(\alpha_{xy}, \alpha_{yz}, \alpha_{zx})$

T_d	E	$8C_3$	$3C_2$	$6S_4$	$6\sigma_d$		
A_1	1	1	1	1	1		$(\alpha_{xx} + \alpha_{yy} + \alpha_{zz})$
A_2	1	1	1	−1	−1		
E	2	−1	2	0	0		$(\alpha_{xx} + \alpha_{yy} - 2\alpha_{zz}, \alpha_{xx} - \alpha_{yy})$
F_1	3	0	−1	1	−1	(R_x, R_y, R_z)	
F_2	3	0	−1	−1	1	(T_x, T_y, T_z)	$(\alpha_{xy}, \alpha_{yz}, \alpha_{zx})$

T_h	E	$4C_3$	$4C_3^2$	$3C_2$	i	$4S_6^5$	$4S_6$	$3\sigma_h$		
A_g	1	1	1	1	1	1	1	1		$\alpha_{xx} + \alpha_{yy} + \alpha_{zz}$
A_u	1	1	1	1	−1	−1	−1	−1		
E_g	$\begin{Bmatrix} 1 \\ 1 \end{Bmatrix}$	ε ε^*	ε^* ε	1 1	1 1	ε ε^*	ε^* ε	1 1		$(\alpha_{xx} + \alpha_{yy} - 2\alpha_{zz}, \alpha_{xx} - \alpha_{yy})$
E_u	1 1	ε ε^*	ε^* ε	1 1	−1 −1	$-\varepsilon$ $-\varepsilon^*$	$-\varepsilon^*$ $-\varepsilon$	−1 −1		
F_g	3	0	0	−1	3	0	0	−1	(R_x, R_y, R_z)	$(\alpha_{xy}, \alpha_{yz}, \alpha_{zx})$
F_u	3	0	0	−1	−3	0	0	1	(T_x, T_y, T_z)	

O	E	$8C_3$	$3C_2$	$6C_4$	$6C_2'$		
A_1	1	1	1	1	1		$(\alpha_{xx} + \alpha_{yy} + \alpha_{zz})$
A_2	1	1	1	−1	−1		
E	2	−1	2	0	0		$(\alpha_{xx} + \alpha_{yy} - 2\alpha_{zz}, \alpha_{xx} - \alpha_{yy})$
F_1	3	0	−1	1	−1	$(R_x, R_y, R_z)(T_x, T_y, T_z)$	
F_2	3	0	−1	−1	1		$(\alpha_{xy}, \alpha_{yz}, \alpha_{zx})$

O_h	E	$8C_3$	$3C_2$	$6C_4$	$6C_2'$	i	$8S_6$	$3\sigma_h$	$6S_4$	$6\sigma_d$		
A_{1g}	1	1	1	1	1	1	1	1	1	1		$\alpha_{xx} + \alpha_{yy} + \alpha_{zz}$
A_{2g}	1	1	1	−1	−1	1	1	1	−1	−1		
E_g	2	−1	2	0	0	2	−1	2	0	0		$(\alpha_{xx} + \alpha_{yy} - 2\alpha_{zz}, \alpha_{xx} - \alpha_{yy})$
F_{1g}	3	0	−1	1	−1	3	0	−1	1	−1	(R_x, R_y, R_z)	
F_{2g}	3	0	−1	−1	1	3	0	−1	−1	1		$(\alpha_{xy}, \alpha_{yz}, \alpha_{zx})$
A_{1u}	1	1	1	1	1	−1	−1	−1	−1	−1		
A_{2u}	1	1	1	−1	−1	−1	−1	−1	1	1		
E_u	2	−1	2	0	0	−2	1	−2	0	0		
F_{1u}	3	0	−1	1	−1	−3	0	1	−1	1	(T_x, T_y, T_z)	
F_{2u}	3	0	−1	−1	1	−3	0	1	1	−1		

Linear groups

$C_{\infty v}$	E	$2C_\infty$	$\ldots$	$\infty\sigma_v$		
Σ^+	1	1	$\ldots$	1	T_z	$\alpha_{xx}+\alpha_{yy}, \alpha_{zz}$
Σ^-	1	1	$\ldots$	-1	R_z	
Π	2	$2\cos\phi$	$\ldots$	0	$(T_x, T_y); (R_x, R_y)$	$(\alpha_{yz}, \alpha_{zx})$
Δ	2	$2\cos 2\phi$	$\ldots$	0		$(\alpha_{xx}-\alpha_{yy}, \alpha_{xy})$
Φ	2	$2\cos 3\phi$	$\ldots$	0		
	$\ldots$	$\ldots$	$\ldots$	$\ldots$		

$D_{\infty h}$	E	$2C_\infty$	$\ldots$	$\infty\sigma_v$	i	$2S_\infty$	$\ldots$	∞C_2		
Σ_g^+	1	1	$\ldots$	1	1	1	$\ldots$	1		$\alpha_{xx}+\alpha_{yy}, \alpha_{zz}$
Σ_g^-	1	1	$\ldots$	-1	1	1	$\ldots$	-1	R_z	
Π_g	2	$2\cos\phi$	$\ldots$	0	2	$-2\cos\phi$	$\ldots$	0	(R_x, R_y)	$(\alpha_{yz}, \alpha_{zx})$
Δ_g	2	$2\cos 2\phi$	$\ldots$	0	2	$2\cos 2\phi$	$\ldots$	0		$(\alpha_{xx}-\alpha_{yy}, \alpha_{xy})$
$\ldots$	$\ldots$	$\ldots$	$\ldots$	$\ldots$	$\ldots$	$\ldots$	$\ldots$	$\ldots$		
Σ_u^+	1	1	$\ldots$	1	-1	-1	$\ldots$	-1	T_z	
Σ_u^-	1	1	$\ldots$	-1	-1	-1	$\ldots$	1		
Π_u	2	$2\cos\phi$	$\ldots$	0	-2	$2\cos\phi$	$\ldots$	0	(T_x, T_y)	
Δ_u	2	$2\cos 2\phi$	$\ldots$	0	-2	$-2\cos 2\phi$	$\ldots$	0		
$\ldots$	$\ldots$	$\ldots$	$\ldots$	$\ldots$	$\ldots$	$\ldots$	$\ldots$	$\ldots$		

Appendix II
Correlation Tables

C_4	C_2
A	A
B	A
E	$2B$

C_6	C_3	C_2
A	A	A
B	A	B
E_1	E	$2B$
E_2	E	$2A$

	z	y	x
D_2	C_2	C_2	C_2
A	A	A	A
B_1	A	B	B
B_2	B	A	B
B_3	B	B	A

D_3	C_3	C_2
A_1	A	A
A_2	A	B
E	E	$A + B$

			C'_2	C''_2
D_4	C_4	C_2	C_2	C_2
A_1	A	A	A	A
A_2	A	A	B	B
B_1	B	A	A	B
B_2	B	A	B	A
E	E	$2B$	$A + B$	$A + B$

D_5	C_5	C_2
A_1	A	A
A_2	A	B
E_1	E_1	$A + B$
E_2	E_2	$A + B$

		C'_2	C''_2				C'_2	C''_2
D_6	C_6	D_3	D_3	D_2	C_3	C_2	C_2	C_2
A_1	A	A_1	A_1	A	A	A	A	A
A_2	A	A_2	A_2	B_1	A	A	B	B
B_1	B	A_1	A_2	B_2	A	B	A	B
B_2	B	A_2	A_1	B_3	A	B	B	A
E_1	E_1	E	E	$B_2 + B_3$	E	$2B$	$A + B$	$A + B$
E_2	E_2	E	E	$A + B_1$	E	$2A$	$A + B$	$A + B$

		$\sigma(zx)$	$\sigma(yz)$
C_{2v}	C_2	C_s	C_s
A_1	A	A'	A'
A_2	A	A''	A''
B_1	B	A'	A''
B_2	B	A''	A'

C_{3v}	C_3	C_s
A_1	A	A'
A_2	A	A''
E	E	$A' + A''$

		σ_v	σ_d		σ_v	σ_d
C_{4v}	C_4	C_{2v}	C_{2v}	C_2	C_s	C_s
A_1	A	A_1	A_1	A	A'	A'
A_2	A	A_2	A_2	A	A''	A''
B_1	B	A_1	A_2	A	A'	A''
B_2	B	A_2	A_1	A	A''	A'
E	E	$B_1 + B_2$	$B_1 + B_2$	$2B$	$A' + A''$	$A' + A''$

C_{5v}	C_5	C_s
A_1	A	A'
A_2	A	A''
E_1	E_1	$A' + A''$
E_2	E_2	$A' + A''$

		σ_v	σ_d	$\sigma_v \rightarrow \sigma_{(zx)}$			σ_v	σ_d
C_{6v}	C_6	C_{3v}	C_{3v}	C_{2v}	C_3	C_2	C_s	C_s
A_1	A	A_1	A_1	A_1	A	A	A'	A'
A_2	A	A_2	A_2	A_2	A	A	A''	A''
B_1	B	A_1	A_2	B_1	A	B	A'	A''
B_2	B	A_2	A_1	B_2	A	B	A''	A'
E_1	E_1	E	E	$B_1 + B_2$	E	$2B$	$A' + A''$	$A' + A''$
E_2	E_2	E	E	$A_1 + A_2$	E	$2A$	$A' + A''$	$A' + A''$

C_{2h}	C_2	C_s	C_i
A_g	A	A'	A_g
B_g	B	A''	A_g
A_u	A	A''	A_u
B_u	B	A'	A_u

C_{3h}	C_3	C_s
A'	A	A'
E'	E	$2A'$
A''	A	A''
E''	E	$2A''$

C_{4h}	C_4	S_4	C_{2h}	C_2	C_s	C_i
A_g	A	A	A_g	A	A'	A_g
B_g	B	B	A_g	A	A'	A_g
E_g	E	E	$2B_g$	$2B$	$2A''$	$2A_g$
A_u	A	B	A_u	A	A''	A_u
B_u	B	A	A_u	A	A''	A_u
E_u	E	E	$2B_u$	$2B$	$2A'$	$2A_u$

C_{5h}	C_5	C_s
A'	A	A'
E'_1	E_1	$2A'$
E'_2	E_2	$2A'$
A''	A	A''
E''_1	E_1	$2A''$
E''_2	E_2	$2A''$

C_{6h}	C_6	C_{3h}	S_6	C_{2h}	C_3	C_2	C_s	C_i
A_g	A	A'	A_g	A_g	A	A	A'	A_g
B_g	B	A''	A_g	B_g	A	B	A''	A_g
E_{1g}	E_1	E''	E_g	$2B_g$	E	$2B$	$2A''$	$2A_g$
E_{2g}	E_2	E'	E_g	$2A_g$	E	$2A$	$2A'$	$2A_g$
A_u	A	A''	A_u	A_u	A	A	A''	A_u
B_u	B	A'	A_u	B_u	A	B	A'	A_u
E_{1u}	E_1	E'	E_u	$2B_u$	E	$2B$	$2A'$	$2A_u$
E_{2u}	E_2	E''	E_u	$2A_u$	E	$2A$	$2A''$	$2A_u$

		$C_2(z)$	$C_2(y)$	$C_2(x)$	$C_2(z)$	$C_2(y)$	$C_2(x)$	$C_2(z)$	$C_2(y)$	$C_2(x)$	$\sigma(xy)$	$\sigma(zx)$	$\sigma(yz)$
D_{2h}	D_2	C_{2v}	C_{2v}	C_{2v}	C_{2h}	C_{2h}	C_{2h}	C_2	C_2	C_2	C_s	C_s	C_s
A_g	A	A_1	A_1	A_1	A_g	A_g	A_g	A	A	A	A'	A'	A'
B_{1g}	B_1	A_2	B_2	B_1	A_g	B_g	B_g	A	B	B	A'	A''	A''
B_{2g}	B_2	B_1	A_2	B_2	B_g	A_g	B_g	B	A	B	A''	A'	A''
B_{3g}	B_3	B_2	B_1	A_2	B_g	B_g	A_g	B	B	A	A''	A''	A'
A_u	A	A_2	A_2	A_2	A_u	A_u	A_u	A	A	A	A''	A''	A''
B_{1u}	B_1	A_1	B_1	B_2	A_u	B_u	B_u	A	B	B	A''	A'	A'
B_{2u}	B_2	B_2	A_1	B_1	B_u	A_u	B_u	B	A	B	A'	A''	A'
B_{3u}	B_3	B_1	B_2	A_1	B_u	B_u	A_u	B	B	A	A'	A'	A''

				$\sigma_h \to \sigma_v(zy)$			σ_h	σ_v
D_{3h}	C_{3h}	D_3	C_{3v}	C_{2v}	C_3	C_2	C_s	C_s
A_1'	A'	A_1	A_1	A_1	A	A	A'	A'
A_2'	A'	A_2	A_2	B_2	A	B	A'	A''
E'	E'	E	E	$A_1 + B_2$	E	$A + B$	$2A'$	$A' + A''$
A_1''	A''	A_1	A_2	A_2	A	A	A''	A''
A_2''	A''	A_2	A_1	B_1	A	B	A''	A'
E''	E''	E	E	$A_2 + B_1$	E	$A + B$	$2A''$	$A' + A''$

		$C'_2 \rightarrow C'_2$	$C''_2 \rightarrow C'_2$			C'_2	C''_2			C'_2	C''_2	C_2, σ_v	C_2, σ_d	C'_2	C''_2
D_{4h}	D_4	D_{2d}	D_{2d}	C_{4v}	C_{4h}	D_{2h}	D_{2h}	C_4	S_4	D_2	D_2	C_{2v}	C_{2v}	C_{2v}	$.C_{2v}$
A_{1g}	A_1	A_1	A_1	A_1	A_g	A_g	A_g	A	A	A	A	A_1	A_1	A_1	A_1
A_{2g}	A_2	A_2	A_2	A_2	A_g	B_{1g}	B_{1g}	A	A	B_1	B_1	A_2	A_2	B_1	B_1
B_{1g}	B_1	B_1	B_2	B_1	B_g	A_g	B_{1g}	B	B	A	B_1	A_1	A_2	A_1	B_1
B_{2g}	B_2	B_2	B_1	B_2	B_g	B_{1g}	A_g	B	B	B_1	A	A_2	A_1	B_1	A_1
E_g	E	E	E	E	E_g	$B_{2g} + B_{3g}$	$B_{2g} + B_{3g}$	E	E	$B_2 + B_3$	$B_2 + B_3$	$B_1 + B_2$	$B_1 + B_2$	$A_2 + B_2$	$A_2 + B_2$
A_{1u}	A_1	B_1	B_1	A_2	A_u	A_u	A_u	A	B	A	A	A_2	A_2	A_2	A_2
A_{2u}	A_2	B_2	B_2	A_1	A_u	B_{1u}	B_{1u}	A	B	B_1	B_1	A_1	A_1	B_2	B_2
B_{1u}	B_1	A_1	A_2	B_2	B_u	A_u	B_{1u}	B	A	A	B_1	A_2	A_1	A_2	B_2
B_{2u}	B_2	A_2	A_1	B_1	B_u	B_{1u}	A_u	B	A	B_1	A	A_1	A_2	B_2	A_2
E_u	E	E	E	E	E_u	$B_{2u} + B_{3u}$	$B_{2u} + B_{3u}$	E	E	$B_2 + B_3$	$B_2 + B_3$	$B_1 + B_2$	$B_1 + B_2$	$A_1 + B_1$	$A_1 + B_1$

	C_2	C'_2	C''_2	C_2	C'_2	C''_2	σ_h	σ_v	σ_d	
D_{4h}	C_{2h}	C_{2h}	C_{2h}	C_2	C_2	C_2	C_s	C_s	C_s	C_i
A_{1g}	A_g	A_g	A_g	A	A	A	A'	A'	A'	A_g
A_{2g}	A_g	B_g	B_g	A	B	B	A'	A''	A''	A_g
B_{1g}	A_g	A_g	B_g	A	A	B	A'	A'	A''	A_g
B_{2g}	A_g	B_g	A_g	A	B	A	A'	A''	A'	A_g
E_g	$2B_g$	$A_g + B_g$	$A_g + B_g$	$2B$	$A + B$	$A + B$	$2A''$	$A' + A''$	$A' + A''$	$2A_g$
A_{1u}	A_u	A_u	A_u	A	A	A	A''	A''	A''	A_u
A_{2u}	A_u	B_u	B_u	A	B	B	A''	A'	A'	A_u
B_{1u}	A_u	A_u	B_u	A	A	B	A''	A''	A'	A_u
B_{2u}	A_u	B_u	A_u	A	B	A	A''	A'	A''	A_u
E_u	$2B_u$	$A_u + B_u$	$A_u + B_u$	$2B$	$A + B$	$A + B$	$2A'$	$A' + A''$	$A' + A''$	$2A_u$

					$\sigma_h \rightarrow \sigma_{(zx)}$		σ_h	σ_v
D_{5h}	D_5	C_{5v}	C_{5h}	C_5	C_{2v}	C_2	C_s	C_s
A'_1	A_1	A_1	A'	A	A_1	A	A'	A'
A'_2	A_2	A_2	A'	A	B_1	B	A'	A''
E'_1	E_1	E_1	E'_1	E_1	$A_1 + B_1$	$A + B$	$2A'$	$A' + A''$
E'_2	E_2	E_2	E'_2	E_2	$A_1 + B_1$	$A + B$	$2A'$	$A' + A''$
A''_1	A_1	A_2	A''	A	A_2	A	A''	A''
A''_2	A_2	A_1	A''	A	B_2	B	A''	A'
E''_1	E_1	E_1	E''_1	E_1	$A_2 + B_2$	$A + B$	$2A''$	$A' + A''$
E''_2	E_2	E_2	E''_2	E_2	$A_2 + B_2$	$A + B$	$2A''$	$A' + A''$

		C'_2	C''_2			C''_2	C'_2	$\sigma_h \to \sigma(xy)$ $\sigma_v \to \sigma(yz)$			C'_2	C''_2	σ_v	σ_d		
D_{6h}	D_6	D_{3h}	D_{3h}	C_{6v}	C_{6h}	D_{3d}	D_{3d}	D_{2h}	C_6	C_{3h}	D_3	D_3	C_{3v}	C_{3v}	S_6	D_2
A_{1g}	A_1	A'_1	A'_1	A_1	A_g	A_{1g}	A_{1g}	A_g	A	A'	A_1	A_1	A_1	A_1	A_g	A
A_{2g}	A_2	A'_2	A'_2	A_2	A_g	A_{2g}	A_{2g}	B_{1g}	A	A'	A_2	A_2	A_2	A_2	A_g	B_1
B_{1g}	B_1	A''_1	A''_2	B_2	B_g	A_{2g}	A_{1g}	B_{2g}	B	A''	A_1	A_2	A_2	A_1	A_g	B_2
B_{2g}	B_2	A''_2	A''_1	B_1	B_g	A_{1g}	A_{2g}	B_{3g}	B	A''	A_2	A_1	A_1	A_2	A_g	B_3
E_{1g}	E_1	E''	E''	E_1	E_{1g}	E_g	E_g	$B_{2g} + B_{3g}$	E_1	E''	E	E	E	E	E_g	$B_2 + B_3$
E_{2g}	E_2	E'	E'	E_2	E_{2g}	E_g	E_g	$A_g + B_{1g}$	E_2	E'	E	E	E	E	E_g	$A + B_1$
A_{1u}	A_1	A''_1	A''_1	A_2	A_u	A_{1u}	A_{1u}	A_u	A	A''	A_1	A_1	A_2	A_2	A_u	A
A_{2u}	A_2	A''_2	A''_2	A_1	A_u	A_{2u}	A_{2u}	B_{1u}	A	A''	A_2	A_2	A_1	A_1	A_u	B_1
B_{1u}	B_1	A'_1	A'_2	B_1	B_u	A_{2u}	A_{1u}	B_{2u}	B	A'	A_1	A_2	A_1	A_2	A_u	B_2
B_{2u}	B_2	A'_2	A'_1	B_2	B_u	A_{1u}	A_{2u}	B_{3u}	B	A'	A_2	A_1	A_2	A_1	A_u	B_3
E_{1u}	E_1	E'	E'	E_1	E_{1u}	E_u	E_u	$B_{2u} + B_{3u}$	E_1	E'	E	E	E	E	E_u	$B_2 + B_3$
E_{2u}	E_2	E''	E''	E_2	E_{2u}	E_u	E_u	$A_u + B_{1u}$	E_2	E''	E	E	E	E	E_u	$A + B_1$

	C_2'	C_2''	C_2'	C_2	C_2''		C_2	C_2'	C_2''	σ_h	σ_d	σ_v	
D_{6h}	C_{2v}	C_{2v}	C_{2h}	C_{2h}	C_{2h}	C_3	C_2	C_2	C_2	C_s	C_s	C_s	C_i
A_{1g}	A_1	A_1	A_g	A_g	A_g	A	A	A	A	A'	A'	A'	A_g
A_{2g}	B_1	B_1	B_g	A_g	B_g	A	A	B	B	A'	A''	A''	A_g
B_{1g}	A_2	B_2	A_g	B_g	B_g	A	B	A	B	A''	A'	A''	A_g
B_{2g}	B_2	A_2	B_g	B_g	A_g	A	B	B	A	A''	A''	A'	A_g
E_{1g}	$A_2 + B_2$	$A_2 + B_2$	$A_g + B_g$	$2B_g$	$A_g + B_g$	E	$2B$	$A + B$	$A + B$	$2A''$	$A' + A''$	$A' + A''$	$2A_g$
E_{2g}	$A_1 + B_1$	$A_1 + B_1$	$A_g + B_g$	$2A_g$	$A_g + B_g$	E	$2A$	$A + B$	$A + B$	$2A'$	$A' + A''$	$A' + A''$	$2A_g$
A_{1u}	A_2	A_2	A_u	A_u	A_u	A	A	A	A	A''	A''	A''	A_u
A_{2u}	B_2	B_2	B_u	A_u	B_u	A	A	B	B	A''	A'	A'	A_u
B_{1u}	A_1	B_1	A_u	B_u	B_u	A	B	A	B	A'	A''	A'	A_u
B_{2u}	B_1	A_1	B_u	B_u	A_u	A	B	B	A	A'	A'	A''	A_u
E_{1u}	$A_1 + B_1$	$A_1 + B_1$	$A_u + B_u$	$2B_u$	$A_u + B_u$	E	$2B$	$A + B$	$A + B$	$2A'$	$A' + A''$	$A' + A''$	$2A_u$
E_{2u}	$A_2 + B_2$	$A_2 + B_2$	$A_u + B_u$	$2A_u$	$A_u + B_u$	E	$2A$	$A + B$	$A + B$	$2A''$	$A' + A''$	$A' + A''$	$2A_u$

	$C_2 \to C_2(z)$			C_2	C_2'	
D_{2d}	S_4	D_2	C_{2v}	C_2	C_2	C_s
A_1	A	A	A_1	A	A	A'
A_2	A	B_1	A_2	A	B	A''
B_1	B	A	A_2	A	A	A''
B_2	B	B_1	A_1	A	B	A'
E	E	$B_2 + B_3$	$B_1 + B_2$	$2B$	$A + B$	$A' + A''$

D_{3d}	D_3	C_{3v}	S_6	C_3	C_{2h}	C_2	C_s	C_i
A_{1g}	A_1	A_1	A_g	A	A_g	A	A'	A_g
A_{2g}	A_2	A_2	A_g	A	B_g	B	A''	A_g
E_g	E	E	E_g	E	$A_g + B_g$	$A + B$	$A' + A''$	$2A_g$
A_{1u}	A_1	A_2	A_u	A	A_u	A	A''	A_u
A_{2u}	A_2	A_1	A_u	A	B_u	B	A'	A_u
E_u	E	E	E_u	E	$A_u + B_u$	$A + B$	$A' + A''$	$2A_u$

						C_2	C_2'	
D_{4d}	D_4	C_{4v}	S_8	C_4	C_{2v}	C_2	C_2	C_s
A_1	A_1	A_1	A	A	A_1	A	A	A'
A_2	A_2	A_2	A	A	A_2	A	B	A''
B_1	A_1	A_2	B	A	A_2	A	A	A''
B_2	A_2	A_1	B	A	A_1	A	B	A'
E_1	E	E	E_1	E	$B_1 + B_2$	$2B$	$A + B$	$A' + A''$
E_2	$B_1 + B_2$	$B_1 + B_2$	E_2	$2B$	$A_1 + A_2$	$2A$	$A + B$	$A' + A''$
E_3	E	E	E_3	E	$B_1 + B_2$	$2B$	$A + B$	$A' + A''$

D_{5d}	D_5	C_{5v}	C_5	C_2	C_s	C_i
A_{1g}	A_1	A_1	A	A	A'	A_g
A_{2g}	A_2	A_2	A	B	A''	A_g
E_{1g}	E_1	E_1	E_1	$A + B$	$A' + A''$	$2A_g$
E_{2g}	E_2	E_2	E_2	$A + B$	$A' + A''$	$2A_g$
A_{1u}	A_1	A_2	A	A	A''	A_u
A_{2u}	A_2	A_1	A	B	A'	A_u
E_{1u}	E_1	E_1	E_1	$A + B$	$A' + A''$	$2A_u$
E_{2u}	E_2	E_2	E_2	$A + B$	$A' + A''$	$2A_u$

											C_2	C_2'	
D_{6d}	D_6	C_{6v}	C_6	D_{2d}	D_3	C_{3v}	D_2	C_{2v}	S_4	C_3	C_2	C_2	C_3
A_1	A_1	A_1	A	A_1	A_1	A_1	A	A_1	A	A	A	A	A'
A_2	A_2	A_2	A	A_2	A_2	A_2	B_1	A_2	A	A	A	B	A''
B_1	A_1	A_2	A	B_1	A_1	A_2	A	A_2	B	A	A	A	A''
B_2	A_2	A_1	A	B_2	A_2	A_1	B_1	A_1	B	A	A	B	A'
E_1	E_1	E_1	E_1	E	E	E	$B_2 + B_3$	$B_1 + B_2$	E	E	$2B$	$A + B$	$A' + A''$
E_2	E_2	E_2	E_2	$B_1 + B_2$	E	E	$A + B_1$	$A_1 + A_2$	$2B$	E	$2A$	$A + B$	$A' + A''$
E_3	$B_1 + B_2$	$B_1 + B_2$	$2B$	E	$A_1 + A_2$	$A_1 + A_2$	$B_2 + B_3$	$B_1 + B_2$	E	$2A$	$2B$	$A + B$	$A' + A''$
E_4	E_2	E_2	E_2	$A_1 + A_2$	E	E	$A + B_1$	$A_1 + A_2$	$2A$	E	$2A$	$A + B$	$A' + A''$
E_5	E_1	E_1	E_1	E	E	E	$B_2 + B_3$	$B_1 + B_2$	E	E	$2B$	$A + B$	$A' + A''$

S_4	C_2
A	A
B	A
E	$2B$

S_6	C_3	C_i
A_g	A	A_g
E_g	E	$2A_g$
A_u	A	A_u
E_u	E	$2A_u$

S_8	C_4	C_2
A	A	A
B	A	A
E_1	E	$2B$
E_2	$2B$	$2A$
E_3	E	$2B$

T	D_2	C_3	C_2
A	A	A	A
E	$2A$	E	$2A$
F	$B_1 + B_2 + B_3$	$A + E$	$A + 2B$

T_h	T	D_{2h}	S_6	D_2	C_{2v}	C_{2h}	C_3	C_2	C_s	C_i
A_g	A	A_g	A_g	A	A_1	A_g	A	A	A'	A_g
E_g	E	$2A_g$	E_g	$2A$	$2A_1$	$2A_g$	E	$2A$	$2A'$	$2A_g$
F_g	F	$B_{1g} + B_{2g} + B_{3g}$	$A_g + E_g$	$B_1 + B_2 + B_3$	$A_2 + B_1 + B_2$	$A_g + 2B_g$	$A + E$	$A + 2B$	$A' + 2A''$	$3A_g$
A_u	A	A_u	A_u	A	A_2	A_u	A	A	A''	A_u
E_u	E	$2A_u$	E_u	$2A$	$2A_2$	$2A_u$	E	$2A$	$2A''$	$2A_u$
F_u	F	$B_{1u} + B_{2u} + B_{3u}$	$A_u + E_u$	$B_1 + B_2 + B_3$	$A_1 + B_1 + B_2$	$A_u + 2B_u$	$A + E$	$A + 2B$	$2A' + A''$	$3A_u$

T_d	T	D_{2d}	C_{3v}	S_4	D_2	C_{2v}	C_3	C_2	C_s
A_1	A	A_1	A_1	A	A	A_1	A	A	A'
A_2	A	B_1	A_2	B	A	A_2	A	A	A''
E	E	$A_1 + B_1$	E	$A + B$	$2A$	$A_1 + A_2$	E	$2A$	$A' + A''$
F_1	F	$A_2 + E$	$A_2 + E$	$A + E$	$B_1 + B_2 + B_3$	$A_2 + B_1 + B_2$	$A + E$	$A + 2B$	$A' + 2A''$
F_2	F	$B_2 + E$	$A_1 + E$	$B + E$	$B_1 + B_2 + B_3$	$A_1 + B_1 + B_2$	$A + E$	$A + 2B$	$2A' + A''$

					$3C_2$	$C_2, 2C'_2$		C_2	C'_2
O	T	D_4	D_3	C_4	D_2	D_2	C_3	C_2	C_2
A_1	A	A_1	A_1	A	A	A	A	A	A
A_2	A	B_1	A_2	B	A	B_1	A	A	B
E	E	$A_1 + B_1$	E	$A + B$	$2A$	$A + B_1$	E	$2A$	$A + B$
F_1	F	$A_2 + E$	$A_2 + E$	$A + E$	$B_1 + B_2 + B_3$	$B_1 + B_2 + B_3$	$A + E$	$A + 2B$	$A + 2B$
F_2	F	$B_2 + E$	$A_1 + E$	$B + E$	$B_1 + B_2 + B_3$	$A + B_2 + B_3$	$A + E$	$A + 2B$	$2A + B$

O_h	O	T_d	T_h	D_{4h}	D_{3d}
A_{1g}	A_1	A_1	A_g	A_{1g}	A_{1g}
A_{2g}	A_2	A_2	A_g	B_{1g}	A_{2g}
E_g	E	E	E_g	$A_{1g} + B_{1g}$	E_g
F_{1g}	F_1	F_1	F_g	$A_{2g} + E_g$	$A_{2g} + E_g$
F_{2g}	F_2	F_2	F_g	$B_{2g} + E_g$	$A_{1g} + E_g$
A_{1u}	A_1	A_2	A_u	A_{1u}	A_{1u}
A_{2u}	A_2	A_1	A_u	B_{1u}	A_{2u}
E_u	E	E	E_u	$A_{1u} + B_{1u}$	E_u
F_{1u}	F_1	F_2	F_u	$A_{2u} + E_u$	$A_{2u} + E_u$
F_{2u}	F_2	F_1	F_u	$B_{2u} + E_u$	$A_{1u} + E_u$

Index